# THERMAL CONDUCTIVITY
## 23

# THERMAL CONDUCTIVITY 23

Edited by

## Kenneth E. Wilkes
## Ralph B. Dinwiddie
## Ronald S. Graves

Oak Ridge National Laboratory
Oak Ridge, Tennessee

**CRC Press**
Taylor & Francis Group
Boca Raton London New York

CRC Press is an imprint of the
Taylor & Francis Group, an **informa** business

# Thermal Conductivity 23

First published 1996 by Technomic Publishing Company, Inc.

Published 2021 by CRC Press
Taylor & Francis Group
6000 Broken Sound Parkway NW, Suite 300
Boca Raton, FL 33487-2742

ISBN 13: 978-1-56676-477-3 (hbk)

**Visit the Taylor & Francis Web site at**
**http://www.taylorandfrancis.com**

**and the CRC Press Web site at**
**http://www.crcpress.com**

Main entry under title:
  Thermal Conductivity 23

A Technomic Publishing Company book
Bibliography: p.

ISSN 0163-9005

# Table of Contents

*Preface*   xi

**KEYNOTE LECTURES**

**Physical Properties Are Controlled by Structures** . . . . . . . . . . . . . . . . . .3
D. L. McELROY

**Improved Correlations for the Thermal Conductivity
of Propane and *n*-Butane** . . . . . . . . . . . . . . . . . . . . . . . . . . . . . . . . .13
M. L. V. RAMIRES, C. A. NIETO DE CASTRO, L. CUSCO
and R. A. PERKINS

**SESSION 1: TECHNIQUES**

**Automatic Non-Destructive Brake Tester for
Thermal Diffusivity/Conductivity of Aircraft
Brake Disks** . . . . . . . . . . . . . . . . . . . . . . . . . . . . . . . . . . . . . . . . . . .29
R. E. TAYLOR, T. R. GOERZ and J. V. BECK

**Improved Axial Cut-Bar Apparatus** . . . . . . . . . . . . . . . . . . . . . . . . . . .41
P. S. GAAL

**Design of a Subminiature Guarded Hot Plate Apparatus** . . . . . . . . . .46
D. R. FLYNN and R. GORTHALA

**Specific Heat Measurements with the
Hot Disk Thermal Constants Analyser** . . . . . . . . . . . . . . . . . . . . . . . . .56
M. GUSTAVSSON, N. S. SAXENA, E. KARAWACKI
and S. E. GUSTAFSSON

**Optimization of the Thermal Conductivity Measurement
by Modulated DSC™** . . . . . . . . . . . . . . . . . . . . . . . . . . . . . . . . . . . . .66
S. R. AUBUCHON and R. L. BLAINE

v

**Estimation of a Space-Varying Heat Transfer**
**Coefficient or Interface Resistance by Inverse Conduction** . . . . . . . . . . 72
D. MAILLET, A. DEGIOVANNI and S. ANDRÉ

**Kalman Filtering Applied to Thermal Conductivity**
**Estimation in Pulse Experiments** . . . . . . . . . . . . . . . . . . . . . . . . . . . . . . 85
F. RIGHINI, F. SCARPA and G. MILANO

**Using the Discrete Fourier Transform in Thermal**
**Diffusivity Measurement** . . . . . . . . . . . . . . . . . . . . . . . . . . . . . . . . . . . . . 95
J. GEMBAROVIČ and R. E. TAYLOR

**Parameter Estimation Method for Flash Thermal**
**Diffusivity with Two Different Heat Transfer Coefficients** . . . . . . . . . . 107
J. V. BECK and R. B. DINWIDDIE

**Multiple Station Thermal Diffusivity Instrument** . . . . . . . . . . . . . . . . . . 119
H. WANG, R. B. DINWIDDIE and P. S. GAAL

### SESSION 2: COATINGS AND FILMS

**Theoretical Aspects of the Thermal Conductivity**
**in Thin Films** . . . . . . . . . . . . . . . . . . . . . . . . . . . . . . . . . . . . . . . . . . . . . . . 129
F. VOELKLEIN and TH. FRANKE

**Thermal Conductivity of Thin Dielectric Films** . . . . . . . . . . . . . . . . . . . 145
S.-M. LEE and D. G. CAHILL

**Measurement of the Thermal Diffusivity of Diamond**
**with a Modified Angström Method** . . . . . . . . . . . . . . . . . . . . . . . . . . . . . 152
H. ALTMANN, O. NILSSON and J. FRICKE

**An Improved Dynamical Method for Thermal**
**Conductivity and Specific Heat Measurements of**
**Thin Films in the 100 nm-Range** . . . . . . . . . . . . . . . . . . . . . . . . . . . . . . 162
R. SCHMIDT, TH. FRANKE and P. HÄUSSLER

**Phonon Knudsen Flow in GaAs/AlAs Superlattices** . . . . . . . . . . . . . . . 172
P. HYLDGAARD and G. D. MAHAN

**Measurement of Thermal Properties of Thin Films**
**Using Infrared Thermography** . . . . . . . . . . . . . . . . . . . . . . . . . . . . . . . . . 183
H. M. RELYEA, F. BREIDENICH, J. V. BECK and J. J. McGRATH

**Laser-Assisted AC Measurement of Thin-Film Thermal**
**Diffusivity with Different Laser Beam Configurations** . . . . . . . . . . . . . 195
X. Y. YU, L. ZHANG and G. CHEN

## SESSION 3: THEORY

**Thermal Conductivity of Zirconia** ............................209
P. G. KLEMENS

**A Rigid-Ion Lattice Dynamical Model for Calculating
Group Velocities in Simple, Body-Centered, and
Face-Centered Cubic Crystals** ...............................221
A. K. McCURDY

**Phonon-Focusing Effects in Dispersive Body-Centered
Cubic Crystals** ..........................................231
M. G. BELL and A. K. McCURDY

**Phonon-Focusing Effects in Dispersive Face-Centered
Cubic Crystals** ..........................................243
G. S. CAPRIGLIONE and A. K. McCURDY

**Energy-Flow-Reversing $N$ and $U$ Processes
in Dispersive Crystals** ....................................254
A. K. McCURDY

**Thermal Conductivity of Linear Chain
Semiconductor $(NbSe_4)_3I$** ...............................266
A. SMONTARA, K. BILJAKOVIĆ, A. BILUŠIĆ,
D. PAJIĆ, D. STAREŠINIĆ, F. LÉVY and H. BERGER

## SESSION 4: COMPOSITES

**Composition Effects on MMC's Thermal Properties and
Behavior under Transient High Fluxes** ........................279
J. J. SERRA, E. MILCENT and P. F. LOUVIGNÉ

**Effect of Heat Treatment on the Thermal Conductivity of
a Particulate SiC Reinforced 6061 Aluminum Matrix Composite** ....288
H. WANG and S. H. JASON LO

**Method for Measuring the Orthotropic Thermal
Conductivity and Volumetric Heat Capacity in a
Carbon-Carbon Composite** ..................................299
K. J. DOWDING, J. V. BECK and L. EILERS

**Thermal Diffusivity Imaging of Continuous Fiber
Ceramic Composite Materials and Components** ..................311
S. AHUJA, W. A. ELLINGSON, J. S. STECKENRIDER and S. KING

## SESSION 5: INSULATION

**Effective Diffusion Coefficients for CFC-11 by Gravimetric
Depletion from Thin Slices of PIR Foam** ......................325
J. R. BOOTH, R. S. GRAVES and D. W. YARBROUGH

Influence of Moisture and Air Movement in
Porous Materials on Thermal Insulation . . . . . . . . . . . . . . . . . . . . . . .338
H. STOPP, J. GRUNEWALD and P. HÄUPL

Heat and Moisture Transfer in Fiberglass Insulation
under Air Leakage and Frosting Conditions . . . . . . . . . . . . . . . . . . . .350
D. R. MITCHELL, Y.-X. TAO and R. W. BESANT

Low-Pressure Thermophysical Properties of
EPB—Expanded Perlite Board . . . . . . . . . . . . . . . . . . . . . . . . . . . . . .362
G. MILANO, F. SCARPA and G. TIMMERMANS

Thermal Conductivity of Spun Glass Fibers as Filler
Material for Vacuum Insulations . . . . . . . . . . . . . . . . . . . . . . . . . . . . .373
R. CAPS, J. HETFLEISCH, TH. RETTELBACH and J. FRICKE

Thermal Conductivity of Evacuated Insulating Powders
for Temperatures from 10 K to 275 K . . . . . . . . . . . . . . . . . . . . . . . . .383
TH. RETTELBACH, D. SATOR, S. KORDER and J. FRICKE

Precision and Bias of the Large Scale Climate Simulator
in the Guarded Hot Box and Cold Box Modes . . . . . . . . . . . . . . . . . . .395
K. E. WILKES, T. W. PETRIE and P. W. CHILDS

Thermal Conductivity of Resorcinol-Formaldehyde Aerogels . . . . . . .407
TH. RETTELBACH, H.-P. EBERT, R. CAPS, J. FRICKE,
C. T. ALVISO and R. W. PEKALA

Reference Materials and Transfer Standards for Use
in Measurements on Low Thermal Conductivity
Materials and Systems . . . . . . . . . . . . . . . . . . . . . . . . . . . . . . . . . . . . .419
R. P. TYE and U. S. RIKO

The NPL High Temperature Guarded Hot-Plate . . . . . . . . . . . . . . . . .431
D. R. SALMON

Combined Radiation and Conduction in Fibrous
Insulation (from 24°C to 400°C) . . . . . . . . . . . . . . . . . . . . . . . . . . . . .442
P. BOULET, G. JEANDEL, P. DE DIANOUS and F. PINCEMIN

Thermoinsulation of Oil- and Gas-Pipelines with
Foamed Plastics in Russia . . . . . . . . . . . . . . . . . . . . . . . . . . . . . . . . . .455
F. SHUTOV

Effect of Sub-Minute High Temperature Heat Treatments
on the Thermal Conductivity of Carbon-Bonded
Carbon Fiber (CBCF) Insulation . . . . . . . . . . . . . . . . . . . . . . . . . . . . .466
R. B. DINWIDDIE, G. E. NELSON and C. E. WEAVER

## SESSION 6: FLUIDS

**Initial Density Dependence of the Thermal Conductivity of Polyatomic Gases** ........................................... 481
A. BERNNAT, M. PAPADAKI and W. A. WAKEHAM

**Measurement of Thermal Diffusivity and Conductivity of a Multi-Layer Specimen with Temperature Oscillation Techniques** ....................................... 492
W. CZARNETZKI and W. ROETZEL

**Photoacoustic Measurements to Determine Thermal Diffusivities of Gases at Moderate Pressures** .................... 504
J. SOLDNER and K. STEPHAN

## SESSION 7: METALS

**Thermal Conductivities of Liquid Metals** ....................... 519
K. C. MILLS, B. J. MONAGHAN and B. J. KEENE

**Thermophysical Property Measurements on Single-Crystal and Directionally Solidified Superalloys into the Fully Molten Region** ............................... 530
J. B. HENDERSON and A. STROBEL

**Thermophysical Property Measurements for Casting Process Simulation** .......................................... 538
R. A. OVERFELT and R. E. TAYLOR

## SESSION 8: CONTACTS AND JOINTS

**Effect of Interstitial Materials on Joint Thermal Conductance** ...... 553
C. MADHUSUDANA and E. VILLANUEVA

**Analytical Models for Solid-Solid Contact in Thermal Transient States** ................................... 564
A. DEGIOVANNI, S. ANDRÉ and D. MAILLET

## SESSION 9: CERAMICS

**A Model for Thermal Conductivity of Irradiated $UO_2$ Fuel** ........ 579
P. G. LUCUTA, H. J. MATZKE, I. J. HASTINGS and P. G. KLEMENS

**Phonon Scattering in SIMFUEL—Simulated High Burnup $UO_2$ Fuel** ..................................... 590
P. G. KLEMENS and P. G. LUCUTA

**The Thermal Conductivity of "Ceramic Powders"** ............... 598
T. ASHWORTH and E. ASHWORTH

**SESSION 10: ORGANICS**

**Organic Compounds: Correlations and Estimation
Methods for Thermal Conductivity** . . . . . . . . . . . . . . . . . . . . . . . . . . . . . **613**
G. LATINI, G. PASSERINI, F. POLONARA and G. VITALI

**A Thermistor Based Method for Measuring Thermal
Conductivity and Thermal Diffusivity of Moist
Food Materials at High Temperatures** . . . . . . . . . . . . . . . . . . . . . . . . . . . **627**
M. F. van GELDER and K. C. DIEHL

**Measuring Thermal Properties of Elastomers
Subject to Finite Strain** . . . . . . . . . . . . . . . . . . . . . . . . . . . . . . . . . . . . . **639**
N. T. WRIGHT, M. G. DA SILVA, D. J. DOSS and J. D. HUMPHREY

*Author Index*     647

*Subject Index*     649

# Preface

The *23rd International Thermal Conductivity Conference* (*ITCC*) was held at the Sheraton Music City Hotel in Nashville, Tennessee on October 29 to November 1, 1995. The *23rd ITCC* was hosted by the Building Materials Program and the High Temperature Materials Laboratory User Program of the Oak Ridge National Laboratory.

The conference Chairmen were K. E. Wilkes, R. B. Dinwiddie, and R. S. Graves, all of the Oak Ridge National Laboratory. They were assisted by a scientific committee, which consisted of the following members:

| | |
|---|---|
| E. Ashworth (U.S.A.) | S. E. Gustafsson (Sweden) |
| T. Ashworth (U.S.A.) | D. P. H. Hasselman (U.S.A.) |
| J. V. Beck (U.S.A.) | J. B. Henderson (Germany) |
| J. R. Booth (U.S.A.) | P. G. Klemens (U.S.A.) |
| D. G. Cahill (U.S.A.) | D. L. McElroy (U.S.A.) |
| C. J. Cremers (U.S.A.) | R. E. Taylor (U.S.A.) |
| A. O. Desjarlais (U.S.A.) | T. W. Tong (U.S.A.) |
| D. R. Flynn (U.S.A.) | D. W. Yarbrough (U.S.A.) |
| Th. Franke (Germany) | |

The conference opened with greetings to the conference participants by the Chairmen and Dr. T. Ashworth, Chairman of the Governing Board of the International Thermal Conductivity Conferences. The conference was attended by 94 participants, who represented 14 different countries. Two keynote lectures and 60 technical papers were presented. Dr. D. L. McElroy, a preeminent expert on thermophysical properties who has participated in all 23 *ITCCs* and was Chairman of the *3rd ITCC*, gave a keynote lecture titled "Properties Are Controlled by Structures." Dr. M. Ramires, who received the C. F. Lucks Memorial Award at the *22nd ITCC*, gave a keynote lecture titled "Improved Correlations for the Thermal Conductivity of Propane and *n*-Butane." Both keynote lectures and 54 of the technical papers presented at the conference are published in this volume.

The conference banquet was held on October 31, 1995. R. S. Graves and R. B. Dinwiddie welcomed the participants. Roy Taylor and D. W. Yarbrough were presented the International Thermal Conductivity Award. J. Wood was given the C. F. Lucks Memorial Award to recognize him as an outstanding young researcher. Four individuals were named Fellows of the International Thermal Conductivity Conferences. They were: R. Roghini, C. J. Shirtliffe, T. W. Tong, and W. A. Wakeham.

Many people assisted with the organization of the conference. They included the members of the scientific committee and the following individuals: Norma Cardwell, Joy Kilroy, Thomas Kollie, and Therese Stovall. We would like to thank the following organizations for sponsoring the conference:

> *Anter Corporation*
> *Holometrix, Inc.*
> *LaserComp, Inc.*
> *Micropore International Limited*
> *National Physical Laboratory*
> *Netzsch Instruments, Inc.*
> *Sinku-Riko, Inc.*
> *Thermetrol, Inc.*
> *Theta Industries*
> *TopoMetrix*
> *U.S. Department of Energy*
> > *Building Materials Program*
> > *High Temperature Materials Laboratory User Program*

Mr. Peter S. Gaal will serve as the Conference Chairman for the *24th ITCC*. The *24th ITCC* will be held in Pittsburgh, Pennsylvania in October, 1997.

> KENNETH E. WILKES
> RALPH B. DINWIDDIE
> RONALD S. GRAVES
> *Chairmen, 23rd International Thermal*
> *Conductivity Conference*

# KEYNOTE LECTURE I

CHAIRS

R. B. Dinwiddie
R. S. Graves

# KEYNOTE LECTURE II

CHAIRS

T. Ashworth
K. E. Wilkes

# Physical Properties Are Controlled by Structures

D. L. McELROY*

## ABSTRACT

The specific heat of iron and steels depends on temperature, composition and heat treatment. Thermal electromotive force from a temperature gradient test can show thermoelement chemical and mechanical homogeniety.

Corrections are needed/possible/useful for: non-adiabaticity in adiabatic calorimeters; non-radial heat flow in radial heat flow systems; non-longitudinal heat flow in longitudial heat flow systems.

Heat conduction computer programs can be helpful in understanding intentional changes to an apparatus and in designing new test systems. The thermal resistivity of closed cell foams decreases with time due to intrusion of air and loss of foaming gas.

## INTRODUCTION

Welcome to Tennessee! I hope you enjoy visiting Nashville as much as I do. I would like to review some of the contributions to the International Thermal Conductivity Conferences (ITCC) that show structure controls properties. Since 1961 the ITCC has provided a forum for people with experience in the development and application of techniques to determine thermal conductivity and thermal diffusivity of materials over a broad temperature range to present their tests and results. A paper on the history of the International Thermal Conductivity Conferences was the first paper in Thermal Conductivity 22 (1).

─────* Consultant, Oak Ridge National Laboratory, Oak Ridge, TN; present address: 1204 Chason Drive, Birmingham, Alabama 35216, (205) 979-9717. This work was performed at Oak Ridge National Laboratory, managed by Lockheed Martin Energy Research Corp. for the U. S. Department of Energy under contract DE-AC05-96OR22464.

The subject, structure controls properties, is illustrated by a short story about Fort Morgan and Admiral Farragut. Part of this story was told to me by my Uncle W. I. (Buster) McElroy who lives in Mobile, Alabama and is 84. Fort Morgan was built at the mouth of Mobile Bay after the War of 1812 to guard the bay from enemy fleets. During the Civil War such forts were repeatedly defeated by steam-powered warships and this was dramatically demonstrated at Fort Morgan on August 5, 1864, when Union Admiral David G. Farragut led his fleet past the guns of the fort and into the bay with the loss of only one ship. At dawn on August 5, 1864, an 18-ship Union fleet commanded by Admiral Farragut steamed toward Fort Morgan and the entrance to Mobile Bay. The Confederate defenders of the fort opened fire. At 7:30 AM, as cannon fire reached a crescendo, the leading Union monitor, the Tecumseh, struck a mine (known as a torpedo during the Civil War) and sank within a minute, taking most of the crew down with her. This sudden disaster threw the Union fleet into confusion causing them to hesitate under the guns of Fort Morgan. At this critical moment Farragut gave his famous order, "Damn the torpedoes, full speed ahead!", which lead the remaining vessels past the fort through the mine field and into Mobile Bay. Farragut's message has the property of inspiration.

But my Uncle Buster says, the public relations people got the story wrong. According to my Uncle Buster, Farragut actually said, "Damn! Torpedoes! Full Speed Ahead!" This simple change in the structure of his comment yields a message with the property of humor!

## EXAMPLES OF STRUCTURE/PROPERTY CHANGES

Structure can include many technical aspects. For the purposes of this review, it includes the morphology of phases present, the crystal structure of the material, the chemical composition, the scatterers of electrons and phonons, as well as the construction of test equipment. Several examples are discussed below.

The structures (and properties) obtained for iron-carbon alloys, such as steels, can be changed from austenite to pearlite to martensite and these structures depend on temperature, composition, and heat treatment. Studies of this reaction in high purity eutectoid Fe-C and Fe-C-Mn steels have been carried out using an isothermal transformation apparatus, that was developed at ORNL (2-5). As shown in Figure 1, the apparatus consisted of an austenitizing furnace, a stirred lead bath, and a mercury bath, all enclosed in an inert argon atmosphere with a movable device for transferring the specimen from the furnace to the quenching baths. The experimental method of determining the growth rates consisted of heating the specimen into the austenite range and quenching to a predetermined subcritical temperature for a fixed time, after which it was quenched to room temperature. The rate of growth was determined by metallographic examinations of the reacted specimen. These studies showed that the rate of growth of pearlite from austenite is a function of temperature and significantly reduced by the addition of manganese to a eutectoid steel.

Figure 1. The Isothermal Transformation Apparatus.

Thermodynamic data on these steels were obtained using a high temperature adiabatic calorimeter developed at the University of Tennessee by E. E. Stansbury, M. L. Picklesimer, G. E. Elder, R. E. Pawel, and D. L. McElroy (6-8). If adiabatic conditions are assumed, the apparent specific heat, $C_p^a$, is calculated usually over a 20°C temperature interval, $\Delta T$, from the heater power, EI, and the time required, $\Delta t$, for a specimen of mass, m, to heat through the temperature interval.

$$C_p^a = \frac{E \cdot I \cdot \Delta t}{m \cdot \Delta T} \tag{1}$$

However, in any real system, nonadiabatic conditions exist. The true specific heat, $C_p$, can be calculated from $C_p^a$ by measuring the effect of heating rate, R, on the specific heat or by measuring the drift of the specimen temperature when no heater current is applied or by measuring with specimens of significantly different mass.

$$C_p = \frac{C_p^a}{1 - (C_p^a \cdot R/K)} \tag{2}$$

Equation 2, where K is the heating rate for adiabatic condictions, yields specific heats which agree to within 0.5%. Figure 2 shows the specific heat of iron and two steels of eutectoid composition from 80 to 950°C. The results for iron show the sharp rise in

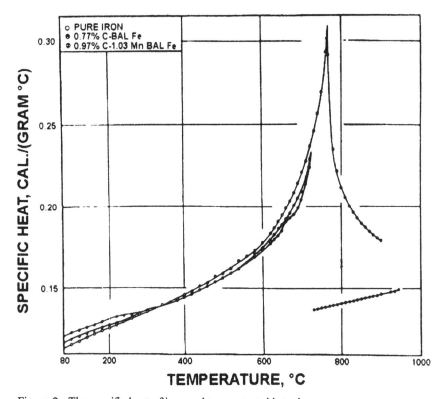

Figure 2. The specific heat of iron and two eutectoid steels.

 the specific heat near the Curie temperature and the drop in specific heat at the alpha-gamma phase transformation at 912°C. The enthalpy of transformation was 215±2 cal/g atom. For the steels, the gamma phase is stable above 728°C, which permitted measurement of the specific heat over a wide temperature range. Although not shown in Figure 2, the specific heat for one steel did not drop rapidly to the values expected for gamma iron above the transformation temperature. Instead agreement was not reached until 850° C, when carbon gradients were removed. After the austenite was homogenized, the specific heat agreed with gamma iron data (8).

Satisfactory operation of an adiabatic calorimeter requires accurate temperature measuring devices. Platinum and platinum-13% rhodium thermocouples provide stable and reproducible thermal emfs and are much superior to available base metal thermocouples for the precise measurements involved, even considering the low thermal potential developed. One way to check for thermoelectric homogeniety is the unilateral temperature gradient furnace (9), which is shown in Figure 3. This apparatus compares the thermal emf of a length of wire passing through a temperature gradient to that of a wire from the same spool or a wire of known homogeneity in a second fixed temperature gradient. An emf-distance trace for as-received Alumel wire had an average offset of about 25 microvolts, and a distinct cycle in the thermal emf was observed for the entering wire. On leaving the furnace, the average was observed to have shifted due to the metallurgical process of recovery, however the cyclic behavior was still apparent. This cyclic behavior lent itself to a more interpretative experiment, i.e., by a wire length calibration at the melting point of lead for an Alumel-platinum thermocouple. Deviation versus distance showed the same cyclic behavior observed in the homogeneity test.

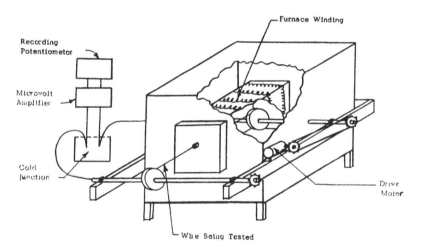

Figure 3. Schematic Drawing of the Unilateral Temperature Gradient Furnace.

## ABSOLUTE RADIAL AND LONGITUDINAL HEAT FLOW SYSTEMS

The design target for one-directional heat flow systems is to minimize unwanted heat flow. Construction and operation for longitudinal systems should avoid radial heat flow, and construction and operation for radial systems should avoid longitudinal heat flow. When the
required quantities are measured directly, the methods are called absolute methods.

"Radial-heat-flow methods offer no panacea for problems of measuring k, and no attempt is made to palliate their shortcomings. They do have some unique theoretical and practical advantages that through careful experimental work can yield good results over wide k and temperature ranges" (10). The ORNL guarded radial heat flow apparatus has been previously described in detail (10). It has a most probable error of ±1.5%, if an isothermal intercomparison is used. This can correct for any calibration differences between the thermocouple wires and for non-radial heat flow. The temperature difference $\Delta T$ used to calculate thermal conductivity was

$$\Delta T = \Delta T \text{ (data)} - \Delta T \text{ (iso)} \qquad (3)$$

where the temperature difference-data (obtained when the core heater is energized) is corrected for the temperature difference-iso (obtained when there is no power to the core heater).

"Of the methods for measuring thermal conductivity of solids (at high temperatures), those employing linear and quasi-linear heat flow are the most important group. For example the Thermophysical Properties Research Center (TPRC) Data books list about 800 references pertaining to the measurement of thermal conductivity of solids, of which about 600 employ essentially linear heat flow. These divide about equally into work below and above room temperature" (11). The ORNL low temperature longitudinal heat flow apparatuses have been described in detail (12-14) and obtain accurate measurements of thermal conductivity, electrical resistivity and Seebeck coefficient between 78 and 400°K on rod specimens 5 to 8 cm long. Obtaining these three properties is essential, if one is to identify the electronic and lattice components of k. Results from the ORNL high accuracy apparatus are with ±0.2% of an equation for Mo.

## RADIAL AND LONGITUDINAL APPARATUSES FOR THERMAL INSULATIONS

A computer program was used to evaluate the system performance of an ASTM C 335 pipe insulation tester for a variety of input conditions (15, 16). This simulation showed the end guards do not have to be in exact balance to obtain accurate thermal conductivity values. Calculations were made on an "ideal" pipe tester to check the computer code being used for simulation purposes. This ideal tester involved two regions, an outer cylinder of pipe insulation and an inner cylinder. Figure 4 shows an unguarded pipe tester modeled after the ideal tester. A crucial element of the design is the selection of a core heater with a low thermal conductivity. This pipe tester confirmed the simulation results for the ideal pipe tester. An unguarded longitudinal heat flow apparatus for measuring the apparent thermal conductivity of insulations from 300 to 330 K, was derived from the unguarded cylindrical screen heater pipe insulation tester, and was adopted as ASTM C 1114 in 1989 (17, 18).

The ORNL heat flow meter apparatus (HFMA) meets C 518 and an interlaboratory comparison with other HFMA apparatuses on foam insulation boards showed a two standard deviation value of 2.2% (19, 20). The ORNL HFMA has been used to measure the thermal performance of board stock foamed with hydrochlorofluorocarbon gases as alternatives to chlorofluorocarbon gases. Three thicknesses of specimens planed from the boardstock, 33, 19 and 10 mm, were aged at 75 and 150° F. The apparent thermal conductivity results from the HFMA at each test time showed the expected order with thickness, i.e.,

$$k \ (32 \ mm) < k \ (19 \ mm) < k \ (10 \ mm).$$

All of the thin-specimen aging results show a nonlinear increase in apparent thermal conductivity with time/thickness$^2$. If one plots ln of the thermal conductivity versus (time)$^{1/2}$/thickness, then the test data for the specimens of three thicknesses describe two distinct linear regions. It is believed that the first linear region corresponds to the influx of air components and the second linear region corresponds to the loss of the blowing agent from the foam. Analysis of this provides a prediction of the thermal performance as a function of time for various foams. ASTM has developed a standard based on this protocol, ASTM C 1303 (21).

## ACKNOWLEDGEMENT

I thank the following people who helped me prepare this presentation: Peter Scofield, DOE, Program Manager; Kenneth E. Wilkes, ORNL, Program Manager; Ralph B. Dinwiddie, ORNL; T. Gordon Godfrey, ORNL; Ronald S. Graves, ORNL and R & D Services; Fred J. Weaver, ORNL; David W. Yarbrough, ORNL, TTU, and R & D Services; and Ray E. Taylor, Purdue University. In addition I thank all of my fellow workers whose contributions I have reported and who have helped me maintain a proper perspective about my efforts in this field.

Figure 4. Unguarded cylindrical screen pipe insulation tester.

## REFERENCES

1. C. F. Lucks and H. J. Sauer, Jr., "International Thermal Conductivity Conferences, 1961-1993: A Historic Profile", pp. 3-29, Thermal Conductivity 22, Ed. T. W. Tong, Technomic Publishing Co., 1994.

2. Marion L. Picklesimer, "The Austenite-Pearlite Reaction in a 1 per cent Manganese High-Purity Eutectoid Steel", M. S. Thesis, University of Tennessee, 1951.

3. David L. McElroy, "Isothermal Transformations on the Austenite Sorbite Reaction in Three High-Purity Eutectoid Steels", M. S. Thesis, Metallurgy, University of Alabama, January, 1953.

4. J. H. Frye, Jr., E. E. Stansbury, and D. L. McElroy, "Absolute Rate Theory Applied to Rate of Growth of Pearlite". Transactions AIME, Journal of Metals, Volume 197, pp. 219-224, February 1953.

5. M. L. Picklesimer, D. L. McElroy, T. M. Kegley, Jr., E. E. Stansbury, and J. H

Frye, Jr., "Effects of Manganese on the Austenite-Pearlite Transformation", Transactions of the Metallurgical Society of AIME, Volume 218, PP. 473-480, June, 1960.

6. E. E. Stansbury, D. L. McElroy, M. L. Picklesimer, G. E. Elder, and R. E. Pawel, "Adiabatic Calorimeter for Metals in the Range 50 to 1000°C", The Review of Scientific Instruments, Vol. 30, No. 2, 121-126, February, 1959.

7. Richard E. Pawel, "The Application of Dynamic Adiabatic Calorimetry to the Copper-Nickel System from 50 to 620°C", University of Tennessee, Ph. D. Thesis, Metallurgy, August, 1956.

8. David L. McElroy, "Applications of Dynamic Adiabatic Calorimetry to the Iron - Iron Carbide System from 50 to 950°C", Ph. D. Thesis, Metallurgy, University of Tennessee, August, 1957.

9. J. F. Potts and D. L. McElroy, "The Effects of Cold Working, Heat Treatment, and Oxidation of the Thermal emf of Nickel-Base Thermoelements", Temperature-Its Measurement and Control in Science and Industry, 243-264, Vol. 3, Part 2, Reinhold Publishing Co., 1963.

10. D. L. McElroy and J. P. Moore, "Radial Heat Flow Methods for the Measurement of the Thermal Conductivity of Solids", Chapter 4 in Volume 1, Thermal Conductivity, R. P. Tye, Ed., Academic Press, New York, 1969.

11. M. J. Laubitz, "Measurement of the Thermal Conductivity of Solids at High Temperatures by using Steady-state Linear and Quasi-linear Heat Flow", Chapter 3 in Volume 1, Thermal Conductivity, R. P. Tye, Ed., Academic Press, New York, 1969.

12. J. P. Moore, D. L. McElroy, and R. S. Graves, "Thermal Conductivity and Electrical Resistivity of High-purity Copper from 78 to 400°K", Can. J. Phys. 45, 3849, 1967.

13. M. J. Laubitz and D. L. McElroy, "Precise Measurement of Thermal Conductivity at High Temperatures (100 - 1200 K)", Metrologia, Vol. 7, No. 1, pp.1-15, January 1971.

14. J. P. Moore, R. K. Williams, and R. S. Graves, "Precision measurements of the thermal conductivity, electrical resistivity, and Seebeck coefficient from 80 to 400 K and their application to pure molybdenum", Rev. Sci. Instrum., Vol. 45, No. 1, pp. 87-95, January, 1974.

15. S. H. Jury, D. L. McElroy, and J. P. Moore, "Pipe Insulation Testers", Thermal Transmission Measurements of Insulations, ASTM STP 660, R. P. Tye, Ed., American Society for Testing and Materials, pp. 310-326, 1978.

16. ASTM C 335-89, <u>Standard Test Method for Steady-State Heat Transfer Properties of Horizontal Pipe Insulation</u>, pp. 70-79, Vol. 04.06, 1994 Annual Book of ASTM Standards (1994).

17. R. S. Graves, D. W. Yarbrough, and D. L. McElroy, "Apparent Thermal Conductivity Measurements by an Unguarded Technique", pp. 339-355, <u>Thermal Conductivity 18</u>, T. Ashworth and David R. Smith, Ed., Plenum Publishing Corporation, 1985.

18. ASTM C 1114-92, <u>Standard Test Method for Steady-State Thermal Transmission Properties by Means of the Thin-Heater Apparatus</u>, pp. 598-606, Vol. 04.06, 1994 Annual Book of ASTM Standards (1994).

19. Ronald S. Graves, David L. McElroy, and Thomas G. Kollie, "Evaluation of a commercial, computer-operated heat flow meter apparatus", <u>Rev. Sci. Instrum.</u> 64(7), pp. 1961-1970, July 1993.

20. ASTM C 518-91, <u>Standard Test Method for Steady-State Heat Flux Measurements and Thermal Transmission Properties by Means of the Heat Flow Meter Apparatus</u>, pp. 150-161, Vol. 04.06, 1994 Annual Book of ASTM Standards (1994).

21. ASTM C 1303-95, <u>Standard Test Method for Estimating The Long-Term Change in The Thermal Resistance of Unfaced Rigid Closed Cell Plastic Foams by Slicing and Scaling Under Controlled Laboratory Conditions</u>. To be published in Vol. 04.06, 1996 Annual Book of ASTM Standards.

# Improved Correlations for the Thermal Conductivity of Propane and *n*-Butane*

M. L. V. RAMIRES, C. A. NIETO DE CASTRO,
L. CUSCO and R. A. PERKINS

## ABSTRACT

New experimental data on the thermal conductivity of propane and n-butane have been reported since the last correlations proposed by Holland et al. [1] in 1979 and Younglove et al. [2] in 1987. These new experimental data, covering a temperature range of 110 K - 700 K and a pressure range of 0.1 MPa - 70 MPa, are used together with the previously available data to develop an empirical correlation for the thermal conductivity of gaseous and liquid propane. The quality of the data is such that the thermal conductivity correlation for propane is estimated to have an uncertainty of about ±5% except near the critical region, where the uncertainty in the equation of state of Younglove et al. [2], increases the uncertainty of the correlation to 8% at the 95% confidence level.

For n-butane, the experimental data for the thermal conductivity is scattered by ±10%. Only a preliminary empirical correlation can be proposed for n-butane, since additional measurements with high accuracy are needed. The thermal conductivity of n-butane can be estimated with this correlation with an uncertainty of about ±5% at the 95% confidence level, except the critical region where the only available set of data shows deviations up to 8%.

M. L. V. Ramires, C. A. Nieto de Castro, Centro de Ciência e Tecnologia de Materiais, Faculdade de Ciências da Universidade de Lisboa, Campo Grande, Edifício C1, Piso 1, 1700 Lisboa, PORTUGAL
L. Cusco, R. A. Perkins, Thermophysics Division, National Institute of Standards and Technology, 325 Broadway, Boulder, CO 80303, USA

## INTRODUCTION

Thermophysical properties of industrially important fluids, such as propane and n-butane, are required for the design of chemical processes and equipment. Recently, the search for environmentally acceptable halocarbons has renewed the idea of using propane and n-butane as refrigerants with zero ozone-depletion potential. The growing demand for information on these fluids motivated an effort to develop empirical correlations based on accurate experimental data, covering the entire fluid region, to represent their thermophysical properties as functions of temperature and pressure (or density).

Two empirical correlations for the thermal conductivity of propane [1,2] and one for n-butane [2] were published based on comprehensive analysis of the data which were available. The first, in 1979, by Holland et al. [1] for propane at temperatures from 140-500 K with pressures up to 50 MPa has an estimated uncertainty of 8% outside the critical region and 15% near the critical point. The second, in 1987, by Younglove et al. [2] for propane and for n-butane at temperatures from 85 - 600 K with pressures up to 100 MPa has an estimated uncertainty of 5% outside the critical region and an uncertainty of 10% near the critical point. New sets of experimental data which cover a wide range of temperatures and pressures with high accuracy have been reported. The new data offer improved coverage in the critical region, where almost no data existed, prompting the development of new empirical based correlations described in this paper.

## METHODOLOGY

The thermal conductivity is represented as a sum of three contributions

$$\lambda(\rho, T) = \lambda_0(T) + \Delta\lambda_e(\rho, T) + \Delta\lambda_c(\rho, T) \tag{1}$$

where $\lambda_0$ is the dilute gas thermal conductivity which is only dependent on the temperature, $\Delta\lambda_e$ is the excess thermal conductivity, and $\Delta\lambda_c$ is the thermal conductivity enhancement in the critical region, both of which are dependent on density and temperature. This representation is useful as it allows these contributions to be analyzed independently using the best available theory. Theory is incorporated into the present correlations for both the dilute gas thermal conductivity and the thermal conductivity enhancement in the critical region.

To analyze the thermal conductivity in terms of density and temperature, the density of the fluid must be evaluated from the temperatures and pressures reported by the authors. The modified Benedict-Webb-Rubin (MBWR) equations of state [2], valid for propane at pressures up to 100 MPa with temperatures between the triple point and 600 K, and for n-butane at pressures up to 70 MPa with temperatures between the triple point and 500 K, have been employed here. The experimental data which are available were generally reported prior to 1990 and so all calculations are based on the IPTS68 temperature scale. Furthermore, the densities are in mol/L and the temperatures are in K.

The available experimental data are grouped into categories of primary and secondary data according to accepted criteria for evaluation of experimental data [3]. However, to fulfill those requirements, all the data prior to 1982 except one point would be considered secondary and only on four sets of data could be considered primary for use in the correlation development. This would restrict the present correlations to a very limited range of temperature and pressure. Consequently, it was decided to exclude only the relative measurements, and to include absolute results with greater uncertainty but which are consistent with other more accurate data. A summary of the primary data for propane and n-butane, together with the ranges in temperature and pressure, experimental technique, and assigned uncertainty is given in Tables I and II respectively.

TABLE I. PRIMARY EXPERIMENTAL DATA FOR THE THERMAL CONDUCTIVITY OF PROPANE

| References | Temperature Range / K Pressure Range / MPa | Technique[a] | Number of points | Estimated Uncertainty % |
|---|---|---|---|---|
| Mann et al. [5] | 275 - 285 ~ 0.1 | SSHW | 6 | 5 |
| Leng et al. [6] | 323 - 413 0.1 - 30 | CC | 83 | 5 |
| Smith et al.[7] | 323 - 423 0.1 | CC | 3 | 5 |
| Carmichael et al. [8] | 277 - 444 0.1 - 35 | SC | 33 | 5 |
| Brykov et al. [9] | 93 - 223 0.1 | HF | 14 | 5 |
| Clifford et al. [10] | 303 0.1 | THW | 1 | 1 |
| Mahesh et al. [11] | 400 - 725 0.1 - 0.6 | CC | 41 | 3 |
| Roder et al. [12,13] | 110 - 300 1 - 70 | THW | 400 | 1.5 |
| Zheng et al.[14] | 323.75 0.1 - 1 | CC | 6 | 3 |
| Tufeu et al. [15] | 298 - 578 1 - 70 | CC | 175 | 2 |
| Prasad et al. [16] | 192 - 320 0.2 - 70 | THW | 128 | 1.5 |
| Yata et al. [17] | 254 - 315 1 - 30 | THW | 16 | 1 |

[a] SSHW, Steady State Hot Wire; CC, Concentric Cylinders, Steady State ; SC Spherical Cell, Steady State ; HF, Heated Filament, Steady State; THW, Transient Hot Wire

TABLE II. PRIMARY EXPERIMENTAL DATA FOR THE THERMAL CONDUCTIVITY OF
n-BUTANE

| References | Temperature Range / K Pressure Range / MPa | Technique[a] | Number of points | Estimated Uncertainty % |
|---|---|---|---|---|
| Mann et al. [5] | 275 - 285 ~ 0.1 | SSHW | 5 | 5 |
| Smith et al.[7] | 323 - 423 0.1 | CC | 5 | 5 |
| Kramer et al. [18] | 348 - 427 0.1 - 100 | CC | 97 | 5 |
| Carmichael et al. [19] | 277 - 444 0.1 - 35 | SC | 45 | 5 |
| R. Kandiyoti [20] | 148 - 252 0.1 | THW | 42 | 2 |
| Brykov et al. [9] | 143 - 273 0.1 | HF | 14 | 5 |
| Parkinson et al. [21] | 323-373 0.1 | SSHW | 2 | 2 |
| Nieto de Castro et al [22] | 298 - 601 0.1 - 70 | CC | 215 | 2 |
| Yata et al. [17] | 258 - 335 1 - 20 | THW | 15 | 1 |
| Cusco et al. [23][1] | 303 - 600 0.2 - 70 | THW | 1295 | 1.5 |

[a] SSHW, Steady State Hot Wire; CC, Concentric Cylinders, Steady State ; SC Spherical Cell, Steady State ; THW, Transient Hot Wire; HF, Heated Filament, Steady State

## CORRELATION PROCEDURES AND RESULTS

### THE ZERO-DENSITY LIMIT

The term zero-density limit refers to those data resulting from mathematical extrapolation of a density series, for a particular transport property at constant temperature, to zero density [4]. Consequently, it does not represent a real physical situation and its value must be inferred from the experimental values of the gas phase thermal conductivity at low densities. Table III summarizes the selected primary data used for correlating the zero density thermal conductivity of propane and n-butane. Some of the reported experimental data do not allow extrapolation to zero density and so the present correlation is based on some data extrapolated to the zero density limit and other data at densities corresponding to atmospheric pressure. The difference between $\lambda_0$ and $\lambda(p=1 \text{ bar})$ is made smaller than the accuracy of the fit, even at subcritical temperatures very near the critical temperature.

---

[1] Preliminary results that will be subjected to a future analysis in a way to avoid the burdening of the fit of the excess thermal conductivity correlation derived from the large number of points

TABLE III. PRIMARY EXPERIMENTAL DATA FOR THE THERMAL CONDUCTIVITY OF PROPANE AND n-BUTANE IN THE DILUTE GAS PHASE

| References | Temperature Range / K | Technique | Number of Points | Estimated Uncertainty % |
|---|---|---|---|---|
| PROPANE | | | | |
| Mann et al. [5][#] | 275 - 285 | SSHW | 6 | 5 |
| Leng et al. [6][#] | 323 - 413 | CC | 5 | 5 |
| Smith et al. [7] | 323 - 423 | CC | 3 | 5 |
| Carmichael et al. [8] | 277 - 444 | SC | 6 | 5 |
| Clifford et al. [10] | 303 | THW | 1 | 5 |
| Mahesh et al. [11][#] | 400 - 725 | CC | 8 | 3 |
| Zheng et al. [14][#] | 323.75 | CC | 1 | 3 |
| Tufeu et al. [15][#] | 298 - 578 | CC | 5 | 2 |
| n-BUTANE | | | | |
| Mann et al. [5][#] | 275 - 285 | SSHW | 6 | 5 |
| Smith et al. [7] | 323 - 423 | CC | 3 | 5 |
| Kramer et al. [18] | 348 - 427 | CC | 97 | 5 |
| Carmichael et al. [19] | 277 - 444 | SC | 6 | 5 |
| Parkinson et al. [21] | 323 - 373 | SSHW | 2 | 2 |
| Nieto de Castro et al [22][#] | 298 - 601 | CC | 9 | 2 |
| Cusco et al. [23][#] | 303 - 600 | THW | 1 | 1.5 |

[#] Extrapolated to zero density

The experimental thermal conductivity data have been fitted to a quadratic polynomial in temperature:

$$\lambda_0(T) = A_1 + A_2 T + A_3 T^2 \qquad (2)$$

where $\lambda_0$ is the dilute gas thermal conductivity in W m$^{-1}$ K$^{-1}$ and $T$ the absolute temperature in K. The data have been fitted to this equation using the method of least squares with weighting factors reflecting the accuracy of the data as given in Table II.

Table IV contains the coefficients $A_i$ together with their standard deviations. The maximum deviation of the primary experimental data is 4% for propane and 5% for n-butane. The deviations of the primary data from the correlation are presented in Fig. 1, and Fig. 2 respectively for propane and n-butane, where it is clearly shown that almost all the data are reproduced within their assigned experimental uncertainty.

TABLE IV COEFFICIENTS FOR THE THERMAL CONDUCTIVITY IN THE LIMIT OF ZERO DENSITY, EQ. (2), WITH THE RESPECTIVE STANDARD DEVIATIONS

| | PROPANE | n-BUTANE |
|---|---|---|
| $A_1$ | $-1.121 \times 10^{-2} \pm 1.08 \times 10^{-3}$ | $-3.380 \times 10^{-3} \pm 1.05 \times 10^{-3}$ |
| $A_2$ | $8.17 \times 10^{-5} \pm 5.3 \times 10^{-6}$ | $2.39 \times 10^{-5} \pm 5.3 \times 10^{-6}$ |
| $A_3$ | $6.39 \times 10^{-8} \pm 6.3 \times 10^{-9}$ | $1.38 \times 10^{-7} \pm 6.7 \times 10^{-9}$ |

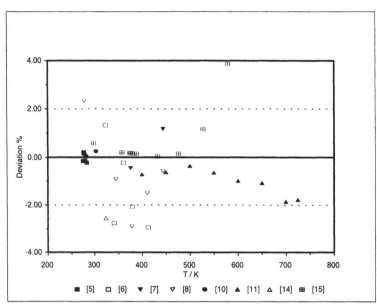

**Figure 1:** Deviations of the primary experimental data in the dilute gas phase from Equation 2 for propane

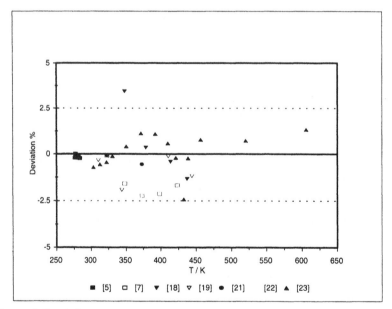

**Figure 2:** Deviations of the primary experimental data in the dilute gas phase from Equation 2 for n-butane

## THE EXCESS THERMAL CONDUCTIVITY AND THE CRITICAL ENHANCEMENT

The thermal conductivity of fluids exhibits an anomalous enhancement over a wide range of densities and temperatures around the critical point, becoming infinite at the critical point itself. This behavior can be accurately described by mode-coupling crossover equations for the thermal conductivity, but application of this theory requires a knowledge of the background thermal conductivity away from the critical point [24]. The density dependent terms for the thermal conductivity can be grouped according to Equation (3)

$$\lambda(\rho,T) = \overline{\lambda}(\rho,T) + \Delta\lambda_c(\rho,T) \tag{3}$$

where $\overline{\lambda}(\rho,T) = \lambda_0(T) + \Delta\lambda_e(\rho,T)$ is the background contribution.

In order to assess the critical enhancement either theoretically or empirically, the background contribution, which is the sum of the dilute gas and the excess thermal conductivity, must be characterized. The scheme adopted in this analysis was to use ODRPACK V. 2.01 [25] to fit all the primary data simultaneously for the excess thermal conductivity and the critical enhancement, fixing the parameters already obtained for the thermal conductivity of the dilute gas phase. This was not possible in the past with less robust nonlinear fitting programs.

The excess thermal conductivity was represented with a polynomial in temperature and density of the form:

$$\Delta\lambda_e(\rho,T) = \sum_{i=1}^{4}\sum_{j=0}^{2}\frac{B_{i,j}\rho^i}{T^j} \tag{4}$$

with $B_{3,2}=B_{4,2}$, $\rho$ expressed in mol/L and T in K.

For engineering purposes the empirical expression of the critical enhancement is given by Equation 5. This equation does not require accurate information on the compressibility and specific heat of the fluids in the critical region as does the theory of Olchowy and Sengers [24, 25].

$$\Delta\lambda_c(\rho,T) = \frac{C_1}{C_2 + \Delta T_c^2}\exp\left[-\left(C_3 \times \Delta\rho_c\right)^2\right] \tag{5}$$

where $\Delta T_c = \dfrac{T}{T_c} - 1$ and $\Delta\rho_c = \dfrac{\rho}{\rho_c} - 1$. Tables V and VI contain the coefficients for Equations 4 and 5 together with their standard deviations.

To demonstrate consistency with theory the thermal conductivity of propane and n-butane the critical enhancement is calculated with the theoretically based crossover equations proposed by Olchowy and Sengers [23, 25], fixing the parameters $B_{ij}$ obtained in the empirical fit of Equation 4. Figures 3 and 4 show a plot of deviations of the experimental data from the total correlation using this

TABLE V: COEFFICIENTS FOR THE REPRESENTATION OF THE EXCESS THERMAL
CONDUCTIVITY, EQ. (4), WITH THE RESPECTIVE STANDARD DEVIATIONS

| | i = 1 | i = 2 | i = 3 | i = 4 |
|---|---|---|---|---|
| | **PROPANE** | | | |
| j = 0 | $0.00745 \pm 0.00182$ | $-0.00013 \pm 0.00034$ | $-4.6 \times 10^{-5} \pm 2.95 \times 10^{-5}$ | $6.49 \times 10^{-6} \pm 8.78 \times 10^{-7}$ |
| j = 1 | $-4.783 \pm 1.4278$ | $0.61750 \pm 0.19929$ | $-0.04479 \pm 0.01463$ | $0.001190 \pm 0.00045$ |
| j = 2 | $1089.9 \pm 273.31$ | $-75.3401 \pm 22.00$ | $0.04361 \pm 0.021254$ | $0.04361 \pm 0.021254$ |
| | **n-BUTANE** | | | |
| j = 0 | $-0.01458 \pm 0.00152$ | $0.00250 \pm 0.00042$ | $-0.00014 \pm 5.33 \times 10^{-5}$ | $1.63 \times 10^{-5} \pm 2.61 \times 10^{-6}$ |
| j = 1 | $15.5476 \pm 1.2455$ | $-1.6254 \pm 0.2496$ | $-0.08484 \pm 0.02415$ | $0.00686 \pm 0.00114$ |
| j = 2 | $-3351.3 \pm 255.1$ | $509.00 \pm 32.05$ | $-1.2398 \pm 0.0736$ | $-1.2398 \pm 0.0736$ |

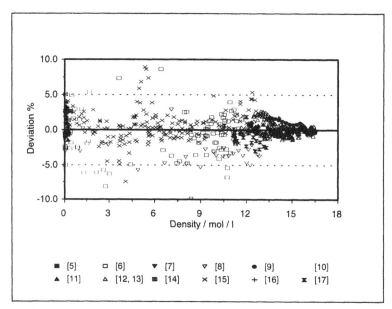

**Figure 3:** Deviations of the primary experimental data from the present correlation
(Eq. 1) using the Olchowy and Sengers Theory [24] for the critical enhancement for
propane

TABLE VI: COEFFICIENTS FOR THE REPRESENTATION OF THE CRITICAL
ENHANCEMENT OF THE THERMAL CONDUCTIVITY, EQ. (5), WITH THE RESPECTIVE
STANDARD DEVIATIONS

| | PROPANE | n-BUTANE |
|---|---|---|
| $C_1$ | $2.81 \times 10^{-5} \pm 8.33 \times 10^{-7}$ | $1.48 \times 10^{-5} \pm 5.61 \times 10^{-7}$ |
| $C_2$ | $1.03 \times 10^{-3} \pm 6.17 \times 10^{-5}$ | $4.76 \times 10^{-4} \pm 2.35 \times 10^{-5}$ |
| $C_3$ | $-3.0602 \pm 5.56 \times 10^{-2}$ | $-3.1226 \pm 3.80 \times 10^{-2}$ |

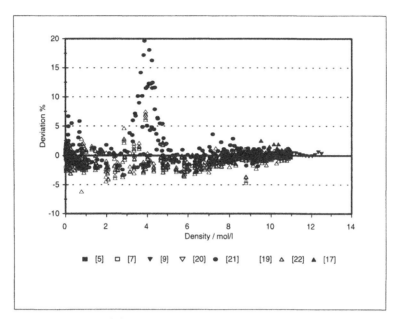

**Figure 4:** Deviations of the primary experimental data from the present correlation (Eq. 1) using the Olchowy and Sengers Theory [24] for the critical enhancement for n-butane

theory for the representation of the critical enhancement. It can be seen that in the critical region, and in spite of the lack of a crossover equation of state, the data are reproduced to within ± 10 % for propane and 20 % for n-butane, though a systematic trend is observed in that region. However, outside this region almost all data are within a band of ± 5 %. It should be noted that only one set of experimental data is available for n-butane in the critical region so the correlation is heavily influenced by this data set.

## DISCUSSION

The final equations for the thermal conductivity are given by Equation (1), with the individual contributions given by Equations (2), (4) and (5) respectively. This representation is valid in the temperature range from 192 K $\leq$ T $\leq$ 725 K over the density range $0 \leq \rho \leq 17$ mol/L for propane and 125 K $\leq$ T $\leq$ 600 K over the density range $0 \leq \rho \leq 13$ mol/L for n-butane.

The deviations of the primary data from the total correlation are presented in Figure 5 and Figure 6, respectively, for propane and n-butane, where it is clearly shown that, outside the critical region, almost all data are reproduced within ± 5 %.

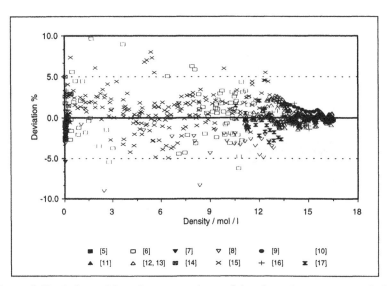

**Figure 5:** Deviations of the primary experimental data from the present correlation (Eq. 1) for propane

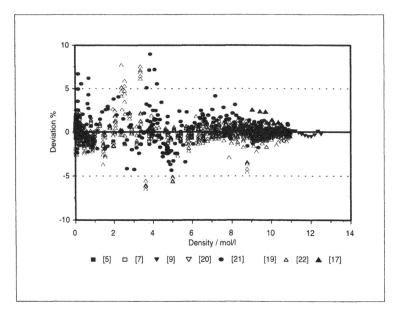

**Figure 6:** Deviations of the primary experimental data from the present correlation (Eq. 1) for n-butane

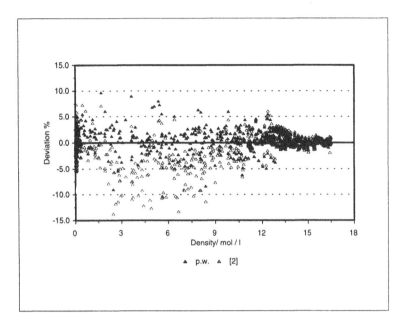

**Figure 7:** Deviations of the primary experimental data from the present correlation (Eq. 1) (base line) and from the correlation of Ref. 2

An overall comparison between the experimental data and our correlation and from the correlation of ref. 2 is shown in figure 7 for propane. The departures are small for densities below 2 mol/L, that is in the low density region, and higher than 8 mol/L. However, at moderate densities, near the critical density, the correlation of ref. 2 is systematically lower than the present. The largest deviations between the two correlations are found in the density range where new, more accurate, data have become available.

**CONCLUSIONS**

Improved empirical correlations are presented for propane (192 K $\leq$ T $\leq$ 725 K; 0 $\leq$ $\rho$ $\leq$ 17 mol/L) and n-butane (125 K $\leq$ T $\leq$ 600 K; 0 $\leq$ $\rho$ $\leq$ 13 mol/L). For temperatures higher than 600 K, these correlations must be considered with caution, since the equation of state employed is valid only to 600 K for propane and 500 K for n-butane and any extrapolation beyond this is subject to error. Nevertheless, an uncertainty of $\pm$ 5% is estimated for the thermal conductivity values obtained with this present correlation in regions where $\rho \leq 0.75 \times \rho_c$ and $\rho \geq 1.25 \times \rho_c$ and $\pm$ 10% for those values within $0.75 \times \rho_c < \rho < 1.25 \times \rho_c$, which is satisfactory for many engineering purposes.

## ACKNOWLEDGMENTS

One of us (M. L. V. Ramires) thanks the Faculty of Sciences of the University of Lisbon for a leave of absence, the Thermophysics Division of the National Institute of Standards and Technology for support as guest researcher and JNICT, Portugal for an additional grant.

## REFERENCES:

1. P. M. Holland, H. J. M. Hanley, K. E. Gubbins and J. M. Haile, *J. Phys. Chem. Ref. Data*, **8**, 559 - 575, (1979)
2. B. Younglove and J. F. Ely, *J. Phys. Chem. Ref. Data*, **16**, 577 - 798, (1987)
3. C. A. Nieto de Castro, S. F.Y. Li, A. Nagashima, R. D. Trengove and W. A. Wakeham, *J. Phys. Chem. Ref. Data* **15**, 1073 (1986)
4. J. Millat, V. Vesovic and W. A. Wakeham, Transport Properties of Dilute Gases and Gaseous Mixtures, Chap 4. in *Transport Properties of Fluids. Their Correlation, Prediction and Estimation*, Capítulo 16, Ed. J. Millat, J.H. Dymond, C.A. Nieto de Castro, Cambridge University Press, Oxford (1994), in Press
5. W. B. Mann., B.G. Dickins, *Proc. Roy. Soc. (London)* **134A**, 77 (1932)
6. D. E. Leng and E. W. Comings, *Industrial and Engineering Chemistry*, **49**, 2042 (1957)
7. W. J. Smith, L. D. Durbin and R. Kobayashi, *J. Chem. Eng. Data*, **5**, 316 (1960)
8. L. T. Carmichael, J. Jacobs, and B. H. Sage, *J. Chem. Eng. Data*, **13**, 40 (1968)
9. V. P. Brykov, G. Kh. Mukhamedzyanov and A. G. Usmanov, *J. Engr. Phys. (USSR)*, **18**, 62 (1970)
10. A. A.Clifford, E. Dickinson, , and P.Gray, *J. Chem Soc. Faraday. Trans. I*, **72**, 1997 (1976)
11. Mahesh C. Aggarwal and George S. Springer, *J. Chem. Phys.* **70**, 3948 (1979)
12. H.Roder, C. A. Nieto de Castro, *J. Chem. Eng. Data*, **27**, 12 (1982)
13. H. Roder, *Nat. Bur. Stand (U.S.)* NBSIR 84-3006 May (1984)
14. Xi-Yin Zheng, S. Yamamoto and H Yoshida, H. Masuoka and M. Yorizane, *J. Chem Eng. of Japan*, **17**, 237 (1984)
15. R. Tufeu and B. Le Neindre, *Int. J. of Thermophysics*, **8**, 27 (1987)
16. R. C. Prasad, G. Wang and J. E. S. Venart, *Int. J. of Thermophysics*, **10**, 1013 (1989)
17. J. Yata, M. Hori and T. Hagiwara, *Proc. Of the Fourth Asian Thermophysical Properties Conference*, Tokyo (1995)
18. F. R. Kramer and E. W. Comings, *J. Chem. Eng. Data*, **5**, 462 (1960
19. L. T. Carmichael, and B. H. Sage, *J. Chem. Eng. Data*, **9**, 511 (1964)
20. R. Kandiyoti, E. McLaughlin and J. F. T. Pittman, *J. Chem. Soc. Faraday Trans. I*, **68**, 860 (1972)
21. C. Parkinson and P. Gray, *J. Chem Soc. Fara. Trans. I*, **68**, 1065 (1972)
22. C. A. Nieto de Castro, R. Tufeu and B. Le Neindre, *Int. J. of Thermophysics*, **4**, 11 (1983)

23. L. Cusco , R. A. Perkins, and M. L. V. Ramires (to be published)

24. Vesovic, W. A. Wakeham, G. A. Olchowy, J. V. Sengers, J. T. R. Watson and J. Millat, *J. Phys. Chem. Ref. Data*, **19**, 3 (1990)

25. P. T. Boggs, R. H. Byrd, J. E. Rogers and R. B. Schnabel., NISTIR 4834 (1992)

26. G. A. Olchowy and J. V. Sengers, *Phys. Rev. Lett.* **61,** 15 (1988)

27. G. Vines, and L. A. Bennett, *J. of Chem. Phys.*, **22**, 360 (1954)

28. H. Senftlben, *Z. Angew. Phys.*, **17**, 86 (1964)

29. N. I. Ryabtsev, and V. A. Kazaryan, *Gazov. Prom*, **14**,46 (1969)

30. H Ehya, F. M. Faubert and G. S. Springer, *J. Heat Trans.*, **94**, 262-5 (1972)

**SESSION 1**

---

# TECHNIQUES

SESSION CHAIRS
T. W. Tong
S. E. Gustafsson

# Automatic Non-Destructive Brake Tester for Thermal Diffusivity/Conductivity of Aircraft Brake Disks

R. E. TAYLOR, T. R. GOERZ and J. V. BECK

## ABSTRACT

This apparatus is a computerized turn-key device capable of non-destructive testing of aircraft brake disks at eight locations simultaneously. The automatic non-destructive tester uses a step heating method to measure the thermal diffusivity (and hence thermal conductivity) of carbon-carbon brake disks. The test requires about five minutes. The basic idea is that one flat surface of the disk is subjected to a reasonably constant uniform heat flux and the resulting temperature response of the opposite face is monitored. Thermal diffusivity values are calculated from the temperature rise curves. The method used to calculate diffusivity values from the data generated with the non-destructive tester involves parameter estimation techniques.

Several apparatuses have been built, tested and supplied to users. Reproducibility of averaged diffusivity results is routinely within 3% on regular disks. The average diffusivity values agree within several percent with those obtained on small samples using our standard laser flash apparatus.

## INTRODUCTION

The thermal diffusivity/conductivity of carbon/carbon materials is a sensitive measure of fiber fraction and orientation and also of

Raymond E. Taylor and Thomas R. Goerz, Properties Research Laboratory, 2595 Yeager Road, West Lafayette, IN 47906
James V. Beck, Dept. of Mechanical Engineering, Michigan State University, East Lansing, MI 48824-1225

delaminations. Consequently, it is very usseful as a quality control check, especially when somewhat localized measurements at a number of locations is involved. The automatic non-destructive tester, which uses a step heating method to measure the thermal diffusivity (and hence thermal conductivity) of carbon-carbon brake disks was designed to do this. The basic idea is that one flat surface of the disk is subjected to a constant uniform heat flux from a bank of lamps and the resulting temperature response of the opposite face is monitored. Thermal diffusivity values can be calculated from the temperature rise curve. The units for thermal diffusivity are length squared/time. The length is that traveled by the heat pulse, which for the disk is its thickness. The time is that required for the temperature rise to reach some criterium. Actually, for a temperature rise curve, many diffusivity values could be calculated. The method used to calculate diffusivity values from the data generated with the non-destructive tester uses parameter estimation techniques [1]. This methodology has proven to be substantially better than conventional methods such as those described in standard texts like Carslaw and Jaeger [2]. The automatic tester will determine the thermal diffusivity values at eight locations simultaneously and store the results in such a fashion that calculated diffusivity values are easily accessed later.

## PARAMETER ESTIMATION ANALYSIS

In the determination of the thermal diffusivity, two basic things are needed. One is the data set of measurements of temperatures at the unheated side of the specimen. The other is a mathematical model for the temperature at this location. The specimen is heated by constant heat flux by the heating lamps and is effectively insulated at the back surface. Since the temperature rise is small, the assumption of temperature independent thermal properties is valid. The one dimensional analytical solution for an isotropic plate insulated on the bottom and with a constant heat flux applied to the top is given as [1],

$$T(L,t,\alpha,\beta_2) = T_i + \beta_2 \left( -\frac{1}{6} + \frac{\alpha t}{L^2} + \frac{2}{\pi^2} \sum_{n=1}^{\infty} \frac{1}{n^2} e^{-n^2\pi^2\alpha t/L^2}(-1)^n \right) \qquad (1)$$

where $\alpha$ is thermal diffusivity, t is time from the beginning of heating and L is the thickness of the specimen. Temperature $T(L,t)$ is measured as a function of t and the initial temperature, $T_i$, is known. Parameters found from the data are $\alpha$ and $\beta_2$ which is given by

$$\beta_2 = \frac{q_0 L}{k} \tag{2}$$

The net heat flux, $q_0$, is used in the model but it is not known because the absorptivity and emissivity of the specimen are unknown. However, the magnitude of $q_0$ is not needed since the data and the model allow the estimation of both parameters simultaneously; $\beta_2$ is called a "throw away" parameter because it is not needed.

The estimation procedure can be done for $\alpha$ and the group $q_0 L/\rho c$ instead of for $\alpha$ and $\beta_2$. However, little difference in the estimated value of $\alpha$ is noted whatever group of two parameters ($\alpha$ and $q_0 L/\rho c$ or $\alpha$ and $\beta_2$) is used.

The parameter estimation if accomplished by minimizing a sum of squares of the measurement errors,

$$S = \sum_{n=1}^{N} \left(Y_n - T(L, t_n; \alpha, \beta_2)\right)^2 \tag{3}$$

with respect to the parameters $\alpha$ and $\beta_2$. A method for doing this is discussed in Chp. 7 [1]. The recommended method uses a sequential procedure so that the model can be checked when additional measurements are used.

## APPARATUS DESCRIPTION

Figure 1 shows a simplified overview of the automatic brake disk tester. The testing unit is mounted in a cabinet and the computer and printer are on an adjacent table. Starting at the top of the cabinet, there is a cooling fan which is turned on before the run while the lamps are warming up, is turned off during the run while the shutter is open, and is turned back on after the measurements are completed. The lamp assembly consists of ten 500 Watt bulbs, each mounted at the focus of a parabolic reflector. The shutter consists of two thick cloths, separated by about one inch. The top cloth is a brown PBI synthetic cloth, medium/heavy duty plain weave 0.072 inches thick. Some of the intense light can shine between the weaves, but most of the light energy is

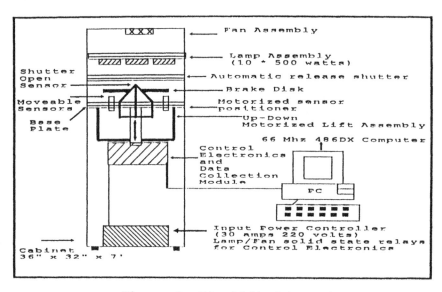

Figure 1. Simplified Overview

absorbed. The second layer is an aluminized silica cloth, with the aluminized surface facing upward to reflect the light energy which penetrates the upper cloth. This lower cloth does not increase in temperature by more than one or two degrees prior to the measurement and prevents any energy impinging on the sample. The upper cloth does not reflect much energy back into the lamps, so the lamp power is the same whether the shutter is in the open or closed position. Dead weights are attached to the rear end of the cloths and gravity is used to open the shutter (Figure 2).

Carbon/carbon brake disks are usually 12 to 22 inches OD, 5 to 12 inches ID and 0.5 to 1.5 inches thick, depending on the type of plane involved. Also, the disks have lugs on the ID or OD depending upon whether it is a stator or rotor. Thus, the tester must be capable of accepting a wide variety of shapes. The brake disk is placed on the cone-shaped motorized lift assembly so that the disk ID rests on the edges of four thin triangular riser plates. Before the lift assembly is lowered, the rear face temperature sensors are automatically positioned at the radius midway between the disk OD and ID (or at some other desired radius).

Control electronics, motors and data collection modules are mounted on the lower surface of the base plate. The input power and switches are located near the bottom of the cabinet. The lamp assembly, base plate

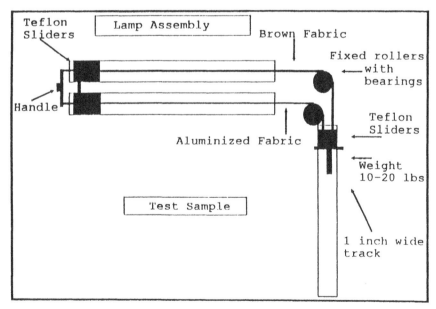

Figure 2. Shutter Mechanism

and power units are each mounted on separate slide rails so that each can be slid out of the cabinet for easy maintenance.

## OPERATING SOFTWARE

Before inserting a brake disk, the computer is turned on and then the tester is powered up. The home menu on the monitor consists of: Perform Material Testing; Reanalyze/Edit Saved Data; Diagnostics; Transfer Material Testing Data; Erase Material Testing Data and Exit to DOS.

Operator input includes choices of data storage subdirectory; sample identification; technicians initials, carbon type, disk type; brake configuration and comments; duration of experiment in seconds; number of sections to be tested; inside diameter; outside diameter; material thickness, bulk density (gm cm$^{-3}$) and specific heat (W s gm$^{-1}$K$^{-1}$). The bottom surface of the disk is wiped of loose material and placed on the lift assembly. It should be reasonably centered and in a horizontal position with the identity number upright and facing the front of the tester. When the space bar is pressed the thermocouples will position themselves at the mid-radius positions and the lift assembly will lower the pad onto the Teflon supports, spring-loading the thermocouples in the process. The lights will come on for a pre-determined time in order for them to reach steady-state. Then the shutters will open and the transient data will be collected. The light will shut off. The motorized lift assembly will rise and the thermocouples will move to the center. The diffusivity/conductivity results will be printed out.

A diffusivity value which looks suspicious may be checked by rotating the disk 45 degrees and retesting to see if the suspicious value follow the rotation. If the suspicious reading always occurs at the same sensor location, the sensor should be checked.

There is an optimum time for the lamps to irradiate the sample. The rear surface temperature rise should be about two degrees. If the rise is less than this, the irradiation time should be increased. If the rise is more than this, the irradiation time should be decreased. The heating rate is determined by the sample's emissivity and mass. Therefore, selected irradiation times may be different for different disks. Too small a temperature increase causes sensitivity and noise problems. Too large a temperature rise causes significant heat losses which are not taken into account in the mathematical analysis. However, there is a fair latitude in satisfactory temperature rises.

Another consideration is important besides the two degree temperature rise, which relates to the signal to noise ratio. This consideration is to make the optimal experiment duration $t_{max}$, and thickness L of the specimen such that roughly

$$\alpha \frac{t_{max}}{L^2} = 1.3 \tag{4}$$

is satisfied. The group could be any value from 0.7 to 1.8 and the accuracy would not suffer greatly. However, much smaller and much larger values can results in less accuracy. An example for a poor experiment is for values of this parameter less than 0.1, when the body is effectively semi-infinite and little T rise occurs at the insulated surface. Another example is for finding $\alpha$ of a "thin" specimen of high thermal diffusivity; here the body has negligible temperature gradients (a "lumped" body) compared to the maximum temperature rise. In this example the group is relatively large compared to 1.3.

The data may be re-analyzed. The options include: text printout of all channels; bar-graph of all channels; plots of individual channels, creating files of selected channels and editing a file to change information. Data files may be transferred to and from floppy disks.

## RESULTS

Because of the highly competitive nature of the carbon/carbon brake disk manufacturers and the proprietary nature of the data, all test results reported here have been normalized to the average value for each disk and the disk numbers have been changed to 1, 2, 3, etc. A typical temperature rise curve for a particular channel is shown as Figure 3, and a sanitized output for one experiment using all eight channels is shown as Figure 4. Two disks were measured on Brake Tester One in 1993. They were remeasured in 1995 on Brake Tester Three. The results are summarized in Table 1. In order to make a valid comparison, both sets of data have been normalized to the average values of the 1995 tests. At the time the tests were made in 1993, data acquisition speed permitted testing only four channels at one time. Consequently, the disk was rotated and the tests repeated in order to cover all eight locations. The average results for the two sets of tests for Disk 1 were only 1% different and those for Disk 2 were 3% different. Although the maximum difference in the average values at any location was 12% (Position 3 on Disk 1, 1993 tests), the maximum difference in the average of all eight positions in the 1995 tests was less than 3%.

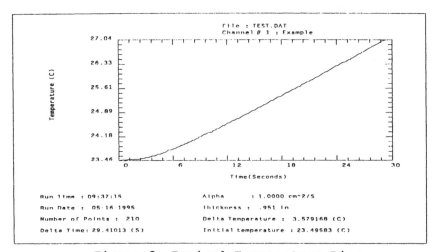

Figure 3. Typical Temperature Rise

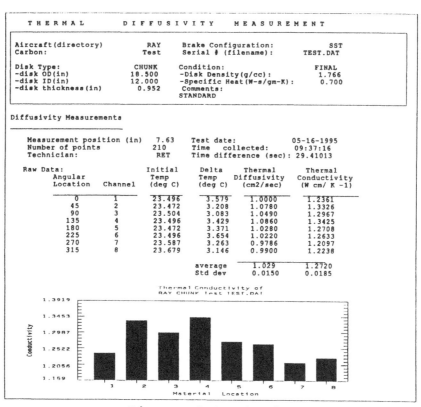

Figure 4.Typical Output

TABLE I. Comparison of 1993 and 1995 Thermal Diffusivity Test on Two Disks (Normalized to Average Values for 1995 Results)

| Disk | RUN | 1 | 2 | 3 | 4 | Position 5 | 6 | 7 | 8 | AVE | Std Dev |
|---|---|---|---|---|---|---|---|---|---|---|---|
| 1 | 1-1 | 0.959 | 0.948 | 0.948 | 1.011 | | | | | 0.967 | 0.030116 |
| 1 | 1-2 | 0.959 | 1.021 | | | | | 1.061 | 1.004 | 1.011 | 0.042240 |
| 1 | 1-3 | | | | | 0.982 | 0.965 | 1.027 | 1.066 | 1.010 | 0.045585 |
| 1 | 1-4 | | | 0.908 | 0.931 | 1.016 | 1.027 | | 1.035 | 0.971 | 0.059802 |
| 1 | AVE | 0.959 | 0.985 | 0.928 | 0.971 | 0.999 | 0.996 | 1.044 | 1.035 | 0.990 | 0.038271061 |
| 1 | Std Dev | 0.000000 | 0.051619 | 0.028284 | 0.056569 | 0.024042 | 0.043841 | 0.024042 | 0.043841 | 0.024381 | |
| | | | | | | | | | | | |
| 1 | 3-1 | 0.948 | 0.976 | 0.999 | 1.021 | 1.038 | 0.999 | 0.931 | 0.987 | 0.987 | 0.035500 |
| 1 | 3-2 | 0.925 | 1.016 | 1.027 | 0.999 | 1.055 | 1.049 | 0.959 | 0.982 | 1.002 | 0.044760 |
| 1 | 3-3 | 0.965 | 0.959 | 1.061 | 1.055 | 1.089 | 1.004 | 0.982 | 0.987 | 1.013 | 0.048963 |
| 1 | 3-4 | 0.965 | 0.942 | 1.083 | 1.083 | 1.049 | 0.959 | 0.999 | 0.971 | 1.006 | 0.057279 |
| 1 | AVE | 0.951 | 0.973 | 1.043 | 1.040 | 1.058 | 1.003 | 0.968 | 0.982 | 1.002 | 0.039984595 |
| 1 | Std Dev | 0.018945 | 0.031700 | 0.037036 | 0.037036 | 0.021991 | 0.036827 | 0.029477 | 0.007544 | 0.010783 | |
| | | | | | | | | | | | |
| 2 | 1-1 | 1.078 | 1.083 | 1.109 | 1.113 | | | | | 1.096 | 0.017802 |
| 2 | 1-2 | 1.003 | 1.047 | | | | | 0.977 | 0.994 | 1.005 | 0.029848 |
| 2 | 1-3 | | | | | 0.994 | 0.991 | 0.959 | 1.012 | 0.989 | 0.022045 |
| 2 | 1-4 | | | 1.025 | 0.968 | 1.056 | 1.083 | | 1.003 | 1.033 | 0.049390 |
| 2 | AVE | 1.041 | 1.065 | 1.025 | 1.041 | 1.025 | 1.037 | 0.968 | 1.003 | 1.031 | 0.032638715 |
| 2 | Std Dev | 0.053033 | 0.025456 | 0.059397 | 0.102530 | 0.043841 | 0.065054 | 0.012728 | 0.012728 | 0.046987 | |
| | | | | | | | | | | | |
| 2 | 3-3 | 0.991 | 1.003 | 1.003 | 1.043 | 1.025 | 1.096 | 0.991 | 1.038 | 1.024 | 0.035459 |
| 2 | 3-2 | 0.919 | 0.977 | 0.977 | 1.065 | 0.991 | 1.021 | 0.987 | 0.981 | 0.990 | 0.041500 |
| 2 | 3-3 | 0.937 | 0.932 | 1.016 | 1.096 | 0.994 | 1.021 | 0.932 | 1.003 | 0.991 | 0.056795 |
| 2 | AVE | 0.949 | 0.971 | 0.999 | 1.068 | 1.003 | 1.046 | 0.970 | 1.007 | 1.002 | 0.039990522 |
| 2 | Std Dev | 0.057470 | 0.035921 | 0.019858 | 0.026627 | 0.018824 | 0.043301 | 0.032970 | 0.026746 | 0.019178 | |

Note: Under Run 1 - 1993 Tests Using Tester 1
Under Run 3 - 1995 Tests Using Tester 3

38

To get better statistics on the latest and most improved tester, nine runs were made on Disk 3 over a two day period. These runs were interspersed among other tests and the disk was removed from the tester between each run. No attempt was made to control the temperature of the test chamber or the initial temperature of the disk, except to give it a chance to come to equilibrium. These conditions are the same as those which the tester operates on the quality control line. The results indicated that those runs measured when the test chamber was the hottest (runs 7 and 8) yielded higher values, indicating that a significant portion of the 7% spread in the average values was associated with variations in the test chamber temperature. An examination of revealed that Positions 4 and 8 consistently had higher values than those measured at the other locations and the values at Position 6 were usually the smallest. These results could be caused by either local differences in the disk or bias with the sensors. We noted this trend after the fourth run. Consequently, we interchanged the sensors at Positions 4 and 6 after the fourth run. The tendency for Position 4 to have large values and Position 6 to have low values was not appreciably altered. Thus, the tester is capable of detecting small variations in the properties at different positions on the disk, when used in a research mode. When used in a mass production quality control mode in a room with normal air conditioning, it can reproduce average values within 7%. Under better controlled conditions, it can reproduce values within 4%.

In order to determine the accuracy (as opposed to reproducibility) of the values, a disk was measured with the automatic tester. A section of the disk was then cut out and samples machined for our regular step-heat apparatus [3] and our laser flash apparatus [4]. The results were all within 3% of each other.

## CONCLUDING REMARKS

Carbon/carbon brake disks are made in batch processes so that a large number of disks are completed at one time. A method for rapid non-destructive thermal testing to replace the existing procedures of limited destructive testing of selected disks or laborious testing of a limited portion of a few disks was desired. The automatic brake disk tester has been shown to fulfill the requirement of non-destructive quality control thermal testing of the disks. Tests at eight locations on an individual disk can be completed within 5 minutes. The tester is also useful in research and development. Unfortunately for the authors, the number of carbon/carbon brake disk manufacturers is very limited and the market is approaching saturation.

**REFERENCES**

[1]    Beck, J.V. and Arnold, K.J., 1977, **Parameter Estimation in Engineering and Science,** Wiley

[2]    Carslaw, H.S. and Jaeger, J.C., 1959, **Conduction of Heat in Solids,** Oxford, U.P. England

[3]    Rooke, S.P. and Taylor, R.E., Feb. 1988, "Transient Experimental Technique for the Determination of Thermal Diffusivity of Fibrous Insulation", J. of Heat Transfer, 110, pp. 270-273

[4]    Taylor, R.E., 1979, "Heat Pulse Thermal Diffusivity Measurements", High  Temp. - High Press., 11, pp. 43-58

# Improved Axial Cut-Bar Apparatus

P. S. GAAL

## ABSTRACT

A comparative cut-bar apparatus with novel mechanical and thermal design features is described. As a departure from conventional designs, the guard surrounding the cylindrical sample and reference stack is not a single piece, tubular body, but a split one made up of two half shells. Instead of granular or powder insulation between the guard and the stacks, solid insulating sections are used that are vacuum formed from fibrous stock. The device is also equipped with a unique dual heatsink which allows uninterrupted testing over the entire RT to 1000°C range.

## INTRODUCTION

Comparative cut-bar systems have been amply described both from the standpoints of underlying theory and general execution by numerous excellent investigators [1], [2], [3], [4], and [5]. The purpose of this paper is not to rehash what has already been said, but to concentrate on several aspects of the thermal and mechanical design of a cut-bar system that alleviates problematic or just cumbersome operating aspects of conventional designs.

Historically, everybody, including commercial manufacturers [6], have used a stack made up of ceramic parts, references, heaters, heatsinks and a tubular outer guard around it with loose insulation filled in the annular space in between and quite often even around the guard. Due to the constraints imposed by the geometry of the

Peter S. Gaal, Anter Corporation, 1700 Universal Road, Pittsburgh, PA 15235

tubular guard, all thermocouples were running axially along the stack to their respective levels where they entered into grooves or holes. In some cases, care was exercised to thermally "temper" each thermocouple by wrapping it around the sample partially or fully before allowing to run in the axial direction; in others, no precautions were taken. As a result, heat leaks along the thermocouples became a problem when low thermal conductivity samples were tested. (Six thermocouples represent 15% of the thermal conductance of a 2 inch diameter sample made out of a material having a conductivity of 1 W/mK.)

The handling of bulk (powdered or granular) thermal insulation is a constant source of aggravation. Powders tend to pack, get into crevices, and are very difficult to keep out of the sample/reference interfaces. Granules, especially the mica based ones (vermiculite, etc.), dust excessively producing a layer of debris over the entire laboratory in no time. Due to their airborne nature and health hazards associated with them, technicians are required to wear goggles and respirators while handling the material, but it still exposes others in the vicinity to the dust and hazards.

Another problem inherent to the conventional configuration used is the need to vary the thermal resistance between the bottom stack heater and the heat sink as the test temperature goes up. With little thermal resistance present, it is impossible to raise the heater temperature to 6-700°C against the water cooled heat sink. Conversely, if a thick thermal resistance layer is placed in between, it will not allow testing at low temperatures. It is therefore necessary for all conventionally built machines to break a test up into segments, allowing a full cool down in between, and to insert different thermal resistance spacers into the stack. To do that, one has to remove the bulk insulation and the guard completely, which is a major operational headache and is considered a serious drawback.

## DISCUSSION

Having no alternative methods to take the place of the comparative cut-bar, a critical evaluation of the previously mentioned short-comings led to a markedly different mechanical design and a drastically improved thermal design for the system under discussion (Figure 1).

To eliminate the use of bulk insulation, a split-shell guard was constructed using molded insulation. While it takes about two hours to remove and replace bulk insulation from a conventional machine, it only takes 20 seconds to swing open and close the currently described guard shell. The split-guard design also allows an entirely new approach to thermocouple management, and eliminates the earlier mentioned thermal shunting of the stack by thermocouples. With the split configuration, the thermocouples exit the stack radially and also pass through the guard radially, through the gap naturally provided by the split. Thus, each thermocouple is in a quasi-isothermal plane and therefore carry little or no heat between the stack and the guard. (The guard is closely matched to the stack, therefore any plan perpendicular to the axis is nearly isothermal.) With the radially

FIGURE 1

placed thermocouples, sample replacement and reference exchange became a very simple matter.

To virtually eliminate the problem of changing spacers with increasing temperature, the new design incorporates dual heat sinks that allow continuous testing over the entire temperature range without interruption. The test starts out using the heat sink located right below the bottom stack heater. At a predetermined temperature, coolant flow is switched to a second heat sink below it, while the upper one is allowed to drain fully. The two heat sinks are separated by a layer of insulation equivalent to the earlier mentioned spacer. Since the switch occurs without mechanical change to the stack, the test proceeds without interruption.

Several other areas of the design employ parts that make servicing easier and less costly. For example, the stack heaters have almost always been complex bobbin-type devices, custom configured and quite expensive to make. The current design employs commercially available clam-shell heaters which are clamped around solid shafts that are part of the stack. This allows heater replacement to be a simple chore while it also ensures that a well-defined solid path for heat flow is provided along the stack. Using this type of heaters and judicious choice of materials for stack components, routine operation to 1000°C is no longer a problem. With special alloy heaters, 1300°C upper limit is quite achievable.

Another area the new design addresses is the method by which the stack is held together. In some previous designs, rods running along the guard were used to support a cross bar on top which in turn was used to load the stack. Again, due to the tubular guards, the load could not be applied before the guard was in place, and therefore afforded no opportunity to inspect the alignment of the stack after load was applied. It is not unusual for disks to slide when pressure is applied. That in turn reduces interface area. The current design completely eliminates this problem. Since the split guard when opened up completely exposes the stack, stack alignment can be checked after load is applied and before the guard is closed back. Thermocouple placement and replacement has also been improved with the use of very fine sheath thermocouples that are quite durable and enter only from one side. Thermocouple wells approximately 30 times the diameter of the sheath are used to ensure good thermal contact and minimal cooling of the junction by conduction along the wire. Further reducing this parasitic effect is the path the thermocouples are led through, as discussed earlier. Since the guard closely matches the thermal gradient of the stack, only very small temperature differences will be present along any thermocouple.

The instrument is configured to be a stand alone device having a small footprint so as to conserve laboratory space. Its control electronics is totally integrated within the front cabinet. Thermocouples are amplified either in the differential mode or as absolute ones depending on the locations. These signals are scanned and measured to 17 bits accuracy. All heaters are controlled from software without the use of discrete temperature controllers. The Windows™ based operating software resides in a separate PC which is connected to the instrument via a serial port. In addition to the control functions, the software also takes care of all data acquisition, data reduction and plotting functions in real time. Extensive safety features round out the control functions performed by the computer.

**COMPARISON DATA**

Even though this paper deals primarily with mechanical and thermal design concepts, the performance of the device in its final configuration will be ultimate measure of its success. The new design has been tested on numerous medium and low conductivity materials with good results. Commercial 304 stainless steel was found to be within a ±5% band of literature values. Additionally, pyroceram was tested and the data fell within a $\pm$4% band around the generally accepted values of that material.

## SUMMARY

In conclusion, the device employing a split cylindrical hinged guard has proven to be easy to operate without sacrificing accuracy, and with the aid of a dual heatsink arrangement, it allows covering the complete temperature range between ambient and 1000°C without interruption.

## REFERENCES

1. J.P. Moore, D.L. McElroy, and R.S. Graves, <u>A Technique for Determining Thermal and Electrical Conductivity and Absolute Seebeck Coefficient Between 300 and 1000 K</u>, ORNL-4986 (1974).

2. J.P. Moore, R.K. Williams, and R.S. Graves, "Thermal Conductivity, Electrial Resistivity, and Seebeck Coefficient of High-Purity Chromium from 280 to 1000 K," <u>J. Appl. Phys.</u> 48:610 (1977).

3. An Axial Heat-Flow Apparatus for Determining Thermal Conductivity of Solids from 20° to 600° K, B.L. Rhodes, et al, Proceedings of the 3rd Conference on Thermal Conductivity, <u>V1</u>, p. 203 (Oak Ridge National Laboratory, 1963).

4. Heat Losses in Cut Bar Apparatus: Experimental Analytical Comparisons, M. Minges, Thermal Conductivity, Proceedings of the Seventh Conference, National Bureau of Standards, Special Publication 302 (1967), p. 197.

5. A longitudinal Symmetrical Heat Flow Apparatus for the Determination of the Thermal Conductivity of Metals and their Alloys, H. Chang and G. Blair, Thermal Conductivity, Proceedings of the Seventh Conference, National Bureau of Standards, Special Publication 302 (1967), p. 355.

6. Improved High Temperature Cut Bar Apparatus, G.R. Chusener, Proceedings of the Fourteenth International Conference on Thermal Conductivity, The University of Connecticut, (1975).

7. Technical Brochure COM-800 Instrument, Holometrix, Inc., Bedford, MA.

# Design of a Subminiature Guarded
# Hot Plate Apparatus

D. R. FLYNN and R. GORTHALA

## ABSTRACT

Under a Phase I Small Business Innovative Research contract from the National
Institute of Standards and Technology (NIST), we are designing a subminiature
guarded hot plate apparatus that can be used to determine the thermal conductivity
or thermal resistance of very small specimens (1 to 3 cm square) of thermal
insulation materials over a temperature range of (at least) −40 °C to +100 °C, with
the capability to carry out measurements in air, selected gases, or vacuum. In this
paper, we present the performance criteria and specifications for the apparatus,
alternative measurement approaches that were reviewed, the overall design approach
that was selected, and the current stage of development of the detailed design of the
prototype apparatus to be built during Phase II of this contract. The apparatus to
be developed under this project will be of direct value to manufacturers of
insulation materials and polymers, particularly for use in characterizing experimental
products that are only available in very small sample sizes.

## INTRODUCTION

Most measurements of the thermal conductivity or thermal resistance of building
insulations, other than pipe insulation, are made using a guarded hot plate apparatus
or a heat flow meter apparatus that has been calibrated with specimens whose
R-value is known from guarded hot plate tests. Most such apparatus requires
specimen sizes in the range 20 to 60 cm across and, say, 2 to 20 cm thick. The
NIST line-heat-source guarded hot plate takes 1-meter diameter specimens. Most
guarded hot plates in the United States and Canada are designed to accept a
matched pair of specimens, although some apparatus can be operated in a one-sided
mode. If an insulation company wishes to make a small quantity of an experimental
product, it frequently would not be feasible to make specimens large enough for the
thermal conductivity equipment that is available. Companies would like to acquire
data rapidly and with sufficient accuracy to make meaningful comparisons of the

Daniel R. Flynn and Ravi Gorthala, DRF R&D, 18777C N. Frederick Ave., Gaithersburg, MD 20879 thermal performance of different materials or of different formulations of the same material. There is no commercial apparatus, and no custom apparatus of which we are aware, that is suitable for carrying out rapid, reliable measurements of thermal conductivity or resistance on specimens smaller than 10 cm across, let alone on specimens with a cross sectional area of the order of a few square centimeters.

Under the Small Business Innovative Research Program, NIST is funding the development of an apparatus capable of carrying out thermal conductivity measurements on small samples of experimental material. The objectives of the Phase I development effort are to: develop performance criteria and specifications for the apparatus to be developed; review alternative measurement approaches, select the design, and refine it to the point where it can be subjected to detailed analysis; develop numerical models of the apparatus that will enable reliable prediction of apparatus performance and measurement accuracy as functions of the design parameters; study and analyze alternative approaches to the most crucial design features of the apparatus; and develop a detailed design of the prototype apparatus to be built in Phase II.

## PERFORMANCE CRITERIA AND SPECIFICATIONS

| | |
|---|---|
| Specimen Size: | 3-cm-square test specimen *or* 1-cm-square test specimen with 3-cm-square guard specimen |
| Maximum Specimen Thickness: | 1.3 cm |
| Maximum Specimen Anisotropy: | 3:1 (lateral-to-longitudinal conductivity) |
| Thermal Conductivity: | Primary range is 0.02 to 0.05 W/m·K (conventional insulations), and if possible 0.05 to 0.35 W/m·K (solid organics) and <0.005 to 0.02 W/m·K (superinsulations) |
| Thermal Resistance: | 0.01 to 1.0 $m^2$·K / W |
| Environmental Control: | Atmospheric pressure (low humidity), gas at controlled pressure, and vacuum |
| Low Temperature Limit: | −40 °C (with capability to −195 °C) |
| High Temperature Limit: | +100 °C (goal: 150 °C) |
| Accuracy: | ≤ 5 percent (goal: ≤ 2 percent near 25 °C) |
| Repeatability: | ≤ 2 percent (goal: ≤ 0.5 percent) |

## ALTERNATIVE MEASUREMENT APPROACHES

NIST did not place restrictions on the type of apparatus other than to indicate that a probe or needle apparatus would not be suitable for insertion into friable materials. After reviewing alternative measurement approaches, we believe strongly that the best type of apparatus to be developed for this project is a Miniature Guarded Hot Plate Apparatus, meeting the requirements of ASTM C177 [1] and ISO 8302 [2]. The following discussion indicates our rationale for that conclusion.

### STEADY STATE VERSUS TRANSIENT APPROACH

Since the data taken with this apparatus will need to be compared with data taken on larger specimens using guarded hot plates or heat flow meter apparatus, thermal conductivity, not diffusivity, needs to be measured. When there is significant radiative heat transfer, as is the case for most thermal insulations, thermal diffusivity is a rather dubious concept at best. For many thermal insulations, one would not obtain correct thermal conductivity or thermal resistance values from thermal diffusivity, specific heat, and density data.

The "hot wire" method is a transient technique that yields values for thermal conductivity rather than thermal diffusivity. There are, however, several problems with such line-heat-source methods that make them unsuitable for the present project: (1) data analysis procedures are based upon pure conduction, not conduction plus radiation, so that the thermal conductivity values obtained may be seriously in error for materials that transmit significant thermal radiation; (2) even if radiation effects were not important, line-heat-source methods are sensitive only to heat transfer within a rather small region surrounding the line heater, so that for materials with even moderately large pores, the data obtained would not represent any sort of meaningful average over a representative volume of material; (3) these techniques assume that the thermal properties of the specimen are independent of angle, which would not be the case for many materials; and (4) probe, or needle, techniques would require insertion of the probe into the specimen, a process that would be unacceptable for delicate or friable materials. In summary, line-heat-source methods can yield valid thermal conductivity data on very fine powders but are really quite questionable for most other forms of thermal insulation.

### GEOMETRY

For isotropic materials or loose-fill insulations, thermal conductivity measurements utilizing radial heat flow in a cylindrical geometry can provide excellent results. However, for board, batt, or blanket insulations, radial-heat-flow methods would require difficult specimen preparation, would require larger amounts of specimen material than is desirable, and would not be suitable for anisotropic materials. Therefore we conclude that the apparatus to be developed should use axial heat flow (one-dimensional heat flow in a Cartesian coordinate system).

## ABSOLUTE VERSUS COMPARISON METHODS

For determinations of the thermal conductivity or thermal resistance of building insulation materials, the workhorse of the industry is the heat flow meter apparatus. Such equipment is rapid and easy to use. It must, however, be calibrated using specimens of known thermal resistance, specimens that are almost invariably tested in, or traceable to, a guarded hot plate apparatus. While it certainly would be possible to develop a heat flow meter apparatus that could accommodate specimens as small as those of interest for this project, there are not suitable reference standards available for calibrating such a heat flow meter apparatus. In principal, one might calibrate a heat flow meter apparatus in air and use the same calibration for measurements carried out with another gas, or vacuum, present. In practice, one should have serious concerns about the validity of data obtained in this manner. Our bottom line on comparative versus absolute methods is that one should not use comparative methods, including heat flow meter apparatus, unless there are absolute methods available that enable accurate testing of calibration specimens of the same size, and under the same environmental conditions, as are required in the comparison apparatus.

## GUARDED HOT PLATE APPARATUS

Having rejected the other alternatives, we have concluded that a subminiature guarded hot plate apparatus is the preferred approach for this project. A guarded hot plate apparatus consists of a heated metering plate, which may be square or circular, separated by a narrow insulating gap from a surrounding coplanar guard plate. Typically, similar specimens are placed on either side of the hot plate; the outside surfaces of the specimen are held between constant temperature cold plates. In operation, the electrical power input to the guard plate is adjusted, usually automatically, so that a multiple-junction differential thermocouple spanning the guard gap has zero output, indicating that there is no temperature difference across the guard gap. Thus the electrically generated heat input to the metering plate flows perpendicularly from both sides of the plate through the specimens to the cold plates. The average thermal conductivity of the two specimens is determined from the measured power to the metering plate, the temperature drops across the specimens, and the geometry. In using a guarded hot plate, measurement errors can arise from imperfect guarding, inaccurate determination of temperature differences, or failure to achieve thermal equilibrium, as well as from possible errors in the instrumentation, in measuring specimen thickness, or in thermocouple calibrations.

While is it relatively straightforward to measure accurately the electrical power dissipated in the metering plate heater, it is not at all easy to ascertain that all of this heat is going where it is supposed to. The hot plate must be well designed in order that there be no significant heat exchange between the metering and guard plates. The thermal resistance across the guard gap should be large, the differential thermocouple junctions must be installed so as to appropriately sample the temperature on either side of the gap, and the power to the guard heater must be controlled so as to maintain a sufficiently small temperature difference across the guard gap. If the conductivity of the specimen is high, a corresponding gap may be cut part way or all the way through the specimens. If the guard is too small, or its surfaces are not isothermal, the guarding will not be adequate. In order to ensure adequate guarding, it is common to use edge insulation around the periphery of the guard plate and specimens and/or to effectively increase the area of the guard plate by using a second (outer) guard plate or by controlling the ambient temperature around the apparatus to a value near the mean temperature of the specimens.

Most guarded hot plate apparatus in the United States and Canada is designed to take two specimens, one on either side of the hot plate. Many two-sided guarded hot plates can be operated with one of the cold plates held at the same temperature as the hot plate, so that there is essentially no heat flow through that "dummy" specimen and the meter power input flows entirely through the test specimen across which there is a substantial temperature difference, The U.S. [1] and International [2] standards, however, allow the use of a one-sided guarded hot plate and some equipment, particularly in Europe, is designed to accommodate only a single specimen. We have been working closely with Robert Zarr, at NIST, on updating the ASTM Standard [3] that addresses single-sided operation of a guarded hot plate. For the present project, it was concluded that the apparatus should be a single-sided guarded hot plate. Such equipment requires only a single specimen and, as described below, can be made quite compact, easy to operate, and accurate.

One additional choice needs to be made, i.e., whether to use a traditional type of guarded hot plate apparatus with a separate guard and meter section or to use a thin foil or screen heater such as described in ASTM C1114 [4]. This latter type of heater has much to comment it, at least in terms of simplicity of construction. However, proper operation of such designs requires that the specimen material in the "guard region" have the same thermal resistance as the material in the "meter" region. A traditional design allows the power input to the guard to be controlled separately from the power input to the meter section, thus allowing the use of a "guard specimen" that differs from the test specimen. Another problem with a thin foil or screen heater apparatus is the difficulty of making accurate temperature measurements of such small thin hot plates as would be required for a miniature version of such apparatus. Accordingly, we have elected to use a design with the guard section being separate from the meter section.

## DESIGN FOR A SUBMINIATURE GUARDED HOT PLATE APPARATUS

A cross section of the proposed design for the subminiature guarded hot plate apparatus is shown in Figure 1. The minimum specimen size is 1-cm square, surrounded by a guard specimen that is 3-cm square. If the material to be tested can be obtained as a 3-cm square sample, its outer portion serves as the guard specimen. Otherwise, the guard specimen can be a 3-cm square mask with a 1-cm square hole to accommodate the test specimen, or can be comprised of several pieces of material, e.g., four 1 × 2 cm pieces or eight 1-cm-square pieces.

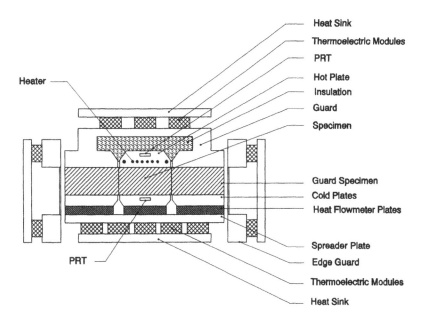

Figure 1. Thermal configuration of the hot plate, lateral and back guard, cold plate, and edge guard for the subminiature guarded hot plate apparatus.

The guarded hot plate assembly (as opposed to the entire guarded hot plate apparatus) consists of a 1-cm square meter section surrounded by a 3-cm square guard section. The guard section includes both a lateral guard, with its face coplanar with that of the meter plate, and a back guard that is extended to prevent heat flow to or from the back side of the meter section. The meter section is supported, on low-thermal-conductance legs (not shown), from the back guard portion, thus avoiding the need for structural members to cross the guard gap. In operation, the guard is controlled at the desired hot-side temperature, by Peltier cooling or heating, and the meter section is controlled to match the guard temperature.

The cold plate assembly is divided into a central "meter" section and a surrounding "guard" section, each of which is backed by a heat flow meter to measure the heat flow arriving through the specimen. In practice, the meter-section heat flow meter is calibrated, when the apparatus reaches steady-state conditions, by the measured power input to the meter-section hot plate. The purposes of the heat flow meters are two-fold. By measuring both the heat flow into the specimen and the heat flow out of the specimen, the steady-state value for the thermal conductivity of the specimen can be predicted earlier and more reliably than would be the case if only the hot-side heat flow were known. In addition, while the electrical power input to the meter-section heater would be the fundamental heat flow measurement, the measurement of heat flow leaving the specimen serves as a quality control check on the overall heat flow measurement. The cold plate assembly is also controlled at the desired temperature by Peltier elements.

The apparatus is provided with an edge guard that uses Peltier elements to create a longitudinal temperature profile that closely matches that across the test specimen and the guard specimen. The upper end of the linear edge guard is positioned even with the hot plate assembly (meter and guard sections). In order to allow various specimen thicknesses, the lower end of the guard extends past the cold plate assembly, with the end temperatures of the linear guard controlled to achieve the desired temperatures at the locations opposite the faces of the hot and cold plates. Since the temperature dependence of thermal conductivity is known for the material (tentatively, stainless steel) from which the edge guard is made, the deviation from linearity of the edge guard temperature profile can be computed. After the thermal conductivity of the guard specimen (or the test specimen if separate guard specimens are not needed) is measured as a function of temperature, the temperature profile along the outer edge of the guard specimen can be computed for each test and (small) corrections can be made for any residual heat flows due to the slight mismatches between the edge guard temperature profile and the guard specimen temperature profile.

The meter-section hot plate and cold plate temperatures are measured using miniature calibrated platinum resistance thermometers (we also are exploring the possibility of using diode thermometers). Other temperatures in the system are measured using thermocouples. All temperature controllers are implemented in software.

Figure 2 shows an overall view of the apparatus, including the supporting structure and the base plate and bell jar which provide the environmental enclosure. The cold plate assembly is supported on low-conductance legs above the cold-side base. A rigid column (to the right of the system) acts as a guide for the edge guard, which can be slid down against the cold-side base to allow access for specimen installation. This same column supports a hot-side base, from which the hot plate assembly, consisting of the meter plate, primary guard (lateral guard plus back guard), and thermoelectric cooling system, is suspended via an electromechanical actuator which is used to position the hot plate. The actuator is provided with a built-in displacement transducer so that the specimen thickness can be determined automatically.

The entire apparatus rests on a base plate and is contained inside a bell jar. As shown schematically, air can be drawn through holes in the base plate for conditioning to the desired humidity. Normally, one carries out thermal conductivity measurements under very low humidity conditions so that the air would be dried by a desiccant or a dehumidifier. In the interest of achieving rapid thermal equilibrium, the hot plate, cold plate, and edge guard serve as the environmental chamber for the specimen so that the base plate, bell jar, hot-side base, cold-side base, and support column can remain near room temperature. The only time this approach would create a problem would be if it were desired to carry out tests at an elevated temperature with the humidity being sufficiently high that condensation on the outer enclosure or the support structure might occur. If it were desired to

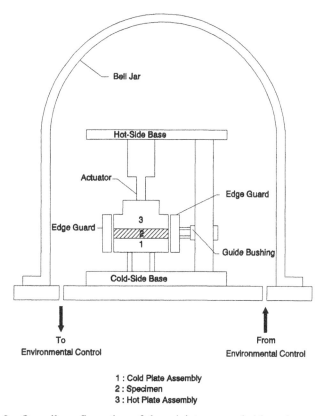

Figure 2. Overall configuration of the miniature guarded hot plate apparatus.

carry out such high humidity tests, the base plate and bell jar can be heated to prevent condensation. The apparatus also will be provided with a vacuum pump and gaging and with the capability of backfilling the system with other gases at a controlled pressure.

Currently, we are carrying out a series of analyses in order to refine the design and minimize measurement uncertainties. These analyses include calculation of temperature distributions throughout the apparatus, computation of unwanted heat exchanges between different portions of the apparatus, and estimation of the uncertainties associated with all power, temperature, and dimensional measurements.

Different alternatives are being considered before the detailed apparatus design can be finalized. As an example, Figure 3 shows two possible alternatives for design of the heater in the meter section of the guarded hot plate. In the left-hand design (a), the meter plate would be fabricated from a high-thermal-conductivity ceramic (e.g., aluminum oxide or aluminum nitride) extrusion, which would be extruded with holes for the heater wires and heater leads. A piece of this extrusion would be fabricated into a square plate and grooves would be cut into the ends to accommodate the loops in the heater winding. After the heater was installed, the grooves would be filled with refractory cement. An advantage of this design is that it does not matter if the heater wires are significantly hotter than the surrounding plate and the (upper) surface temperature can be computed from a temperature measurement made on the backside of the plate. Disadvantages of this design are that construction is somewhat labor intensive and the cost of the meter plate might be higher than it would be for some other designs.

In the right-hand design (b) in Figure 3, the temperature sensor is near the front surface of the plate and the heater is made using thick film electronic circuit fabrication techniques. A cover plate is placed over the heater to provide protection and to prevent the heater element, which would be somewhat hotter than the rest of the plate, from being "seen" by the back guard. If many heaters were being made, this approach would have a large cost advantage; if only a few heaters are being made, however, the unit cost might be rather high. With this approach, the cover plate also might be significantly hotter than desired.

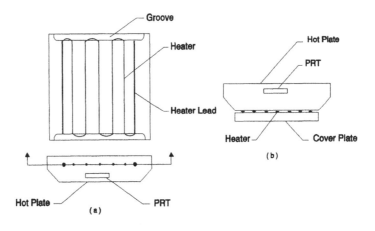

Figure 3. Two alternative designs for the heater in the
meter section of the guarded hot plate.

During the current Phase I effort, we are using computer models and analyses to refine the entire apparatus design and then iterate on design parameters in order to optimize the design in terms of speed, accuracy of operation, production feasibility, and cost. The overall detailed design, for Phase II fabrication and checkout, will include all necessary instrumentation and sensors, as well as the apparatus proper and its environmental enclosure and control system. The design to be developed will meet all the requirements of the relevant standards [1-3] and will address the precautions and concerns expressed by various experts [5-8] on the design of guarded hot plate apparatus.

## REFERENCES

[1] ASTM C177-85, Standard Test Method for Steady-State Heat Flux Measurements and Thermal Transmission Properties by Means of the Guarded-Hot-Plate Apparatus.

[2] ISO 8302, Determination of Steady-State Thermal Resistance and Related Properties — Guarded Hot Plate Apparatus.

[3] ASTM C1044-90, Standard Practice Using the Guarded-Hot-Plate Apparatus in the One-Sided Mode to Measure Steady-State Heat Flux and Thermal Transmission Properties.

[4] ASTM C1114-92, Standard Test Method for Steady-State Thermal Transmission Properties by Means of the Thin-Heater Apparatus.

[5] Hahn, M.H., Robinson, H.E., and Flynn, D.R., "Robinson Line-Heat-Source Guarded Hot Plate Apparatus," in Heat Transmission Measurements in Thermal Insulations, ASTM STP 544 (American Society for Testing and Materials, Philadelphia, PA, 1974), pp. 167-192.

[6] Klarsfeld, S., "Guarded Hot-Plate Method for Thermal Conductivity Measurements," *Compendium of Thermophysical Property Measurement Methods, Vol. 1, Survey of Measurement Techniques* (Maglić, K.D., Cezairliyan, A. and Peletsky, V.E., eds.), Plenum, New York, 1984, pp. 169-230.

[7] De Ponte, F., Langlais, C., and Klarsfeld, S., "Reference Guarded Hot Plate Apparatus for the Determination of Steady-State Thermal Transmission Properties," *Compendium of Thermophysical Property Measurement Methods, Vol. 2, Recommended Measurement Techniques and Practices* (Maglić, K.D., Cezairliyan, A. and Peletsky, V.E., eds.), Plenum, New York, 1992, pp. 99-131.

[8] Salmon, D.R., "The NPL 610 mm Guarded Hot-Plate/Heat Flow Meter Standard Apparatus for Thick Insulations," *Thermal Conductivity 22* (Tong, T.A., ed.), Technomic Publishing Co., Lancaster, PA, 1994, p. 834.

# Specific Heat Measurements with the Hot Disk Thermal Constants Analyser

M. GUSTAVSSON, N. S. SAXENA, E. KARAWACKI
and S. E. GUSTAFSSON

## ABSTRACT

Hot Disk measurements, based on the TPS-(Transient Plane Source)-technique, is being used for determining the thermal conductivity and the thermal diffusivity of solids. The specific heat can be calculated from the ratio of these two transport coefficients. This kind of measurements presuppose an initially isothermal sample with a total temperature increase that varies between 0.1 K and 1.0 K.

In this work solid samples are placed inside a thermally insulated holder made of a material with high thermal conductivity. The holder with the sample is exposed to a constant output of power from a Hot Disk sensor attached to the holder. The experiment is carried out over periods of time which are long compared with the time it takes to establish non-varying temperature gradients inside the holder and sample assembly. During the heating period the temperature increase of the sensor is continuously recorded by following its resistance increase. Indications from a series of measurements on metallic and ceramic samples at room temperature are that it is possible to conveniently estimate the specific heat of a solid with an accuracy of a few percent.

## INTRODUCTION

In certain transient measurements of thermal transport properties of materials, it is possible to determine both the thermal conductivity and thermal diffusivity. In these cases one can also estimate the specific heat per unit volume, of the sample material, from the ratio of these two coefficients. A case in point being the Transient Plane Source (TPS) or the Hot Disk method [1].

Mattias Gustavsson, Department of Thermo and Fluid Dynamics, Chalmers Institute of Technology, S-412 96 Gothenburg, Sweden
N S Saxena, Department of Physics, University of Rajasthan, Jaipur, India
Ernest Karawacki and Silas E Gustafsson, Department of Physics, Chalmers Institute of Technology, University of Gothenburg, S-412 96 Gothenburg, Sweden

The solution most commonly used when applying the TPS method presupposes that the heat wave created by the output of power in the Hot Disk sensor must not reach the outside boundaries of the sample, which means that the thermal conductivity equation has been solved with the assumption that the sensor is located in a sample of infinite extension.

In some experiments Hot Disk sensors have been used for direct determination of the specific heat capacity of materials by transient heating of thermally insulated samples over time periods which are long compared to the time it takes to establish non-varying temperature gradients inside the sample [2].

Irrespective of the way the sample is insulated, there will always appear heat losses through thermal radiation and thermal conduction to the surroundings and through the leads to the Hot Disk sensor. In order to control the conditions for the heat losses mentioned above, the sample is in the present method placed in a holder made of a material with high thermal conductivity and designed so that the holder can be closely attached to the plane Hot Disk sensor. The selection of a material with high thermal conductivity for the sample holder is made in order to minimise the time it takes to establish a constant difference between the mean temperature of the sample holder and that of the Hot Disk sensor.

In order to estimate the heat losses to the surroundings and the specific heat of the holder, a separate experiment is initially being performed with the empty holder. With the information from this initial experiment, it is possible to deduce the specific heat of the sample material, from a recording with the sample inside the holder. In these experiments, we have made sure that the temperature increase versus time is essentially the same as in the recording with the empty holder.

If we assume that the holder with the sample is heated over a comparatively long period of time, and that it is placed in an environment such that the heat loss from the holder to the surroundings is kept at a minimum, it turns out that the temperature increase of the holder and sample after a short initial transient will be an almost linear function of time, cf. figure 1. This linear temperature increase appears when the temperature gradients inside the holder and the sample tend to constant values and there is a constant difference between the temperature recorded with the sensor and the average temperature of the sample.

## THEORY

If we assume that the holder and the sample have infinitely large thermal conductivities and are placed inside a perfect insulator, it is obvious that the temperature recorded with the Hot Disk sensor gives the average temperature over the whole time period, and the only calibration needed is to determine the specific heat capacity of the holder. However, in the normal experimental situation, it is necessary to consider deviations from this ideal model.

From the definition of specific heat capacity we can write for the holder without the sample

$$P_1(t) = \left(mc_p\right)_{\text{holder}} \frac{d\overline{T}_1}{dt}(t) + \left|Q_1(t)\right| \tag{1}$$

and for the holder together with the sample we have

$$P_2(t) = \left\{\left(mc_p\right)_{\text{holder}} + \left(mc_p\right)_{\text{sample}}\right\} \frac{d\overline{T}_2}{dt}(t) + \left|Q_2(t)\right| \tag{2}$$

where $P$ is input of power to the system from the sensor, $t$ is the time, $\overline{T}(t)$ is the average temperature of the holder (and sample) and $|Q|$ is the heat losses to the surroundings.

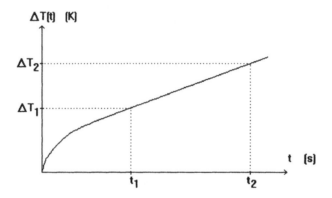

Figure 1.   Temperature versus time recording for holder with or without sample keeping the output of power from the Hot Disk sensor constant.

With reference to Figure 1 we define the average value of the time derivative of the temperature in the time interval $\left[t_1, t_2\right]$ as:

$$\overline{\left(\frac{dT}{dt}\right)} = \frac{1}{t_2 - t_1}\int \frac{dT}{dt}dt = \frac{1}{t_2 - t_1}\int dT = \frac{\Delta T}{t_2 - t_1} \tag{3}$$

With this expression it is possible to calculate the specific heat capacity since the output of power in the Hot Disk sensor is well known provided the heat losses summarised by the notation Q is known. This can in principle be done by performing a separate experiment with the holder without the sample.

Let us express the heat losses in the two experiments with and without the sample inside the holder in the following way:

$$\left|Q_1\right| = f_1(t)\frac{d\overline{T}_1}{dt}(t) \tag{4}$$

$$|Q_2| = f_2(t)\frac{d\overline{T}_2}{dt}(t) \tag{5}$$

We can then write:

$$P_1(t) = \left\{ \left(mc_p\right)_{\text{holder}} + f_1(t) \right\}\frac{d\overline{T}_1}{dt}(t) \tag{6}$$

and $$P_2(t) = \left\{ \left(mc_p\right)_{\text{holder}} + \left(mc_p\right)_{\text{sample}} + f_2(t) \right\}\frac{d\overline{T}_2}{dt}(t) \tag{7}$$

Over the time period, when the temperature versus time is essentially a linear function, we can write the average temperature increase as the sum of a constant part and a time dependent part is such a way that the time average of the time dependent part is zero:

$$\frac{d\overline{T}_1}{dt}(t) = \overline{\left(\frac{dT}{dt}\right)}_1 + \left(\frac{d\overline{T}_1}{dt}\right)'(t) \tag{8}$$

and similarly for the average temperature increase with a sample.

We now make the assumption, that there exists a unique function $f^*(t)$, for our holder, insulation and sample configuration (but different for different holders, insulations etc.), such that:

$$|Q(t)| = f^*(t + \Delta t_{corr})\overline{\left(\frac{dT}{dt}\right)} \tag{9}$$

inside the time interval $[t_1, t_2]$ of interest. We have here included a small time correction $\Delta t_{corr}$ to compensate for the initial transients at the beginning of the experiments, during which the holder with or without the sample establishes constant spatial temperature gradients.

The detailed validity of this assumption has eventually to be tested experimentally. However, an increase in the output of power $\overline{P}$ from the sensor means a corresponding increase in the averaged temperature increase and it seems reasonable to suppose a proportionate loss of heat to the surroundings.

With the above assumptions we can now write:

$$|Q_1(t)| = f^*(t + \Delta t_{corr1})\overline{\left(\frac{dT}{dt}\right)}_1 \tag{10}$$

and $$|Q_2(t)| = f^*(t + \Delta t_{corr2})\overline{\left(\frac{dT}{dt}\right)}_2 \tag{11}$$

from these equations it follows that:

$$P_1(t) = \left(mc_p\right)_{\text{holder}} \frac{d\overline{T_1}}{dt}(t) + f^*\left(t + \Delta t_{corr1}\right)\overline{\left(\frac{dT}{dt}\right)_1} \tag{12}$$

and $$P_2(t) = \left\{\left(mc_p\right)_{\text{holder}} + \left(mc_p\right)_{\text{sample}}\right\}\frac{d\overline{T_2}}{dt}(t) + f^*\left(t + \Delta t_{corr2}\right)\overline{\left(\frac{dT}{dt}\right)_2} \tag{13}$$

If we now take the time average of equations (12) and (13) over $\left[t_1, t_2\right]$, neglecting the influence on the average values of $f^*$ from the difference in time corrections - or

$$f^*\left(t + \Delta t_{corr1}\right) \cong f^*\left(t + \Delta t_{corr2}\right) \text{ when } t \in \left[t_1, t_2\right] - \tag{14}$$

we obtain the final equations accordingly:

$$\overline{P_1} = \left(mc_p\right)_{\text{holder}}\overline{\left(\frac{dT}{dt}\right)_1} + \overline{f^*}\overline{\left(\frac{dT}{dt}\right)_1} = \left\{\left(mc_p\right)_{\text{holder}} + \overline{f^*}\right\}\overline{\left(\frac{dT}{dt}\right)_1} \tag{15}$$

and $$\overline{P_2} = \left\{\left(mc_p\right)_{\text{holder}} + \left(mc_p\right)_{\text{sample}} + \overline{f^*}\right\}\overline{\left(\frac{dT}{dt}\right)_2}. \tag{16}$$

The final expression from which to determine the specific heat of the sample then becomes:

$$\frac{\overline{P_2}}{\delta_2} - \frac{\overline{P_1}}{\delta_1} = \left(mc_p\right)_{\text{sample}}, \tag{17}$$

where we have defined $\delta = \overline{\left(\frac{dT}{dt}\right)}$. $\tag{18}$

## ESTIMATION OF ERRORS

During the whole experiment the temperature is measured as a function of time keeping the total driving voltage over the bridge - particularly over the sensor and a series resistance - constant. This makes it in particular possible to estimate the deviations from the linearity of the temperature versus time curve.

It is clear from what is shown above that the analytical error in $\left(mc_p\right)_{\text{sample}}$, according to the suggested treatment of the experimental data (cf. equations 15 and 16), is equal to the error in $\overline{f^*}$. The relative error in determining $\left(mc_p\right)_{\text{sample}}$ is obtained from the following relations:

Following equations (4) and (5) we can write

$$\left|\left(\frac{d\bar{T}}{dt}(t)\right)'\right| \le \varepsilon \left|\overline{\left(\frac{dT}{dt}\right)}\right|, \tag{19}$$

where we define $\varepsilon$ as the smallest positive number making the inequality hold in the considered time interval $[t_1, t_2]$. A straight forward estimation of the loss of heat from the holder indicates that:

$$|Q(t)| \le \varepsilon \cdot \left\{(mc_p)_{\text{holder}} + (mc_p)_{\text{sample}}\right\}\overline{\left(\frac{dT}{dt}\right)}. \tag{20}$$

Using the equations (19) and (20) we get the final estimation of the relative error as

$$\text{Error} \le \frac{f_{\max}^* \cdot \varepsilon}{(mc_p)_{\text{sample}}} \le \frac{\left\{(mc_p)_{\text{holder}} + (mc_p)_{\text{sample}}\right\} \cdot \varepsilon^2}{(mc_p)_{\text{sample}}} \le 0.7\% \tag{21}$$

The estimation of a relative error less than 0.7 % from equation (21) is based on the fact that the specific heat of the holder was in all our measurements more than ten times larger that of the sample and a direct estimation of the deviation from linearity of the temperature versus time curve indicates that $\varepsilon \approx 2.5\%$.

Equation (21) indicates rather well the possibilities to improve accuracy of the measurements by working with a more optimal value of the ratio of the specific heat of the holder to that of the sample - $(mc_p)_{\text{holder}}\big/(mc_p)_{\text{sample}}$ - and at the same time improve the linearity of the temperature versus time curve.

It should be noted that we are in these experiments dealing with the following three errors.
1. Errors related to the assumptions and approximations in the theoretical model treated above.
2. Statistical errors of the measurements.
3. Inaccuracy related to the output of power, temperature coefficient of resistivity, failure to reproduce the surroundings of the sample holder, with or without the sample, etc.

## EXPERIMENTAL ARRANGEMENTS

In the present experiments the holder was made of copper with the following inside dimensions: diameter 20 mm and height 5 mm. The walls of the holder was on the average 3 mm thick and it was attached to a Hot Disk sensor with a diameter of 20 mm. The total time of the experiments was chosen to 80 sec with a approximate output of power in the sensor of 1 Watt, which gave a reasonable time scale for the temperature measurements and a total temperature increase of around 2 K.

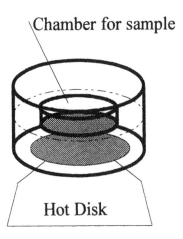
Chamber for sample

Hot Disk

Figure 2. Sample holder with the Hot Disk sensor attached.

The reason for using a rather massive holder was to reduce the transient period at the beginning of the experiment and also to easily arrange for the same temperature versus time graph with or without the sample placed inside the holder. A high thermal diffusivity in the holder is consequently desirable in order to reduce $t_1$, or the time when the spatial temperature gradients in the holder and sample become approximately constant in time.

There are a series of conditions that obviously should be considered when designing these kind of experiments and we will be listing a few below:

1. *The holder should be designed so that the transient period at the beginning of the experiment is minimised.*
2. *It should be possible to easily attach a plane Hot Disk sensor to the holder.*
3. *The insulation material around the holder should have as low a thermal conductivity as possible.*
4. *The surface area of the holder should be selected so that the heat loss is minimised.*
5. *It is desirable that the temperature versus time graphs for the holder with and without the sample are as close as possible or that* $\overline{\left(\dfrac{dT}{dt}\right)_1} \cong \overline{\left(\dfrac{dT}{dt}\right)_2}$ *over the time window* $\left([t_1, t_2]\right)$.

Some comments should be made as to the possibilities to meet the above mentioned requirements, which in some cases are quite incompatible. It should for instance be remembered that it is possible to work with more than one Hot Disk sensor attached to the holder. If we still assume that the holder is flat one can easily arrange with one sensor on both side of the holder and in that way further reduce the time $t_1$.

To estimate the time $(t_1)$ it takes to establish non-varying temperature gradients in both the holder and the sample using only one sensor, we have derived the following formula for the case depicted in Figure 3.

$$t_1 \geq \left( \frac{L_{\text{sample}}}{\sqrt{\kappa_{\text{sample, approx.}}}} + \frac{L_{\text{holder}}}{\sqrt{\kappa_{\text{holder}}}} \right)^2 \tag{22}$$

The transient development of the temperature profile in the sample can be considered closely one-dimensional, since the major heat flux comes from the underlying contact area, whilst, due to the more complex geometry, the transient development in the holder is purely 3-dimensional. The characteristic lengths for the formula have been chosen with this in mind.

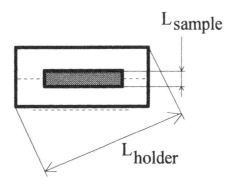

Figure 3.    Characteristic lengths of the sample and holder.

## DETERMINATION OF THE OUTPUT OF POWER $\overline{P}$

It is important to determine as accurately as possible the output of power during the time period when the temperature is increasing linearly with time. The Wheatstone bridge with the sensor and a series resistance in one of the arms gives a fairly constant output of power. The driving voltage over the sensor and the series resistance is kept constant throughout the experiment. However, in these measurements it appears important to determine the average output of power as precisely as possible.

The resistance increase in the sensor is

$$R(t) = R_0 \left( 1 + \alpha \Delta T(t) \right), \tag{23}$$

and $P(t) = R(t)I(t)^2$. $\tag{24}$

The power to be entered into equation (17) is following some further treatment calculated from the expression

$$\overline{P} = \frac{1}{t_2 - t_1} \int_{t_1}^{t_2} P(t)dt \qquad\qquad (25)$$

giving an estimated accuracy better than 0.1 %.


## RESULTS AND DISCUSSION

The results of measurements on six different materials is presented in Table 1. On each material three different measurements were performed and the average of these measurements deviates only 2.4 % from recommended values.


Table 1:    Results from measurements of the specific heat capacity on five materials with different sample sizes. The average deviation from recommended values [3,4] was 2.4 %. Each value corresponds to a single recording over 80 sec. The stainless steel was of the SIS 2343 quality and CEC stands for a cordierite-based sintered material provided by Lafarge Company, France, called Cecorite 130P

|  | Copper | Aluminium | Nickel | Silver | Stainless Steel | CEC |
|---|---|---|---|---|---|---|
| **Mass of sample [g]** | 8.34 | 3.64 | 3.77 | 3.80 | 5.12 | 3.08 |
| **Specific heat [J/kgK]** | 368 348 370 | 911 948 956 | 473 458 472 | 236 233 244 | 434 443 452 | 731 684 721 |
| **Mean value** | 362 | 938 | 468 | 237 | 443 | 712 |
| **Rec. value** | 387 | 940 | 444 | 237 | 440 | 728 |
| **Deviation** | 6.5% | 0.2% | 5.0% | 0% | 0.6% | 2.2% |


It is important to note that the data presented in this paper refer to ongoing work with a view to develop an experimental technique to independently determine the specific heat capacity of a sample material using the same experimental set-up - the Hot Disk Thermal Constants Analyser- which is presently being used for measuring the thermal conductivity and thermal diffusivity of materials. It should also be noted that we have not yet designed an optimal sample and holder configuration. This optimal design will probably reduce the errors accordingly.

The method presented here will as a matter of fact be complementary to Hot Disk measurements particularly when studying small samples with high thermal conductivity. In those cases the time period for the measurements is very limited, which makes it difficult to directly extract the thermal diffusivity from the experimental data. With this new method it is possible to get independent information on the specific heat and use these data when evaluating thermal conductivity experiments. In this way it has turned out to be possible to work with substantially smaller high conducting samples than what has been possible before.

The work that is presently going on with reference to this experimental technique includes an investigation of the energy storage ratio between the holder and the sample and temperature range limitations. It should also be noted that this technique should be possible to apply also to fluids.

## REFERENCES

1. Gustafsson Silas E., 1991. "Transient plane source techniques for thermal conductivity and thermal diffusivity measurements of solid materials." *Review of Scientific Instruments*, 62:797-804.

2. Gustavsson Mattias, Ernest Karawacki, and Silas E. Gustafsson, 1994. "Thermal conductivity, thermal diffusivity, and specific heat of thin samples from transient measurements with hot disk sensors." *Review of Scientific Instruments*, 65:3856-3859.

3. Cabannes Francois and M. L. Minges, 1989. "Thermal conductivity and thermal diffusivity of a corderite-based ceramic. Results of a CODATA measurements program." *High Temperatures - High Pressures*, 21:69-78.

4. Bolz R. E. and G. L. Tuve, editors, 1976. *CRC Handbook of tables for Applied Engineering Science*, 2nd Edition. CRC Pres Inc. Boca Raton, Florida.

# Optimization of the Thermal Conductivity Measurement by Modulated DSC™

S. R. AUBUCHON and R. L. BLAINE

## ABSTRACT

One of the many benefits of Modulated DSC (MDSC™) is the direct measurement of heat capacity. This capability is of great interest because heat capacity and thermal conductivity are related properties. Through the use of MDSC, a technique has been developed in which the thermal conductivity of semi-insulating materials can be measured directly.

Investigations into the applicability of this technique in a variety of polymer systems have been performed. In addition, research has centered on the measurement of thermal conductivity as a function of temperature, optimization of the calibration factors, and the comparison of experimental values to those obtained from the literature. This work has helped in the development of a secondary reference material kit for thermal conductivity measurements. Salient results will be discussed.

## INTRODUCTION

One of the benefits of MDSC is the direct measurement of heat capacity. This capability is of interest in this discussion because heat capacity and thermal conductivity are related properties. In the thermal conductivity measurement, the heat capacity ($C_p$) of a sample is measured by encapsulating a thin sample in a crimped aluminum pan and subjecting it to an isotherm, with a modulation amplitude of $\pm$ 1°C over a period (P) of 80 seconds. The apparent heat capacity (C) of the sample is then measured by forming the sample into a right circular cylinder of length (L) and diameter (d), having parallel end faces. The mass (M) of this specimen is recorded, then the heat capacity is measured with the specimen resting directly on the constantan disc in the MDSC cell. A thin layer of aluminum foil wetted with silicone oil acts as an efficient heat transfer path between the disk and sample.

Steven R. Aubuchon & Roger L. Blaine, TA Instruments, Inc., 109 Lukens Drive, New Castle, DE 19720

The thermal conductivity of the sample is then calculated from the experimentally determined C and $C_p$ values by the following equation:

$$K_o = (8 \, L \, C^2) \, / \, (C_p \, M \, d^2 \, P) \tag{1}$$

Several additional assumptions are necessary for this calculation to be valid:

- The face of the specimen at the heat source follows the applied temperature modulation.
- The heat flow through the opposing face is zero.
- There is no heat flow through the side of the specimen.

Practical application demonstrates that the accuracy of this method declines with decreasing thermal conductivity, most likely owing to the loss of thermal energy through the sides of the test specimen under flowing purge gas conditions. Modeling the premise of heat loss through the sides of the test specimen creates a thermal conductivity calibration constant (D) which may be used to correct for this effect. This constant may be incorporated with the previously calculated value to obtain a more accurate observed thermal conductivity value via the following equation:

$$K = [K_o - 2D + (K_o^2 - 4Dk_o)^{0.5}]/2 \tag{2}$$

The accurate and precise measurement of thermal conductivity by MDSC is dependent on the optimization of experimental conditions. Proper calibration and sample preparation are vital to the successful thermal conductivity measurement. Recommended experimental techniques and parameters are discussed in the following sections.

## DISCUSSION

### CELL CALIBRATION

Through MDSC, the direct quantitative measurement of heat capacity can be performed. It is this unique ability which allows for the direct measurement of thermal conductivity. In order for the heat capacity measurement (and hence the thermal conductivity measurement) to be accurate, the MDSC system must be calibrated for heat capacity. For routine heat capacity measurements on polymers, a single-temperature heat capacity calibration, usually utilizing high density polyethylene at a temperature above the melt is sufficient. The heat capacity constant, or $K(C_p)$, is calculated as the ratio of the measured $C_p$ to the literature $C_p$ at the chosen temperature, usually 140°C. Calibration at a single point is adequate for most MDSC studies, in that the MDSC heat capacity constant will not vary considerably over the operating temperature range of the instrument. However, the precision and accuracy can be improved by determining $K(C_p)$ at multiple

determining $K(C_p)$ at multiple temperatures over the range of interest. For that purpose, sapphire is recommended. Figure I shows typical results for a 60 mg sapphire disk. Note that the variation in $K(C_p)$ over 40°C is only 1%.

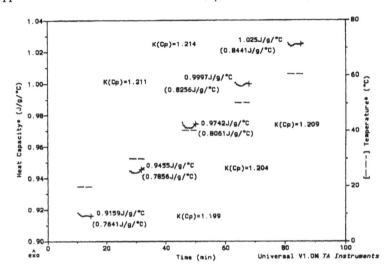

Figure I: $K(C_p)$ at Varying Temperatures. Literature Values are in Parentheses

## THERMAL CONDUCTIVITY CALIBRATION

The accuracy of the thermal conductivity measurement is improved through the utilization of the thermal conductivity calibration constant, "D". This constant is used to correct for heat loss through the sides of the specimen during the measurement. The calibration constant is determined by measuring the thermal conductivity of a material, and comparing the calculated value with the literature reference. The absolute value of D is then determined from the following equation:

$$D = (K_o \bullet K_r)^{0.5} - K_r \qquad (3)$$

where $K_o$ is the calculated value of thermal conductivity and $K_r$ is the reference value. Polystyrene is an excellent calibrant, based on its low thermal conductivity, the ease of fabrication and handling, and the abundance of thermal data available for this material. The calibration constant (D) is valid for a sample of specific dimensions. Therefore, the sample specimen should be fabricated to match the dimensions of the calibrant disk as closely as possible. Studies in our laboratory have also shown that the calibration constant "D" is also somewhat temperature sensitive. Figure II illustrates this effect.

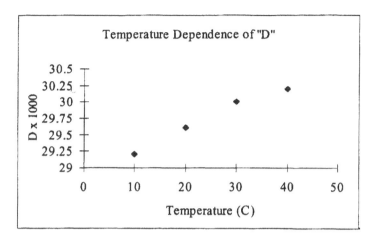

Figure II. Temperature Dependence of the Thermal Conductivity Calibration
Constant (D)

For this reason, we recommend that a unique value of D is calculated for each measurement temperature. If a dynamic scan is desired, the calibration constants can be averaged, however this will decrease the accuracy and precision of the measurement.

For isothermal measurements, the heat capacity calibration constant $[K(C_p)]$ and the thermal conductivity calibration constant (D) can be simultaneously employed to improve the quality of the experimental result.


## SAMPLE PREPARATION

Perhaps the most important (and frequently overlooked) aspect of the DSC experiment involves sample preparation. For a technique as sensitive as DSC, improper preparation or placement of the specimen can introduce experimental error which can invalidate results. Modulated DSC results can also be affected by improper sample preparation. When MDSC is applied to a technique such as the thermal conductivity measurement, potential errors arising from poor sample preparation can be greatly pronounced. Therefore, proper sample pretreatment and handling is vital to the success of the thermal conductivity measurement.

The first step of the thermal conductivity measurement involves the determination of the specific heat capacity of a thin specimen of the material of interest. This measurement is performed via the standard MDSC technique, therefore traditional sample preparation guidelines should be used.

1.    *Maximize the contact area between the sample and the pan.*

Keep the sample as thin as possible in order to cover as much of the pan as possible. Do not use large irregular chunks of sample.

2.    *Use lids on the DSC pans to keep the sample flat and pressed
      against the bottom of the pan.*

      If hermetic pans are used, flatten the lid before crimping to
      force the sample to the bottom of the pan and to minimize its
      ability to move during the experiment.

3.    *Use samples of 10-15 mg for polymers and keep them as thin
      as possible.*

      Although 10-15 mg is larger than typically used for traditional
      DSC experiments, it is recommended for MDSC in order to
      provide accurate heat capacity and to maximize the size of the
      total and signals that can be smaller than typical DSC
      measurements. This is due to the much lower average heating
      rates used with MDSC.

Unique to the thermal conductivity experiment is the measurement of the
apparent heat capacity of a more massive, right-circular cylinder of the specimen.
Ideally, the sample should be cut from a preformed 1/4" diameter circular rod of
the material. The suggested height of the sample 3-3.5 mm. Alternate
dimensions can be used, but the shape of the specimen should remain constant.
The height of the sample should be constrained so as to not interfere with the
inner lid placement on the DSC cell. It is important to fashion a sample of large
enough dimensions such that a temperature gradient is induced during the
measurement. If a preformed rod is not available, the sample can be cut from a
block of the material using a cork bore or low-speed hollow drill bit. Care should
be taken to avoid imparting any thermal history to the sample during the
fabrication process.

For softer, more compliant samples, it may be necessary to cast the
material into the desired shape. A 1/4" diameter hole drilled into a 3.5 mm thick
piece of polished aluminum has been used in our laboratory as a mold. The
pliable sample is pressed into the hole, and the excess material is scraped from
the surface of the mold. The formed cylinder is then carefully removed from the
hole, taking care not to distort the shape. The sample can then be weighed, and
the dimensions determined from the size of the mold. Since the calculation is
dependent upon sample dimension, it is very important to minimize any
geometric distortion when removing the sample from the mold. In some cases,
the material may be so pliable that removing the sample from the mold is very
difficult. It is possible to cool a sample in liquid nitrogen while in the mold,
lowering the sample temperature below its glass transition temperature.
Extraction is then more feasible. However, the sample should be allowed to
slowly warm to room temperature in a dessicator, to minimize condensation. It is
advised to analyze the material previous to this procedure to ensure that
subambient temperatures do not impart thermal history to the sample.

Measurement of the apparent heat capacity of the large specimen is performed employing the same experimental and modulation conditions as for the thin sample. The sample cylinder is placed directly onto the DSC constantan disc, to maximize heat transfer. A thin layer of aluminum foil wetted on both sides with silicone oil serves as an efficient thermal buffer between the sample and the constantan disk. The aluminum foil disk should be initially wetted on both sides with the oil, then carefully placed on the sample position. (An identically prepared disk should also be placed on the reference position.) The sample cylinder is then carefully placed onto the aluminum disk, ensuring sufficient contact across the entire face of the cylinder. The measurement can then be performed.

## SUMMARY

The utility of Modulated DSC is illustrated in the direct measurement of thermal conductivity. This technique provides for the facile determination of accurate and precise values of thermal conductivity for many polymeric samples. As with most thermal analysis techniques, experimental procedure can affect the quality of the results. With correct calibration and sample preparation, high quality determination of thermal conductivity can routinely be achieved by MDSC.

# Estimation of a Space-Varying Heat Transfer Coefficient or Interface Resistance by Inverse Conduction

D. MAILLET, A. DEGIOVANNI and S. ANDRÉ

## ABSTRACT

Two regularization methods are compared in this article. These methods are used to calculate a heat transfer coefficient distribution on a cylindrical pipe located in a transverse flow, starting from internal temperature measurements. The first inversion method corresponds to a Tikhonov regularization (applied to surface heat flux density). The direct problem is solved by separating the variables in the usual space domain (two-dimensional steady-state problem). The second method uses an inversion in a transformed domain, the space Fourier domain (experimental transforms of the temperature distribution, which relies on the thermal quadrupole method for solving the direct problem). This second method requires the setting of only one parameter: the desired number of harmonics. The agreement between these two inversion techniques is very good. It is also shown that the second method (regularization by spectrum truncation) constitutes a singular value decomposition of the operator of the direct problem, with analytical singular values.

## NOTATION

| | | | |
|---|---|---|---|
| $A, B, C, D$ | coefficients of the quadrupole matrix ($2 \times 2$) | $h(s)$ | local heat transfer coefficient (Wm$^{-2}$K$^{-1}$) |
| A | operator | J | merit function to be minimized |
| **A** | matrix produced by discretizing operator A ($m \times n$) | **K** | quadrupole matrix of order $2 \times 2$ (after Fourier transformation) |
| Arg | argument | $m$ | number of measurement points |
| $b_i(s)$ | base function (on the $[0\ \pi]$ interval) | **M** | quadrupole matrix ($2 \times 2$) |
| **b** | vector of base functions | $n$ | number of desired harmonics for $\varphi$ |
| **e** | error vector | | |
| g | function | | |

Denis Maillet, Alain Degiovanni, Stéphane André, Laboratoire d'Energétique et de Mécanique Théorique et Appliquée, URA CNRS 875,
Institut National Polytechnique de Lorraine - Université Henri poincaré Nancy I
2, avenue de la Forêt de Haye - 54516 Vandoeuvre-lès-Nancy cedex - France

| N | norm | $\gamma$ | regularization |
|---|------|----------|----------------|
| $r$ | polar radius    (m) | | coefficient |
| R | penalization term in J | | $( = \mu/(1-\mu) )$ |
| $R$ | thermal resistance of | $\Gamma$ | contour |
| | a unit length of the | $\mu$ | compromise |
| | pipe $(KmW^{-1})$ | | parameter |
| **R** | penalization vector | $\lambda$ | thermal conductivity |
| | $(1 \times m)$ | | $(Wm^{-1}K^{-1})$ |
| $s$ | curvilinear abscissa | $\lambda_i$ | singular value |
| | (m) | $\Lambda$ | diagonal matrix |
| T, $T$ | temperature function | | $(m \times n )$ of the |
| | and temperature (K) | | singular values of **A** |
| **T** | temperature vector | $\rho$ | normalized radius |
| | $(m \times 1)$ | | $( = r/r_e)$ |
| $T_e$ | temperature of the | | current value of |
| | external air flow (K) | | parameter vector $\varphi$ in |
| **U, V** | orthogonal matrices | | criterion J |
| | $(n \times n$ and $m \times m )$ | $\theta$ | temperature difference |
| **u, v** | column vectors of **U** | | $(= T_1 - T)$     (K) |
| | and **V** (truncated to $n$ | | |
| | components) | *Subscripts* | |
| $x$ | angular abscissa | 1, 2 | relative to internal (1) |
| **x** | measurement position | | and external (2) radii |
| | vector | | of the pipe, or to the |
| **X** | matrix of the position | | first and second layers |
| | vectors of all the | exp | relative to transforms |
| | measurements | | of  the experimental |
| **Y** | vector of measured | | temperatures |
| | temperatures ( $m \times 1$) | i | relative to the input of |
| $w$ | weighting term in the | | the quadrupole, or |
| | calculation of the | | subscript of a flux |
| | experimental | | component, or |
| | transforms | | of a singular value |
| | of temperature | k | subscript of a function |
| $\varepsilon$ | measurement noise | | or of a harmonic |
| | (K) | o | relative to the output |
| $\boldsymbol{\varepsilon}$ | noise vector ( $m \times 1$) | | of the quadrupole |
| $\varphi$ | linear density of flux | $\varphi$ | relative to $\varphi$ |
| | $(Wm^{-1})$ | $\mu$ | relative to $\mu$ |
| $\boldsymbol{\varphi}$ | column vector ($n$ | | |
| | components of $\varphi (s)$ | *Superscripts* | |
| | in base **b**) | $\sim$ | Fourier cosine |
| $\phi$ | product of the linear | | transform |
| | density of flux and the | $\wedge$ | estimated value |
| | half-perimeter | $'$ | component in the new |
| | $(= \pi r \varphi)$ (W) | | base of $\varphi$ or **Y** |
| $\Phi$ | heat flux through a unit | t | matrix transpose |
| | length of pipe (W) | | |

# INTRODUCTION

This article compares two different inversion techniques for estimating the heat transfer coefficient distribution over a cylinder in crossflow: Tikhonov regularization and singular value decomposition. They were both implemented using the same experimental temperature measurements. Results for the first technique have already been published. Conception and implementation of the second technique, which is based on an integral transform method for solving direct problems (thermal quadrupoles), are presented here for the first time.

One of the most popular methods used nowadays for solving multidimensional steady-state inverse conduction problems is the regularization method with penalization (for example: Tikhonov regularization). In this modified least squares technique, where a surface temperature or flux distribution is desired, knowing internal temperature measurements, two parameters have to be set more or less arbitrarily. These parameters are the number of pieces that will be used for defining the function to be estimated (parametrization problem) and the level of the coefficient that weights the penalization term with respect to the least squares sum. The role of this weighting coefficient is to make the inverse problem more well-posed, i.e. to produce a solution that will be less sensitive to the measurement noise, especially when the number of desired pieces becomes large.

This problem will be presented first in a rather general way, based on this penalization method. Another regularization method will then be presented: the regularization by truncation of the spectrum of singular values.

When the desired function is a spatial distribution of heat flux density in a two-dimensional problem , the temperature $T$ at any point P of coordinate vector $x$ (vector with two components) in the medium is:

$$T\ (\mathbf{x})\ =\ \int_{\Gamma}\ A\ (\mathbf{x},s)\ \varphi\ (s)\ ds\ =\ (A\ \varphi)\ (\mathbf{x}) \tag{1}$$

where $\varphi\ (s)$ is the spatial distribution of heat flux density at the medium boundary $\Gamma$, defined by its curvilinear abscissa $s$, and A $(\mathbf{x}\ ,\ s)$ the Kernel of the operator that transforms the wall heat flux density $\varphi$ to internal temperature $T$ *(figure 1)*.

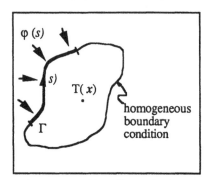

Figure 1. Wall heat flux estimation

If the temperature is desired at $m$ points $x_1$ to $x_m$ , T (x) becomes a vector function **T** (**X**), where the coordinates of these $m$ points have been ordered in a vector **X**. Kernel A $(\mathbf{x}\ ,\ s)$ then becomes a vector function A $(\mathbf{X}\ ,\ s)$ of both **X** and $s$, and operator A becomes a vector operator (**A**):

$$\mathbf{T}\ (\mathbf{X})\ =\ \int_{\Gamma}\ \mathbf{A}\ (\mathbf{X},s)\ \varphi\ (s)\ ds\ =\ (\mathbf{A}\ \varphi)\ (\mathbf{X}) \tag{2}$$

It is sometimes possible to calculate the temperature using equation (1), for particular forms of function $\varphi(s)$. In the general case, this function has to be parametrized, i.e. replaced by a column vector $\varphi$ of initially infinite length:

$$\varphi(s) = \sum_{i=1}^{\infty} \varphi_i \, b_i(s) = \mathbf{b}^t(s) \, \boldsymbol{\varphi} \tag{3}$$

where functions $b_i(s)$ form a base of the function space where we want to find $\varphi(s)$, and where superscript "t" designates the transposition of the column vector $\mathbf{b}$ that is built starting with these $b_i(s)$ functions. These functions may for example be constant functions over a given interval (zero outside this interval) if a constant piecewise function is wanted, or "hat" functions and so forth. In practice, vector $\varphi$ has to be truncated after considering only a finite number $n$ of components. Thus the dimension of the estimation problem becomes finite and the temperatures at the $m$ points that are of interest for the inversor (measurement points) are calculated using the discretized form of equation (2). This form is:

$$\mathbf{T} = \mathbf{A} \, \boldsymbol{\varphi} \quad \text{with} \quad \mathbf{T} = \mathbf{T}(\mathbf{X}) \quad \text{and} \quad A_{ij} = \int_{\Gamma} A(x_i, s) \, b_j(s) \, ds \tag{4}$$

where $\mathbf{A}$ is an ($m \times n$) matrix and $\mathbf{T}$ and $\varphi$ are column vectors of lengths $m$ and $n$ respectively. In the previous direct problem, $\varphi$ is given and $\mathbf{T}$ is calculated, using equation (4). If the inverse problem is now considered, $\mathbf{T}$ is measured, with an additive measurement noise $\boldsymbol{\varepsilon} = [\varepsilon_1, \varepsilon_2, ..., \varepsilon_m]^t$, and $\varphi$ is looked for in the form of an estimation $\hat{\boldsymbol{\varphi}}$. In this second problem, integral equation (2) is never solved. It is the matrix form (4) of this second problem that is solved, because the number $m$ of measurements being finite, the number $n$ of "components" (or pieces) that represent the flux function shall be less or equal to it at the most, otherwise the problem becomes indeterminate. Let $\mathbf{Y} \, (= \mathbf{T} + \boldsymbol{\varepsilon})$ be the vector of measured temperatures. The most "natural" estimation of $\varphi$ is then the one that minimizes the norm of the differences between measurements and model (4).

$$\hat{\boldsymbol{\varphi}} = \text{Arg}\left[\min\left(J(\boldsymbol{\xi})\right)\right] \quad \text{with} \quad J(\boldsymbol{\xi}) = \left\| \mathbf{Y} - \mathbf{A}\,\boldsymbol{\xi} \right\|^2 \tag{5a}$$

Usually the norm that is used for merit function J to be minimized is the sum of the squares of the components of the vector considered. In the non-trivial case where matrix $\mathbf{A}$ is regular, of rank $n$ here, the solution of this linear estimation problem is the estimator of the ordinary least squares [1], which is the solution of the following equation in $\boldsymbol{\xi}$.

$$(\mathbf{A}^t\mathbf{A}) \, \boldsymbol{\xi} = \mathbf{A}^t \, \mathbf{Y} \tag{5b}$$

or:

$$\hat{\boldsymbol{\varphi}} = (\mathbf{A}^t\mathbf{A})^{-1} \, \mathbf{A}^t \, \mathbf{Y} \tag{5c}$$

As soon as the number $n$ of desired harmonics for $\varphi$ becomes large enough, this solution becomes unstable, as any noise $\boldsymbol{\varepsilon}$, even weak, produces a very large error $\mathbf{e}_\varphi \, (= \hat{\boldsymbol{\varphi}} - \boldsymbol{\varphi})$ in the heat flux density. Regularization therefore becomes necessary. Two types of regularization are possible.

In the regularization through penalization technique, the preceding least square sum is modified by adding the square of the norm of a penalization term R ($\xi$) [2, 3].

$$\hat{\varphi}_{\mu} = \text{Arg}\left[\min\left(J_{\mu}(x)\right)\right] \quad \text{with} \quad J_{\mu}(\xi) = (1-\mu)\left\|Y - A\,\xi\right\|^2 + \mu\left\|R(\xi)\right\|^2 \quad (6a)$$

R ($\xi$) being either a deviation ($\xi$ - $\varphi_{ref}$) from a reference heat flux density $\varphi_{ref}$ (possibly equal to zero) or a discretized form of the first or second derivative, $\dfrac{d\varphi}{ds}$ or $\dfrac{d^2\varphi}{ds^2}$, of function $\varphi$ ($s$). For these three cases, where the penalizing term R ($\xi$) is linear with respect to $\xi$ and can therefore be written **R** $\xi$, where **R** is a $m$ component column vector, the regularized solution $\hat{\varphi}_{\mu}$ is the solution of the following equation which replaces equation (5b).

$$(\text{A}^t\text{A} + \gamma\ \text{R}^t\text{R})\ \xi = \text{A}^t\ \text{Y} \qquad\qquad (6b)$$

$\gamma$ ( = $\mu/(1-\mu)$ ) is the regularization coefficient, which has to be adjusted to obtain the necessary compromise between the precision of the estimation (if $\mu = \alpha = 0$, in the absence of noise) and the stability of the solution (if $\mu = 1$ or $\gamma =\infty$, that is R($\hat{\varphi}$)=0).

In the regularisation through truncature technique, the case considered here corresponds to $n$ smaller than or equal to $m$ (fewer unknowns than measurements), with matrix **A** (the discretization of the A operator) having full rank $n$. The singular values $\lambda_i$ (for i = 1 à $n$) of **A** ($m$ x $n$) are used, which are the strictly positive square roots of the eigenvalues of the matrix **A**$^t$**A**. It is then possible to use a Lanczos decomposition  (or singular value decomposition) of **A** [4, 5].

$$\text{A} = \text{V}\ \Lambda\ \text{U}^t \qquad \text{with:} \qquad \Lambda = \begin{bmatrix} \lambda_1 & 0 & \cdots & 0 \\ 0 & \lambda_2 & \cdots & 0 \\ \vdots & 0 & \ddots & \vdots \\ 0 & 0 & \cdots & \lambda_n \\ \hline & & \mathbf{0} & \end{bmatrix} \qquad (7a)$$

- where **U** ($n$ x $n$) and **V** ($m$ x $m$ ) are two orthogonal matrices,
- and $\Lambda$ is a matrix that can be partitioned into a diagonal matrix, diag ( $\lambda_1$, $\lambda_2$,... , $\lambda_n$) (singular values ordered according to  decreasing values, taking into account possible multiplicities), and  a zero matrix (size $(m-n)$ x $n$ ):

This corresponds to one change of base for the flux density vector, and another change of base for the measurement vector, using matrices **U** and **V** for these changes. Using the prime notation (') to denote these two vectors in the new bases, we have:

$$\xi = \text{U}\ \xi' \quad \text{and} \quad \text{Y} = \text{V}\,\text{Y}' \qquad\qquad (7b)$$

The flux density function $\varphi\,(s)$ is parametrized with a column vector $\varphi'$ in the new function base $\mathbf{b}'\,(s)$, with:

$$\varphi\,(s) = \mathbf{b}'^{t}(s)\,\varphi' \quad \text{with} \quad \mathbf{b}'(s) = \left[b_1{}'(s)\ b_1{}'(s)....b_n{}'(s)\right]^{t} = U\,\mathbf{b}(s) \quad (7c)$$

Using equations (5b, 7a and 7b), it can easily be shown that the estimator $\hat{\varphi}'$ of the flux density in the new base is the solution of:

$$(\Lambda^t\Lambda)\ \xi' = \Lambda^t\ Y' \tag{8a}$$

$$\text{or} \quad \hat{\varphi}'_i = y'_i\ /\ \lambda_i \quad \text{for} \quad i = 1\ \grave{a}\ n \tag{8b}$$

Returning to the original base of the flux density yields a new expression of its estimator, an expression that is strictly equivalent to equation (5c).

$$\hat{\varphi} = U\ \hat{\varphi}' = \sum_{i=1}^{n} \frac{(\mathbf{v}_i^t Y)}{\lambda_i}\ \mathbf{u}_i \quad \text{with:}\ U = [\mathbf{u}_1.... \mathbf{u}_n]\ \text{and}\ V = [\mathbf{v}_1.... \mathbf{v}_m] \tag{8c}$$

where $\mathbf{u}_i$ and $\mathbf{v}_i$ designate the first $n$ and $m$ columns of matrices $U$ and $V$.

An obvious advantage of estimator (8c), when compared to the estimator produced by solving equation (6b), is that only one parameter needs to be set - the number $n$ of components of the flux density - and not both $n$ and the regularization parameter $\gamma$ (or $\mu$) in the second case. If $n$ is too large, the inversion will become unstable, because the estimation error, in the original base of the $n$ functions $b_i\,(s)$, is expressed:

$$e_\varphi = \hat{\varphi} - \varphi = \sum_{i=1}^{n} \frac{(\mathbf{v}_i^t \varepsilon)}{\lambda_i}\ \mathbf{u}_i \tag{9a}$$

while in the "singular" bases of functions $b_i'(s)$ defined by $U$ it is:

$$\left(e_{\varphi'}\right)_i = \varepsilon'_i\ /\ \lambda_i \quad \text{for} \quad i = 1\ \text{to}\ n \tag{9b}$$

This means that singular values (of large rank) that are too low will disproportionately amplify the measurement noise. In this case the objectives will be more realistic if a parametrization of function $\varphi\,(s)$ is considered with a lower number of $b_i\,(s)$ functions. In the following part of this work, two regularization methods will be compared: regularization through penalization (with fixed $n$ and optimized $\gamma$) and the spectrum truncation technique (variable $n$). This last method stems directly from the quadrupole method used in a transformed space [6, 7, 8].

## DIRECT PROBLEM

We want to find the internal temperature at a point P with polar coordinates $(r, x)$, inside a pipe defined by two cylindrical surfaces of radii $r_1$ et $r_2$ (*figure 2*). A uniform temperature is given at the internal radius $r_1$, along with the transfer coefficient $h$, which varies with angle $x$ and determines the boundary condition at the external wall $r_2$ (the variation of this coefficient will be estimated in the inverse problem). It is further assumed that $h$ varies symmetrically with respect to the plane $x = 0$. The temperature is thus the solution of the following system of equations.

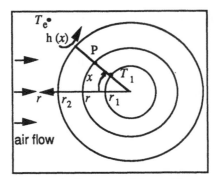

Figure 2. Geometry of the problem

$$\frac{\partial^2 T}{\partial r^2} + \frac{1}{r}\frac{\partial T}{\partial r} + \frac{1}{r^2}\frac{\partial^2 T}{\partial x^2} = 0 \qquad (10a)$$

$T$ is an even function in $x$                               (10b)

$T$ is a periodical function of period $2\pi$ in $x$     (10c)

$T = T_1$                      at $r = r_1$           (10d)

$$-\lambda\frac{\partial T}{\partial r} = h(x)\left[T - T_e\right] \qquad \text{at } r = r_2 \qquad (10e)$$

In order to linearize the inverse problem that will be considered later on, it is advantageous to replace the third-kind condition (10e) with a second-kind condition for which the wall heat flux is initially set.

$$-\lambda\frac{\partial T}{\partial r} = \varphi(x) \qquad\qquad \text{at } r = r_2 \qquad (10f)$$

Once the functions $h(x)$ and $\varphi(x)$ are parametrized, and the formal solution of $T$ is found in terms of $\varphi$, it is possible to express $\varphi$ in terms of $h(x)$ by setting the right-hand members of equations (10e) and (10f) equal to each other.

The solution of the system (10a,b,c,d,f), that can be obtained by the <u>method of separation of variables</u> [9, 10], corresponds, before parametrization, to the form given by equation (1), replacing $T(\mathbf{x})$ by $T_1 - T(r, x)$, $s$ by angle $x$, contour $\Gamma$ by the angular integral $[0\ \pi]$, and Kernel $A(\mathbf{x}, s)$ by $A(r, x, x')$, with:

$$A(r, x, x') = \frac{2\,r_2}{\pi\,\lambda} \sum_{k=0}^{\infty} g_k(r/r_2)\,\cos(kx)\,\cos(kx') \qquad (11a)$$

with  $g_o(\rho) = \frac{1}{2}\,\ell n\,(\rho/\rho_1)$     $g_k(\rho) = \frac{1}{k}\,\frac{1}{\rho^k}\,\frac{\rho^{2k} - \rho_1^{2k}}{1 + \rho_1^{2k}}$   for $k > 0$  (11b)

and                      $\rho = r/r_2$            $\rho_1 = r_1/r_2$

By parametrizing function $\varphi(x)$ in the form of a piecewise constant function, the coefficients of matrix $\mathbf{A}$ defined by equation (4) can be calculated analytically for any number $m$ of points where the temperature has to be calculated. In that case functions $b_i$ $(s = x)$ of equation (3) have a constant value of unity on each angle interval of width $\pi/n$, and are equal to zero outside.

In the quadrupole method, a Fourier cosine transform of temperature T on the $[0 \ \pi]$ interval is now considered, with its associated explicit inversion equation.

$$\tilde{T} \ (r,k) \ = \ \int_0^\pi T \ (r,x) \ \cos \ (kx) \ dx \quad \text{with} \quad k: \text{natural integer} \quad (12a)$$

$$T \ (r,x) \ = \ \sum_{k=0}^{\infty} \frac{\cos \ (kx)}{N_k} \ \tilde{T} \ (r,k) \quad \text{with} \quad N_k \ = \ \int_0^\pi \cos^2 \ (kx) \ dx \ = \begin{cases} \pi & \text{if} \ k = 0 \\ \dfrac{\pi}{2} & \text{if} \ k > 0 \end{cases}$$

$$(12b)$$

The $\cos \ (kx)$ functions, where $k$ is a non-negative natural integer, are the eigenfunctions of the problem in the $x$ variable (they can be found by separation of variables, with the symmetry condition (10b) determining a zero tangential flux density at $x = 0$ and $x = \pi$). Multiplication of equation (10a) by $\cos \ (kx)$, followed by integration of the resulting equation between 0 et $\pi$ [11], yields the following equation (after integration by parts that takes into account the zero tangential gradient at $x = 0$ and $x = \pi$ ).

$$\frac{d^2\tilde{T}}{dr^2} + \frac{1}{r} \frac{d\tilde{T}}{dr} - \frac{k^2}{r^2} \ \tilde{T} \ = \ 0 \qquad (13a)$$

The cosine Fourier transform of the product of the flux density by the half perimeter (at radius $r$) is now considered.

$$\tilde{\phi} \ (r,k) = \int_0^\pi \phi \ (r,x) \ \cos \ (kx) \ dx \ = -\lambda \ \pi r \ \frac{d\tilde{T}}{dr} \quad \text{with} \quad \phi \ (r,x) \ = \left( - \lambda \ \frac{\partial T}{\partial r} \right) \ \pi r$$

$$(13b)$$

Integration of equations (13a and b) allows the establishment of a linear relationship between the transforms of the temperature-flux vectors at the entry $(r = r_i)$ and at the exit $(r = r_o)$ from a cylindrical layer of thickness $r_o - r_i$ (with $r_1 \leq r_i \leq r_o \leq r_2$ ), and for a fixed harmonic of order $k$ [11].

$$\begin{bmatrix} \tilde{T} \ (k) \\ \tilde{\phi} \ (k) \end{bmatrix}_{r_i} = \ \mathbf{M} \ (r_i, \ r_o, \ k) \ \begin{bmatrix} \tilde{T} \ (k) \\ \tilde{\phi} \ (k) \end{bmatrix}_{r_o} = \begin{bmatrix} A \ (k) & B \ (k) \\ C \ (k) & D \ (k) \end{bmatrix}_{r_i,r_o} \begin{bmatrix} \tilde{T} \ (k) \\ \tilde{\phi} \ (k) \end{bmatrix}_{r_o} \quad (14a)$$

Coefficients $A, B, C$ et $D$ of the quadrupole matrix $\mathbf{M}$ are defined by:

$$A \ (0) \ = \ D \ (0) \ = \ 1; \quad B \ (0) \ = \ \frac{1}{\pi \ \lambda} \ \ln \ (r_o \ / \ r_i); \quad C \ (0) \ = \ 0 \quad (14b)$$

$$\left. \begin{aligned} A \ (k) \ &= \ D \ (k) \ = \ \frac{1}{2} \left[ (r_o \ / \ r_i)^k + (r_i \ / \ r_o)^k \right] \\ B \ (k) \ &= \ \frac{1}{2\pi \ \lambda k} \left[ (r_o \ / \ r_i)^k - (r_i \ / \ r_o)^k \right] \\ C \ (k) \ &= \ \frac{\pi \ \lambda k}{2} \left[ (r_o \ / \ r_i)^k - (r_i \ / \ r_o)^k \right] \end{aligned} \right\} \qquad \text{for} \ k \ > 0 \qquad (14c)$$

Let us note that the determinant ($AD - BC$) of this matrix is equal to unity for any value of $k$. If $\mathbf{M_1}$ is the matrix of the first layer defined from $r_i = r_1$ to $r_0 = r$, $\mathbf{M_2}$ the matrix of the second layer defined from $r_i = r$ to $r_0 = r_2$, and $\mathbf{M}$ that of the whole pipe defined by $r_i = r_1$ and $r_0 = r_2$ (and if the coefficients of these matrices have the corresponding subscripts, just omitting argument $k$), we can say that for the whole pipe :

$$
\begin{bmatrix} \tilde{T}_1 \\ \tilde{\phi}_1 \end{bmatrix} = \mathbf{M} \begin{bmatrix} \tilde{T}_2 \\ \tilde{\phi}_2 \end{bmatrix} \quad \Rightarrow \quad \tilde{T}_2 = (\tilde{T}_1 - B\,\tilde{\phi}_2)/A \tag{15a}
$$

and for layer (2):

$$
\begin{bmatrix} \tilde{T} \\ \tilde{\phi} \end{bmatrix} = \mathbf{M_2} \begin{bmatrix} \tilde{T}_2 \\ \tilde{\phi}_2 \end{bmatrix} \Rightarrow \tilde{T} = A_2\,\tilde{T}_2 + B_2\,\tilde{\phi}_2 = (A_2\,\tilde{T}_1 - B_1\,\tilde{\phi}_2)/A \tag{15b}
$$

using the fact that $\mathbf{M} = \mathbf{M_1 M_2}$. This equation allows the calculation of the $k^{\text{th}}$ harmonic $\tilde{T}$ of internal temperature in terms of the $k^{\text{th}}$ harmonic $\tilde{\phi}_2$ of the surface heat flux density and of the corresponding harmonic of temperature (at $r = r_1$). Boundary conditions (10d and 10f) are included this way. A Fourier inversion according to equation (12b) allows the calculation of the original $T(r, x)$ in the initial domain for any point in the cylindrical pipe.

## INVERSE PROBLEM

Measurements have been made in a wind tunnel [9, 10], under forced convection conditions (Reynolds number 64230, $\lambda = 0.224$ W/mK; $r_1 = 10$ mm; $r_2 = 16$ mm; $m = 19$ internal thermocouple measurements for an angle $x$ varying between 0 and 180° with a 10° step at a radius of 14.02 mm ; $T_1 = 65°C$ ; $T_e = 25°C$).

These measurements were inverted using regularization through penalization (Tikhonov) on the flux density (the penalizing term R of equation (6a) corresponding to the second derivative of the flux density), with the direct model being solved either by the analytical method presented above (separation of variables with 18 flux pieces), or by the boundary element method [10]. The value of the regularization coefficient was optimized ($\gamma = 10^{-6}$) by minimization of the residual $J_\mu$ - equation (6a) - for a simulated internal temperature distribution (corresponding to a given transfer coefficient distribution) that has been polluted by an additive noise. Measurements are shown in Figure 3a (circles). The estimated transfer coefficient distribution using this inversion method is shown in Figure 3b (circles).

Regularization through truncation can also be considered here: when $m$ measurements $y_i$ of temperatures $T_i$ corresponding to regularly spaced angles (at a constant radius $r$) are available, it is possible to estimate ($n+1$) components $\tilde{T}(k)$ of the spectrum of function $T(x, \text{constant } r)$ by simple quadrature.

$$
\tilde{T}_{\text{exp}}(k) = \sum_{i=1}^{m} y_i\, w_i \cos(kx_i) \tag{16}
$$

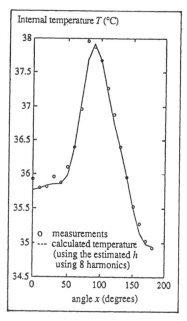

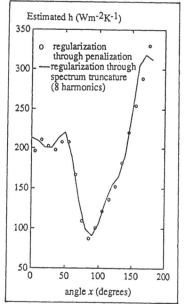

a - Measured temperatures
and recalculated temperatures
(quadrupole method, r=14,02mm)

b - Estimated transfer coefficients

Figure 3. Inversion through penalization and inversion through truncature
(8 harmonics)

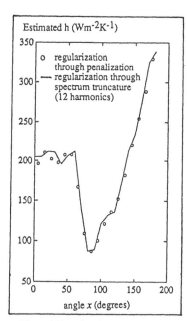

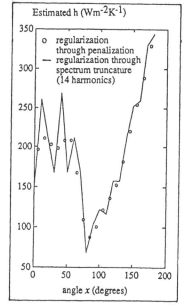

Figure 4. Inversion through penalization
and inversion through
truncature (12 harmonics)

Figure 5. Inversion through penalization
and inversion through
truncature (14 harmonics)

**81**

with $\quad w_1 = w_m = \dfrac{\pi}{2\,(m-1)} \quad$ and $\quad w_i = 2\,w_1 \quad$ if $\quad i \neq 0 \quad$ and $\quad i \neq m$

In practice, one can evaluate a maximum of $n = m$ harmonics this way - two points for each period, considering $2m$ points symmetrized with respect to the symmetry axis of the problem - in order to prevent any aliasing when the frequencies are too high (Gibbs's phenomenon). The preceding matrix approach can be adopted again, but now $\tilde{T}$ is given (calculated from measurements - equation (16)) and $\tilde{\phi}$ must be calculated. So if equation (14a) is written for layer (1), it is possible to express the transform of $\phi$ (at radius $r$) as a function of the transforms of $T_1$ and of $T$. The same equation can then be written for layer (2), which yields the transforms of $T_2$ and of $\phi_2$ (on the outside wall, at radius $r_2$). The final result is:

$$\tilde{\phi}_2 = (A_2\,\tilde{T}_1 - A\,\tilde{T})\,/\,B_1 \quad \text{and} \quad \tilde{T}_2 = (-B_2\,\tilde{T}_1 + B\,\tilde{T})\,/\,B_1 \qquad (17a)$$

These two transforms can be estimated, harmonic by harmonic, by replacing $\tilde{T}$ with $\tilde{T}_{exp}\,(k)$ and $\tilde{T}_1$ with $\pi\,T_1$ (if $k = 0$) and with $0$ else. The outside surface temperature $T_2\,(x)$ and that of the corresponding surface flux density $\varphi\,(x)$ can be estimated by Fourier inversion according to equation (12b), both spectra being truncated at $n$ harmonics. The angular variation of the estimated heat transfer coefficient is then:

$$\hat{h}\,(x) = \frac{\hat{\varphi}\,(x)}{\hat{T}_2\,(x) - T_e} = \frac{\hat{\phi}_2\,(x)}{\pi\,r_2\left[\hat{T}_2\,(x) - T_e\right]} \qquad (17b)$$

The transfer coefficient distributions estimated by this method are presented in the solid line graphs shown in Figure 3b ($n = 8$), in Figure 4 ($n = 12$) and in Figure 5 ($n = 14$). The estimated distribution corresponding to the above penalization technique has also been plotted in the same figures (circles). The agreement between the two inversion techniques are very good, except for the last case where an instability begins to develop for low values of $x$. The interesting point in this last method is the low variability of the distribution with the number of chosen harmonics. This number is the only adjusting parameter. Furthermore, the 12-harmonic curve (12 functions $b_i\,(x) = \cos\,(ix)$) is still stable. This would probably not be the case for a distribution with 12 constant pieces, which would very likely require the introduction of a penalizing term . Temperatures recalculated from eight harmonics are set on the experimental plot in Figure 3a. Here also the agreement is quite satisfactory.

## EXPRESSION OF THE SINGULAR VALUES

As the temperature $T_1$ on the inside surface is constant, only the zero order harmonic is different from zero ($\tilde{T}_1\,(0) = \pi T_1$). If one uses the change of function $\theta = T_1 - T$, equation (15b) becomes:

$$\tilde{\phi}_2\,(k) = \frac{A\,(k)}{B_1\,(k)}\,\tilde{\theta}_{exp}\,(k) \quad \text{with} \quad \tilde{\theta}_{exp}\,(k) = \pi T_1 - \tilde{T}_{exp}\,(k) \quad (18a)$$

For $k = 0$, one gets $\quad \Phi = (T_1 - \bar{T})\,/\,R \quad$ with $\quad R = \dfrac{1}{\pi\lambda}\,\ell n\,(r\,/\,r_1) \quad (18b)$

where $\Phi$ is the total flux transferred between the inside ($r = r_1$) and outside surfaces ($r_2$), $\overline{T}$ being the average temperature at radius $r$ and $R$ the thermal resistance of the first layer (between $r_1$ and $r$). The first equation (18a) has exactly the same form as equation (8b). One can easily derive the singular values of operator A whose kernel is defined by equation (11a), by identification: $\lambda_i = B_1(i) / A(i)$ — for $i = 0, 1, 2, \ldots\infty$. At that point, the condition number of matrix A, the standard variation of each harmonic produced by measurement noise, for different numbers of harmonics, can be easily calculated.

## CONCLUSION

It has been shown here that an integral transformation can be applied to measured distributions of internal temperature in order to estimate a surface heat flux distribution using an inverse heat conduction procedure. This technique, based on the idea of thermal quadrupoles, is justified each time the geometry is simple and the heat transfer linear. This approach has been applied in estimating the heat transfer coefficient in external flow over a cylinder starting from internal measurements at a constant radius. The profiles produced by this spectral method, as well as the corresponding residuals, have been compared to those obtained earlier, starting from the same measurements but using a Tikhonov regularization with a second order penalization term (direct model produced by the method of separation of variables). The agreement is excellent and the implementation is quite easier.

The link between this inversion technique, using experimental transforms, and the method of regularization using spectrum truncation (spectral method based on singular value decomposition of the operator of the inverse problem) has been stated. The singular values have an explicit analytical expression in this last case.

A similar technique can be applied in estimating an internal interface resistance distribution starting from surface temperature measurements: such a technique is presented in [12] for a heat pulse excitation in planar geometry (quadrupoles based on a Laplace time transformation and a Fourier cosine space transformation).

## REFERENCES

1. Beck, J.V. and K.J.Arnold, 1977, *Parameter Estimation in Engineering and Science*, John Wiley and sons, New York.
2. Beck, J.V., B. Blackwell and C.R. St-Clair Jr., 1985, *Inverse Heat Conduction - Ill-Posed Problems*, John Wiley and sons, New York.
3. Tikhonov, A.N.and V.Y. Arsenine, 1977, *Solution of Ill-Posed Problems*, V.H. Winston and Sons, Washington, D.C.
4. Golub, G.H. and C.F. Van Loan, 1983, *Matrix Computations*, Johns Hopkins University Press, 1983.
5. Linz, P., Feb. 1994, "A new numerical method for ill-posed problems", *Inverse Problems*, 10(1): L1-L6.
6. Degiovanni, A., 1988, "Conduction dans un mur multicouche avec sources: extension de la notion de quadripôle", *Int. J. Heat Mass Transfer* 31(3):553-557.
7. Leturcq, Ph., J.M. Dorkel, F.E. Ratolojanarhary and S. Tounsi, 1993, "A two-port network formalism for 3D heat conduction analysis in multilayered media", *Int. J.Heat Mass Transfer*, 36(9): 2317-2326.
8. André, A., J.C. Batsale, A. Degiovanni, D. Maillet and C. Moyne, *Thermal Quadrupoles - An efficient Method for solving the Heat Equation through Integral Transforms*, John Wiley and Sons, to be published, end 1996.

9. Maillet, D. and A. Degiovanni, July 1989., "Méthode analytique de conduction inverse appliquée à la mesure du coefficient de transfert local sur un cylindre en convection forcée", *Revue de Physique Appliquée*, 24 :741-759.

10. Maillet, D., A. Degiovanni and R. Pasquetti, august 1991, "Inverse heat conduction applied to the measurement of heat transfer coefficient on a cylinder: comparison between an analytical and a boundary element technique", *Journal of Heat Transfer*, 113,:549-557.

11. Batsale, J.C., D. Maillet and A. Degiovanni, 1994, "Extension de la méthode des quadripôles thermiques à l'aide de transformations intégrales - Calcul d'un transfert thermique au travers d'un défaut plan ", *Int. J. Heat Mass Transfer*, 37(1):111-127.

12. Bendada, A., J.C. Batsale, A. Degiovanni and D. Maillet, 1994, "Interface Resistances: the Inverse Problem for the Transient Technique" in *Inverse Problems in Engineering Mechanics,* H.D. Bui, M. Tanaka, M. Bonnet, H. Maigre, E. Luzzato and M. Reynier, eds. Rotterdam: Balkema, pp 347-354.

# Kalman Filtering Applied to Thermal Conductivity Estimation in Pulse Experiments

F. RIGHINI, F. SCARPA and G. MILANO

## ABSTRACT

A new dynamic technique for the determination of the thermal conductivity of electrical conductors is being implemented at the CNR Istituto di Metrologia "G. Colonnetti" (IMGC, Italy). A tubular specimen undergoes a subsecond Joule heating and in the following free cooling period the temperature profiles are measured via microsecond scanning pyrometry. Thermal conductivity is reconstructed in form of polynomial approximation by solving the associated nonlinear inverse conductive problem with the use of the Kalman filter. This stochastic technique provides a way to consider the various noises and disturbances affecting the measured signals. Some simulated and preliminary experimental results on niobium are presented and discussed. From a single cooling experiment the above procedure is able to reconstruct the temperature dependent thermal conductivity in the range 1300-2300 K.

## INTRODUCTION

Thermal conductivity is an important thermophysical property difficult to measure directly at high temperatures. A new dynamic technique for its measurement is being implemented at the CNR Istituto di Metrologia "G. Colonnetti" (IMGC, Italy). The experiment [1] consists in the heating of a tubular specimen with a current pulse of subsecond duration and in the accurate measurement via high-speed scanning pyrometry [2] of the temperature profiles established on the specimen during the free cooling period. A schematic diagram of this dynamic experiment to measure thermal conductivity is presented in figure 1. Similar measurements are also in development at the National Institute of Standards

Francesco Righini, CNR Istituto di Metrologia "G. Colonnetti", Strada delle Cacce 73, I 10135, Torino - ITALY.
Federico Scarpa and Guido Milano, Dipartimento di Termoenergetica e Condizionamento Ambientale, Università di Genova, Via All'Opera Pia 15 A, I 16145, Genova - ITALY.

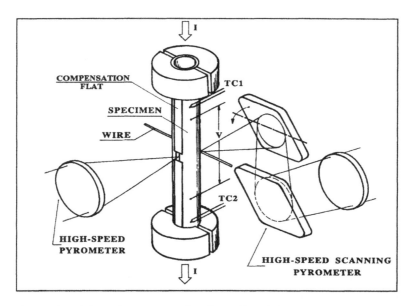

**Figure 1** - Schematic representation (from Ref. [1]) of the new dynamic technique for thermal conductivity measurements. I, current passing through the specimen; V, voltage drop across the central part of the specimen; TC1 and TC2, chromel-alumel thermocouples.

and Technology (NIST, USA), by using a scanning pyrometer based on an array detector [3]. This new measurement technique is a dynamic version of the classical determinations of thermophysical properties using direct heating methods [4].

The evaluation of the thermal conductivity from the temperature profiles measured in dynamic experiments is a difficult problem. At the moment no completely satisfactory solutions have been found. A method was developed at IMGC opening a vertical or horizontal window on the profiles and attempting to transform the set of differential heat equations representing the central point of the window into a set of linear equations of the unknown thermal conductivity [5]. A different analytical method was tried at NIST using a numerical solution of the second order time-dependent partial differential equation for heat conduction [6]. In both cases, after applying the methods to the data of preliminary experiments, it was not possible to establish the accuracy of the predicted thermal conductivity results. This is due to a series of factors, involving the preliminary nature of the performed experiments, the rather large quantity of data, the difficulty of establishing the accuracy of derivatives computed from experimental data, the different accuracy of temperatures in various ranges, and the possibility that the model does not completely reflect the physical reality of the experiment.

The present paper describes the possible application of an alternative estimation method, based on the Kalman filter, to identify the temperature dependent thermal conductivity from a large set of experimental temperature profiles obtained in dynamic experiments. The method inherently yields estimates of the standard deviation of all the identified parameters and, unlike other inverse techniques, takes into account the uncertainty related to initial and boundary conditions. Moreover, the modular structure of the algorithm permits to discriminate easily between possible mismatches in the assumed physical model.

## PHYSICAL MODEL AND ESTIMATION ALGORITHM

We regard the specimen as a homogeneous, isotropic thin rod subjected to one-dimensional transient cooling from an initially measured temperature distribution. The thermal conductivity of the material is the temperature dependent unknown function to be identified, while the thermal field and the time temperature histories at two arbitrary boundaries are the measured quantities. We assume the volumetric heat capacity and the total hemispherical emissivity to be known in the temperature range of interest. Geometrical quantities are also known and used without considering thermal expansion effects, by performing computations in the so-called "tube-space" [5].

The resulting one-dimensional, nonlinear, heat transfer equation, also known as "long thin rod approximation", can be written in the form:

$$C(T)\frac{\partial T}{\partial t} = \frac{\partial}{\partial z}\left[k(T)\frac{\partial T}{\partial z}\right] - \sigma\varepsilon_{ht}(T)\frac{p_e}{S}\left[T^4 - T_{amb}^4\right]; \quad T \equiv T(z,t) \tag{1}$$

$$z_1 < z < z_2 \quad 0 < t \le t_m$$

where $k(T)$, $C(T)$, and $\varepsilon_{ht}(T)$ are the temperature dependent thermal conductivity, volumetric heat capacity and total hemispherical emissivity. S is the cross-sectional area, $p_e = 2\pi r_e$ the external perimeter, $z$ the spatial co-ordinate, $t$ the time, $T$ the temperature, $\sigma$ the Stefan-Boltzmann constant. $z_1$ and $z_2$ represent the left and right limits of the investigated portion of the specimen and $t_m$ is the duration of the considered part of the experiment. Eq. (1) is coupled to the following initial and boundary conditions (first kind):

$$T(z,0) = T_0(z); \quad T(z_1,t) = u_1(t); \quad T(z_2,t) = u_2(t) \tag{2}$$

The parameterization of the unknown function is obtained by a polynomial approximation as follows:

$$k(T) = \sum_i k_i(T - T_{ref})^i \tag{3}$$

where $k_i$ are the unknown parameters to be determined and $T_{ref}$ is a suitable reference temperature. The cooling process is represented in a discretized form and the temperature distribution, at the generic time instant $t_k$, is replaced by the vector $T(t_k)$. The numerical scheme adopted to discretize Eqs. (1-2) must have a form suitable to be incorporated into the Kalman algorithm; the simplest way to meet this requirement is the use of an explicit recursive formula as follows:

$$T(t_k) = f[T(t_{k-1}), u_f(t_{k-1})]; \qquad T \equiv [T_1, .., T_{Nz-1}]^t \tag{4}$$

$$0 < k < N_t$$

$$T(t_0) = T_0 \qquad\qquad \text{known} \tag{5}$$

$$u_f(t_k) \qquad\qquad \text{known} \qquad u_f \equiv [u_1, u_2]^t \tag{6}$$

where $T(t_k)$ is the time dependent temperature vector and $N_z$ and $N_t$ are the total number of discretized intervals in the spatial and time directions. The above formulation is often referred as the "state-space model" or the "internal representation" of a physical process.

The Kalman algorithm [7] is essentially based on a sequence of recursive predictor-corrector calculations by which the current estimate is continuously improved as a new set of measurements becomes available from the experiment:

*Prediction (dynamic update)*

$$x_{k|k-1} = f[x_{k-1|k-1}, u_{k-1}] \tag{7}$$

$$P_{k|k-1} = A_{k-1} \cdot P_{k-1|k-1} \cdot A^t_{k-1} + B_{k-1} \cdot R_{w\,k-1} \cdot B^t_{k-1} \tag{8}$$

*Correction (static update)*

$$x_{k|k} = x_{k|k-1} + K_k \cdot (y_{m\,k} - H \cdot x_{k|k-1}) \tag{9}$$

$$P_{k|k} = [I - K_k \cdot H_k] \cdot P_{k|k-1} \tag{10}$$

where $x$ is the state estimate vector (temperature) and $y_m$ refers to the measured signals. The notation $x_{k|i}$ stands for "estimate of $x_{\text{true}}(t_k)$ by means of the information available at $t_i$". $[\,\cdot\,]^t$ denotes the transposition operator; $f[\,]$ is a non-linear vector function; $A$ and $B$ are the following jacobian matrices:

$$A = \left[\frac{\partial f_i}{\partial x_j}\right] \quad B = \left[\frac{\partial f_i}{\partial u_j}\right] \tag{11}$$

Equations (7) and (9) represent the evolution of the state estimate, while (8) and (10) are the associated covariance equations.
The Kalman gain, $K$, is given by

$$K_k = P_{k|k-1} \cdot H^t_k \cdot [H_k \cdot P_{k|k-1} \cdot H^t_k + R_{v\,k}]^{-1} \tag{12}$$

and **H** represents the measurement matrix (in our case the unit matrix). $\mathbf{R_v}$ is the diagonal covariance matrix of the measurement noise, while $\mathbf{R_w}$ takes into account the noise associated to the measured boundary conditions.

Since some parameters of the process model are unknown, we modify the Kalman algorithm by augmenting the sought for parameters into the state. In this work we are involved with the estimation of the temperature dependent function $k(T)$, so the vector $\beta$ of the unknown parameters and the augmented state vector $x$ result respectively:

$$\beta = [\, k_i \,]^{\,t}; \qquad x \equiv [T^{\,t}, \beta^{\,t}]^{\,t} \tag{13}$$

On the basis of the above formulation the Kalman filter discretized equations, reported in detail in [8], have been implemented by a computer program which has been set-up and tested for a wide class of transients and materials [9,10]. Moreover it has been shown [11] that the fully stochastic Kalman estimator, for this kind of application, provides results more reliable than those given by other standard semi-deterministic methods like OLS (Ordinary Least Squares), MAP (Maximum A Posteriori), etc.

## SIMULATION RESULTS

The ability of the Kalman algorithm to identify the thermal conductivity from typical dynamic measurements of temperature profiles has been proven by means of computer simulations. Using realistic data of the thermophysical properties of a material (niobium), we have computed by software simulation a set of temperature profiles as they would be created by a typical dynamic experiment. The processing of these simulated profiles enable us to address two different kinds of problems. At first, if the simulation model and the filter model are the same, we can put in evidence the possible presence of errors in the structure of the algorithm or in its software implementation. For the case of niobium we have been able to reconstruct the assumed thermal conductivity to better than 0.001%. This is a clear indication that the software introduced no bias and that there are no round-off errors in the complex computations.

In a second phase, the use of a more sophisticated simulator, sometimes called a *truth model*, can be used to study the effect of the simplifications introduced in the model. A better understanding of the limits of the assumptions made during the preliminary modeling phase is thus possible.

Due to its one-dimensional formulation, the long thin rod approximation holds in case of very small and constant thickness. The real specimen does not strictly adhere to these requirements on account of the presence of the small black-body hole (necessary to refer all surface temperature measurements to blackbody conditions) and of the cross-section flats (necessary to compensate for the presence of the blackbody hole, see Fig. 1) on the same side of the blackbody hole. Furthermore, the radiative exchange in the interior of the tubular specimen is not considered and this last item has turned out to be an important cause of modeling error.

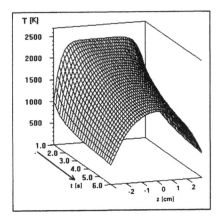 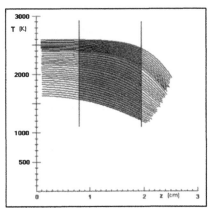

**Figure 2** - Typical temperature history during the cooling phase (simulation).

**Figure 3** - Analyzed region of the specimen (experimental profiles)

We ensure that a set of realistic measurements is produced by using a complete 3-D simulator that takes into account the effective geometry of the specimen and the inner and outer radiative boundary conditions. A typical simulated temperature vs. time field during the cooling period is shown in Fig. 2 but in the real case we analyze only a limited portion of the specimen (see, for example, Fig. 3). The more appropriate space window to be analyzed is chosen by considering that the central region of the specimen should be avoided because there the temperature profiles are nearly flat and very little thermal conduction takes place. Also the ends of the temperature profiles are not used on account of the higher noise affecting the lower temperature signals.

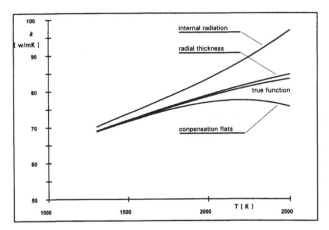

**Figure 4** - Influence of various phenomena on the reconstructed thermal conductivity.

Figure 4 shows the effects of the three major contributions to the conductivity distortion. The specimen thickness is responsible for a small error in the estimated values while both the internal radiation and the presence of the compensation flats greatly disturb the reconstruction process, but produce effects in opposite directions. For niobium in the temperature range from 1300 to 2500 K the average error induced by these two factors on the estimated thermal conductivity is of the order of 6% and 2.5%, respectively. To avoid these drawbacks we incorporate in the filter model the internal radiation transfer as follows:

$$C(T)\frac{\partial T}{\partial t} = \frac{\partial}{\partial z}\left[k(T)\frac{\partial T}{\partial z}\right] - \frac{p_e}{S}\sigma\varepsilon_{\mathrm{ht}}\left[T^4 - T_{amb}^4\right] - \frac{p_i}{S}\frac{\varepsilon_{\mathrm{ht}}}{(1-\varepsilon_{\mathrm{ht}})}\left[\sigma T^4 - J(z,T)\right] \quad (14)$$

where $J$ is the internal radiosity of an elementary annulus and $p_i = 2\pi r_i$ is the internal perimeter. By means of this enhanced model the error due to the inner radiation is reduced to about 0.1%. This residual mismatch is due to the modeling of only a portion of the specimen. In fact, if we consider the radiosity definition, that is:

$$J_i = \rho_i G_i + \varepsilon_i E_{\mathrm{n}\,i} \quad (15)$$

we can see that the term $G_i$, the incident radiation, depends on the whole specimen and so the radiosity. Since we do not take into account the radiation coming from outside the selected region of the specimen, the model will be partially incorrect.

The compensation flats cause a circumferential flux that can be fully taken into account only by means of a 2-D model and, because of the covariance propagation equation (8), this is for the moment being beyond the capability of our hardware. Nevertheless, if the analysis is limited to a smaller temperature interval (1300-2300 K), the global error in the reconstructed conductivity function can be reduced to less than 1%. A simplified model to take into account the effect of the axial non-symmetry without 2-D extensions is under development. Another interesting approach would be a radical change in the geometry of the specimen along with a redesign of the identification technique used in the heating phase to estimate various properties such as the total hemispherical emissivity, etc. Also the use of strip specimens could greatly simplify the experimental arrangement, provided that accurate measurements of the normal spectral emissivity of the strips could be performed.

Concerning the value of the relative mean standard deviation provided by the filter and defined as

$$\overline{\sigma}_k = \frac{1}{T_{max} - T_{min}} \int_{T_{min}}^{T_{max}} \frac{\sigma_k(T)}{k(T)}\, dT \quad (16)$$

we achieve $\overline{\sigma}_k < 0.1\%$, showing the potential of the technique at the level of measurement noise typical of these experiments.

## EXPERIMENTAL RESULTS

The experiment consists in the rapid heating, by a current pulse of subsecond duration, of a tubular specimen (Fig. 1) with the following dimensions: internal and external radii, $r_i$ =2.705 mm, $r_e$=3.59 mm ; length $L$= 5.87 cm. Two types of experiments are performed on these specimens. At first very fast (subsecond) heating experiments are performed, using large currents (>1000 A). In these conditions the central portion of the specimen has at all times a flat temperature distribution and therefore represents a specimen at uniform high temperature. The heat capacity and the hemispherical total emissivity as functions of temperature are obtained from measurements of the true temperature inside the blackbody, of the voltage drop across the central part of the specimen (see Fig. 1), and of the current in the specimen [12]. The thermal conductivity is determined with a second set of experiments (performed on the same specimen without opening the experimental chamber) by measuring the temperature profiles in the free cooling phase while the heat content is dissipated by radiation and conduction. This second set of experiments can be performed with different heating rates, obtaining different initial conditions (the first temperature profile at the beginning of cooling) for the thermal conductivity determinations. The temperature profiles are measured by a high speed scanning pyrometer [2] and recorded every 25 ms. In the present setup every profile consists of 600 temperature measurements (sampling frequency: 500 kHz, A/D resolution: 16 bit).

The presence of a cold wire placed across the central part of the specimen causes a marked dip in the pyrometer output and defines a reference "zero" position. Furthermore, the identification algorithm automatically takes care of any residual error since the reference positions at every time are considered as further parameters to be estimated.

The scanning pyrometer measures temperatures at fixed positions in space, but from one profile to another, the measured temperature refers to a different location on the specimen on account of thermal expansion effects. This phenomenon is accurately compensated as described in [5]. The finite time spent by the rotating mirror in front of the specimen is accounted for and the integrating effect of the pyrometer target (diameter 0.8 mm) is also corrected.

Besides the thermal conductivity, the temperature history in the filter model depends on the measured values of the heat capacity and total hemispherical emissivity, therefore errors in these quantities will affect the quality of the reconstruction process. Moreover, this transient technique is based on accurate temperature measurements: the signals by the pyrometer must be transformed into true temperatures and any error in this conversion will cause bias in the estimated thermal conductivity. So, both a great care in the identification of the above properties during the heating phase and an accurate calibration of measurement instrumentation are mandatory.

Fig. 5 shows the results obtained by the elaboration of a preliminary set of eight experiments performed on niobium under different conditions. In the same figure some literature data [13, 14] are also shown.

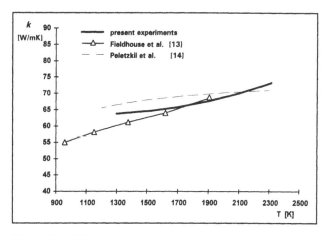

**Figure 5** - Estimated thermal conductivity as a function of temperature (four unknown parameters). Comparison between the present technique and literature values.

## CONCLUSIONS

An estimation technique able to exploit the potential of high temperature dynamic experiments for the measurement of thermal conductivity has been presented.

The effectiveness of the proposed approach has been proven by means of extensive simulations while preliminary applications in true experiments give results, over a large temperature range, similar to those reported in literature. The problem of establishing the accuracy of the predicted thermal conductivity is inherently addressed by the technique, but the temperature dependent confidence regions provided by the filter are reliable only in the case of accurate and complete modeling of the phenomenon under investigation. Further work is planned both to improve the modeling and to perform experiments with different geometries to investigate the optimal conditions for dynamic experiments for the determination of thermal conductivity at high temperatures.

## REFERENCES

1. Righini, F., G. C. Bussolino, A. Rosso, and R. B. Roberts, 1990. "Thermal Conductivity by a Pulse-Heating Method: Theory and Experimental Apparatus", *International J. Thermophysics*, **11**: 629-641.

2. Righini, F., A. Rosso, and A. Cibrario, 1985. "Scanning Pyrometry with Microsecond Time Resolution", *High Temperatures High Pressures*, **17**: 153-160.

3.  Cezairliyan A., R. F. Chang, G. M. Foley, and A. P. Miiller, 1993. "High-Speed Spatial Scanning Pyrometer", *Rev. Sci. Instrum.*, **64**: 1584-1592.

4.  Taylor R. E., 1984. "Thermophysical Property Determinations Using Direct Heating Methods" in *Compendium of Thermophysical Property Measurement Methods*, Vol. **1**, K. D. Maglic, A. Cezairliyan and V. E. Peletskii Eds., New York, Plenum Press, pp. 125-167.

5.  Bussolino, G. C., J. Spišiak, F. Righini, A. Rosso, P. C. Cresto, and R. B. Roberts, 1993. "Evaluation of Thermal Conductivity from Temperature Profiles", *International J. Thermophysics*, **14**: 525-539.

6.  MacDonald, R. A., 1995. "An Analytical Method for Determining Thermal Conductivity from Dynamic Experiments", *International Journal of Heat and Mass Transfer*, **38**: 2549-2556.

7.  Candy, J. V., 1988. *Signal Processing: The Modern Approach*, New York, McGraw Hill, p. 322.

8.  Scarpa, F., and G. Milano, 1993. "Identification of Temperature Dependent Thermophysical Properties from Transient Data: a Kalman Filtering Approach", Internal Report EGR/21, Dipartimento di Termoenergetica e Condizionamento Ambientale, Università di Genova, Italy, pp.1-16.

9.  Milano G., Scarpa F., and Timmermans G., 1994. "Thermophysical Properties of EPB-Expanded Perlite Board: a Transient Measurements Technique", in *Thermal Conductivity 22,* T. W. Tong Ed., Lancaster, Penn., USA, Technomic Publishing Co, pp. 263-274.

10. Scarpa F., G. Milano, 1994. "Influence of the Boundary Conditions on the Estimation of Thermophysical Properties from Transient Experiments", *High Temperatures-High Pressures*, **26**: 163-175

11. Scarpa, F., G. Milano, and D. Pescetti, 1993. "Thermophysical Properties Estimation from Transient Data: Kalman Versus Gauss Approach", in *Inverse Problems in Engineering*, N. Zabaras, K. A. Woodbury, and M. Raynaud Eds., Palm Coast, Florida, ASME book no. 100357, pp. 109-116.

12. Cezairliyan, A., 1984. "Pulse Calorimetry" in *Compendium of Thermophysical Property Measurement Methods*, Vol. **1**, K. D. Maglic, A. Cezairliyan and V. E. Peletskii Eds., New York, Plenum Press, pp. 643-668.

13. Fieldhouse, I. B., J. C. Hedge, and J. I. Lang, 1958. "Measurements of Thermophysical Properties", WADC Tech. Rept., 58-274, in *Handbook of Thermophysical Properties of Solid Materials*, MacMillan: NewYork, 1961, Vol. **1**, p. 467.

14. Peletskii, V. E., A. P. Grishchuk, E. B. Zaretskii, and A. A. Zolotukhin, 1987. "Study of a Set of Thermophysical Properties of Niobium in the High-Temperature Region" High Temperature (USSR) **25**: 205-211.

# Using the Discrete Fourier Transform in Thermal Diffusivity Measurement

J. GEMBAROVIČ and R. E. TAYLOR

## ABSTRACT

This paper presents a new way to reduce data in the flash method and the step method of measuring thermal diffusivity. Experimental temperature vs. time data are first properly periodized, then transformed by using the Discrete Fourier Transform (DFT), and the real part of the second term of the image temperature is then fitted with a theoretical formula derived in this paper. The main advantage of this procedure is that the thermal diffusivity calculation actually does not depend on the temperature level before the flash, and can be applied also in the case when the measured temperature signal of the sample is superimposed with an arbitrary linearly rising or falling signal, with no need to know the parameters of this imposed signal. The second advantage is that the procedure allows the treatment of perturbed signals, even in the case when the data are partly statistically correlated, i.e. the noise is not Gaussian. The data reduction procedure has been tested for a correction of the effect of an electronic noise imposed on the temperature vs. time signal in the flash, as well as in the step method. The results show that the reproducibility of our procedure is good, and the accuracy is comparable with other data reduction methods.

## INTRODUCTION

Today commonly used thermal diffusivity measurement methods are based on comparison of experimentally gained transient temperature of a sample with a theoretically derived one, represented by an analytical formula. Integral transformations are frequently used for finding the analytical formula as a solution of heat conduction equation. The temperature of the sample in a real thermal

Jozef Gembarovič, University of Education, Department of Physics, Trieda A. Hlinku 1, Nitra, 949 74, Slovakia
Raymond E. Taylor, Thermophysical Properties Research Laboratory Inc., 2595 Yeager Road, West Lafayette, IN 47906, USA

95

diffusivity measurement is monitored either in one, or in few points and recorded in time. This temperature vs. time curve (or curves) is then fitted with the theoretical one in order to find the desired thermal diffusivity as one of the parameters in the analytical formula. The problem of data reduction lies in the fact that the analytical formulae are usually complicated infinite series of transcendental functions, and their 'shape' strongly depends on boundary conditions of the heat conduction equation used for their calculation.

The main idea of our approach for finding thermal diffusivity is not to fit temperature vs. time curve with the theoretical one, but a transformed temperature vs. time curve with the correspondent transformed theoretical one. Integral transformation, as a mathematical operation, eliminates one of the independent variables (e.g. time) in the theoretical model, so the resulting analytical formula for the transformed temperature could be expressed in less complex way that the non-transformed one. This principal advantage can lead to other 'side' effects, which can either make the calculation simpler and faster, or make the algorithm more universal. As already described [1], e.g. using the Laplace transformation in data reduction in the flash method [2] leads to substantial reduction in complexity of the analytical formula for temperature vs. time curve after pulse, and the algorithm can be effectively applied on data affected by the finite pulse time effect. Similarly, using the Fourier transformation offers full advantages with a noise management, signal filtering, etc.

In this paper we focus on using the Discrete Fourier Transform (DFT) for data reduction in the flash method and in the step method [3] of measuring thermal diffusivity. First mentioned is one of the most popular and the most frequently used method of measurement of thermal diffusivity for a wide range of homogeneous materials from cryogenic temperatures into molten region, as well as many heterogeneous ones including certain layered, dispersed and fiber reinforced composites. Instantaneous laser pulse, which is used to test the samples in the flash method may sometimes cause a problem with overheating of the exposed sample surface, especially when it is applied on highly insulating materials, or thick samples of large grain heterogeneous materials, where the grain size is of the order of the usual sample thickness used (order of mm). In these cases it is reasonable to use the step method, in which the instantaneous pulse is replaced by step-wise continuous heating of the sample front surface, which enable the use of relatively larger specimens than those used in the flash method. Another important advantage of the step method is the relatively low intensity level of the imposed constant heat flux, compared with the instantaneous heat pulse necessary with the flash method. Specimens are less likely to begin a phase transition, decomposition or another irreversible process as a result of a sudden rise in temperature of the front surface caused by a heat pulse. But, in comparison with the flash method, in which tens of data reduction procedures for computing the thermal diffusivity were proposed, the step method experimenter has a simpler choice, because the only one was described, till today, is in the original work of Bittle & Taylor [3]. The purpose of this work is also to show that data from the step method (properly converted) can be treated by the same data reduction procedures as those from the flash method. In mathematical sense, these two methods are similar, and the theoretical temperature rise for the flash method can be expressed as a derivative of the step method temperature with respect to time.

Using the Discrete Fourier Transformation (DFT) for data reduction in the flash method and in the step method requires two steps. First the experimental temperature vs. time data are properly periodized, and transformed by using the Fast Fourier Transform (FFT) algorithm of computing DFT. Then, in order to find the thermal diffusivity of the sample, the real part of the second term of the transformed temperature is fitted with an appropriate theoretical formula .

The main advantage of our procedure is that the thermal diffusivity calculation does not depend on the baseline temperature level, and can be applied also in the case where the measured signal corresponding to the sample temperature is superimposed with an arbitrary linearly rising or falling signal, with no need to know the parameters of this imposed signal. It is inherent that our procedure simply ignores the presence of a component $At + B$ (where $t$ is the time, and $A$, $B$ are arbitrary finite real constant numbers) in the temperature signal.

The proposed procedure allows the treatment of perturbed signals, even in the case when the data are partly statistically correlated, i.e. the noise is not Gaussian.

## THEORETICAL MODEL

The theoretical formula needed for the data reduction procedure can be derived as a discrete Fourier transform of the solution of the heat conduction equation which corresponds to the model considered in the respective method of measuring thermal diffusivity.

### DISCRETE FOURIER TRANSFORM

The DFT is a relatively new transform. A major development of this transform appeared in 1965, when J. W. Cooley and J. W. Tukey [4] described a very fast algorithm for computation, known as the Fast Fourier Transform (FFT). Thanks to this algorithm the DFT became the most widespread means of numerical computation of the Fourier transform and of numerical harmonic analysis.

Direct DFT of a sequence of N finite complex numbers $x_i, i = 0,1,2,...,N-1$, is defined [5] as a sequence of values $X_k$ given by

$$X_k = \sum_{i=0}^{N-1} x_i \exp(-jik2\pi / N), \qquad k = 0,1,...,N-1 \tag{1}$$

where $j = \sqrt{-1}$. We shall call the sequence $X_k$ the image of the sequence $x_i$.

If $X_k, k = 0,1,2,...,N-1$ is a sequence of $N$ finite complex numbers, then its inverse DFT is a sequence of $N$ numbers expressed by the relation

$$x_i = \frac{1}{N}\sum_{k=0}^{N-1} X_k \exp(jik2\pi / N), \quad i = 0,1,2,...,N-1. \tag{2}$$

In the case where a sequence $X_k$ is an image of a sequence $x_i$, Equation expresses the original sequence $x_i$ and is inverse to the form given by Equation (1).

## FLASH AND STEP METHOD MODEL

Our considerations are based on the most simple (ideal) model which assumes that the sample used in both methods is isotropic, homogeneous, opaque to the used radiation, thermally insulated, and that the heat source is instantaneous in the flash method and continuous step like in the case of the step method. Energy of the heat source is uniformly absorbed in the front surface of the sample. The temperature excursion in the rear face of the sample (for $x = L$, and $t > 0$,) can be expressed by the known formula [2]

$$T_F(L,t) = T_0 \left[ 1 + 2 \sum_{n=0}^{\infty} (-1)^n \exp\left( -\pi^2 n^2 \frac{at}{L^2} \right) \right] \qquad (3)$$

for the flash method, and

$$T_S(L,t) = T_0 \left[ \frac{at}{L^2} - \frac{1}{6} - \frac{2}{\pi^2} \sum_{n=0}^{\infty} \frac{(-1)^n}{n^2} \exp\left( -\pi^2 n^2 \frac{at}{L^2} \right) \right] \qquad (4)$$

for the step method [3].

In the case of the flash method $T_0$ is the steady state temperature in the sample after the heating pulse given by

$$T_0 = \frac{Q}{\rho c L} , \qquad (5)$$

where $Q$ is the total amount of heat absorbed through unit surface of the sample, $\rho$ is the density, $c$ is the heat capacity of the sample material, $L$ is the sample thickness, $t$ is time, and $a$ is the thermal diffusivity.

In the case of the step method $T_0$ is

$$T_0 = \frac{F_0 L}{k} , \qquad (6)$$

where $F_0$ is the constant and uniform heat flux, and $k$ is the thermal conductivity of the sample.

## DFT OF TEMPERATURE RESPONSE

Experimental points of measured temperature data are in the ideal model represented by a sequence of $M$ real numbers $T_i$ given by Equation (3) for the flash method:

$$T_i = T_0\left[1 + 2\sum_{n=0}^{\infty}(-1)^n \exp\left(-\pi^2 n^2 \frac{a}{L^2}(i+1)\vartheta\right)\right], \qquad i = 0, 1, 2, \ldots, M-1 \quad (7)$$

or Equation (4) for the step method

$$T_i = T_0\left[\frac{a(1+i)\vartheta}{L^2} - \frac{1}{6} - \frac{2}{\pi^2}\sum_{n=0}^{\infty}\frac{(-1)^n}{n^2}\exp\left(-\pi^2 n^2 \frac{a}{L^2}(1+i)\vartheta\right)\right], \quad i = 0, 1, 2, \ldots, M-1$$

$$(8)$$

respectively. Here $\vartheta$ represents time between two consecutive temperature points.

Before computing the formula for the theoretical ideal image of the temperature response using Equation (1), we will define the sequence $x_i$ as a composition of two sequences - original $T_i$ and the $T_i$ in reverse order :

$$x_i = \begin{cases} T_i, & i = 0, 1, \ldots, M-1 \\ T_{N-1-i}, & i = M, M+1, \ldots, N-1 \end{cases} \qquad (9)$$

where $N = 2M$.

This 'periodization' of temperature response enables the use of this data reduction method also in the case when the response is superimposed with a linearly rising or decreasing signal. Typical temperature vs. time data from both methods after the periodization are shown in Figure 1.

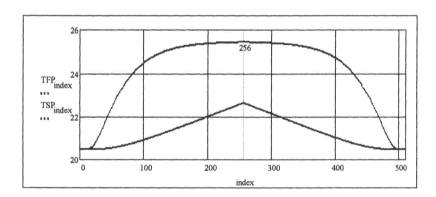

Figure 1. Typical temperature vs. time data (256 points) for the flash method (the upper curve - TFP), and for the step method (the lower curve - TSP) after periodization given by Equation (9).

After inserting periodized Equation (7) into Equation (1) and using the formula for summation of a geometrical sequence, the real part of the second term of the image sequence $X_k$ for the flash method can be written in the form

$$\mathrm{Re}\left\{X_2(N,a)\right\} = 2T_0(1+\mathrm{C}(N))\sum_{n=1}^{m}(-1)^n \frac{\mathrm{W}(a,n)\left[1-\mathrm{W}(a,n)\right]\left[1-\mathrm{W}^{N/2}(a,n)\right]}{1+\mathrm{W}^2(a,n)-2\,\mathrm{C}(N)\,\mathrm{W}(a,n)} \quad (10)$$

where

$$\mathrm{W}(a,n) = \exp\left(-\frac{\pi^2 n^2 a\vartheta}{L^2}\right), \qquad \mathrm{C}(N) = \cos\left(\frac{4\pi}{N}\right) \quad (11)$$

Analogically for the step method the real part of the second term can be expressed as:

$$\mathrm{Re}\left\{X_2(N,a)\right\} = -\frac{2T_0}{\pi^2}(1+\mathrm{C}(N))\sum_{n=1}^{m}\frac{(-1)^n}{n^2}\frac{\mathrm{W}(a,n)\left[1-\mathrm{W}(a,n)\right]\left[1-\mathrm{W}^{N/2}(a,n)\right]}{1+\mathrm{W}^2(a,n)-2\,\mathrm{C}(N)\,\mathrm{W}(a,n)}$$

$$(12)$$

As can be seen from Equations (10) - (12), the term is an unique function of $a$, and can be used for computing the desired value of thermal diffusivity.

It should be noted, that although the theoretical formulae derived for the image temperature in this paper is for the most simple ideal models, it is easy to derive similar ones for more complicated models, including heat losses from the sample, or for multi-layer samples.

## DATA REDUCTION PROCEDURE

The aim of the data reduction procedure is to calculate thermal diffusivity values from the experimental temperature vs. time data.

While in the case of the flash method there exist tens of data reduction procedures (see e.g. review in [6]), in the step method there is only one described in the Bittle and Taylor's papers [3, 7]. Their procedure is based on solving so called working equation, defined as

$$\frac{T(t_1)}{T(t_2)} = \frac{\dfrac{at_1}{L^2} - \dfrac{1}{6} - \dfrac{2}{\pi^2}\sum_{n=0}^{\infty}\dfrac{(-1)^n}{n^2}\exp\left(-\pi^2 n^2 \dfrac{at_1}{L^2}\right)}{\dfrac{at_2}{L^2} - \dfrac{1}{6} - \dfrac{2}{\pi^2}\sum_{n=0}^{\infty}\dfrac{(-1)^n}{n^2}\exp\left(-\pi^2 n^2 \dfrac{at_2}{L^2}\right)}, \quad (13)$$

where on the left side is a ratio of the observed rear face temperature changes in two different times ($t_2 > t_1$), and on the right side is the ratio of the theoretical temperatures in the same times. Once the experimental temperature response is obtained, the only unknown left in Equation (13) is the thermal diffusivity $a$.

It can be seen from Equation (3) and (4), that the theoretical temperatures are linearly dependent, when the step method temperature is differentiated with respect to time :

$$\frac{dT_s(L,t)}{dt} \propto T_F(L,t) \quad (14)$$

This fact can be used in data reduction process. Temperature vs. time curves gained in the step method can be simply converted to those from the flash method and then all appropriate data reduction procedures for computing thermal diffusivity can be used from the flash method.

The experimental temperature data can be represented by a sequence of $M$ numbers $E_i$, where $E_i$ is the temperature in time $t = (i + 1)\vartheta$, and $i = 0, 1, ..., M - 1$. The same periodization of $E_i$ has to be done before the transformation, as the one given by Equation (9). If the set of periodized experimental data has $N = 2^K$ temperature points, where $K$ is a positive integer number (usually greater then 6), then the response signal can be transformed by using the well-known standard FFT algorithm.

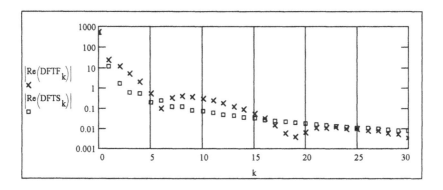

Figure 2. Amplitudes of the real parts of the image temperatures - the first 30 terms. Symbols (x) correspond to the flash method data (DFTF), and (□) to the step method data (DFTS). The original sequences are from Figure 1.

Because $E_i$ is a sequence of $N$ real numbers, the image $\overline{E}_k(N, a)$ can be expressed as a sequence of $(N/2 - 1)$ complex number. As an example, the real parts of the first 30 terms of the image temperature for both methods (the original sequences are in Fig. 1.) are shown in Figure 2. Due to a fact, that the magnitudes exponentially fall with k, and the higher terms are more sensitive to noise, for data reduction it is reasonable to choose as low a term as possible. The zero term of the image temperature (for $k = 0$) is affected by presence of a shift of the response curve up or down as a whole (by adding a constant to the signal), while the first term can be distorted by the presence of a superimposed linear rising or falling signal. This is why we will use for the fitting procedure the real part of the second term of the image temperature $\overline{E}_2(N, a)$.

In order to compute thermal diffusivity, and to eliminate the second unknown parameter, temperature $T_0$ in Equation (8) or (10) respectively, we introduce a function F , defined as

$$F(a) = \frac{\text{Re}\{\overline{E}_2(N,a)\}}{\text{Re}\{\overline{E}_2(N/2,a)\}} - \frac{\text{Re}\{X_2(N,a)\}}{\text{Re}\{X_2(N/2,a)\}} \, . \tag{15}$$

$F(a)$ has one simple root $a_0$, which is the desired thermal diffusivity of the sample. In order to avoid singularities, $F(a)$ can be written also in the form

$$F(a) = \text{Re}\{\overline{E}_2(N,a)\}\,\text{Re}\{X_2(N/2,a)\} - \text{Re}\{\overline{E}_2(N/2,a)\}\,\text{Re}\{X_2(N,a)\} \tag{16}$$

With respect to FFT procedure used for transformation of experimental data, the second time interval in $F(a)$ was chosen to be half of the first one. The problem of finding $a$ reduces to solving an equation $F(a) = 0$. Typical curve of $F(a)$ for the step method is shown Fig. 3.

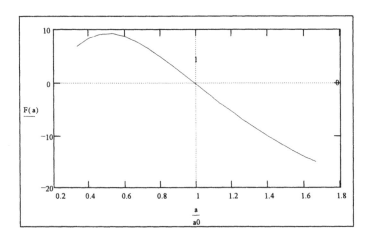

Figure 3. Typical curve of the function $F(a)$ given by Equation (16) for the step method. The root $a_0$ is the desired value of the thermal diffusivity. Similar shape has the function $F(a)$ in the case of the flash method.

Because of slow convergence of the sum for theoretical amplitude given by Equation (10) - (12), the upper limit of the summation index $m$ has to be equal or greater then 30.

### PRACTICAL DEMONSTRATION

In order to demonstrate that our procedure for thermal diffusivity calculation does not depend on knowing the baseline temperature level and can be applied also in the case of noisy signal, two sets of data were tested, with different degrees of noise on the response signal from the flash, as well as the step method. Each set consists of the five sequences of temperature vs. time points.

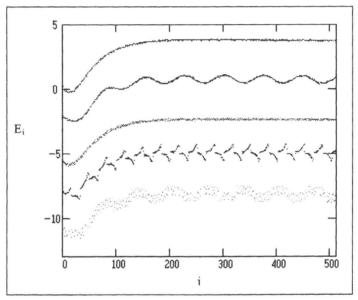

Figure 4. Experimental temperature response curves gained by the flash method, with different electronic noise superimposed on the signal. (The data curves are consequently numbered from the top of the graph.)

A set of temperature curves for the flash method was generated with different degrees of superimposed electronic noise. All the temperature response curves shown in Fig. 4 were obtained using the same sample, at the same temperature and other experimental conditions except that different noises were superimposed on the response signal. The sample thickness was 0.105 cm, and sampling frequency ($1/\vartheta$) of the signal was 1530 Hz. 512 and 256 points after the flash were used in data reduction calculations. As can be seen from Table 1 the results for the calculated thermal diffusivity values are within $\pm$ 2.1 % from that for the first data set, which represents the smoothest data.

The second set (graphs are given in Fig. 5.) for the step method was generated by a computer using the Equation (8), where $a = 9.00 \times 10^{-6}$ $m^2.s^{-1}$, $L = 3.00 \times 10^{-2}$ m, $T_0 = 15$, $\vartheta = 40/1012$ s. The noise was superimposed on an ideal signal with using a random number generator, as well as sinusoidal low frequency 'hum'. The relative differences (see Table 1.) between the desired and the computed value of thermal diffusivity remain again in reasonable limits of $\pm$ 3%.

The reproducibility of the data reduction procedure is negligibly affected by noise, provided the noise to useful signal ratio is small, and the noise has frequencies $f_n$ which are much higher than the frequencies of the useful part of the signal, i.e. the condition $f_n \gg 1/(N\vartheta)$ is fulfilled [8].

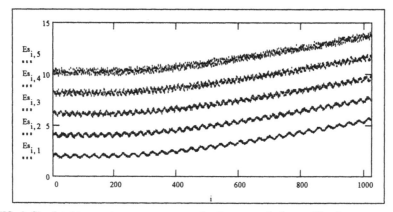

FIG. 5. Simulated temperature response curves for the step method test. (The data curves are consequently numbered from the bottom of the graph.)

TABLE I. Results of the test of reproducibility (the value $\Delta$ is calculated as a relative difference between the computed and the reference value of thermal diffusivity $a$ . In the case of the Flash Method the reference value is the first value in the column - for Data No. 1, and for the Step Method it is the desired value of thermal diffusivity equal to $9.000 \times 10^{-6}$ m$^2$s$^{-1}$.)

| | Flash Method | | Step Method | |
|---|---|---|---|---|
| Data No. | $a.10^6\left[m^2 s^{-1}\right]$ | $\Delta[\%]$ | $a.10^6\left[m^2 s^{-1}\right]$ | $\Delta[\%]$ |
| 1 | 3.833 | 0.0 | 9.223 | 2.483 |
| 2 | 3.802 | -0.8 | 8.983 | -0.194 |
| 3 | 3.859 | 0.7 | 9.249 | 2.770 |
| 4 | 3.753 | -2.1 | 9.176 | 1.951 |
| 5 | 3.883 | 1.3 | 8.876 | -1.378 |

## DISCUSSION

Our preliminary testing of the data reduction procedure on both methods shows that the number of points $N$ used in the computation should be at least equal or higher that $2^8$. The upper boundary of the time interval, $(N\vartheta)$ for the step method should be within a range $\frac{aN\vartheta}{L^2} \in (0.3, 1.0)$. The lower limit is given by the onset of the transient rise in temperature vs. time curve for the step method which is about $\frac{aN\vartheta}{L^2} \approx 0.12$, and the upper limit is given by the possibility to meet the ideal adiabatic boundary condition during a real experiment. For the flash method the range of the time limits is about the same, but the lower limit is approximately half of this in the step method. It is due to the nature of the temperature response curve in the flash method, where earlier onset of a transient rise is observed.

An example of the above mentioned conversion of the data between the step and the flash method is given on Fig. 6. The step method data $ES_i$ (256 points) were generated by computer with using Equation (8), where $a = 3.500 \times 10^{-6}$ $m^2.s^{-1}$, $L = 1.2345 \times 10^{-3}$ m, $T_0 = 5$, $\vartheta = 0.3048$ s. The noise was superimposed on an ideal signal with using a random number generator.

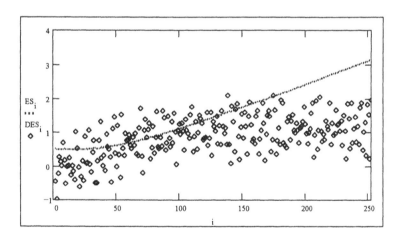

Figure 6. Simulated sequence of 256 temperature vs. time points in the step method (ES), and its derivative with respect to time (DES).

In order to compute a differentiation with respect to time from ESi, the standard formula for numerical differentiation was used

$$DES_i = \frac{ES_{i-2} - 8ES_{i-1} + 8ES_{i+1} - ES_{i+2}}{12t_i}, \qquad (17)$$

where $DES_i$ is the derivative of $ES_i$ with respect to time in the time point $t_i$.

As can be seen from Fig. 6. DES$_i$ points are much more 'scattered' than the original ones. This operation is very sensitive to presence of the noise. But even with this fact, the results of our data reduction procedure applied on these two sequences are quite optimistic, because of the step method gives $a = 3.473 \times 10^{-6}$ m$^2$s$^{-1}$ (relative difference = - 0.766 %), and the flash method procedure gives $a = 3.506 \times 10^{-6}$ m$^2$s$^{-1}$ ( 0.158 %). If the noise to signal ratio rises above a level of few percent, the derived points became chaotic and the reproducibility of our data reduction procedure becomes poor.

The other data reduction procedures used in flash method, especially those which use one or two points from the curve to determine the thermal diffusivity, can be applied to the data obtained from the step method by a numerical differentiation with respect to time, only after an effective smoothing operation.

## REFERENCES

1. Gembarovič, J., and R.E. Taylor, 1993, "Use of the Laplace Transformation in the Flash Method of Measuring Thermal Diffusivity" *International Journal of Thermophysics* 14: 297- 311.

2. Parker, W.J., R.J. Jenkins, C.P. Butler, and G.L. Abbott, 1961, "Flash Method for Determining Thermal Diffusivity, Heat Capacity, and Thermal Conductivity" *Journal of Applied Physics* 32(9):1679 - 1684.

3. Bittle R.R. and R.E. Taylor, 1984, "Step Heating Technique for Thermal Diffusivity Measurements of Large-Grained Heterogeneous Materials" *Journal of American Ceramic Society*, 67(3): 186 -190.

4. Cooley, J.W., and J.W. Tukey, 1965, "An Algorithm for Machine Calculation of Complex Fourier Series" *Math. of Computation*, 19: 297 - 301.

5. Čížek, V. 1986: *Discrete Fourier Transforms and their Applications*, Bristol: Adam Hilger Ltd., p.51

6. Balageas, D.L., 1988, *Proceedings of the 11th ETCP*, Umea, p. 85.

7. Bittle, R.R. and R.E. Taylor, 1985 "Thermal Diffusivity of Heterogeneous Materials and Non-Fibrous Insulators", in *Thermal Conductivity 18*, T. Ashworth and D. R. Smith, eds. New York: Plenum Publishing Corp., pp. 379 - 390.

8. Gembarovič, J., and R.E. Taylor, 1994, "A New Data Reduction Procedure in the Flash Method of Thermal Diffusivity Measuring" *Review of Scientific Instruments* 65(11): 3535 - 3539.

# Parameter Estimation Method for Flash Thermal Diffusivity with Two Different Heat Transfer Coefficients

J. V. BECK and R. B. DINWIDDIE

## ABSTRACT

Determining thermal diffusivity using flash diffusivity tests at high temperatures is investigated using parameter estimation. One aspect is the development of a method for determining two different heat transfer coefficients, one at the heated face and one at the opposite face. Both simulated exact and experimental data are used to illustrate the procedure. Although the heat transfer coefficients are different, assuming the heat transfer coefficients in the estimation process are the same does not significantly affect the estimates of the thermal diffusivity.

Insight into the estimation of thermal diffusivity and other parameters is obtained from a study of the sensitivity coefficients. Although the thermal diffusivity is the primary parameter of interest, a measured signal proportional to the temperature rise also depends on the heat transfer coefficients and energy input, which are called nuisance parameters (if they are not of interest). As the temperatures increase above 1500°C, the heat losses become very large and greatly influence the temperature response. By using insights from the study of the sensitivity coefficients for each of these parameters, the thermal diffusivity can be estimated despite the large heat losses.

## INTRODUCTION

Flash diffusivity methods have been used to determine the thermal diffusivity of solids from low to elevated temperatures [1-6]. However, as the temperature increases, the heat losses from the specimen surfaces rapidly increase, resulting in more difficult analysis of the data. The heat loss from the specimen can be caused by free convection and radiation. If the specimen is in a vacuum, only radiation losses are possible. In both cases, the heat losses can be described by heat transfer coefficients.

James V. Beck, Mechanical Engineering, Michigan State University, E. Lansing, MI 48824
Ralph Dinwiddie, Oak Ridge National Laboratory, Bldg. 4515, MS 6064, Oak Ridge, TN 37831

For tests at elevated temperatures (greater than 1500°C), the heat losses can be large and the surface temperatures on either face are quite different. The heat transfer coefficients can also be different in magnitude. However, the heat transfer coefficients on both faces of the specimen are commonly assumed to have the same value[1-6].

This paper investigates the simultaneous estimation of the thermal diffusivity, two heat transfer coefficients (one at $x = 0$ and the other at $x = L$, see Fig. 1) and the input power. The analysis is intended for elevated temperatures and the associated large heat losses. The main parameter of interest is the thermal diffusivity but sometimes the three other parameters must be simultaneously estimated; they are termed nuisance parameters. Parameter estimation techniques are used to estimate these parameters and are described for this problem. Estimating all these parameters simultaneously is deceptively difficult. The correlation between parameters can be very high, which means that simultaneous determination of such parameters can be both difficult and inaccurate. Fortunately, it also means that the number of parameters can be reduced.

An outline of the remainder of the paper is now given. First a mathematical model for this problem is given and followed by the analytical solution. Next the parameter estimation concepts are given and a case with exact data is investigated. The method is then applied to analyze a set of experimental data. The paper ends with conclusions.

## MATHEMATICAL MODEL AND SOLUTION

The specimen is modeled as a flat plate of thickness $L$. A signal proportional to the temperature rise at $x = L$ is measured using noncontact, averaging radiation sensors. In the experimental data used herein, the specimen is about 1 mm in thickness and about 25 mm in diameter. For these conditions, the one-dimensional plate model as shown in Figure 1 is appropriate. The surface at $x = 0$ is assumed to be heated with an instantaneous heat flash at time $t = 0$. (The analysis can be readily modified to treat a finite duration of the pulse.) Heat transfer coefficients of $h_1$ and $h_2$ are at the faces at $x = 0$ and $L$, respectively. The ambient temperature is assumed to be the constant value of $T_\infty$. These heat transfer coefficients can be used to describe the heat loss for both free convection and radiation.

The mathematical model and boundary conditions are

$$\alpha \frac{\partial^2 T}{\partial x^2} = \frac{\partial T}{\partial t} \tag{1}$$

$$-\lambda \frac{\partial T}{\partial x}\Big|_{x=0} = q_0 \delta(t) - h_1[T(0,t) - T_\infty] \tag{2}$$

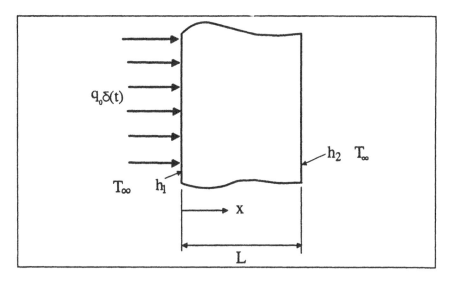

Figure 1. Diagram of a specimen heated by a flash at $x = 0$ and with heat transfer coefficient $h_1$ at $x = 0$ and heat transfer coefficient $h_2$ at $x = L$.

$$-\lambda\frac{\partial T}{\partial x}\Big|_{x=L} = + h_2[T(L,t) - T_\infty] \tag{3}$$

The initial temperature is $T_\infty$. The symbol $\delta(t)$ is the Dirac delta function that is zero everywhere except $t$ near zero and its integral over $t$ is equal to one. The units of the energy input, $q_0$, are J/m². Implicit in eq. (1) is the assumption that the thermal conductivity, $\lambda$, does not vary significantly over the temperature range of a particular flash experiment, although it can vary greatly one experiment to another.

An analytical solution of the above problem is a Green's function [7],

$$T(L,t) = T_\infty + \frac{2q_0}{\rho cL}\sum_{m=1}^{\infty} e^{-\eta_m^2 \alpha t/L^2}\frac{C_m}{N_m} \tag{4}$$

$$C_m = \eta_m(\eta_m\cos\eta_m + B_1\sin\eta_m) \tag{5}$$

$$N_m = (\eta_m^2 + B_1^2)[1 + \frac{B_2}{\eta_m^2 + B_2^2}] + B_1 \tag{6}$$

The $\eta_m$'s are eigenvalues found from $tan\,\eta_m = \eta_m(B_1 + B_2)/(\eta_m^2 - B_1 B_2)$ where $B_1 = h_1 L/\lambda$ is the Biot number at $x = 0$ and $B_2 = h_2 L/\lambda$ is the Biot number at $x = L$; $\lambda$ is the thermal conductivity.  The thermal diffusivity, the parameter of interest, is denoted $\alpha$.

One important decision is determining which parameters or groups can be estimated from transient signals at the $x = L$ face.  The above solution shows that the temperature rise at $L$ can be expressed as a function of four parameters denoted $\beta_1$, $\beta_2$, $\beta_3$, and $\beta_4$ where

$$\beta_1 = \alpha, \quad \beta_2 = \frac{q_0}{\rho cL}, \quad \beta_3 = B_1 = \frac{h_1 L}{\lambda}, \quad \beta_4 = B_2 = \frac{h_2 L}{\lambda} \tag{7}$$

For $\beta_2$, three unknowns are included: energy input $q_0$, density $\rho$ and specific heat $c$; however, only the parameter $\beta_2 (= q_0 /\rho cL)$ is needed.  Since the $T$ rise is proportional to $\beta_2$, only a signal proportional to the temperature rise must be measured. The last two parameters, $\beta_3$ and $\beta_4$, also involve the ratio of unknown quantities, such as $h_1$ over $\lambda$. Although only the thermal diffusivity, $\alpha$ , is often the single desired parameter, the other three groups (or parameters) must also be simultaneously estimated.

## PARAMETER ESTIMATION

In these experiments, the measurement errors in the temperature rise (or a signal proportional to it) can be considered additive and unbiased  and to have a constant variance. A cost function for these assumptions is the sum of squares.  Since the sensitivity coefficients for $\beta_3$ and $\beta_4$ are correlated, Tikhonov regularization [8] is used, resulting in the sum of squares function for $j = 1, 2, . . , J$ measurements,

$$S = \sum_{j=1}^{J} (Y_j - T_j)^2 + \alpha_{Tik}(\beta_3^2 + \beta_4^2) \tag{8}$$

where $Y_j$ and  $T_j$ are the  measured and calculated temperatures at time $t_j$ and $x = L$;  $T_j$ is calculated using the model given by eq. (4). The second term in this equation is called a zeroth order Tikhonov regularization term. The Tikhonov regularization parameter, $\alpha_{Tik}$, is made sufficiently small that the estimates of the  diffusivity are little affected but $\alpha_{Tik}$ is made big enough to allow convergence.  Some examples of selecting $\alpha_{Tik}$ is given later.

Estimates of the four parameters are obtained by minimizing eq. (8) by taking the first derivative of $S$ with respect to the parameters $\beta_i$ ($I = 1, 2, 3, 4$) and setting each equation equal to zero (see chap. 7, [9] for a complete discussion),

$$\frac{\partial S}{\partial \beta_i} = 2\sum_{j=1}^{J} (Y_j - T_j)(-\frac{\partial T_j}{\partial \beta_i}) + 2\alpha_{Tik}\beta_i \Delta = 0 \tag{9}$$

where $\Delta = 0$ for $i = 1$ and 2 and $\Delta = 1$ for $i = 3$ and 4.  Four simultaneous nonlinear algebraic equations are obtained from eq. (9).  The partial derivatives, $\partial T/\partial\beta_i$, in eqs. (9) are called sensitivity coefficients; see [10] for explicit expressions.

Determination of the confidence intervals for the thermal diffusivity is found using the classical statistical procedure with some assumptions regarding the measurement errors. The covariance matrix of the estimates of the parameters is calculated using eq. (7.7.1) of ref. 9. The diagonal term associated with the thermal diffusivity is the variance of the estimated value. Its square root is the estimated standard deviation of the estimated thermal diffusivity. The estimated confidence region is calculated as shown in Sect. 7.7 of ref. 9.

The values of the covariance matrix depend upon the assumptions that are valid for the measurement errors. These assumptions used herein are that the errors are additive, have zero mean (that is, are unbiased), have a constant variance and are first order autoregressive. A method of treating the first order autoregressive errors is given in Sect. 6.9 of ref. 9. These assumptions should be checked by examining the residuals which are simultaneously obtained with the parameter estimates.

Another basic assumption is that the model is correct. If it is not, then a systematic variation (a characteristic bias or "signature") will occur in the residuals that is repeated from test to test. One such imperfection in the model might be the lack of treatment of thermal penetration of the laser flash, causing the initial temperature distribution to be nonuniform.

Ideally uncorrelated measurements errors would be obtained and revealed by the residuals; unfortunately measurement errors are frequently either correlated or biased. Nevertheless it can be stated that the confidence intervals of the thermal diffusivity are certain values, provided the assumptions are valid.

## EXACT DATA EXAMPLES

An example with simulated temperatures (correct to six significant figures) is first given. The thickness is 1.0 and the initial temperature is 0.0. The true values of the parameters are $\alpha$ $(=\beta_1)$ equals 1; $q_0/\rho cL$ $(=\beta_2)$ equals 1, $B_1$ $(=\beta_3)$ equals 0.5 and $B_2$ $(=\beta_4)$ equals 0.1. The temperature curve is shown as the upper one in Figure 2. Forty data points are used with dimensionless time steps of $\alpha \Delta t/L^2 = 0.05$. Two sets of initial "guesses" are used. For the first three rows of Table I, all the starting parameter values are correct except the second one which is 0.7 while the true value is 1.0. The last three rows of Table I use the initial "guesses" of 1, 0.7, 0.5 and 0.5. Estimated parameters are denoted $b_i$ and results are shown in Table I for values of the Tikhonov regularization parameter $\alpha_{Tik}$ from $10^{-16}$ to $10^{-8}$. In each convergent case the estimated

TABLE I. RESULTS OF ESTIMATING PARAMETERS USING EXACT DATA

| Initial Values of Parameters | $\alpha_{Tik}$ | Estimated | Parameters | | | Std. Dev. | Thermal Diffusivity |
| | | $b_1$ | $b_2$ | $b_3$ | $b_4$ | $s$ | Confidence Interval |
| --- | --- | --- | --- | --- | --- | --- | --- |
| 1, 0.7, 0.5, 0.1 | $10^{-8}$ | 1.0035 | 0.9893 | 0.2810 | 0.2810 | 0.167E-4 | 1.0027 to 1.0042 |
| 1, 0.7, 0.5, 0.1 | $10^{-12}$ | 1.0002 | 0.9994 | 0.4932 | 0.1046 | 0.257E-6 | 0.99982 to 1.00056 |
| 1, 0.7, 0.5, 0.1 | $10^{-16}$ | 1.0000 | 1.0000 | 0.5000 | 0.1000 | 0.268E-8 | 0.99996 to 1.00004 |
| 1, 0.7, 0.5, 0.5 | $10^{-8}$ | 1.0035 | 0.9893 | 0.2810 | 0.2810 | 0.166E-4 | 1.00270 to 1.0042 |
| 1, 0.7, 0.5, 0.5 | $10^{-12}$ | 1.0034 | 0.9994 | 0.2859 | 0.2761 | 0.109E-5 | 1.0032 to 1.0036 |
| 1, 0.7, 0.5, 0.5 | $10^{-16}$ | NONCONVERGENT | | | | | |
| Exact Parameter Values: | | 1 | 1 | 0.5 | 0.1 | | |

$\alpha$ (that is, $b_1$) is very near the true value of 1.0. The confidence regions for this parameter are also given by the last pair of numbers. The quantity denoted $s$ is the estimated standard deviation of the measurements, which is an estimate of the standard deviation of the simulated measurements (about $10^{-6}$).

For the first row of Table I ($\alpha_{Tik} = 10^{-8}$), the estimated values of the Biot numbers ($b_3$ and $b_4$) are both about 0.28, which is near the average of the true values of 0.1 and 0.5. For even smaller values of $\alpha_{Tik}$ shown in the second and third rows of Table I, estimates $b_3$ and $b_4$ are quite accurate, indicating that the computational procedure is correct with extremely accurate data and quite different values of $\beta_3$ and $\beta_4$. The last three rows of Table I show that the estimation process is much more difficult if the initial guesses for $\beta_3$ and $\beta_4$ are the same value. However, for the cases that do converge (rows 4 and 5) the parameter of interest, the thermal diffusivity, is negligibly affected by the estimates of the last two parameters. Consequently in many cases, it is satisfactory to estimate the thermal diffusivity with the assumption that the two heat transfer coefficients are equal. A reason why there is a tendency for $b_3$ and $b_4$ to approach the same value is because the sensitivity coefficients for $\beta_3$ and $\beta_4$ tend to be correlated. See the below discussion of Figures 2 and 3.

The choice of the Tikhonov parameter may require some experimentation. One concept is to make it as small as possible and yet obtain convergence. Another concept is to choose $\alpha_{Tik}$ so that the estimated standard deviation of the temperatures, $s$, is about the expected value, which is about $10^{-6}$. Reference to Table I shows that $\alpha_{Tik} = 10^{-12}$ satisfies this condition.

The dimensionless temperature rise, Fourier number, and dimensionless modified (by multiplying by $\beta_i$) sensitivity coefficients $Z(i)$, i = 1,2,3, 4 are defined by

$$T^+ = \frac{T - T_\infty}{\beta_2}, \quad t^+ = \frac{\alpha t}{L^2}, \quad Z(i) = \frac{\beta_i}{\beta_2} \frac{\partial T}{\partial \beta_i}, \quad i = 1,.,4 \tag{10}$$

Dividing $Z(i)$ by $\beta_2$ eliminates the dependence of $Z(i)$ upon $\beta_2$. Multiplication of $Z(i)$ by $\beta_i$ gives the modified sensitivities and permits comparison with the temperature rise. Notice that $T^+$ is equal to $Z(2)$.

Figure 2 displays results for $\beta_3 = 0.5$ and $\beta_4 = 0.1$; $Z(2)$ reaches a maximum value about 0.7 and then starts to decrease with dimensionless time, $t^+$. The dimensionless sensitivity for the thermal diffusivity, $Z(1)$, is relatively large at early times, reaching a maximum about 0.53 and then decreases to negative values. The Biot number sensitivity coefficients ($\beta_3$ and $\beta_4$) are smaller in magnitude and correlated (i.e., have the same shape). Since $\beta_3$ and $\beta_4$ are nuisance parameters, these conditions of small sensitivies and correlation need not significantly affect the estimation of $\beta_1$ $(= \alpha)$. It suggests setting $\beta_3$ equal to $\beta_4$ (and estimating only $\beta_3$) will not significantly affect the estimation of $\beta_1$. This can be seen by examining the results of Table I.

Since it may not be necessary to estimate independently two Biot numbers, in the next case the Biot numbers are assumed to be equal. Figure 3 shows results for $\beta_3 = \beta_4 = 10$. The magnitude of the $\beta_1$ sensitivity coefficient tends to be larger than that of the temperature rise and the $\beta_2$ sensitivity coefficient. That is advantageous for estimating $\beta_1$. The correlation between $Z(2)$ and $Z(3)$ (that is, $Z(2)/Z(3)$ is nearly a constant) indicates that the simultaneous estimation of $\beta_1$, $\beta_2$ and $\beta_3$ may be difficult. However, regularization may be used to improve the convergence for $\beta_1$.

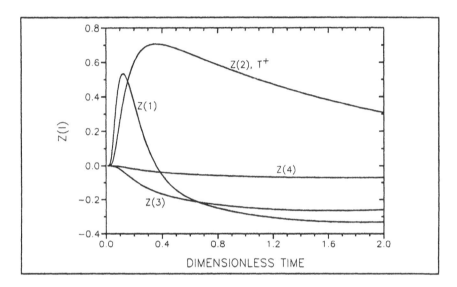

Figure 2. Exact temperature rise and modified sensitivity coefficients for $\alpha = 1$, $\beta_2 = 2q_0/\rho cL = 1$, $\beta_3 = B_1 = 0.5$ and $\beta_4 = B_2 = 0.1$. Values plotted versus dimensionless time, $\alpha t/L^2$.

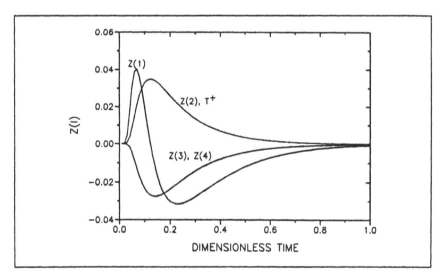

Figure 3. Exact temperature rise and modified sensitivity coefficients for $\alpha = 1$, $\beta_2 = 2q_0/\rho cL = 1$, $\beta_3 = B_1 = 10$ and $\beta_4 = B_2 = 10$. Values plotted versus dimensionless time, $\alpha t/L^2$.

## EXPERIMENTAL DATA EXAMPLE

Transient temperatures for carbon bonded carbon fiber insulation (CBCF) at 2000°C are shown in Figure 4. The temperature response is given in "volts" units. An analysis was performed for estimating the four parameters, (diffusivity, energy input, and the two Biot numbers). Because of the high correlation between the two Biot numbers, Tikhonov regularization using eq. (8) was needed. Using all 472 data points, a range of $\alpha_{Tik}$ values was chosen for the initial estimates of the Biot numbers of $\beta_3 = 12$ and $\beta_4 = 8$. For each $\alpha_{Tik}$ value shown in Table II, the converged values of the two Biot numbers are equal, though different as $\alpha_{Tik}$ is varied. The minimum regularization for $\alpha_{Tik}$ is about $10^9$ which is a large numerical value because the magnitude of the "volts" in Figure 4 is large. For smaller $\alpha_{Tik}$ values and the same initial estimates of parameters, the procedure has difficulty converging. The important point is that the $\alpha$ estimates, denoted $b_1$, are relatively insensitive to changes in the Tikhonov parameter; for example, increasing $\alpha_{Tik}$ by a factor of 1000 increases $b_1$ by only 15%.

The reason that the two Biot numbers converge to the same values in Table II for a specified $\alpha_{Tik}$ value is the very high correlation in the sensitivity coefficients. Since the two Biot numbers coalesce to the same values, it is reasonable to estimate only three parameters, ($\alpha$, energy input, and the same Biot number for both surfaces). The estimated parameters are 0.006524 cm²/s, 579,900 and 10.645 for $\alpha$, energy input and Biot number, respectively. The parameter estimates can be plotted sequentially with time. The sequential values are those that would be obtained if the number of measurements that are used increased one by one until all the data is used. In well-designed experiments and when an appropriate number of parameters are estimated, the sequential estimates should be nearly constant for at least the last half of the experiment duration. For this case, very large variations with time of the sequential parameters is found. The sequential values change so greatly because the energy input and Biot number ($\beta_2$ and $\beta_3$) sensitivity coefficients are correlated, as indicated by Figure 3. (Figure 3 is for the four parameters but for the case of identical Biot values at $x = 0$ and $L$, the sensitivity coefficient for the same Biot number on both surfaces is just a factor of two larger than that shown for $\beta_3$.) Correlation between two sensitivity coefficients can be determined by dividing one by the other and plotting the result as a function of time. If the ratio is nearly constant, then high correlation exists and fewer parameters should be estimated. The ratio of the second and third sensitivity coefficients is almost a constant in this case. This suggests estimating only two parameters with the Biot number given a few values.

TABLE II. RESULTS OF ESTIMATING PARAMETERS USING EXPERIMENTAL DATA FOR INITIAL VALUES OF $\beta_3 = 12$ AND $\beta_4 = 8$. DATA NOT FILTERED.

| | Estimated | Parameters | | | | Std. Dev. |
|---|---|---|---|---|---|---|
| $\alpha_{Tik}$ | $b_1$ | $b_2$ | $b_3$ | $b_4$ | $S_{min}$ | $s$ |
| $10^{12}$ | 0.00738 | 267600 | 6.363 | 6.363 | 0.217E+9 | 680.9 |
| $10^{10}$ | 0.00676 | 450700 | 9.058 | 9.058 | 0.120E+9 | 506.4 |
| $10^9$ | 0.00662 | 519500 | 9.928 | 9.928 | 0.1079E+9 | 480.2 |
| $10^8$ | NONCONVERGENT | | | | | |

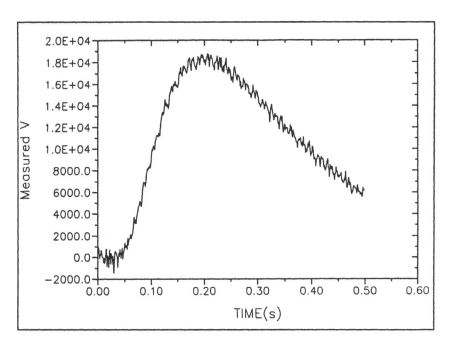

Figure 4. Measured temperature rise (in arbitrary volt units) versus time for CBCF at 2000°C.

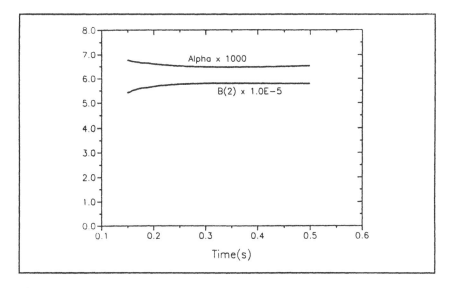

Figure 5. Sequential parameter estimates for CBCF at 2000°C for two parameters: thermal diffusivity and input energy; Biot number = 10.65.

TABLE III. RESULTS OF ESTIMATING $\beta_1$ and $\beta_2$ FOR SPECIFIED VALUES OF $\beta_3$. DATA FILTERED, FIRST TWO MEASUREMENTS DROPPED AND INITIAL TEMPERATURE CORRECTION.

| $\beta_3$ | $b_1$, cm$^2$/s | $b_2$ | $S_{min}$ | $s$ | Thermal Diffusivity Confidence Interval |
|---|---|---|---|---|---|
| 7 | 0.00720 | 305700 | 0.607E+8 | 359 | 0.00648 to 0.00791 |
| 9 | 0.00677 | 444800 | 0.413E+8 | 296 | 0.00636 to 0.00717 |
| 10.65 | 0.00652 | 578500 | 0.379E+8 | 284 | 0.00618 to 0.00687 |
| 12 | 0.00637 | 700600 | 0.386E+8 | 285 | 0.00602 to 0.00671 |
| 15 | 0.00611 | 1013000 | 0.439E+8 | 304 | 0.00572 to 0.00649 |

Figure 5 shows sequential results for estimating only two parameters, thermal diffusivity and energy input. These results are for the Biot number, $\beta_3$, equal to the converged value for three parameters which is 10.65. Before discussing this plot, several points are made. First, the data was filtered at each time by simply using the average of five previous, the measured value at that time and five subsequent measured temperatures. Also the first two measurements were omitted since they seemed to be high and a small correction for a non-zero initial temperature was added. The net effect of the filtering, etc. was negligible upon the parameter estimates shown by Figure 5. Second, filtering is very reasonable to reduce the effects of the periodic noise. The period of this noise is about eleven data points, hence the choice of the filtering region. Third, the estimated values shown in Figure 5 are not greatly affected by the value of the fixed value of $\beta_3$. The fourth and final point is that the estimated value of the thermal diffusivity in Figure 5, 0.00652 cm$^2$/s, is more properly given with a confidence region. Using standard statistical methods and assuming that the measurement errors are first order autoregressive[9] yields the confidence region of 0.00618 to 0.00687 cm$^2$/s. If the same procedure were used for $\beta_3$ equal to 7 to 14, the associated confidence intervals include the above value of 0.00652, shown in Table III.

Returning now to a discussion of Figure 5, the most obvious and satisfactory feature is that the thermal diffusivity and the energy input are nearly constant over a very large time range. This is in contrast to the case of estimating three parameters. The three parameters have large sequential variations because the second and third parameters are highly correlated. See Figure 3 which is for about the same Biot number. Figure 3 also shows that the first two parameters are quite uncorrelated, which is one reason that the sequential values in Figure 4 are nearly constant. One of the difficulties of this analysis for two parameters is that an estimate of $\beta_3$ is needed. However, a 114% increase in $\beta_3$ from 7 to 15 causes only a 15% drop in the estimated thermal diffusivity. If the $\beta_2$ and $\beta_3$ parameters were perfectly correlated then changes in $\beta_3$ would not affect the thermal diffusivity ($\beta_1$). As it is, there is a slight change in the thermal diffusivity. The confidence region of 0.00618 to 0.00687 cm$^2$/s (or ±5%) is reasonable for measurements at 2000°C if the main source of errors is in the random temperature measurement. For lower temperatures, the Biot numbers are smaller and the correlation between the power and Biot number is decreased. This makes estimation of the three parameters ($\beta_1$, $\beta_2$ and $\beta_3$) easier.

Another important aspect is the examination of the residuals which are the differences between the measured and calculated temperatures. Because of space limitations, they cannot be shown. However, it is sufficient to describe them as increasing from zero at t = 0 to 550 at 0.08s, decreasing to -500 at 0.15s increasing to 550 at 0.3s and finally going down to -600. There is a little fluctuation in the residuals and the data was filtered before analysis . Two observations are made. The residuals are relatively small, with the maximum magnitude about 3% of the maximum temperature rise. This indicates that the model is good. The second observation is that the residuals tend to be correlated and the residuals just less than 0.1s are more significant because the temperatures at those times are small.


## CONCLUSIONS

Methods to estimate the thermal diffusivity using data from flash diffusivity tests at elevated temperatures are discussed and illustrated using simulated and experimental data. At elevated temperatures the heat losses from the specimen faces are unequal (caused by a much larger temperature rise at $x = 0$ than at $x = L$) and large (Biot numbers >> 1), making determination of the thermal diffusivity more difficult. For specimens in a vacuum the heat losses are by radiation, which can be described by radiation heat transfer coefficients, one on each side of the specimen. The estimation of the thermal diffusivity for tests having two heat transfer coefficients and an unknown energy input is discussed. A method is given for the simultaneous estimation of these four parameters. (Actually it is more convenient to estimate the thermal diffusivity, energy input divided by the volumetric heat capacity multiplied by the thickness, and two Biot numbers which are proportional to the heat transfer coefficients.)

Tikhonov regularization is needed in a sum of squares function to find the four parameters. The sum of squares function is minimized with respect to the parameters. Tikhonov regularization is needed because the two heat transfer coefficients are highly correlated. Methods for determining the regularization constant are discussed.

A physical understanding of the estimation problem can be obtained by examining the sensitivity coefficients for each of the parameters. The sensitivity coefficients are the first derivatives of the calculated temperature; a modified coefficient is a derivative multiplied by the appropriate parameter. If these modified coefficients are proportional over time or one is small compared to the others, the simultaneous estimation of all the parameters is very difficult because the minimum is poorly defined. However, if either of these conditions are true, the number of parameters being estimated can be reduced. Plots of the sensitivity coefficients show that the two Biot numbers are highly correlated, indicating that estimating a single Biot number is satisfactory and will not greatly affect the estimates of the thermal diffusivity. The values of Biot numbers are not important since they are nuisance parameters.

For elevated temperatures (> 1500°C), not only are the Biot numbers for the two faces highly correlated but the input energy and the Biot numbers are correlated. Thus if a reasonable estimate of the Biot number is available, only two parameters can be simultaneously estimated, namely, the thermal diffusivity and the input energy.

Several additional concepts are helpful. For oscillatory measurement errors, filtering of the data can improve the estimates. Sequential estimation (for adding one measurement after another) can yield much insight into the adequacy of the model. It is important to examine the residuals. For the experimental data examined, which is for CBCF at 2000°C, the residuals are shown to be relatively small, indicating that the model is satisfactory. Confidence regions for the measurements are given which are about ±5% for the data considered.

## ACKNOWLEDGMENTS

Research sponsored by the Space and National Security Programs, Office of Nuclear Energy, Science and Technology, and the Assistant Secretary for Energy Efficiency and Renewable Energy, Office of Transportation Technologies, as part of the High Temperature Materials Laboratory User Program, Oak Ridge National Laboratory, managed by Lockheed Martin Energy Research Corp. for the U.S. Department of Energy under contract DE-AC05-96OR22464.

## REFERENCES

1.  Parker, W. J., R. J. Jenkins, C. P. Butler, and G. L. Abbott, 1961. "Thermal Diffusivity Measurements Using the Flash Technique." *J. Applied Physics*, 32:1679.
2.  Cape, J. A. and G. W. Lehman, 1963. "Temperature and Finite Pulse-Time Effects in the Flash Method for Measuring Thermal Diffusivity." *J. Applied Physics*, 34:1909.
3.  Cowan, R. D., 1963. "Pulse Method of Measuring Thermal Diffusivity at High Temperatures.", *J. Applied Physics*, 34:926.
4.  Clark, L. M., III and R. E. Taylor, 1975. "Radiation Loss in the Flash Method for Thermal Diffusivity." *J. Applied Physics*, 46: 714.
5.  Koski, J. A. 1981. "Improved Data Reduction Methods for Laser Pulse Diffusivity Determination with the Use of Minicomputers," *Proc. 8th Symposium of Thermophysical Prop.*, II: 94-103.
6.  Raynaud, M., J. V. Beck, R. Shoemaker, and R. E. Taylor, 1989. "Sequential Estimation of Thermal Diffusivity for Flash Tests," in *Thermal Conductivity, 20*, D. P. H. Hasselman and J. R. Thomas, Plenum Publishing Corp., 20:305-321.
7.  Beck, J., K. Cole, A. Haji-Sheikh, and B. Litkouhi, 1992. *Heat Conduction Using Green's Functions*, Washington, D.C.: Hemisphere.
8.  Beck, J. V., B. Blackwell, and C. R. St. Clair, Jr., 1985. *Inverse Heat Conduction: Ill-Posed Problems*, New York: Wiley-Interscience.
9.  Beck, J. V. and K. J. Arnold, 1977. *Parameter Estimation in Engineering and Science*, New York, NY: Wiley.
10. Beck, J. V. , 1994. "Solution for the Eigenvalues for the X33 Case," Memo No. 94-2 to R. Dinwiddie, Aug. 5, 1994.

# Multiple Station Thermal Diffusivity Instrument

H. WANG, R. B. DINWIDDIE and P. S. GAAL

## ABSTRACT

A multiple furnace laser flash thermal diffusivity system has been developed. The system is equipped with a movable Nd:Glass laser unit, two IR detectors and four furnaces for precise measurements of thermal diffusivity over the temperature range from -150°C to 2500°C. All furnaces can operate in vacuum and inert gas; the environmental effects furnace also supports oxidizing and reducing environments. To increase testing speed the graphite and aluminum furnaces are both equipped with six-sample carousels. Thermal diffusivity measurements of three standard reference materials show excellent results over the entire temperature range.

## INTRODUCTION

Since its introduction in 1961[1], the laser flash technique has become a standard testing method for thermal diffusivity measurements of solids [2,3]. Traditionally, a single furnace is used for a certain temperature range and a few classes of materials. The rapid development of new materials often requires a laser flash system to cover a wider temperature range and to operate with different environments. This is particularly true for materials research and development. However, it is virtually impossible to find a single furnace type that can service the entire temperature domain of practical use, and to accommodate every material to be tested. Throughput in testing is a paramount factor even for fully automated systems.

The system which is the subject of this paper is based on a traversing laser, multiple furnaces and two photovoltaic IR detectors. To suit a variety of testing requirements, the system is equipped with an ultra-high temperature graphite furnace (500°C to 2500°C), a high temperature Kanthal Super™ furnace (RT to 1700°C), a low temperature aluminum Monoblock™ furnace (-150°C to 500°C), and a high

H. Wang and R.B. Dinwiddie, MS 6064, Bldg. 4515 Oak Ridge National Laboratory, Oak Ridge, TN 37830-6064

P.S. Gaal, Anter Corporation, 1700 Universal Road, Pittsburgh, PA 15235

119

speed quench furnace (RT to 1200°C). It also provides the ability to add to and/or reconfigure an existing system as needs change. High throughput leads to lower unit cost per test and higher level of utilization. Although the laser flash technique is much faster than static methods, a single sample test over a moderate temperature range could still take a whole day in traditional systems. Concurrent testing of multiple samples was made available on this system. The graphite and aluminum furnaces are both equipped with six-sample carousels. They provide an opportunity to study five or six specimens concurrently under completely identical conditions.

The purpose of this paper is to summarize the system capabilities and report results for tests on standard reference specimens. Armco iron, stainless steel and AXM-5Q graphite were tested and compared with data in the literature. The tests showed excellent results over the entire temperature range.

## LASER FLASH THERMAL DIFFUSIVITY SYSTEM

### ULTRA HIGH TEMPERATURE SUBSYSTEM

A schematic of the system is shown in Fig.1. The ultra-high temperature graphite furnace is designed to operate from 500°C to 2500°C in inert gas environment and up to 2000°C in vacuum. It employs a water cooled stainless steel outer shell, molded fibrous carbon insulation, and pyrolitic graphite reflector inside. The graphite heating element is free standing with both power connectors on the bottom. The sample support structure, also of graphite, is an integral part of the furnace. A precision slide mechanism allows the sample holder to be lifted out, co-

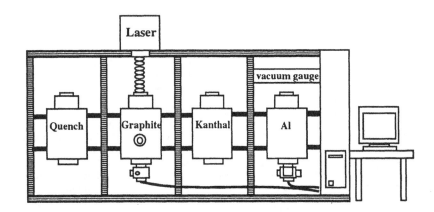

Figure 1. The thermal diffusivity testing system.

linearly with the optical axis. Proper closure is ensured with a sliding seal that engages without a jolt to the sample. A graphite multi-sample carousel allows up to six samples to be loaded and tested during one run. The sample holder is located in the uniform temperature zone of the furnace, and the temperature of the sample holder is monitored by a one-color optical pyrometer.

## HIGH TEMPERATURE SUBSYSTEM

The high temperature Kanthal Super™ furnace is designed to operate between room temperature and 1700°C. The furnace cavity has a 4 inch long hot zone ensuring excellent uniformity and stability in the sample region. The high purity alumina muffle and sample support components allow operation in high vacuum, ($10^{-6}$ torr), oxidizing and reducing atmospheres. The alumina muffle also prevents the Kanthal elements from direct exposure to the operating gas. The shell is air-cooled and is no more than warm to the touch at maximum operating temperature. Temperature control and sensing is via Type S thermocouples, one close to the heaters and another adjacent to the sample. To operate with hydrogen atmosphere, special valves and a built-in hydrogen sensor were incorporated into the gas controls to monitor the muffle and the furnace cavity in case the muffle tube cracks. The alumina sample holder is configured for a single sample. With judicious use of opaque shields in the sample holder, the shine-by of the laser beam due to transparency of alumina is kept at an acceptable level.

## MODERATE TEMPERATURE SUBSYSTEM

The operating range of the low temperature aluminum Monoblock™ furnace is from -150°C to 500°C. The design allows the furnace to run in high vacuum, ($10^{-6}$ torr) and inert gas environment. Measurement below room temperature is made possible by introducing liquid nitrogen to cooling channels within the furnace block. The sample holder is a six-sample carousel for multiple testing. It is made out of aluminum and stainless steel and has similar design to the graphite furnace carousel.

## QUENCH FURNACE

The fourth furnace is a high speed quench furnace. The radiantly heated sample holder structure will be operable between ambient and 1200°C. Cooling is provided by a blast of helium once the heat source is turned off. Through various timed and tracked signals, the lower quenched temperature can be programmed. The furnace is designed to measure the thermal diffusivity of a sample at a certain high temperature and then quench the sample at about 200°C per second to a secondary temperature to test it again without first having to go ambient. This furnace will be installed and tested during the first quarter of 1996.

## OPTICAL SUBSYSTEMS

<u>Laser Unit</u>: The laser unit has a Nd:Glass laser with pulse energy up to 35 Joule. The laser head can be moved along a rigid rail above each of the four furnaces. The front face of the laser is equipped with a full complement of key-locked manual controls for diagnostic testing. An interlocked flexible shield connects the laser and the furnace. Full locking is checked by actuated safety switches before firing may occur. Once the beam paths are shielded and interlocked, Class 1 operation is ensured and thus there is no need for safety goggles in the vicinity. The laser power can be varied by changing either the charging voltage (1200 to 2800V) or the number of high voltage capacitors in parallel connection (1 to 5). The pulse-width may be adjusted by varying the number of parallel capacitors. In addition, the pulse-width may be stretched by the addition of inductors in the discharge circuit. Several optical components were used to aid the alignment of the laser. A helium/neon alignment laser is located on the laser rail in the same housing. It is used for the

alignment of the primary laser, and during thermal diffusivity tests to confirm the presence of a sample in the sample holder.

Detection System:   Two different photovoltaic detectors are incorporated into the system for versatility.  They are mounted underneath the furnaces and monitor the temperature change of the rear surface of the specimen.  A high temperature silicon photodiode detector with adjacent preamplifier is shared by the graphite and Kanthal furnaces.  Its properties are well suited for use with wavelengths common in high temperature testing.  A fiber optic link connects the furnace port and the detector, located 6 feet away.  A cryogenically cooled InSb detector with built-in preamplifier is designed for the low temperature aluminum furnace.

A secondary detector subsystem is incorporated into the design to determine the exact shape of the laser pulse.  The system maps every laser shot to obtain the precise pulse shape and apply the various pulse width corrections.

OTHER SUBSYSTEMS

The vacuum system consists of a mechanical roughing pump and a turbo molecular pump for high vacuum.  All the furnaces share the same vacuum system. In practical use, only one furnace at a time is under vacuum.

Purging gas may be directed to an individual furnace during testing.  Two-stage bubblers are used for the exhaust of all the furnaces.   The bubbler has an oil reservoir located at the bottom of the assembly and a second reservoir located directly above the first.  The design has a ball seal and isolation valve to prevent any air from being pulled into the system during vacuum operation or rapid cool down.

The power supply for the system consists of a water cooled step-down transformer controlled by a low noise controller. Current limiting, water flow sensing, etc., are fully implemented.

SAMPLES

The system is designed to adapt to a variety of samples.  Typical samples are disk-shaped or square-shaped, 6 to 12.5mm in diameter and 1 to 6mm thick. In addition to bulk specimens, powder and molten metal samples can also be tested. Special containment capsules of quartz or sapphire, are used to test materials through the melt.


# OPERATING SOFTWARE AND DATA ANALYSIS

The operation of the system is fully computer controlled and automatic.  The following functions are provided by the application software:

• Sample identification, test parameters and test setup information.

• Choice of filters and their effect in reducing noise.

• Detailed tabulated corrections by the various methods for each test shot at each temperature.

• Rear temperature vs. time plot for each shot as it occurs. Full display of laser pulse shape, and rear surface temperature excursion.

• Tabulated summary of diffusivity averaged over the multiple measurements at each temperature.

The data analysis methods of the laser flash technique have been well documented [2,3]. The first analysis method was developed by Parker, et. al. [1].

Thermal diffusivity, $\alpha$, was calculated from the thickness, D, and half rise time, $t_{0.5}$, of the rear surface temperature:

$$\alpha = CD^2/t_{0.5} \qquad (1)$$

where C is a dimensionless parameter. In Parker's method, C=0.1388 under adiabatic conditions. Cowan [4] later developed a mathematical model to include radiation heat losses. Temperature changes at $5t_{0.5}$ or $10t_{0.5}$ are compared with the temperature change at $t_{0.5}$ to determine the parameter C. Clark and Taylor [5] proposed a new algorithm for analyzing the radiation heat losses from the temperature vs. time plot before the maximum temperature had been reached. Koski [6] and Heckman [7] incorporated a parameter, L, for heat loss from the front to the rear of the sample for each condition of Clark and Taylor as well as Cowan's method.

The system software incorporates all of the data analysis techniques mentioned. One technique can be chosen before the test as a primary one. After the test, thermal diffusivity values obtained from the various methods are given in a table.

Two layer and three layer materials can also be tested in the system. The analysis is based on the technique developed by Lee and Taylor [8]. Specific heat, density, thickness of each layer and thermal diffusivity of the substrate material are required for the calculation. The application software can also calculate the contact resistance between two layers with known diffusivity[8].

## STANDARD MATERIALS TESTING

Several standard materials, SRM 8425 graphite, SRM 1462 stainless steel and ARMCO iron, were used to test the performance of the system. All the samples were disks 12.5mm in diameter. Thermal diffusivity of each material was measured as a function of temperature and compared with reference values. The tests were repeated three times to check the reproducibility of the system. Thermal diffusivity data of standard materials are available in the TPRC Data Series from Purdue University [9] and NIST Research Material data sheet.

SRM 8425 graphite samples were tested in the aluminum furnace from 100°C to 500°C and continued in the graphite furnace from 700°C up to 1900°C. Reference data were calculated from the thermal conductivity values provided on the NIST Research Material data sheet, specific heat values existing in the literature [10] and a density for graphite of 1.73 g/cm$^3$. The testing results of SRM 8425 graphite obtained from the system agreed within ±5% with the reference curve as shown in Fig. 2. The standard deviation of a single test temperature determined by multiple measurements (multiple laser shots) during each test and among the three repeated tests are within ±3% of the experimental values shown in Figure 2.

Stainless steel, SRM 1462, was tested from 100°C to 1050°C in both furnaces and compared with reference data [9,11]. The results shown in Figure 3 exhibit less than 5% deviation from the average reference values. An ARMCO iron sample was tested in the low temperature furnace under vacuum condition. The results from 100 to 500°C are shown in Figure 4. They showed very good agreement with Taylor and Clark's data [12] and are within ±5% of the recommended reference curve [9]. For all the three reference materials, the estimated uncertainties of the reference data values are ±5% for test temperatures above 300K.

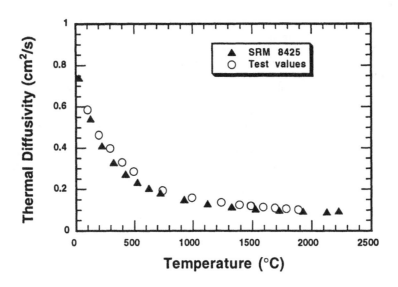

Figure 2. Thermal diffusivity test on SRM 8425 graphite standard sample.

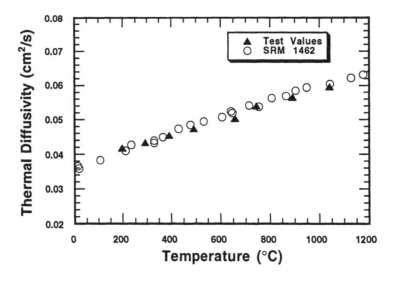

Figure 3.  Thermal diffusivity measurement on SRM 1462 stainless steel sample.

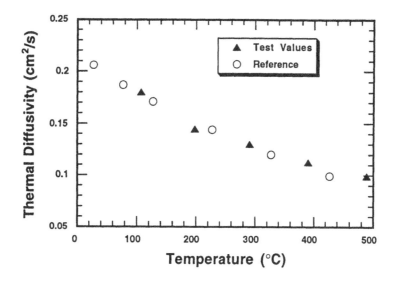

Figure 4. Thermal diffusivity measurement of ARMCO iron standard.

## SUMMARY

A system has been developed for thermal diffusivity measurements over a wide temperature range and in different environments. The multiple station and concurrent testing concepts have been proven successful for materials characterization. Testing of standard reference materials also showed exceptionally good agreement with literature values; in all cases the experimental values were essentially equal to the reference values. Further tests on layered samples, melts and the quench furnace are planned.

## ACKNOWLEDGMENT

The experimental work was supported by the U.S. Department of Energy, Assistant Secretary for Energy Efficiency and Renewable Energy, Office of Transportation Technologies, as part of the High Temperature Materials Laboratory User Program under contract DE-AC05-96OR22464, managed by Lockheed Martin Energy Research Corporation. This research work was also supported in part by an appointment of H. Wang to the ORNL Postdoctoral Research Associates Program administered jointly by ORISE and ORNL.

## REFERENCES

1. Parker, W. J., Jenkins, Butler, C. P. and Abbott, G. L., 1961. "Flash Method of Determining Thermal Diffusivity, Heat Capacity and Thermal Conductivity," *Journal of Applied Physics*, 32(9): 1679-1684

2. ASTM Designation E 1461, 1992. "Standard Test Method for Thermal Diffusivity of Solids by the Flash Method," 933-940

3. Taylor, R.E., 1979. "Heat-Pulse Thermal Diffusivity Measurements," *High Temperature - High Pressure*, 43(11): 43-58

4. Cowan, R.D., 1963. "Pulse Method of Measuring Thermal Diffusivity at High Temperatures," *Journal of Applied Physics*, 34(4): 926-927

5. Clark, L. M. and Taylor, R. E., 1975. "Radiation Loss in the Flash Method for Thermal Diffusivity," *Journal of Applied Physics*, 46: 714

6. Koski, J. A., 1981. "Improved data Reduction Methods for Laser Pulse Diffusivity Determination with the use of Minicomputers," Proceeding of the 8th Symposium of Thermalphysical Properties, Vol. II: 94-103

7. Heckman, R. C., 1974. "Error Analysis of the Flash Thermal Diffusivity Technique," *Thermal Conductivity 14*, Klemens, P. G. and Chu, T. K., Eds., Plenum Publishing Co. NY: 491-498

8. Taylor, R. E., Lee, T. Y. R. and Donaldson, A. B., 1978. "Determination of Thermal physical Properties of Layered Composites by Flash Method," *Thermal Conductivity 15*, Mirkovich, V. V. ed., Plenum Publishing Co. NY: 135-148

9. "Thermal Diffusivity," 1973. Thermalphysical Properties of Matter, Purdue University, Vol.10

10. Deshpande, M. S. and Bogaard, 1983. "Evaluation of Specific Heat Data for Poco Graphite and Carbon-Carbon Composites," *Thermal Conductivity 17*, Hust, J. G. ed., 17: 45-54

11. Maglic, K. D., Perrovic, N. and Zivotic, Z., 1980. "Thermal Diffusivity Measurements on Standard Reference Materials," *High Temperature-High Pressure*, 12: 555-560

12 Taylor, R. E. and Clark, L. M., 1974. "Finite Pulse Time Effect in Flash Diffusivity Method," *High Temperature - High Pressure*, Vol. 2:65-72

# COATINGS AND FILMS

SESSION CHAIRS
Th. Franke
C. J. Cremers

# Theoretical Aspects of the Thermal Conductivity in Thin Films

F. VOELKLEIN and TH. FRANKE

## ABSTRACT

We discuss theoretical models for the analysis of the thermal conductivity in thin films. Crucial features of thin films in regard of the thermal transport properties are their polycrystalline structure and small film thickness. These imperfections lead to additional charge carrier and phonon scattering processes, which are not present in monocrystalline bulk materials. We analyze the influence of grain boundary and surface scattering of electrons and phonons on the thermal conductivity. For thin metal films the theoretical validity of the Wiedemann-Franz law is discussed in the framework of the Fuchs-Sondheimer and Mayadas-Shatzkes scattering models. Experimental results of thin Aluminium films are compared with the theoretical investigations. For two-band semimetal films like Bismuth and Antimony we demonstrate, how the several components of the thermal conductivity can be determined. The dependence of the charge carrier contribution, the bipolar and the lattice thermal conductivity on thickness and grain size is analyzed. A model of the lattice thermal conductivity of polycrystalline films based on phonon grain boundary and surface scattering is presented and compared with experimental results of Antimony and Bismuth films.

## INTRODUCTION

The investigation of the thermal conductivity of thin films is of great interest both, for the understanding of the charge carrier and phonon transport mechanisms and for technical application of thin films (e. g. in thin film devices such as sensors). For the interpretation of the electrical resistivity of thin polycrystalline films the Fuchs-Sondheimer (FS) [1-3] and the Mayadas-Shatzkes (MS) model [4] is well established. The aim now is a theoretical description of the thermal conductivity of thin films based on the FS and MS models of surface and grain boundary scattering. Theoretical results for thin metal and bipolar semimetal films are presented and compared with experimental investigations of Aluminium, Bismuth and Antimony films.

F. Voelklein, Technical College, Department of Technical Physics, Am Brueckweg 26, D-65428 Ruesselsheim, Germany
Th. Franke, Technical University Chemnitz-Zwickau, Institute of Physics, D-09107 Chemnitz, Germany

## MODEL OF CHARGE CARRIER AND PHONON TRANSPORT IN THIN POLYCRYSTALLINE FILMS

### DESCRIPTION OF THE MODEL

Charge carrier and phonon scattering occurs within the crystallites and at the film surfaces and grain boundaries. At the film surface partial diffuse reflection (scattering parameter $0 \leq p_e \leq 1$ as fraction of the specularly reflected carriers) is assumed. According to the MS model, the grain boundaries are considered as planes, oriented perpendicular to the direction of the external fields ($E_x$, $\partial T/\partial x$) and described by $\delta$-shaped scattering potentials of strength S at the positions $x_n$, which are distributed according to a Gaussian (see Figure 1).

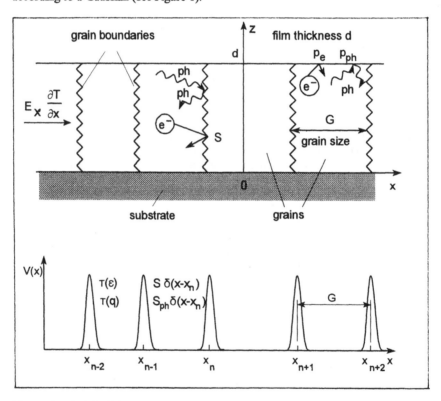

Figure 1 : Model of charge carrier and phonon transport in thin polycrystalline films

For the crystallites the properties of the bulk material and its band structure are assumed. The scattering processes within the grains (background scattering) can be described by relaxation times $\tau(\varepsilon)$ and $\tau(q)$, where $\varepsilon$ is the charge carrier energy and q is the phonon (lattice) mode. There is no preferential orientation of the films in the x-y-plane. Therefore, the x- and y-axes of the coordinate space are arranged arbitrarily. The external fields shall be oriented parallel to the x-axis.

## BOLTZMANN EQUATION, CURRENT DENSITY, HEAT-FLUX DENSITY AND GENERAL TRANSPORT COEFFICIENTS

The calculation of the distribution function for electrons (or holes)

$$f(r, k, t) = f_0 + f_1 \quad, \tag{1}$$

where $f_0$ is the equilibrium distribution function, given by the Fermi-Dirac distribution and $f_1$ is a small deviation, starts from the Boltzmann equation.

There shall be electrical fields $E_x$ and thermal gradients $\partial T/\partial x$ within the film:

$$v_x \frac{\partial f_0}{\partial x} + v_z \frac{\partial f_1}{\partial z} - eE_x \frac{\partial f_0}{\partial k_x} = -\frac{f_1}{\tau(\varepsilon)} - \int_{V_k} P(k, k') \left(f_1(k) - f_1(k')\right) d^3 k' \tag{2}$$

The term $v_z \, \partial f_1 / \partial z$ is due to the scattering at the film surfaces and P (k, k') is the transition probability for scattering by grain boundaries. It is calculated by means of the perturbation theory, considering the scattering potential of the grain boundaries V(x) as perturbation and averaging over a Gaussian distribution of the grain boundary distances. Thus,

$$P(k, k') = \frac{m S^2 (1 - e^{-4k_x^2 s^2}) \delta(k_y - k_y') \delta(k_z - k_z') \delta(k_x + k_x')}{\hbar^3 |k_x| G (1 + e^{-4k_x^2 s^2} - 2e^{-2k_x^2 s^2} \cos 2k_x G)} \tag{3}$$

is obtained, where G is the average grain-boundary distance, s is the standard deviation and m is the effective mass. With F(r) as the Fermi energy and assuming FS boundary conditions [1-3] of the film surfaces the distribution functions

$$f_1^+ = \tau^*(\varepsilon) v_x \left( -\frac{\partial f_0}{\partial \varepsilon} \right) \left[ (\mp e) E_x - \frac{(\varepsilon - F)}{T} \frac{\partial T}{\partial x} \right] \left( 1 - \frac{(1 - p_e) e^{-z/\tau^*(\varepsilon) v_z}}{1 - p_e e^{-d/\tau^*(\varepsilon) v_z}} \right), (v_z > 0)$$

$$f_1^- = \tau^*(\varepsilon) v_x \left( -\frac{\partial f_0}{\partial \varepsilon} \right) \left[ (\mp e) E_x - \frac{(\varepsilon - F)}{T} \frac{\partial T}{\partial x} \right] \left( 1 - \frac{(1 - p_e) e^{(d - z)/\tau^*(\varepsilon) v_z}}{1 - p_e e^{d/\tau^*(\varepsilon) v_z}} \right), (v_z < 0) \tag{4}$$

are obtained with -e for electrons, e for holes, and the relaxation time with

$$\frac{1}{\tau^*(\varepsilon)} = \frac{1}{\tau(\varepsilon)} + \frac{2m S^2 (1 - e^{-4k_x^2 s^2})}{\hbar^3 |k_x| G (1 + e^{-4k_x^2 s^2} - 2e^{-2k_x^2 s^2} \cos 2k_x G)} \tag{5}$$

The current and heat-flux densities in x-direction j and w, respectively, averaged over the film thickness, are calculated with (4):

$$j = -\frac{1}{4\pi^3 d} \int\limits_0^d \int\limits_{V_k} e \, v_x \, f_1 \, d^3k \, dz \tag{6}$$

$$w = -\frac{1}{4\pi^3 d} \int\limits_0^d \int\limits_{V_k} (\varepsilon - F) \, v_x \, f_1 \, d^3k \, dz \tag{7}$$

Thus, the general transport coefficients $K_r$ (d, G), which are functions of the film thickness d and the grain size G, are obtained :

$$j = e^2 K_0(d, G) \, E_x + \frac{e}{T} \, K_1(d, G) \, \frac{\partial T}{\partial x} \tag{8}$$

$$w = -e K_1(d, G) \, E_x + \frac{1}{T} \, K_2(d, G) \, \frac{\partial T}{\partial x} \tag{9}$$

with

$$K_r(d,G) = \frac{1}{4\pi^3 d} \int\limits_0^d \int\limits_{V_k} \varepsilon^r \tau^*(\varepsilon) v_x^2 \left(-\frac{\partial f_0}{\partial \varepsilon}\right) \left[ \left(1 - \frac{(1-p_e)e^{-z/\tau^* v_z}}{1 - p_e e^{-d/\tau^* v_z}}\right)_{v_z > 0} \right.$$
$$\left. + \left(1 - \frac{(1-p_e)e^{(d-z)/\tau^* v_z}}{1 - p_e e^{d/\tau^* v_z}}\right)_{v_z < 0} \right] d^3k \, dz \tag{10}$$

where r = 0, 1, 2 .

## THERMAL CONDUCTIVITY IN POLYCRYSTALLINE METAL FILMS

Presumptions of the MS model and of our calculations of the thermal conductivity of metal films are:
i)  fully degenerate electron gas :  $-\partial f_0 / \partial \varepsilon = \delta(\varepsilon - F)$
ii) spherical Fermi surfaces (isotropy of the transport coefficients)

Using equs. (8) and (10) we calculate the electrical conductivity

$$\sigma_e(d, G) = e^2 K_0(d, G) = \sigma_{e, bulk} \, g_e(d, G), \tag{11}$$

where $\sigma_{e, bulk}$ is the bulk electrical conductivity and the function $g_e$ (d, G) depends on film thickness, grain size and temperature. An useful approximation of this function is

$$g_e \ (d,G) \approx \left( 1 + \frac{3(1-p)l}{8d} + \gamma(S) \ \frac{l}{G} \right)^{-1} , \qquad (12)$$

where l is the mean free path of the carriers (in the bulk material) and
$\gamma(S) = 3m \ S^2 / 2 \ \hbar^2 F$.

The thermal conductivity $\lambda_e$ of the electrons can also be deduced from equs.
(8) - (10).

$$\lambda_e \ (d,G) = K_2 \ (d,G) \ / \ T = \lambda_{e,bulk} \ g_e \ (d,G) \qquad (13)$$

Analogous to the electrical conductivity $\lambda_e$ is given by the bulk value and the same function $g_e$ (d, G). Therefore, the quotient of the thin film thermal and electrical conductivities is equal to the quotient of the bulk :

$$\frac{\lambda_e \ (d,G)}{\sigma_e \ (d,G)} = \frac{\lambda_{e,bulk}}{\sigma_{e,bulk}} = L_0 \ T . \qquad (14)$$

This theoretical result demonstrates, that the Wiedemann-Franz law should be valid even for very thin polycrystalline metal films with the Lorenz-Number $L_0$ of the bulk material.

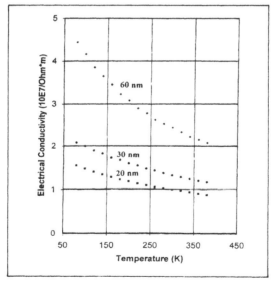

Figure 2 : Electrical conductivity of Aluminium films

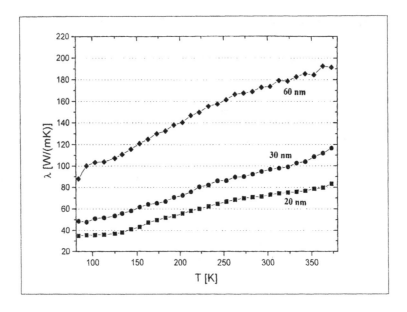

Figure 3 : Thermal conductivity of Aluminium films

An experimental verification of the theory is represented in Figures 2 and 3, which show results of the electrical and thermal conductivity measurements on thin Aluminium films, which were carried out on identical films immediately after the deposition in a vacuum chamber. The thermal conductivity was determined by a microelectromechanical sensor (MEMS) and with a method, described in [7, 8 ].

Table 1 shows the calculated Lorenz-Numbers for three film thicknesses at 300 K.

TABLE 1:  ELECTRICAL AND THERMAL CONDUCTIVITIES AND LORENZ-
                NUMBERS OF ALUMINIUM FILMS AT 300 K

| Thickness d [nm] | $\sigma_e$ [$10^7$/$\Omega$m] | $\lambda_e$ [ W/ mK] | $L = \lambda_e / (\sigma_e T)$ [ $10^{-8}$ $V^2$ / $K^2$ ] |
|---|---|---|---|
| 20 | 1,00 | 72 | 2,40 |
| 30 | 1,30 | 95 | 2,44 |
| 60 | 2,44 | 173 | 2,38 |

## THERMAL CONDUCTIVITY IN BIPOLAR SEMIMETAL FILMS WITH ANISOTROPIC TRANSPORT COEFFICIENTS

As an example of the thermal conductivity in bipolar semimetal films with anisotropic transport coefficients we discuss the properties of Bismuth and Antimony. Their electrical and thermal transport coefficients can satisfactorily be interpreted only

by means of a two-band model. At temperatures of 80 to 400 K Bismuth with its small band overlap is within the intermediate range between degeneracy ( $F > E_c + 5 k_B T$ and $F < E_v - 5 k_B T$ ) and non-degeneracy ( $F > E_c - k_B T$ and $F > E_v + k_B T$ ).
$E_c$ and $E_v$ are the edges of the conduction and valence band, respectively.

For Antimony, the bulk material has been analyzed in the framework of a partially degenerate two-band many-valley-model [5]. In this model with a nearly temperature-independent band overlap the Fermi level is pinned between the two band edges. Therefore, the presumption $(-\partial f_0/\partial \varepsilon) = \delta (\varepsilon - F)$ of the MS model is not justified.

Calculating the transport coefficients, the integration over the carrier energies $\varepsilon_i$ (with i = n, p for electrons and holes, respectively) has to be carried out with the specific energy derivatives of the distribution functions $(-\partial f_{0i}/\partial \varepsilon_i)$.

The band structure of Bismuth and Antimony crystallites can be described by elliptical Fermi surfaces (many-valley-model) [5,6]. The grains are generally columnar and oriented perpendicular to the substrate with their c-axis. This leads to an anisotropy of the transport coefficients.

According to this model the general transport coefficients can be deduced by the procedure demonstrated above. We obtain the current and heat-flux densities :

$$j_i = e^2 K_{0i}(d,G)E_x \mp \frac{eF_i}{T} K_{0i}(d,G) \frac{\partial T}{\partial x} \pm \frac{e}{T} K_{1i}(d,G) \frac{\partial T}{\partial x} \qquad (15)$$

$$w_i = \mp e K_{1i}(d,G)E_x + \frac{F_i}{T}K_{1i}(d,G)\frac{\partial T}{\partial x} - \frac{K_{2i}(d,G)}{T} \frac{\partial T}{\partial x} - \frac{E_i}{e} j_i \qquad (16)$$

with $E_i = E_c$ , $F_i = F - E_c$ for electrons and $E_i = E_v$ , $F_i = E_v - F$ for holes (the upper sign for i = n, the lower for i = p).

In the grains, we assume charge carrier scattering by acoustic phonons. Then, with the general transport coefficients equ. (10) we can calculate the results for the
i)  electrical conductivity :

$$\sigma = e^2 (K_{0n}(d,G) + K_{0p}(d,G)) = en\mu_n \, g_{0n}(d,G) + ep\mu_p g_{0p}(d,G) = \sigma_n + \sigma_p \qquad (17)$$

ii)  Seebeck coefficient :

$$\alpha = ( \sigma_n \alpha_n + \sigma_p \alpha_p ) / \sigma \qquad (18)$$

$$\alpha_i = (\mp)\frac{1}{eT} \left( \frac{K_{1i}(d,G)}{K_{0i}(d,G)} - F_i \right) = (\mp)\frac{k_B}{e} \left[ \frac{2\mathscr{F}_1(I_i)}{\mathscr{F}_0(I_i)} \frac{g_{1i}(d,G)}{g_{0i}(d,G)} - I_i \right] \qquad (19)$$

iii)　thermal conductivity (charge carrier contribution) :

$$\lambda_e = L_n \, \sigma_n \, T + L_p \, \sigma_p \, T + (\alpha_p - \alpha_n)^2 \, \sigma_n \, \sigma_p \, T / \sigma \qquad (20)$$
$$= \lambda_n \quad + \quad \lambda_p \quad + \quad \lambda_{bi}$$

$$L_i = \left( \frac{k_B}{e} \right)^2 \frac{3 \mathscr{F}_2(I_i) \mathscr{F}_0(I_i) g_{2i}(d,G) g_{0i}(d,G) - [2\mathscr{F}_1(I_i) g_{1i}(d,G)]^2}{[\mathscr{F}_0(I_i) g_{0i}(d,G)]^2} \qquad (21)$$

Here, $\mu_n$ and $\mu_p$ are the carrier mobilities in the bulk material, n and p are the electron and hole concentration, respectively, and $I_i = F_i / k_B T$. $\mathscr{F}_r(I_i)$ are the well known Fermi integrals with r = 0, 1, 2.

The functions $g_{ri}(d, G)$ are not identical for electrons and holes, respectively and take into consideration the anisotropy of the Fermi surfaces. These functions can be approximated by

$$g_{ri}(d,G) \approx \frac{\int_0^\infty \left( -\frac{\partial f_{0i}}{\partial \varepsilon_i} \right) \varepsilon_i^{r+3/2} \, \tau_i(\varepsilon_i) \left( 1 + \frac{3}{2} \gamma_i(S_i) + \frac{3(1-p_i)}{8K_i(\varepsilon_i)} \right)^{-1} d\varepsilon_i}{\int_0^\infty \left( -\frac{\partial f_{1i}}{\partial \varepsilon_i} \right) \varepsilon_i^{r+3/2} \, \tau_i(\varepsilon_i) \, d\varepsilon_i} \qquad (22)$$

where 
$$\gamma_i(S_i) = \frac{\sqrt{2m_i} \, S_i^2 \, \tau_i(\varepsilon_i)}{\hbar^2 \, G \, \sqrt{\varepsilon_i}} \quad , \qquad K_i(\varepsilon_i) = \frac{d \, \sqrt{m_{3i}}}{\tau_i(\varepsilon_i) \, \sqrt{2\varepsilon_i}} \qquad \text{and}$$

$$\frac{1}{m_i} = \frac{1}{2} \left( \frac{1}{m_{1i}} + \frac{1}{m_{2i}} \right) \; .$$

$(m_{1i})^{-1}$, $(m_{2i})^{-1}$ are the axes of the effective mass tensor in the x-y-plane, $(m_{3i})^{-1}$ is the axis perpendicularly to this plane. $L_i$ are the Lorenz numbers for the two bands.

We analyze the charge carrier thermal conductivity $\lambda_e$ with the help of the equs. (17) to (21). The quantitative analysis requires the knowledge of the partial conductivities $\sigma_n$, $\sigma_p$ and of the partial Seebeck coefficients $\alpha_n$, $\alpha_p$. However, measured quantities are the total electrical conductivity $\sigma$, the Seebeck coefficient $\alpha$ and the total thermal conductivity $\lambda$. With these quantities, $\sigma_n, \sigma_p, \alpha_n$ and $\alpha_p$ cannot be determined. However, the theoretical analysis of the functions $g_{ri}(d,G)$ show, that the quotients $g_{1i}(d,G) / g_{0i}(d,G)$ and $g_{2i}(d,G) / g_{0i}(d,G)$ can be approximated by unity for film thicknesses larger than 30 nm. Therefore, we can use for $\alpha_n$, $\alpha_p$ and $L_n$, $L_p$ the bulk values. With $\alpha_{i,bulk}$ and the measuring results of $\sigma$ and $\alpha$, the partial conductivities $\sigma_n$, $\sigma_p$ are determined. Then, using equ. (20) the charge carrier thermal conductivity $\lambda_e$ is deduced by the calculation of the Wiedemann-Franz contribution $\lambda_n$, $\lambda_p$ and the bipolar term $\lambda_{bi}$. With the total thermal conductivity $\lambda$, the lattice thermal conductivity $\lambda_{ph}$ (phonon contribution) is determined by $\lambda_{ph} = \lambda - \lambda_e$. Applying this procedure, the thermal conductivities of Bismuth and Antimony were studied for film thicknesses ranging from 30 to 400 nm in the temperature region 80-400 K. The results of measurements of the electrical conductivity and the Seebeck coefficient for Antimony films are shown in Figures 4 and 5 and published in [9].

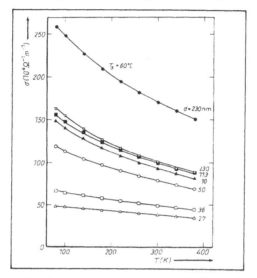

Figure 4:  Electrical conductivity σ versus temperature T for Antimony films of several thicknesses d, prepared at a substrate temperature $T_s$=25°C and for a film with the thickness d=230 nm and $T_s$=60°C

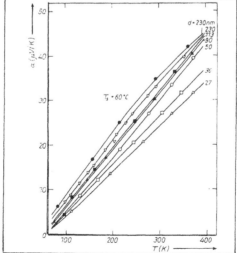

Figure 5:  Seebeck coefficient α versus temperature for Antimony films of several thicknesses d, prepared at a substrate temperature $T_s$=25°C and for a film with the thickness d = 230 nm and $T_s$ = 60°C

Figure 6 demonstrates the thermal conductivity of an Antimony film with a thickness d=230 nm and the results of the above mentioned procedure. The calculated charge carrier contribution $\lambda_e$ is represented as a function of temperature. The bipolar thermal conductivity $\lambda_{bi}$ (as a component of $\lambda_e$) is a crucial part of $\lambda_e$ at room temperature and for higher temperatures.

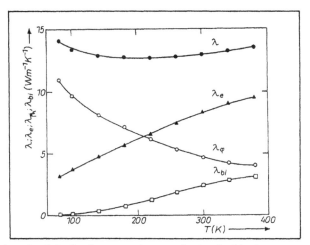

Figure 6:    Thermal conductivity $\lambda$, charge carrier contribution $\lambda_e$, bipolar component $\lambda_{bi}$ and lattice thermal conductivity $\lambda_{ph}$ versus temperature T for an Antimony film with thickness d = 230 nm

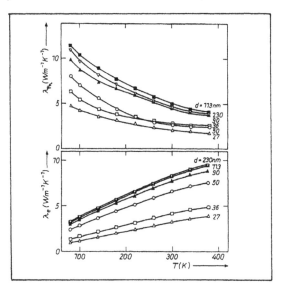

Figure 7:    Lattice thermal conductivity $\lambda_{ph}$ and charge carrier contribution $\lambda_e$ of Antimony films with various film thicknesses d as a function of temperature T

Figure 7 shows the results of our measurements and analysis for Antimony films with various film thicknesses.

Results for Bismuth films are represented in Figures 8 and 9. Here, Figure 8 shows the thermal conductivity with film thicknesses ranging from 28 nm to 350 nm.

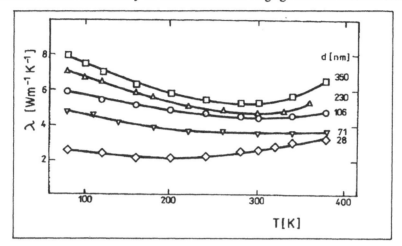

Figure 8:   Thermal conductivity of Bismuth films as a function of temperature T

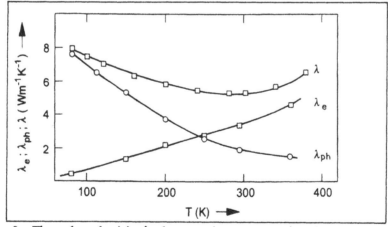

Figure 9:   Thermal conductivity $\lambda$, charge carrier component $\lambda_e$ and lattice thermal conductivity $\lambda_{ph}$ of a Bismuth film with the film thickness d=350 nm

In Figure 9 the charge carrier component $\lambda_e$ and the lattice thermal conductivity $\lambda_{ph}$ are represented together with the total thermal conductivity $\lambda$ of the film with d=350 nm. The components $\lambda_e$ and $\lambda_{ph}$ were calculated according to the above mentioned procedure. The results for the electrical conductivity $\sigma$ and the Seebeck coefficient $\alpha$ of these films were published in [10].

The lattice thermal conductivity of Antimony films and Bismuth films as a function of film thickness is given in Figures 10 and 11, respectively.

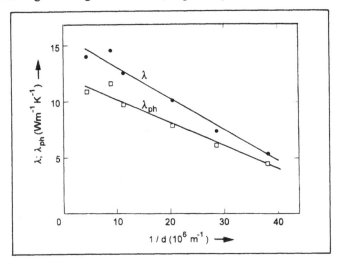

Figure 10: Thermal conductivity $\lambda$ and lattice thermal conductivity $\lambda_{ph}$ of Antimony films as a function of film thickness d at T=80 K

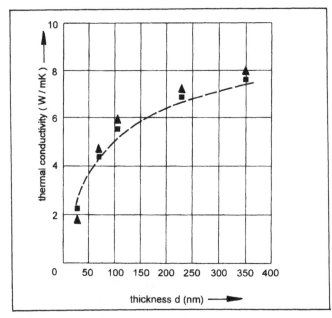

Figure 11: Lattice thermal conductivity of Bismuth films as a function of film thickness d at T=80 K (triangles: experimental; squares and dashed line: theoretical)

The question arose, how these dependences on film thickness can be deduced theoretically. Therefore, in the next section we present a model, which describes the lattice thermal conductivity in thin polycrystalline films by means of phonon scattering on surfaces and grain boundaries.

## THEORY OF THE LATTICE THERMAL CONDUCTIVITY IN THIN POLY-CRYSTALLINE FILMS

We consider the lattice thermal conduction as a transport of phonons (analogous to the charge carrier transport in polycrystalline films according to Figure 1). Appart from the background scattering of phonons within the crystallites (e. g. by phonon-phonon umklapp scattering) the phonons are scattered by the film surfaces and grain boundaries. Partially specular reflection is possible at the surfaces, where the probability $p_{ph}$, that a phonon will be specularly reflected is in the range $0 < p_{ph} < 1$. The grain boundaries, considered as planes perpendicular to the direction of the phonon transport, are represented by $\delta$-potentials $V(x) = S_{ph} \Sigma_n \delta(x-x_n)$.

Arranging the coordinate system as shown in Figure 1 and with the temperature gradient only in the x direction (i. e. any direction of the film plane) we obtain the Boltzmann equation for the calculation of the phonon occupation number $N(q, r, t)$:

$$v_z \frac{\partial n(q)}{\partial z} + v_x \frac{\partial N_0}{\partial T} \frac{\partial T}{\partial x} = \frac{n(q)}{\tau_q(\omega)} - \int_{v_q} P(q,q')\{n(q) - n(q')\} dq' \qquad (23)$$

where

$$n(q) = N(q,r,t) - N_0(q) \qquad (24)$$

q is the lattice mode and $N_0(q) = \{\exp[h\omega(q)/k_B T]-1\}^{-1}$ is the equilibrium occupation number. The background scattering processes within the crystallites are described by the isotropic relaxation time $\tau_q(\omega)$.

The term $v_z \, \partial n(q)/\partial z$ of equ. (23) takes the scattering effect of the surfaces into account; the integral describes the grain boundary scattering, where $P(q,q')$ is the transition probability for a phonon in state q to be scattered to q' by the grain boundaries. $P(q,q')$ is calculated using perturbation theory, in which the grain boundary scattering potential $V(x)$ is considered as a perturbation on the hamiltonian and the square of the matrix element $|\langle q|V(x)|q'\rangle|^2$ is averaged over a Gaussian distribution of the grain boundary distances.

The solution of equ. (23) with regard to the FS boundary conditions [1-3] can be applied to the calculation of the phonon heat flux density

$$w_{ph}(d, G) = \frac{1}{(2\pi)^3 d} \int_0^d \int_{v_q} n(q) \hbar w v_x d^3q \, dz = -\lambda_{ph}(d, G) \frac{\partial T}{\partial x} \qquad (25)$$

We obtain the lattice thermal conductivity $\lambda_{ph} = \lambda_{ph,bulk} \, g_{ph}(d,G)$, where the first factor is the lattice thermal conductivity of the bulk material without grain boundaries, given by the following equation:

$$\lambda_{ph,bulk} = \frac{k_B}{6\pi^2 v} \left( \frac{k_B T}{\hbar} \right)^3 \int_0^{\Theta/T} \frac{x^4 e^x}{(e^x - 1)^2} \tau_q(x) \, dx \qquad (26)$$

The function $g_{ph}(d,G)$ describes the dependence of $\lambda_{ph}(d,G)$ on the thickness d, the grain size G and the temperature T:

$$g_{ph}(d,G) \approx \frac{\int_0^{\Theta/T} \frac{x^4 e^x}{(e^x - 1)^2} \tau_q(x) \left\{ 1 + \frac{3}{2}\gamma + \frac{3(1-p_{ph})}{8K} \right\}^{-1} dx}{\int_0^{\Theta/T} \frac{x^4 e^x}{(e^x - 1)^2} \tau_q(x) dx} \qquad (27)$$

with $\qquad \gamma = \frac{2 S_{ph}^2 \tau_q(x)}{\hbar^2 v G}$ , $\quad x = \frac{\hbar \omega}{k_B T}$ , $\quad K = \frac{d}{\tau_q(x) v}$ .

Here v is the sound velocity, $\Theta$ is the Debye temperature and $p_{ph}$ is the parameter of surface scattering. If there are several phonon branches with various velocities of sound, which contribute to the total lattice thermal conductivity, we have to add the contribution of each branch.

## COMPARISON OF THEORETICAL AND EXPERIMENTAL RESULTS

Phonon-phonon umklapp scattering is the essential phonon scattering mechanism in bulk Bismuth and Antimony in the temperature range 80-400 K [11,12]. Klemens [13] derived the relaxation time for this case $\tau_q(\omega) = \tau_{u0}/k_B T \omega^2$. Applying it to equ. (27) we obtain

$$g_{ph}(d,G) \approx \int_0^{\Theta/T} \frac{x^2 e^x}{(e^x - 1)^2} \left( 1 + \frac{\gamma_1}{x^2} + \frac{\gamma_2}{x^2} \right)^{-1} dx \, \bigg/ \int_0^{\Theta/T} \frac{x^2 e^x}{(e^x - 1)^2} dx \qquad (28)$$

where $\qquad \gamma_1 = \frac{3 S_{ph}^2 \tau_{u0}}{v G (k_B T)^3}$ , $\quad \gamma_2 = \frac{3(1-p_{ph}) v \tau_{u0} \hbar^2}{8 d (k_B T)^3}$

Experimental results for $\lambda_{ph}$ (d,G) of Bismuth films are represented in Figure 11 together with the fitted theoretical curve $\lambda_{ph,bulk}$ $g_{ph}$ (d,G). For $\lambda_{ph,bulk}$ we used the values of [11,14]. The grain size G of Bismuth films is a function of film thickness d. G was determined experimentally, the results were reported in [15]. Fitting the experimental values of Figure 11 with the function $g_{ph}$ (d,G) and with $p_{ph}$ = 0 we obtain the parameters $S_{ph}$ = 2,5·$10^{-32}$ Wsm and $\tau_{u0}$ = $10^{-5}$ W. With $\tau_{u0}$ the relaxation time at the Debye frequency $\tau_q(\omega_D)$ = 4,1·$10^{-11}$s for T=80K and in the same way $\tau_q(\omega_D)$ = 1,1·$10^{-11}$ s for T=300K are calculated. The experimental results for Antimony films can also be interpreted with this model. A complete interpretation is published in [9].

## CONCLUSIONS

The proposed models of surface and grain boundary scattering of charge carriers and phonons seems to be useful for the interpretation of the charge carrier and lattice thermal conductivity in thin polycrystalline films. The Wiedemann-Franz law, which was deduced theoretically for thin metal films, has been verified experimentally for thin Aluminium films.

The models were used succesfully for the analysis of the thermal conductivity in thin bipolar semimetal films. With the phonon scattering model, the investigation of the lattice thermal conductivity as a function of thickness enables the calculation of the mean free path of phonons analogously to the determination of the electron mean free path by the analysis of the electrical resistivity of thin films.

## REFERENCES

1. Fuchs, K. 1938, The Conductivity of Thin Metallic Films According to the Electron Theory of Metals, Proc. Cambridge Phil. Soc. 34, 100
2. Sondheimer, E. H. 1950, The Influence of a Transverse Magnetic Field on the Conductivity of Thin Metallic Films, Phys. Rev. 80, 401
3. Sondheimer, E. H. 1952, The Mean Free Path of Electrons in Metals, Adv. Phys. 1,1
4. Mayadas, A. F., M. Shatzkes 1970, Electrical-Resistivity Model for Polycrystalline Films: The Case of Arbitrary Reflection at External Surfaces, Phys. Rev. B 1, 1382
5. G. A. Saunders, Oektu, Oe., 1968, The Seebeck coefficient and the Fermi Surface of Antimomy Single Crystals, J. Phys. Chem. Solids 29, 327
6. Abeles, B., S. Meiboom 1956, Galvanomagnetic Effects in Bismuth, Phys. Rev. 101, 544
7. Voelklein, F. 1990, Thermal Conductivity and Diffusivity of a Thin Film SiO$_2$-Si$_3$N$_4$ Sandwich System; Electronic and Optics, Thin solid films 188, 27
8. Voelklein, F., T. Staerz, H. Schmidt 1995, Microsensor for In Situ Thermal Conductivity Measurements of Thin Films, Sensors and materials 7, 6, 395
9. Voelklein, F., E. Kessler 1990, Thermal Conductivity and Thermoelectric Figure of Merit of Thin Antimony Films, phys. stat. sol. (b) 158, 521
10. Voelklein, F., E. Kessler 1986, Analysis of the Lattice Thermal Conductivity of

Thin Films by means of a modified Mayadas-Shatzkes Model: The Case of Bismuth Films,Thin solid films 142, 169

11. Gallo, C. F., B. S. Chandrasekhar, P. H. Sutter 1963, Transport Properties of Bismuth Single Crystals, J. Appl. Phys. 34, 144

12. Uher, C., H. J. Goldsmid 1974, Separation of the Electronic and Lattice Thermal, phys. stat. sol. (b) 65, 765

13. Klemens, P. G., in R. P. Tye (ed.) 1969, Theory of the Thermal Conductivity of Solids, Acad. Press London 1,2

14. Ertl, M. E. 1966, Thermal Resistivity due to Mass Fluctuation Scattering in Bismuth-Antimony Alloys, phys. stat. sol. 17, K 63

15. Voelklein, F., E. Kessler 1986, Temperature an Thickness Dependence of Electrical and Thermal Transport Coefficients of $Bi_{1-x}$ $Sb_x$ films in an anisotropic, non-degenerate two-band model, phys. stat. sol. (b) 134, 351

# Thermal Conductivity of Thin Dielectric Films

S.-M. LEE and D. G. CAHILL

## ABSTRACT

The $3\omega$ method is used for the thermal conductivity measurement of ultra thin dielectric films of $SiO_2$ and MgO. The films are deposited on a small opening of Si substrate covered with 200 nm thick insulation layer. This method greatly reduced the probability of getting short-circuited between the heater and Si substrate. With this improvement, the $3\omega$ method has been used to measure the thermal conductivity of 20 - 300 nm thick $SiO_2$ and MgO films deposited on Si substrate by plasma enhanced chemical vapor deposition and reactive sputtering method, respectively. Our results are consistent with the several previous reports obtained by different methods. The thermal resistance at the film interfaces is $\sim 2 \times 10^{-8}$ $m^2 K W^{-1}$ for these films. We suggest that the thermal conductivity of these films is intrinsically thickness-independent for the measured film thickness range.

## INTRODUCTION

It is well established fact that the thermal conductivity of amorphous solids is close to its minimal value calculated based on the Debye model [1]. It is also shown that the thermal conductivity of micron thick oxide films can be explained within the minimal thermal conductivity concept [2]. However, there have been many reports showing anomalous thickness dependence in the thermal conductivity of amorphous $SiO_2$ and $SiN_x$ films [3]. Since the density and the acoustic property of the films would show no serious thickness dependence in the investigated thickness range, the observations are inconsistent with the minimal thermal conductivity concept. Although the effect has been explained assuming unknown thermal boundary resistance at the film interfaces, due to large discrepancy among the measured data, the origin of the extra thermal resistance is not well understood.

Department of Materials Science and Engineering, Coordinated Science Laboratory, University of Illinois, 1101 W. Springfield, Urbana, IL 61801

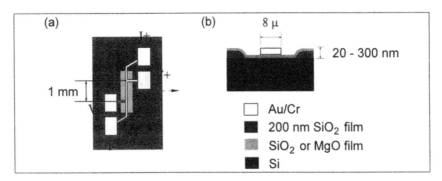

**Figure 1**. (a) Heater pattern. (b) Cross sectional view of the sample geometry.

## EXPERIMENTAL

The measured $SiO_2$ films were deposited by PECVD(plasma enhanced chemical vapor deposition) method and MgO films were deposited by reactive DC magnetron sputtering.

We used the $3\omega$ method which has been used for the thermal conductivity measurement of various insulating materials. With extended analysis of the sample geometry, the method also has been successfully applied to dielectric thin films[2]. However, for films deposited on metal substrate, the practical application of this method has been limited by the electrical insulation problem caused by the pinholes of the dielectric film. In this work, we solved the problem by double deposition of the dielectric films.

The sample geometry is shown in Fig.1. First, a small rectangular window was made by photolithography and wet etching of a 200 nm thick $SiO_2$ film. Then 20 - 300 nm thick films were deposited on the windowed substrates. The double deposition greatly improved the insulation of the heater pattern from the substrate. The resistance between the heater and the substrate was larger than 40 M$\Omega$.

The film thermal conductivity($\Lambda_{film}$) of $SiO_2$ and MgO film is much smaller than the thermal conductivity of $Si(\Lambda_{Si})$ and the heater width(w) is much larger than the film thickness(d). With these conditions, the temperature oscillation amplitude, $\Delta T$, at the heater is a simple superposition of the substrate thermal response($\Delta T_{Si}$) and the film thermal response($\Delta T_{film}$)[4]:

$$\Delta T = \Delta T_{Si} + \Delta T_{film}$$
$$= \frac{P}{\pi \Lambda_{Si}} \left\{ -\frac{1}{2}\ln(2\omega) + \text{const} \right\} + \Delta T_{film} \quad , \qquad (1)$$

where P is the supplied ac power per unit length and

$$\Delta T_{film} = \frac{P}{\Lambda_{film}} \frac{d}{w} \quad . \qquad (2)$$

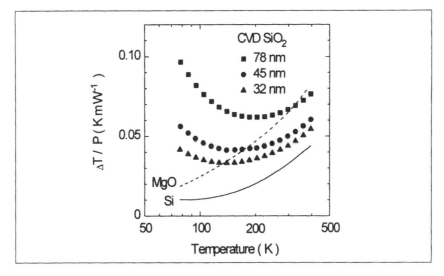

**Figure 2**. Measured temperature oscillation of the heater normalized by P - supplied power per unit length. The heating frequency was 1 kHz.

$\Delta T$ was measured in the heating frequency range of 20 Hz - 20 kHz. In most cases, $\Delta T_{film}$ was calculated around the heating frequency of 1 kHz. Typical data of $\Delta T$ divided by P are shown in Fig. 2 as a function of temperature.

As the film thickness decreases, the thermal response is dominated by the substrate. In Fig. 2 we also plotted the thermal response of MgO crystal substrate, which is commonly used as insulating substrate. The figure shows that Si substrate gives smaller temperature offset, which is better for the thermal conductivity measurement of very thin films.

### RESULTS AND DISCUSSION

The measured thermal conductivity data of $SiO_2$ and MgO films are shown in Fig. 3 and Fig. 4, respectively. The low density of the films compared to bulk is responsible for low film thermal conductivity. For $SiO_2$ films, the results are in good agreement with recent report by Goodson[3], Käding[5] and Swartz[6]. The values are much higher than other reports[7,8]. All the films show decreasing thermal conductivity as the film thickness gets smaller. The tendency is more pronounced for MgO film whose micron thick film has four times larger thermal conductivity than $SiO_2$.

To calculate the film thermal conductivity from the measured $\Delta T$, we have assumed that the most of the temperature shift, $\Delta T_{film}$, is produced by the dielectric film. For ultrathin films, however, extra thermal resistance is relatively significant.

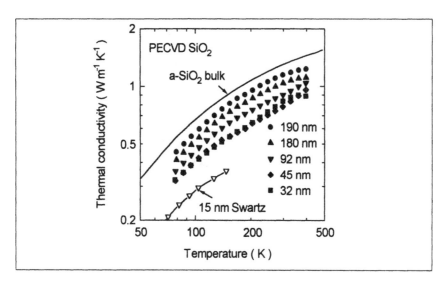

**Figure 3**. Thermal conductivity of PECVD SiO$_2$ films

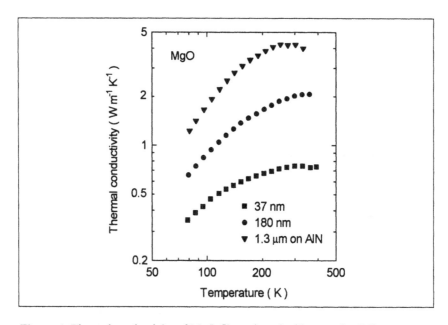

**Figure 4**. Thermal conductivity of MgO films deposited by reactive DC magnetron sputtering method.

In view of the extra thermal resistance, we list several origins for the observed thickness dependence of the measured film thermal conductivity as follows.

As previously noted, any intrinsic thickness dependence of the physical property of film is unusual for the measured film thickness range. Instead, we may suggest that the physical property, e.g. density, is noticeably different at the early deposition due to the existence of film/substrate interface as well as the inequilibrium of deposition parameters. The roughness layer at the film surface also can be considered effectively as intermediate film layer. The structural difference between the film and the substrate is another obvious origin of the thermal resistance, which has been understood in view of acoustic or diffuse mismatch models.[6] These two factors can be combined as extra thermal resistance at the film/substrate interface.

The interface between the sample film and the heater film is also an evident source of extra thermal resistance. According to the measurement of Swartz[6], the thermal boundary resistance between metal film and dielectric substrate is about $10^{-8}$ m$^2$K/W. Since we used ~ 3 nm thick Cr layer to improve the adhesion of Au film, CrO$_x$ can be formed during the deposition or by reaction with the oxide film to give additional unknown thermal resistance. A 300 nm thick Au film also behaves as thermal resistance in our sample geometry. Assuming that Au film has bulk thermal conductivity, the effect can be calculated straightforward; ~ $2 \times 10^{-9}$ m$^2$W/K. Only few of the above origins can be estimated accurately while most of them have to be understood. Hence the measurement of thin film thermal conductivity provides valuable information for the understanding of the heat transport at the interface.

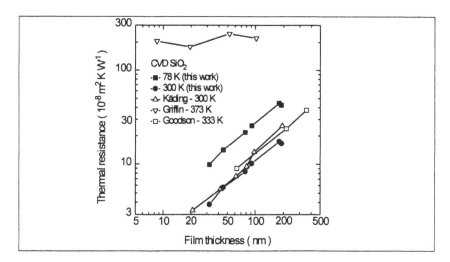

**Figure 5.** Apparent thermal resistance of SiO$_2$ film as a function of ilm thickness.

Practically, the listed thermal resistance may be different for each sample preparation and each film thickness. To get the estimation of extra thermal resistance, however, we analyzed the data assuming that the extra thermal resistance is constant for all the films we measured. Then the measured thermal conductivity is related to the thickness:

$$\frac{d}{\Lambda_{film}} = \frac{d}{\Lambda_{intrinsic}} + R_0 \;. \tag{3}$$

In Fig. 5, we plotted the apparent thermal resistance, $d/\Lambda_{film}$, of $SiO_2$ film as a function of the film thickness. At room temperature, the linear fit to Eq.(3) gave $R_0$ ~ $2\times10^{-8}$ $m^2W/K$. The fit to our data of CVD $SiN_x$ film, which are not presented here, also gave the same answer. The value is essentially the same as previously reported data with different measurement methods. We suppose that the above estimation is the lower limit for our metal-dielectric-substrate sample geometry.

## CONCLUSION

We used the $3\omega$ method to measure the thermal conductivity of thin $SiO_2$ and MgO films as a function of the film thickness. We estimated the extra thermal resistance at the film interfaces as $2\times10^{-8}$ $m^2W/K$, which is assumed to be close to the lower limit for our sample geometry. It is suggested that the thermal conductivity of these dielectric films should have no intrinsic thickness dependence.

## ACKNOWLEDGEMENTS

This work was supported by National Science Foundation Grant No. CTS-9421089. Part of this work was performed at the microelectronics facility of the Materials Research Laboratory, University of Illinois, which is supported by the U. S. Department of Energy under Grant No. DEFG02-91-ER45439.

## REFERENCES

1. David G. Cahill and R. O. Pohl, 1988. "Lattice vibrations and heat transport in crystals and glasses. " *Annual Review of Physical Chemistry*, 39:93-121.

2. S.-M. Lee and David G. Cahill, 1995. "Thermal conductivity of sputtered oxide films." *Physical Review B*, 52(1):253-257.

3. K. E. Goodson, M. I. Flik, L. T. Su, and D. A. Antoniadis, 1993. "Annealing-Temperature Dependence of the Thermal Conductivity of LPCVD Silicon-

Dioxide Layers." *IEEE Electron Device Letters*, 14(10):490-492.

4. David G. Cahill, M. Katiyar, and J. R. Abelson, 1994. "Thermal conductivity of *a*-Si:H thin films." *Physical Review B*, 50(9):6077-6081.

5. O. K. Käding, H. Skurk, and K. E. Goodson, 1994. "Thermal conduction in metallized silicon-dioxide layers on silicon." *Applied Physics Letter* 65(13):1629- 1631.

6. E. T. Swartz and R. O. Pohl, 1988. "Thermal resistance at interfaces." *Applied Physics Letter*, 51(26):2200-2203.

7. J. C. Lambropoulos, M. R. Jolly, C. A. Amsden, S. E. Gilman, M. J. Sinicropi, D. Diakomihalis, and S. D. Jacobs, 1989. "Thermal conductivity of dielectric thin films." *Journal of Applied Physics* 66(9):4230-4242.

8. A. J. Griffin, Jr., F. R. Brotzen, and P. J. Loos, 1994. "Effect of thickness on the transverse thermal conductivity of thin dielectric films." *Journal of Applied Physics* 75(8):3761-3764.

# Measurement of the Thermal Diffusivity of Diamond with a Modified Angström Method

H. ALTMANN,* O. NILSSON** and J. FRICKE*

## ABSTRACT

The measurements of the thermal diffusivities of several freestanding CVD-diamond films performed in the frame of a round robin test are presented in detail. The measurements were done at room temperature using a modified version of the dynamic Ångström method, developed for studies of freestanding films. It was found that thermal diffusivities and conductivities could be correlated with the densities and IR-transmission spectra of the different specimens. The discrepancies with the average thermal conductivity results of the round robin test could not be explained from the uncertainty of our method.

## INTRODUCTION

With the improvements in CVD-technology, diamond films produced by this technique has become technically interesting. One of the many interesting properties of diamond is its very high thermal conductivity $\lambda$. In order to test the reliability of different measurement techniques and laboratories a round robin test [1] was performed by the National Institute of Standards and Technology (NIST, Gaithersburg, MD, USA), in which we took part. The large discrepancies between the results revealed the difficulties of measuring the very high thermal conductivities ($\lambda \approx 500 - 1500\ W\ m^{-1}\ K^{-1}$) of diamond films.

We want to present our results on the diamond films in detail in this paper with a discussion of the possible errors. Since there were indications that the CVD-films have a conductivity gradient in the direction of growth, we also performed a numerical simulation on the heat transfer in a layer consisting of a material with a conductivity gradient.

*Physikalisches Institut der Universität Würzburg
Am Hubland, D - 97074 Würzburg, Germany

**Bavarian Center for Applied Energy Research e.V. (ZAE Bayern)
Am Hubland, D - 97074 Würzburg, Germany

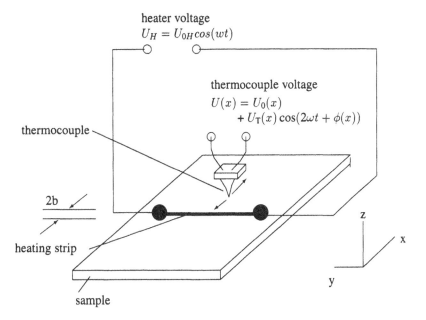

Figure 1: *Principle of the measurements with a modified version of the dynamic Ångström temperature wave method.*

## PRINCIPLE OF MEASUREMENT

Thermal diffusivities $a$ were measured at room temperature using a modified version of the dynamic Ångström temperature-wave method, similar to the method proposed by Hatta et al [2]. Figure 1 shows the principle of the measurement. A silver heating strip of width $2b = 0.2\ mm$ is deposited onto the surface of the specimen in an evaporation process. In the experiment an electric current $I_H(t)$ with angular frequency $\omega$ is fed into the metal strip by applying a heater voltage of $U_H(t) = U_{0H}cos(\omega t)$. The ohmic resistance R of the strip causes a Joule heating of power $P_H(t)$ with a periodicity of $2\omega$,

$$P_H(t) = U_H(t)I_H(t) = \frac{1}{2}\frac{U_{0H}^2}{R}(1 + \cos 2\omega t)\,, \qquad (1)$$

which causes a time-dependent temperature distribution in the sample. Since the thickness of the metal strip ($30 - 60\ nm$) is small compared to the sample thickness ($200 - 500\ \mu m$), the heat capacity of the strip and its influence on the lateral heat flow can be neglected. The temperature wave is probed with a movable micro-thermocouple ($25\ \mu m$) which can be moved along the surface of the sample in the x-direction (Figure 1). The generated thermoelectric voltage is then analyzed by a two-phase lock-in amplifier. Phase and amplitude of the $2\omega$ signal give information about the propagation of the thermal wave along the $x$-axis and thus the thermal diffusivity $a$ of the sample.

The general case can be described by the three-dimensional heat-flow equation [3]. In order to get an analytical solution two assumptions are necessary: The extension of the specimen in $x$ and $y$ directions is infinitely large and the thermal diffusion length $(a/\omega)^{1/2}$ is significantly larger than the thickness of the sample. Then the general problem can be reduced to the one-dimensional form of the heat flow equation

$$\rho c_{\mathrm{p}} \frac{\partial T(x,t)}{\partial t} - \lambda \frac{\partial^2 T(x,t)}{\partial x^2} = Q(x,t)\,. \tag{2}$$

where $\quad Q(x,t) \;=\; \begin{cases} Q_0 \cos^2 \omega t & for \;\; |x| \leq b \\ 0 & for \;\; |x| > b \end{cases}$,

$\qquad Q_0 \qquad:\quad$ amplitude of the absorbed heat per unit volume and time,

$\qquad T(x,t) \;:\quad$ temperature,

$\qquad \rho \qquad:\quad$ mass density,

$\qquad c_{\mathrm{p}} \qquad:\quad$ specific heat capacity.

Considering only the oscillating part of the solution for equation (2) the temperature for times large compared to a wave period can be written as

$$T(x,t) = T_0 \exp(-\sqrt{\frac{\omega}{a}}|x|) \cos(2\omega t - \sqrt{\frac{\omega}{a}}|x| + \phi) \; for \; |x| > b \tag{3}$$

By plotting the phase shift $[(\omega/a)^{1/2} + \phi]$ against the position $x$ the thermal diffusivity $a$ can be determined from the slope of the curve. In principle the diffusivity can also be derived from a logarithmic plot of the amplitude versus $x$. Since the amplitude strongly depends on the contact force of the thermocouple, however, only the data from the phase plot are taken into account.

## NUMERICAL SIMULATION WITH FINITE ELEMENTS

The analytical solution for the experimental setup is valid only for homogeneous isotropic materials. Since measurements were performed on CVD-diamond films, which seem to have a gradient in their thermal properties, numerical simulations were performed for such materials similar to the calculations performed by Fournier et al. [4]. For the calculations we used a commercial finite-elements-method program, ANSYS (Swanson Analysis Systems Inc., Houston, PA, USA). For the simulations the sample was considered to be infinitely large in the $y$-direction. In the $z$-direction the sample was divided into 16 equidistant layers with different thermal conductivities, while mass density and specific heat were still considered to be independent of $z$. The thickness of the sample $\Delta z$ was taken to be $400\,\mu m$. In the $x$-direction the sample had an overall length of $\Delta x = 10\,mm$. The following boundary conditions were chosen to match the experimental setup:

- Adiabatic conditions at $x = 0$ due to the symmetry of the problem.

- Isothermic coupling at $x = \Delta x/2$.

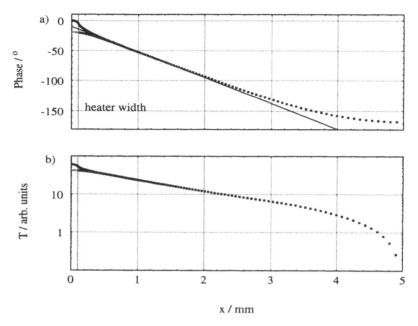

Figure 2: *Simulation of a homogeneous isotropic sample (dots). The solid line shows the one-dimensional analytical solution. Near the heater element the different data points mark results from the top and the bottom of the sample. a) Linear plot of phase versus position x, b) logarithmic plot of amplitude versus x.*

- Adiabatic conditions at $z = 0$ and $z = \Delta z$, i.e. no heat losses to the surroundings by radiation or convection were allowed for.

- At the top ( $z = 0$ ) and for $0 \leq x \leq b$ sinusoidally varying energy is supplied to the sample with a frequency $2\omega = 251 \ rad \ s^{-1}$. The thermal influence of the heater is neglected as in the analytical solution.

Mass density and specific heat were set to $\rho = 3510 \ kg \, m^{-3}$ and $c_p = 502 \ J \, kg^{-1} \, K^{-1}$, respectively. At first the simulations were performed with $\lambda = 403 \ W m^{-1} K^{-1}$ being independent of $z$; this value corresponds to a thermal penetration depth of 750 $\mu m$, which is about 2 times the sample thickness. Thus an influence of the sample thickness is expected to be observed near the heater element. Figure 2 shows the result of the simulation. The data were compared with the one-dimensional analytical solution for the phase. As one would expect, the one-dimensional analytical solution is in good agreement with the numerical data in a region between the end of the heater strip and the region around the heater element. Points near the end of the sample ($x = \Delta x/2$), however, show a significant deviation due to the isothermal coupling.

Then the thermal conductivity $\lambda$ was taken to vary linearly from 800 $W m^{-1} K^{-1}$ at the top of the sample to 1600 $W m^{-1} K^{-1}$ at the bottom (Figure 3). As can be seen the dependence of the phase on $x$ is still linear and the amplitude decays exponentially with $x$. Fitting the data with the one-dimensional analytical solution a value $a = 800 \ mm^2 \ s^{-1}$ is obtained resulting in a mean thermal conductivity of $\lambda = 1400 \ Wm^{-1}K^{-1}$.

The following conclusions can be obtained from the numerical simulations:

- For homogeneous materials the one-dimensional analytical solution gives satisfactory results for the thermal diffusivity if the boundary conditions mentioned above are fulfilled. However, for samples with small lengths $x$ the influence of the coupling at the boundary of the sample must be checked carefully.

- For materials having a constant gradient in the thermal conductivity along the $z$-axis, an effective thermal diffusivity is obtained which is not the arithmetical mean but is closer to the high conductivity value than to the low conductivity value.

- The gradient in $z$-direction is not detectable directly from the measurements with the modified Ångström thermal-wave method.

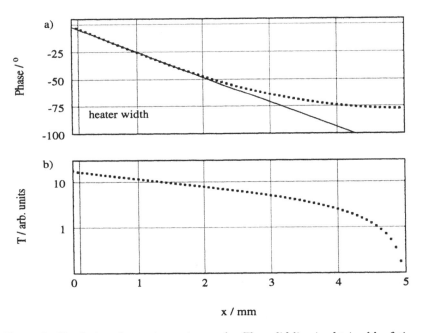

Figure 3: *Simulation of an anisotropic sample. The solid line is obtained by fitting the one-dimensional analytical solution for the linear part $x < 2 \ mm$. a) Linear plot of phase versus position $x$, b) logarithmic plot of amplitude versus $x$.*

## EXPERIMENTAL RESULTS

### THERMAL MEASUREMENTS

First calibration measurements were performed on vitreous silica, on polyimide films and on thin copper plates and the deviations from literature values were < 5% [5]. Those measurements as well as the following were performed under vacuum

TABLE I: *Specimen name, thickness $\Delta z$, mass density $\rho$, mean thermal diffusivity $a$, thermal diffusion length $(a/\omega)^{1/2}$, the determined thermal conductivities $\lambda$ and their mean values from the round robin test $\lambda_{RR}$ [1] for the CVD-diamond samples.*

| sample | $\Delta z$ [mm] | $\rho$ [kg/m³] | $a$ [mm²/s] | $(a/\omega)^{1/2}$ [mm] | $\lambda$ [$10^2 W/mK$] | $\lambda_{RR}$ [$10^2 W/mK$] |
|--------|-----|------|----------|----------|-----|---------------|
| SQ-X   | 0.31 | 3740 | 423 ± 10 | 1.4 ... 2.4 | 8.0 | 14.9 ± 35 % |
| SQ-F   | 0.32 | 3560 | 328 ± 9  | 1.4 ... 1.5 | 5.9 | 11.5 ± 30 % |
| SQ-T   | 0.40 | 3550 | 293 ± 6  | 1.4 ... 1.5 | 5.2 | 12.0 ± 31 % |
| SQ-F-B | 0.38 | 3340 | 247 ± 10 | 1.0 ... 1.6 | 4.1 | 7.6 ± 30 %  |
| SQ-E   | 0.31 | 3300 | 168 ± 7  | 1.0 ... 1.3 | 2.4 | 4.7 ± 21 %  |

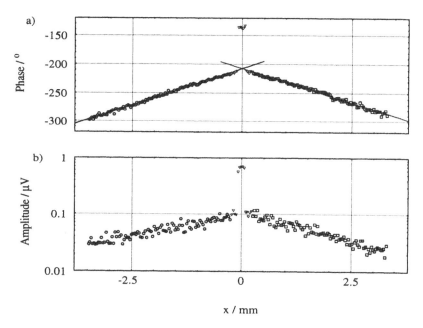

Figure 4: *Phase shift and amplitude of temperatures versus locations $x$ for CVD-diamond sample SQ-F. Frequency of measurement $2\omega$ was 146 $rad\ s^{-1}$. Data points taken into account for the phase fit at the left (O) and right ($\square$) side are drawn black (the gray symbols are omitted).*

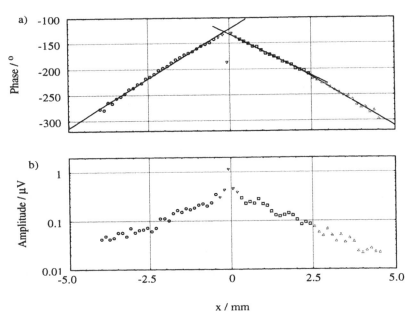

Figure 5: *Phase shift and amplitude R of locations x for CVD-diamond sample SQ-E. Frequency of measurement $2\omega$ was 146 rad $s^{-1}$. For the right-hand side two fits were made, split at $x = 2.5$ mm, to show the detected inhomogeneity of the thermal diffusivity.*

conditions to minimize the influence of the ambient air. The 5 CVD-diamond films were submitted to us as part of a round robin test organized by the National Institute of Standards and Technology [1]. The samples had the dimensions $10\times10\times0.3$ $mm^3$ and were pinned to a copper plate with scotch tape. To have a measuring range in the $x$-direction as large as possible the heater strip was orientated diagonally on the samples. The region beneath the measured area was held self-supporting. Measurements of the thermal diffusivity were performed at different angular frequencies from $2\omega = 111$ rad $s^{-1}$ to $2\omega = 212$ rad $s^{-1}$. From the thermal diffusivities obtained from the fits the thermal diffusion lengths were calculated and it was verified that all criteria for the application of the analytical solution were fulfilled. From all measurements at different frequencies and from both sides of the heater strip mean values for the thermal diffusivity were determined. In addition densities of all samples were determined by weighing and determination of the volume. The values seem to have a systematic error for some are higher than the value for natural diamond (TABLE I, [6]). This is probably due to difficulties in the determination of the thickness by microscopic focusing since the polished specimens still showed some surface roughness.

The diffusivity measurements for each sample showed no dispersion with frequency. Figure 4 shows a plot of the measured phase and amplitude for specimen SQ-F at a frequency $2\omega = 146$ rad $s^{-1}$. Due to the influence of the thermal contact

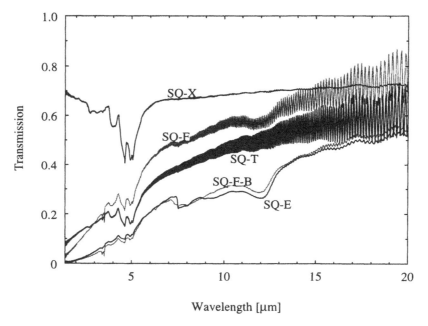

Figure 6: *Directional-directional IR-transmission of the CVD-diamond samples, the oscillations on the right hand side are due to interferences corresponding to the thickness of the films.*

between the thermocouple and the specimen, the measured amplitudes show a larger scattering than the phase values. Figure 5 shows a measurement for specimen SQ-E at the same frequency. For $x > 2.5\ mm$ a drop of the phase curve caused by a smaller thermal diffusivity can be seen. This effect is significantly larger on the right-hand side of the sample. Therfore we performed two fits on this side. The first fit only concerns data points with $x \leq 2.5\ mm$. For the second fit the data points with $x \geq 2.5\ mm$ were used. The thermal diffusivities determined for the split range were 237 $mm^2/s$ and 158 $mm^2/s$, respectively. These results were detected for all frequencies.

The errors for the thermal diffusivities given in TABLE I were calculated from the statistical variations for the used frequencies (typically 5 %). Additional error sources are the uncertainty in the determination of phase ($\leq 1$ %) and location ($\leq 3$ %). The uncertainty in location is caused by the inaccuracy of the inductive position sensor ($\leq 2$ %) and the possibility that the specimen might be tilted. Furthermore, we estimate the influence of surface heat losses on the diffusivity to be less then 5 %. Effects of the finite length of the specimen would be detected from a smaller slope of the phase curve as in figure 2. This was, however, never observed in our experiments. Thus, our results ought to have a maximum error of 10 %.

Another uncertainty in our experiment is the temperature of the specimens since they will be heated up by the dissipated electrical power. Since our thermocouple has

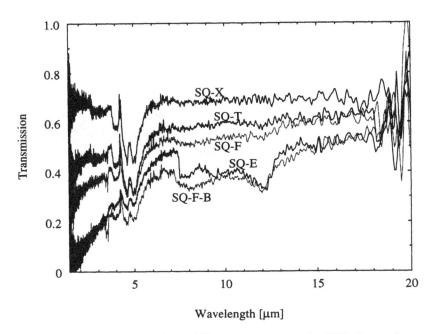

Figure 7: *Directional-hemispherical IR-transmission for the CVD-diamond samples. For sample SQ-X the transmission is similar to the directional-directional IR-transmission, while for the other samples it is higher due to a significant amount of scattered radiation.*

no reference link absolute temperatures could not be measured. For example, at 50 $K$ above room temperature the thermal conductivity of type $IIa$ diamond decreases by 20 % [7], which perhaps partly could explain the low values of our measurements compared to the average value of the round robin test.

## IR OPTICAL MEASUREMENTS

In addition IR-spectroscopic measurements were performed. Figure 6 and figure 7 show the directional-directional and the directional-hemispherical IR-transmission for all samples. Comparing the directional-directional with the densities and the measured thermal diffusivity values, a corelation between those values is obvious, i.e. the higher the density the higher is $\lambda$ and the transmission. In addition sample SQ-X, which shows the largest transmission values, is transparent in the optical range, too. By comparing the directional with the hemispherical transmission measurements, it can be seen that sample SQ-X has nearly identical spectra. This is due to a small amount of diffuse transmitted radiation. For all other samples the hemispherical measurements show higher transmissions. For those samples the diffuse transmitted radiation is about 10 to 20 %.

Therefore a larger amount of diffuse radiation is due to a larger amount of scattering centres correlated with a lower directional-directional IR-transmission and a lower thermal diffusivity. A larger amount of scattering centers is correlated with a lower density, too. This effect ought to be caused by non-crystalline imperfections at the grain boundaries.

Thus, we conclude that our measured thermal diffusivities of the CVD-diamond films are consistent with the determined densities and the IR-transmission spectra. Since we do not know all details about the measurements by the other laboratories in the round robin test, we can only speculate about the remaining discrepancies. However, a second round robin test on specimens as homogeneous as possible would be useful.

## ACKNOWLEDGEMENTS

We are grateful to G. Göbel for the IR-transmission measurements.

## REFERENCES

1. A. Feldman, "Round Robin Thermal Conductivity Measurements on CVD-Diamond." *Application of Diamond Films and related materials: Third International Conference, 1995,* A. Feldman, Y. Tzeng, W. A. Yarbrough, M. Yoshikawa and M. Murakawa ed.

2. I. Hatta, Y. Sasuga, R. Kato, A. Maesono, "Thermal diffusivity measurement of thin films by means of an ac calorimetric method." *Review of Scientific Instruments,* **56** (1985) 1643

3. H.S Carslaw, J.C. Jaeger, *Conduction of Heat in Solids*, Oxford University Press, Oxford 1959

4. D. Fournier, K. Plamann, "Thermal measurement on diamond and related materials." *Diamond and related materials,* **4** (1995) 809-819

5. H. Altmann, O. Nilsson, J. Fricke "An apparatus for the measurement of thermal diffusivities of thin films." to be submitted to: *Review of Scientific Instruments*

6. R. C. Weast (ed.), *CRC Handbook of Chemistry and Physics*, CRC Press, Florida 1985, 65th edition

7. Y. S. Touloukian (ed.), *Thermophysical properties of matter - Volume 2: Thermal Conductivity*, IFI/Plenum, New York (1970)

# An Improved Dynamical Method for Thermal Conductivity and Specific Heat Measurements of Thin Films in the 100 nm-Range

R. SCHMIDT, TH. FRANKE and P. HÄUSSLER

## ABSTRACT

We report on a new experimental technique based on a modified $3\omega$ method that combines two existing techniques. The configuration is insensitive to temperature fluctuations and shows negligible radiation losses. It allows a simultaneous measurement of the thermal conductivity and the specific heat of thin films between 1 K $<$ T $<$ 400 K with high accuracy and reproducibility. A self supporting membrane is used as the substrate. First results demonstrating the applicability of the new technique are presented.

## INTRODUCTION

The knowledge of the thermal conductivity $\lambda$ of thin films is important for a variety of microelectronic devices as well as basic research. For example low-temperature anomalies of amorphous solids are subject of investigations of several groups.

The experimental determination of $\lambda$, especially of thin films, is difficult because small heat flows have to be detected. Problems arise like radiation losses and temperature fluctuations. Thin films are also difficult to prepare. Self supporting thin films are proposed in [1]. Only a few materials with a sufficient low thermal conductivity can be used as substrate.

Subsequently, several methods exist for the determination of thermal properties. Steady state methods [2,3] are time consuming and sensitive to temperature fluctuations as well as radiation losses. But they are useful for measuring thin films.

The second group contains dynamical methods like the heat-pulse technique [4] or the $3\omega$ technique [5,6,7]. The $3\omega$ technique proposed by CAHILL [5] is usefull for dielectric films in the $\mu$m-range. Dynamical methods are insensitive to temperature fluctuations as well as radiation losses.

R. Schmidt, Th. Franke, P. Häussler, Technische Universität Chemnitz-Zwickau, Institut für Physik, Physik dünner Schichten, 09107 Chemnitz, BRD

We use a thermal–conductivity sensor developed by VÖLKLEIN [8], which combines the advantages of both measurement techniques. This configuration is insensitive to temperature fluctuations and is able to measure dielectric as well as metallic thin films. It belongs to the group of dynamic methods.

## THE SENSOR

The sensor (fig.1) is a small chip (5mm × 10mm) produced by standard processes of semiconductor technology. It consists of a silicon wafer with a layered system (membrane) of 200nm $Si_3N_4$, 400nm $SiO_2$ and 200nm $Si_3N_4$. On the rear side the silicon is removed by anisotropic etching on an area of $(g + l) \times b$. In this area the membrane is self supporting and serves as a substrate. A narrow bolometer ($g = 5\mu$m) is evaporated and structured by photolithography onto the top side of the membrane and serves as a heater and thermometer simultaneously. For high–temperature measurements we use a metal and at low temperatures a semiconductor, respectively.

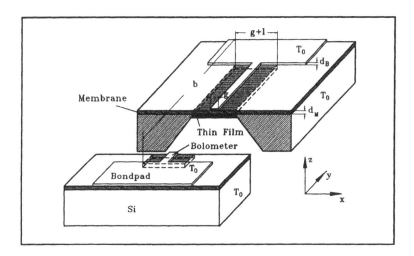

Figure 1. Thermal–conductivity sensor

Typical dimensions are $b = 2$mm, $l + g = 120\mu$m, and $d_M = 500 - 800$nm. The samples under investigation can be evaporated onto the rear side of the membrane. An in–situ preparation by flash evaporation onto liquid helium cooled sensors is possible. The thermal conductivity of the membrane can be calculated from the internal heat balance of the sensor:

$$\frac{\partial^2 \Delta T(x,y)}{\partial x^2} + i2\omega \frac{C_p}{\lambda_M} \Delta T - \frac{2\epsilon_M \sigma(T^4 - T_0^4)}{\lambda_M d_M} = 0 \quad . \tag{1}$$

$2\omega$    frequency of the temperature oszillation
$C_p$    specific heat of the membrane

$\lambda_M$    thermal conductivity of the membrane
$d_M$    thickness of the membrane
$\epsilon_M$    emissive power of the membrane
$\sigma$    Stefan–Boltzmann constant
$T$    temperature
$T_0$    ambient temperature

Because of $b \gg (l+g)$, the heat flow is quasi one–dimensional in $\pm x$–direction. We use

$$\dot{q} = -i2\omega C_p(\omega)\Delta T(x,y) \quad , \tag{2}$$

for describing the heat power per volume with

$$\Delta T(x,y) = T(x,y) - T_0 \tag{3}$$

and the radiation term is approximated by

$$T^4 - T_0^4 \approx 4T_0^3 \Delta T(x,y) \tag{4}$$

because of small temperature gradients. With

$$\mu^2 = -i2\omega \frac{C_p}{\lambda_M} + \frac{8\epsilon_M \sigma T_0^3}{\lambda_M d_M} \tag{5}$$

we simplify (1)

$$\frac{\partial^2 \Delta T(x,y)}{\partial x^2} + \mu^2 \Delta T(x,y) = 0 \tag{6}$$

The solution has to fulfill the boundary conditions

$$\Delta T(x,y) \Big|_{x=\pm \frac{g+l}{2}} = 0 \tag{7}$$

and

$$-\lambda_M \frac{\partial}{\partial x}\Delta T \Big|_{x_B=\pm \frac{g}{2}} = \frac{1}{2}\frac{P(y)}{d_M} \tag{8}$$

with the $y$–dependent heat flow per unitlength

$$\begin{aligned}
P(y) &= \frac{I^2 \varrho(T_0)}{g d_B}\left[1 + \beta \, \Delta T(x_B,y)\right] - 8 \, \epsilon_B \, \sigma \, g \, T_0^3 \, \Delta T(x_B,y) \\
&\quad + \left(\lambda_B d_B + \lambda_M d_M\right) g \, \frac{\partial^2 \Delta T}{\partial y^2} \quad .
\end{aligned} \tag{9}$$

$I$    electric current
$\varrho$    resistivity
$\beta$    temperature coefficient of the resistivity
$d_B$    bolometer thickness
$\lambda_B$    thermal conductivity of the bolometer
$\epsilon_B$    emissive power of the bolometer
$x_B$    half width of the bolometer

The general solution of (1) with (7) and (8) at $x = x_B = \pm\frac{g}{2}$ is given by

$$\Delta T(x_B, y) = \frac{P(y)}{2\lambda_M \mu d_M} \tanh(\mu\frac{l}{2}) \quad . \tag{10}$$

With (9) and (10) we determine the temperature distribution along the bolometer.

We get the average temperature rise of the bolometer by

$$\Delta T_M = \gamma[1 - \frac{2}{\nu b} \tanh(\nu\frac{b}{2})] \tag{11}$$

with

$$\gamma = \frac{\bar{\bar{Q}}}{\nu^2 g b(\lambda_B d_B + \lambda_M d_M)} \tag{12}$$

and

$$\nu^2 = \frac{2\lambda_M d_M b\mu \coth(\mu\frac{l}{2}) + 8\epsilon_B \sigma T_0^3 g b}{g b(\lambda_B d_B + \lambda_M d_M)} \quad . \tag{13}$$

$\bar{\bar{Q}}$ is the average heat power. This leads to

$$\frac{\bar{\bar{Q}}}{\Delta T_M} = \frac{2\lambda_M d_M b\mu \coth(\mu\frac{l}{2}) + 8\epsilon_B \sigma T_0^3 g b}{1 - \frac{2}{\nu b} \tanh(\nu\frac{b}{2})} \quad . \tag{14}$$

Proper dimensions of the membrane ($b$ large, $g$ and $l$ small, $\frac{b}{l} > 8$ and $\frac{b}{g} > 400$) lead to $\frac{2}{\nu b} \tanh(\nu\frac{b}{2}) \ll 1$. Under these conditions the error becomes less than 1 %. The thermal conductance $G = \frac{\bar{\bar{Q}}}{\Delta T_M}$ becomes a sum of the thermal conductance of the membrane

$$G_M(\lambda_M, \epsilon_M) = 2\lambda_M d_M b\mu \coth(\mu\frac{l}{2}) \tag{15}$$

and the bolometer

$$G_B(\epsilon_B) = 8\epsilon_B \sigma T_0^3 g b \tag{16}$$

because of the radiation of the bolometer surface.

If we expand the term $\coth x = \frac{1}{x} + \frac{x}{3}$ in (15) (which requires $|\mu\frac{l}{2}| \ll 1$), we get

$$G_M(\lambda_M, \epsilon_M) = 4\lambda_M d_M \frac{b}{l} + \lambda_M d_M b\mu^2 \frac{l}{3} \quad , \tag{17}$$

and with (5)

$$G_M(\lambda_M, \epsilon_M) = 4\lambda_M d_M \frac{b}{l} - i2\omega\frac{d_M b l C_p}{3} + 8\epsilon_M \sigma T_0^3 \frac{lb}{3} \quad . \tag{18}$$

The third term describing the radiation losses of the membrane is negligible if the area $bl$ is small (micro structure of the membrane). The same is valid for $G_B(\epsilon_B)$, if the bolometer surface $gb$ is small. Thus we get for the total thermal conductance (first term is the D.C.limit, second term describes heating with $\omega$ of the membrane)

$$G = \frac{\bar{Q}}{\Delta T_M} = 4\lambda_M d_M \frac{b}{l} - i2\omega \frac{d_M b l C_p}{3} \quad . \tag{19}$$

$\Delta T_M$ can be determined experimentally, if we use the $3\omega$–method [5]. Through the bolometer flows an electric current

$$I(t) = I_0 \cos(\omega t) \quad , \tag{20}$$

which produces an electric power

$$P_{el}(t) = I^2(t)R_B = \frac{U_\omega^2}{2R_B}(1 + \cos(2\omega t)) = \bar{Q}(1 + \cos(2\omega t)) \quad . \tag{21}$$

$U_\omega$    voltage oscillating with frequency $\omega$
$t$      time
$R_B$   bolometer resistance

The temperature rises independently ($T_{dc}$) on the frequency and oscillates ($\Delta T_M$) according to the heat power by

$$T(t) = T_{dc} + \Delta T_M \cos(2\omega t - \varphi) \quad . \tag{22}$$

A phase shift $\varphi$ occurs. $R_B$ is a function of T. Assuming a linear dependence $R_B(T)$ because of small temperature changes this leads to

$$R_B(t) = R_{dc} + \Delta R_B \cos(2\omega t - \varphi) \quad , \tag{23}$$

and for the voltage drop along the bolometer we get

$$\begin{aligned} U(t) &= I_0 R_{dc} \cos(\omega t) + \frac{I_0 \Delta R_B}{2} \cos(\omega t - \varphi) \\ &+ \frac{I_0 \Delta R_B}{2} \cos(3\omega t - \varphi) \quad . \end{aligned} \tag{24}$$

The second term is negligible regarding the first term ($\frac{\Delta R_B}{R_{dc}} \ll 1$). This leads to

$$U(t) = U_\omega \cos(\omega t) + U_{3\omega} \cos(3\omega t - \varphi) \quad . \tag{25}$$

From the third harmonic $U_{3\omega}$ we determine with (24) $\Delta T_M$

$$U_{3\omega} = \frac{I_0 \Delta R_B}{2} = \frac{I_0 R_B \Delta T_M}{2}\beta = \frac{U_\omega \Delta T_M}{2} \frac{dR_B}{R_B dT} \quad . \tag{26}$$

Exchanging $\frac{\bar{Q}}{\Delta T_M}$ in (19) by (26), it follows

$$\frac{\bar{Q}}{\Delta T_M} = \frac{U_\omega^3 \beta}{4R_B U_{3\omega}} = 4\lambda_M d_M \frac{b}{l} - i2\omega \frac{d_M b l C_p}{3} \tag{27}$$

and we get $\lambda_M$ from the solution for the third harmonic ($U_{3\omega}$ is a complex quantity):

$$U_{3\omega_{rms}} = \frac{U^3_{\omega_{rms}}}{2R_B}\beta \left[ \frac{l}{4\lambda_M d_M b} \frac{1}{1+\omega^2 \left(\frac{d_M C_p l^2}{6 d_M \lambda_M}\right)^2} \right.$$

$$\left. + i \frac{3}{2\omega d_M C_p b l} \frac{1}{1+\frac{1}{\omega^2}\left(\frac{6 d_M \lambda_M}{d_M C_p l^2}\right)^2} \right] \quad . \tag{28}$$

## MEASURING TECHNIQUE

We use a $^4$He cryostat. For the in–situ preparation of the samples a special evaporation source is used. It consists of a rotary tube with an internal thread containing the material. Turning the tube causes a transport of the material to the end of the tube, where it falls onto a hot tungsten filament and gets flash evaporated.

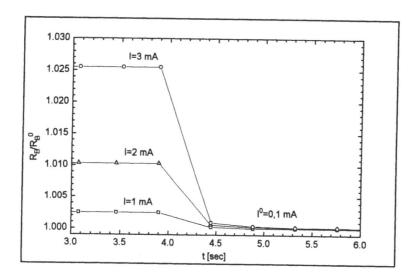

Figure 2. Relaxation time of the sensor after heating

The measuring equipment itself is nearly the same as described in [5,6,7]. The relaxation time of the sensor is shorter than the time constant of the devices necessary to perform measurements with high accuracy. In fig.2 the normalized bolometer resistance $\frac{R_B}{R_B^0}$ is plotted for several heating currents $I$ versus the measuring time $t$. The bolometer temperature is cooled down to $T_0$ and after about 1.5sec the bolometer resistance becomes $R_B^0$ (no heating of the bolometer because of a small current $I^0$).

With $R_B$ and its temperature coefficient we can determine $\lambda$ by a linear regression $\frac{1}{Re(U_{3\omega})}$ versus $\omega^2$ from (28) as shown in fig.3. Only the linear part is used because of $|\mu\frac{l}{2}| \ll 1$ (17). At higher frequencies the thermal penetration depth $|\frac{1}{d_P}| = (\frac{\lambda}{i 2 C_p \omega})^{\frac{1}{2}}$ [7] is smaller and the solution does not fulfill the boundary conditions (7) and (8).

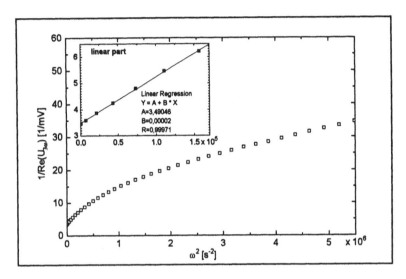

Figure 3. Frequency dependence of the signal. The inset shows the linear part of the chart at low frequencies.

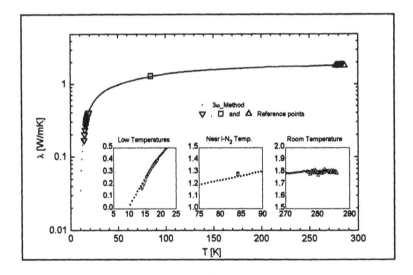

Figure 4. Thermal conductivity of the membrane. Reference points are taken by a method described in [8].

Firstly, we perform a measurement of $\lambda_M$ without any thin film on the rear side of the membrane (fig.4). The thermal conductivity of the whole system (membrane and thin film on the rear side) is measured as

$$\lambda d = \lambda_M d_M + \lambda_F^{eff} d_F \quad .$$

(29)

$\lambda_F^{eff}$   effective thermal conductivity of the film
$d_F$   thickness of the film

Because a thermal resistance $R_{th}$ occures at the interface between the membrane and the film $\lambda_F^{eff}$ is given by [9,10]

$$\lambda_F^{eff} = \lambda_{intr} \frac{d_F}{R_{th}\lambda_{intr} + d_F} \quad .$$

(30)

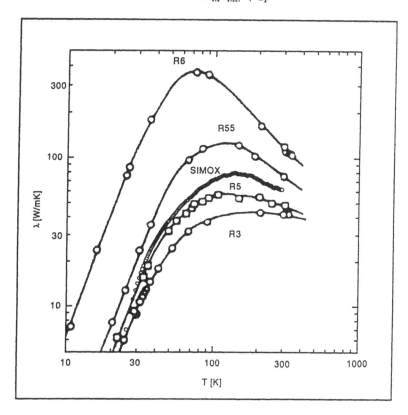

Figure 5. Thermal conductivity of a SIMOX 70 $nm$ silicon film doped with As$^+$. R3, R5, R6 and R55 measured by SLACK [12]

The intrinsic thermal conductivity $\lambda_{intr}$ of the film can be calculated by equation (29) and (30). $R_{th}$ depends on the evaporation method and of course on the evaporated material. For a particular system and a particular evaporation method $R_{th}$ at the interface can be calculated by a linear regression $\frac{d_F}{\lambda_F^{eff}}$ versus $d_F$ (thickness of the thin film not to small) [11]:

$$\frac{d_F}{\lambda_F^{eff}} = \frac{d_F}{\lambda_{intr}} + R_{th} \quad . \tag{31}$$

In fig.5 a plot of $\lambda_F^{eff}$ after correction with $\lambda_M$ of a SIMOX 70nm silicon film (doped with $As^+$, $3 \times 10^{19} cm^{-3}$) is shown. Our data obtained for the silicon film correspond very well with data measured by SLACK [12] for single crystalline silicon bulk material (R5 doped with B, $3 \times 10^{20} cm^{-3}$, R6 doped with P, $2 \times 10^{19} cm^{-3}$, R55 doped with P, $1.7 \times 10^{20} cm^{-3}$). R3 is a polycrystalline silicon sample doped with B, $5 \times 10^{20} cm^{-3}$. At lower temperatures $\lambda_F^{eff}$ is limited by the film thickness. The relaxation time $\tau_d = \frac{d_F}{v}$ ($v$ - sound velocity) resulting from film thickness is smaller than the relaxation time of the main scattering process - the U-phonon process - given by

$$\tau_U(T, \omega_{Ph}) = \frac{\tau_U^0}{k_B T \omega_{Ph}^2} \quad . \tag{32}$$

$\omega_{Ph}$    phonon frequency
$k_B$    Boltzmann constant

It is supposed that the structure of the silicon film is nearly single crystalline and was not affected by the penetration of oxygen ions brought into the single crystalline silicon wafer.


## SUMMARY

Our new technique can be used to determine the thermal conductivity $\lambda$ of thin metallic and nonmetallic films in a temperature range from 1 K to 400 K. The thermal conductivity corresponds very well with data obtained by other techniques.

The equipment is simple and the measurement itself is insensitive to temperature fluctuations and radiation losses. An in-situ preparation as well as an in-situ annealing of thin films is possible. Ex-situ preparation is possible too.


## ACKNOWLEDGEMENT

The authors wish to thank F. Völklein for preparing the thermal-conductivity sensors.

## REFERENCES

1. B.T.Boiko, A.T.Pugachev and V.M.Bratsychin, 1973."Method for the Determination of the Thermophysical Properties of Evaporated Thin Films." *Thin Solid Films*, 17:157-161.

2. F.Völklein and E.Kessler, 1984. "Determination of Thermal Conductivity and Thermal Diffusivity of Thin Foils and Films." *Exp. Technik der Phys.*, 33(4):343-350.

3. F.Völklein and E.Kessler, 1984. "A Method for the Measurement of Thermal Conductivity, Thermal Diffusivity, and Other Transport Coefficients of Thin Films." *phys.stat.sol.(a)*, 81:585-596.

4. S.E.Gustafsson, M.A.Chohan, K.Ahmed and A.Maqsood, 1983. "Thermal Properties of Thin Insulating Layers Using Pulse Transient Hot Strip Measurements." *J.Appl.Phys.*, 55(9):3348-3353.

5. D.G.Cahill, 1989. "Thermal Conductivity Measurement from 30 to 750 K: The $3\omega$ Method." *Rev.Sci.Instrum.*, 61(2):802-808.

6. D.G.Cahill and R.O.Pohl, 1986. "Thermal Conductivity of amorphous solids above the plateau." *Phys.Rev.B*, 35(8):4067-4073.

7. D.G.Cahill, H.E.Fischer, T.Klitsner. E.T.Swartz and R.O.Pohl, 1988. "Thermal Conductivity of thin films: Measurement and Understanding.", *J.Vac.Sci.Technol.A*, 7(3).1259-1266.

8. F.Völklein, 1989. "Thermal Conductivity and Diffusivity of a Thin Film $SiO_2$–$Si_3N_4$ Sandwich System." *Thin Solid Films*, 188:27-33.

9. A.J.Griffin and F.R.Brotzen and P.J.Loos, 1994. "Effect of Thickness on the Transverse Thermal Conductivity of Thin Dielectric Films." *J.Appl. Phys.*, 75(8):3761-3764.

10. A.J.Griffin,Jr., F.R.Brotzen and P.J.Loos, 1994. "The Effective Transverse Thermal Conductivity of Amorphous $Si_3N_4$ Thin Films." *J.Appl.Phys.*, 76(7):4007-4011.

11. J.C.Lambropoulos, M.R.Jolly, C.A.Amsden, S.E.Gilman, M.J.Sinicropi, D.Diakomihalis and S.D.Jacobs, 1989. "Thermal Conductivity of Dielectric Thin Films." *J.Appl.Phys.*, 66(9):4230-4242.

12. G.A.Slack, 1964. "Thermal Conductivity of Pure and Impure Silicon, Silicon Carbide and Diamond." *J.Appl.Phys.*, 35(12):3460-3466.

# Phonon Knudsen Flow in GaAs/AlAs Superlattices

P. HYLDGAARD and G. D. MAHAN

## ABSTRACT

We explain the significant reduction in the in-plane thermal conductivity of GaAs/AlAs superlattices as observed by T. Yao, *Appl. Phys. Lett.* **51**, 1798 (1987.) We estimate that the average bulk-phonon mean free path $l_{mfp} = 115$ nm is significantly longer than the superlattice layer thickness $d = 5$–$50$ nm. The GaAs/AlAs interfaces are generally smooth but does contain a finite density of both microroughness and monolayer island defects which cause an enhanced elastic and anharmonic phonon scattering. We treat this interface scattering as diffusive, i.e., providing a complete phonon thermalization. The resulting phonon Knudsen flow within the individual superlattice layers can account for the observed reduction of the in-plane superlattice thermal conductivity. We also consider the effects of a finite probability for reflection at the interfaces.

## INTRODUCTION

Fast phonon transport in semiconductor heterostructures may allow an improved cooling of densely packed devices such as for example phase-array semiconductor lasers.[1] On the other hand, a significant reduction of the thermal transport may make semiconductor heterostructures attractive possibilities for thermoelectric devices. Either way, an increased understanding of the phonon transport in semiconductor heterostructures is important.

For the in-plane heterostructure phonon transport, T. Yao[2] and G. Chen[3] has observed a significant reduction in equal-layer-thickness GaAs/AlAs superlattices compared with the average, $\kappa_{avg} = 67$ W/Km, bulk-GaAs and bulk-AlAs conductivities. For the smallest layer thickness, $d = 5$ nm investigated in Ref. 2, the superlattice thermal conductivity is reduced down almost to the value of an actual $Al_{0.5}Ga_{0.5}As$ alloy.

Per Hyldgaard and Gerald D. Mahan, Department of Physics and Astronomy, 200 S. College, University of Tennessee, Knoxville, TN 37996; *and* Solid State Division, Oak Ridge National Laboratory, Oak Ridge, TN 37831.

We emphasize, however, that the observed reduction is not due simply to alloy scattering. The molecular-beam-epitaxial grown GaAs/AlAs superlattices are characterized by relatively smooth interfaces and the interdiffusion is limited.[4, 5, 6] However, the interface does contain a finite density of (monolayer) island defects whose in-plane extension can range from below to beyond the exciton Bohr radius,[6] $\sim 10$ nm. At the interface where an GaAs layer is grown on top of an AlAs layer there is also some atomic-scale microroughness.[6]

In this paper we demonstrate that the interface scattering produced by the island defects and microroughness can account for the observed reduction of the superlattice thermal conductivity. Specifically we describe the GaAs/AlAs interfaces and explain how a monolayer island defect can enhance the elastic and anharmonic phonon interface scattering. The average bulk-phonon mean free path $l_{mfp} = 115$ nm is significantly longer than the superlattice layer thicknesses $d = 5$–$50$ nm investigated in Ref. 2. Hence, the interface scattering is important.

Assuming that the interface scattering is diffusive, i.e., causes a complete phonon thermalization, we determine the Knudsen flow within the individual superlattice layers adapting a study by K. Fuchs.[7] We demonstrate in particular that the interface scattering can account for the observed reduction of the in-plane superlattice thermal conductivity. Finally, noting that areas without monolayer steps or islands can extend for more than $\sim 100$ nm at the interface where an AlAs layer is grown on a GaAs layer,[6] we also consider the effect on the thermal conductivity of a finite interface-reflection probability.

## THE MEASURED SUPERLATTICE THERMAL CONDUCTIVITY.

In Ref. 2, T. Yao reports the measurements of the thermal transport in a set of superlattices formed by equally thick layers of GaAs and AlAs as illustrated in the top panel of Fig. 1. The thickness $d$ of these individual layers ranges from $d = 5$ nm and to $d = 50$ nm. As also illustrated in the top panel of Fig. 1, the temperature gradient is applied in the plane of these superlattice layers. We may assume[2, 8] the same temperature in all layers of the superlattices which has a total thickness $\sim 10$ $\mu$m.[2] Hence, the resulting heat transport is given directly by in-plane superlattice thermal conductivity.

The set of diamonds in the bottom panel of Fig. 1 shows this in-plane thermal conductivity as a function of the superlattice layer thickness $d$. For comparison the dashed line in the bottom panel of Fig. 1 shows the average value,[2] $\kappa_{avg} = 67$ W/Km, of the bulk-GaAs and bulk-AlAs thermal conductivities. This average conductivity would provide an appropriate description if the GaAs and AlAs layers contributed independently to the thermal transport and if the phonon mean free path was short enough that one simply had bulk thermal conductance in parallel. It is clear that this simple description is invalid even for a layer thickness of $d = 50$ nm.

The superlattice thermal conductivity is, for layer thicknesses $d \sim 10$ nm, reduced almost to the value,[2] $\kappa_{AlGaAs} = 11$ W/Km — indicated by the dotted line in the bottom panel — of an actual $Al_{0.5}Ga_{0.5}As$ alloy. We emphasize, however, that the observed reduction in the superlattice thermal conductivity can not be explained simply as resulting from an increased bulk alloy scattering. Specifically, the observed factor-of-three reduction of the thermal conductivity

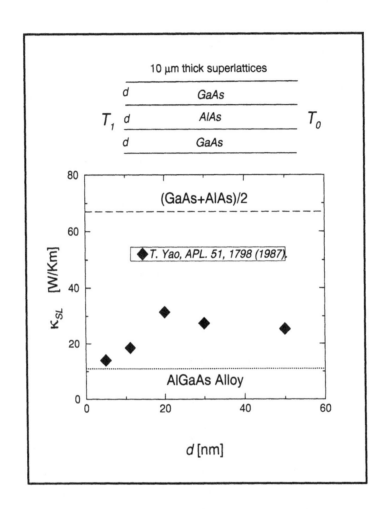

**FIGURE 1.** *Top* panel shows schematics of experimental determination[2] of in-plane thermal conductivity in 10 $\mu$m thick superlattices with GaAs and AlAs layer thicknesses $d = 5$-50 nm. *Bottom* panel compares measured super-lattice thermal conductivities (set of diamonds) with the average bulk-GaAs and bulk-AlAs thermal conductivities (the dashed curve) and with the thermal conductivity of an actual $Al_{0.5}Ga_{0.5}As$ alloy (the dotted curve.)

would — if due to alloy scattering — correspond[ 9] to a 10–20% bulk concentration of GaAs (AlAs) in the AlAs (GaAs) layers. Such a massive interdiffusion is ruled out by high-resolution transmission-electron-microscopy studies as reported for example in Refs. 5,6.

## THE PHONON-INTERFACE SCATTERING

While there is no massive GaAs–AlAs interdiffusion, some disorder does arise at the heterostructure interfaces. As is evident from the high-resolution electron-transmission-microscopy studies in Refs. 5,6 this disorder is confined to within a few monolayers of the interface.

When an AlAs layer is grown on an GaAs layer it is possible to achieve a very smooth interface. There is a finite density of monolayer-island defects[4] but these can have an in-plane extension significantly larger[6] than the exciton Bohr radius $a_B \sim 10$ nm. The top panel of Fig. 2 shows the schematics of such an island defect with a monolayer-region of GaAs protruding into the AlAs layer.

In contrast, the interface where a GaAs layer is grown on top of a AlAs layer is significantly more disordered with so-called microroughness occurring at the atomic scale.[4,5,6] However, with the use of growth interruption it is possible to achieve *additional* longer-range (monolayer) fluctuations with an in-plane extension comparable to the exciton Bohr radius.[6] The GaAs-AlAs interfaces is thus characterized by monolayer defects on several length scales and the interface defects can affect all transport-carrying phonons.

The bottom panel of Fig. 2 illustrates how an interface defect can enhance both the elastic and the anharmonic phonon scattering. Specifically, as shown at the upper interface by the pair of solid arrows, an interface defect breaks the in-plane translational invariance and enhances the elastic scattering. Similarly, since the interface defect in particular softens the stringent momentum-conservation rule limiting the anharmonic phonon scattering,[ 10] monolayer defects will provide an enhanced inelastic scattering. Such an anharmonic scattering event is illustrated at the lower interface with an incoming energetic phonon (longer arrow) decaying to the pair of low-energy phonons (pair of dashed arrows.)

## THE BULK-PHONON MEAN FREE PATH

The interface scattering described above is important because we estimate below a very long average bulk-phonon mean free path, $l_{mfp} \approx 115$ nm. The phonon transport within the heterostructure layers is thus characterized by a finite Knudsen number,[ 11] $(l_{mfp}/d) \geq 2$.

We assume for simplicity an isotropic phonon dispersion with a maximum wave vector, $q_{max} \approx 0.11$ nm$^{-1}$, given by

$$\frac{4\pi}{3} q_{max}^3 = 4(2\pi)^3/a^3, \tag{1}$$

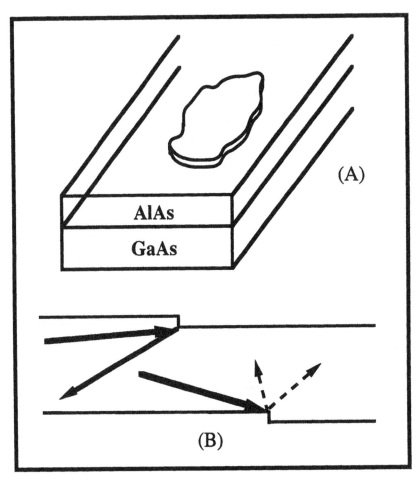

**FIGURE 2.** *Top* panel shows schematics of the interface where an AlAs layer is grown on a GaAs layer. Although this interface is smooth there is still as indicated a finite density of monolayer island defects. The *bottom* panel illustrates how such interface defects enhances both the elastic and anharmonic or inelastic electron scattering as the defect breaks the in-plane translational invariance.

176

where $a = 0.565$ nm denotes the GaAs lattice constant. For the sound velocities of the transverse and longitudinal modes we use the values,[12] $c_t = 3340$ m/s and $c_l = 4730$ m/s.

We also observe that both the transverse and longitudinal modes are relatively soft and is at the zone-boundary (i.e., $q \approx q_{max}$) described by energies $\Theta_t = 100$ K and $\Theta_l = 270$ K.[13] The low value, $\Theta_t = 100$ K, reflects a significant nonlinear dispersion for the transverse modes.[13] There is also a smaller nonlinearity in the longitudinal-mode dispersion.

The nonlinear dispersion reduces the group velocity, $v_g^{t,l} \equiv (\partial \omega_{t,l}/\partial q)$, of both the transverse and longitudinal phonon modes near the zone-boundary and causes a smaller contribution,

$$\kappa_{i=t,l} = \tfrac{1}{3}k_B T^2 \int_0^{q_{max}} q^2 dq \frac{e^{\hbar\omega_i(q)/k_B T}}{(e^{\hbar\omega_i(q)/k_B T} - 1)^2} \times v_g^i(q)[v_g^i(q)\tau_i], \qquad (2)$$

to the bulk thermal conductivity. This transverse- and longitudinal-phonon contribution, Eq. (2), is derived within a phonon-relaxation-time approximation (given by $\tau_i$.) For simplicity we assume a cubic relation,

$$q = \omega(1 + \alpha_i \omega^2)/c_i \qquad (3)$$

between the phonon momentum and frequency within mode $i = t, l$. We determine the value of $\alpha_{i=t,l}$ by the condition, $\hbar\omega_i(q_{max}) = k_B\Theta_i$.

The contribution from phonon mode $(i, q)$ to the bulk thermal conductivity, Eq. (2), is proportional to the mean free path, $[v_g^i(q)\tau_i]$. Assuming for simplicity a mode-independent and constant bulk-phonon mean free path, $l_{mfp}$, we find

$$\kappa_{GaAs} = l_{mfp} \times \frac{k_B^4 T^3}{6\pi^2 \hbar^3}(2\frac{R_t}{c_t^2} + \frac{R_l}{c_l^2}) \qquad (4)$$

where we have introduced the dimensional integrals $(i = t, l)$

$$R_i(T) = \int_0^{\Theta_i/T} dx \frac{x^4 e^x}{(e^x - 1)^2}(1 + \alpha_i(k_B T x/\hbar)^2)^2. \qquad (5)$$

Inserting the *measured* value, $\kappa_{GaAs} = 45$ W/Km, of the bulk-GaAs thermal conductivity in Eq. (4) we obtain an estimate for the averaged bulk-phonon mean-free path, $l_{mfp} \approx 115$ nm. We predict in particular a significant value of the Knudsen number, $l_{mfp}/d \geq 2$, for the phonon transport within the semiconductor layers.

## THE INTERFACE-LIMITED TRANSPORT

The *top* panel of Fig. 3 illustrates how the interface scattering significantly affects the in-plane thermal transport when the average bulk-phonon mean free path $l_{mfp}$ *exceeds* the superlattice layer thickness $d$. The phonon scattering off the interface defects then becomes the dominant source of scattering and causes a reduction in the in-plane thermal conductivity.

The *bottom* panel of Fig. 3 shows as a function of the layer thickness $d$ the estimated resulting superlattice thermal conductivity reduction. The three curves correspond to a set of different assumptions for this scattering as described below.

The simplest possible treatment of interface-scattering results by replacing the relaxation time $\tau$ in Eq. (2) by the layer-traversal time, $\tau_B = d/v_g$, for a phonon moving at right angle to the interfaces. The resulting estimate for the superlattice thermal conductivity is

$$\kappa_{SL}^B \equiv (d/l_{mfp})\kappa_{avg},\qquad(6)$$

expressed relative to the average bulk thermal conductivity, $\kappa_{avg} = 67$ W/Km of GaAs and AlAs. This simple estimate is shown by the dotted curve in the bottom panel of Fig. 3 and is smaller than the measured thermal conductivities (set of diamonds.)

The *solid* curve in Fig. 3 shows the superlattice-thermal conductivity obtained in the phonon-Knudsen-flow[7,11] estimate described further below. We assume in this estimate again a constant bulk-phonon mean free path[14] and that the interface defects causes a diffusive phonon scattering, i.e., causes a complete thermalization of the phonon distribution.

We denote by $\hat{z}$ the direction perpendicular to the superlattice layers (the growth direction) and assume a thermal gradient in direction $\hat{x}$ which induces a change in the thermal distribution,

$$N_{\vec{q}}(z) \equiv N^0(q) + g_{\vec{q}}(z).\qquad(7)$$

Because of the interface scattering (at $z = \pm d/2$) this induced distribution, $g_{\vec{q}}(z)$, depends as indicated on the coordinate $z$. Assuming a constant bulk-phonon mean free path this $z$-variation is described by the linearized Boltzmann equation

$$\left\{l_{mfp}(q_z/q)(\partial/\partial z) + 1\right\}g_{\vec{q}}(z) = (-\partial T/\partial x)Q(q_x/q)$$
$$Q(q_x/q) = l_{mfp}(q_x/q)(\partial N^0(q)/\partial T).\qquad(8)$$

We solve this linearized Boltzmann equation Eq. (8) subject to the boundary conditions mandated by the assumed diffusive interface scattering. Specifically, because phonons *leaving* the interface at $z = d/2$ (and thus traveling with $q_z < 0$) are assumed to be in equilibrium with the interface we must have

$$g_{q_z<0}(z = d/2) \equiv 0.\qquad(9)$$

A similar aguement yields the boundary condition, $g_{q_z>0}(z = -d/2) \equiv 0$, at the upper interface, $z = d/2$.

As a function of the inverse Knudsen number $\eta \equiv d/l_{mfp}$ we then obtain the relative reduction of the Knudsen-flow superlattice thermal conductivity,

$$\frac{\kappa_{SL}}{\kappa_{avg}}(\eta \equiv d/l_{mfp}) = 1 - \frac{3}{8\eta}\left[1 + 4D(\eta)\right]\qquad(10)$$

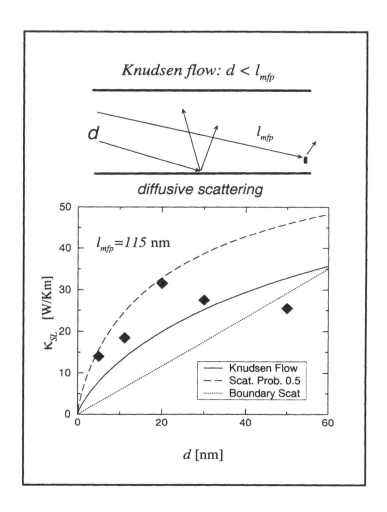

**FIGURE 3.** *Top* panel illustrates how interface scattering restricts the in-plane thermal conductivity because the layer thickness $d$ is shorter than the average bulk-phonon mean free path $l_{mfp}$. *Bottom* panel shows estimates for resulting interface-limited in-plane thermal conductivity. Specifically, the dotted curve shows the boundary-scattering estimate where the layer thickness $d$ simply replaces the average mean free path $l_{mfp}$ in the expression for the bulk thermal conductivity. The solid curve shows the Knudsen-flow estimate where we assume the phonon-interface scattering is diffusive, i.e., provides a complete thermalization. Finally, the dashed curve shows the estimate assuming a finite interface-reflection probability $p = 0.5$.

where

$$D(\eta) \equiv \left( -\frac{\eta^3}{24} + \frac{\eta^2}{24} + \frac{5\eta}{12} - \frac{1}{4} \right) e^{-\eta} + \left( \frac{\eta^4}{24} - \frac{\eta^2}{2} \right) \int_\eta^\infty d\xi \frac{e^{-\xi}}{\xi}. \quad (11)$$

Equation (10) is evaluated with $\kappa_{avg} = 67$ W/Km for the average bulk-GaAs and bulk-AlAs thermal conductivities which yields the phonon Knudsen flow shown by the solid curve in Fig. 3.

Compared with the measured superlattice thermal conductivities (set of diamonds) we find that both the simple boundary-scattering evaluation, Eq. (6) (dotted curve,) and the phonon-Knudsen-flow calculation, Eq. (10) (solid curve,) *overestimates* the observed conductivity reduction from the average bulk conductivity. We conclude that the phonon-interface scattering *can* account for the superlattice conductivity.

## A PARTIAL INTERFACE-REFLECTION

It is no surprise that the Knudsen-flow calculation overestimates the in-plane superlattice thermal conductivity since we have assumed a complete phonon thermalization at both interfaces. At the interface where a AlAs layer is grown on a GaAs layer, however, it is possible to both avoid the microroughness and achieve monolayer islands with a very large an in-plane extension > 100 nm.[6] It is thus likely that the majority of the incoming phonons are simply reflected rather than experiencing an elastic or anharmonic scattering event at this interface. We conclude this paper by a simple estimate of the effect of a finite interface-reflection probability.

In one case we may assume that all phonons approaching the GaAs-to-AlAs interface are simply reflected while all phonons at the AlAs-to-GaAs interface experience a complete thermalization. We can solve the Boltzmann equation in the same manner as for the phonon Knudsen flow but now subject to a reflection boundary condition at the upper ($z = d/2$) interface, i.e., using

$$g_{q_z < 0}(z = d/2) \equiv g_{q_z > 0}(z = d/2) \quad (12)$$

rather than the condition given in Eq. (9).

A related problem arises with the assumption of an *equal* finite reflection probability, $p = 0.5$, at both interfaces. In this estimate we may simply adapt the calculation provided in Ref. 7 and obtain for the in-plane superlattice thermal conductivity shown by the dashed curve in Fig. 3. We note that this estimate is larger than the measured superlattice thermal conductivity.

## CONCLUSION

We have theoretically investigated the in-plane thermal conductivity in GaAs-AlAs superlattices with layer thicknesses $d$ smaller than the estimated average bulk-phonon mean free path, $l_{mfp} \approx 115$ nm. We have described how the defects at the superlattice interfaces enhances both the elastic and anharmonic phonon scattering as they break the in-plane translational invariance. Furthermore, we have demonstrated how this interface scattering can account for the significant reduction of the in-plane thermal conductivity observed by T. Yao[2] in GaAs/AlAs superlattice with layer thicknesses ranging from $d = 5$ nm and to $d = 50$ nm. Finally, we have discusses the effect of a finite reflection probability at the interfaces.

## ACKNOWLEDGEMENTS

The authors appreciate useful discussions with M. Huberman and G. Chen. This work was supported by the University of Tennessee and the Division of Material Science, U. S. Department of Energy under Contract No. DE-AC05-84OR21400 with Lockheed Martin Energy Systems.

## REFERENCES

1. H.-J. Yoo, A. Scherer, J. P. Harbison, L. T. Florez, E. G. Paek, B. P. Van der Gaag, J. R. Hayes, A. von Lehmen, E. Kapon, Y.-S. Kwon, *Appl. Phys. Lett.* **56**, 1198 (1990); P. L. Gourley, M. E. Warren, G. R. Hadley, G. A. Vawter, T. M. Brennan, and B. E. Hammons, *Appl. Phys. Lett.* **58**, 890 (1991).

2. T. Yao, *Appl. Phys. Lett.* **51**, 1798 (1987).

3. G. Chen, "Size and Interface effects on thermal conductivity of superlattices and periodic thin-film structures," preprint.

4. See for example D. Bimberg, D. Mars, J. N. Miller, R. Bauer, and D. Oertel, J. Vac. Sci. Technol. B **4**, 1014 (1986).

5. A. Ourmazd, D. W. Taylor, J. Cunningham, and C. W. Tu, *Phys. Rev. Lett.* **62**, 933 (1989).

6. D. Bimberg, F. Heinrichsdorff, R. K. Bauer, D. Gerthsen, D. Stenkamp, D. E. Mars, and J. N. Miller, J. Vac. Sci. Technol. B **10**, 1793 (1992).

7. K. Fuchs, *Proc. Cambridge Philos. Soc.* **34**, 100 (1938); see also H. B. G. Casimir, *Physica* **V**, 495 (1938).

8. I. Hatta, Y. Sasuga, R. Kato, and A. Maesono, *Rev. Sci. Instrum.* **56**, 1643 (1985).

9. See Fig. 3 in M. A. Afromowitz, *J. Appl. Phys.* **44**, 1292 (1973) showing calculated $Al_xGa_{1-x}As$-alloy thermal conductivity as a function of Aluminum concentration $x$.

10. J. M. Ziman, *Electrons and Phonons* (Oxford University Press, Oxford, 1979,) pp. 128–145.

11. Henrik Smith and H. Højgaard Jensen, *Transport Phenomena* (Clarendon press, Oxford, 1989,) pp. 246–257.

12. H. H. Landolt and R. Börstien, *Numerical Data and Functional Relationships in Science and Technology,* New Series, **17e**, pp. 218–258.

13. See calculated GaAs-phonon dispersion in H. H. Landolt and R. Börstien, *Numerical Data and Functional Relationships in Science and Technology,* New Series, **17e**, p. 528.

14. It is of course straight forward to generalize this Knudsen flow estimate to the case of a $q$- (or frequency-) dependent bulk-phonon mean free path, $l_{mfp}(q)$. The formal results corresponding to Eqs. (10) and (11) — with $\eta(q) \equiv d/l_{mfp}(q)$ — then express the Knudsen-flow reduction from the (average) bulk-conductivity contribution, $\kappa_{avg}(q)$, at phonon momentum $q$. The in-plane superlattice thermal conductivity obtains by a weighted integration of the estimated superlattice contributions, $\kappa_{SL}(q)$, over $q$.

# Measurement of Thermal Properties of Thin Films Using Infrared Thermography

H. M. RELYEA, F. BREIDENICH, J. V. BECK and J. J. McGRATH

## ABSTRACT

One dimensional quasi-steady state heat transfer experiments were performed to measure the thermal diffusivity of thin films. A mathematical model for one dimensional, radial heat conduction in a circular disk was derived; it was implemented in a nonlinear sequential parameter estimation program (NLIN) to estimate the thermal diffusivity. A new experimental configuration was designed which utilized a halogen lamp and ellipsoidal reflector as a heat source. The experimental technique was utilized to determine the thermal diffusivity of CVD-diamond samples, a relatively new material with thermal properties that have not yet been thoroughly quantified. Three samples of CVD-diamond were tested. The results for the thermal diffusivities of the three samples were $3.86 \pm 0.21$ cm$^2$/s, $4.31 \pm 0.27$ cm$^2$/s and $5.69 \pm 0.09$ cm$^2$/s.

## INTRODUCTION/MOTIVATION

Natural diamond, with its high thermal conductivity, is an ideal substance for thin coatings which dissipate large amounts of heat. Diamond films are used in a variety of sophisticated applications including cutting and grinding tools, bearings, speakers, amplifiers, lasers, computer hard disks and integrated circuits. Recent breakthroughs in the process for manufacturing synthetic diamond, known as chemical vapor deposition (CVD), have transformed diamond from a rare, natural substance into a more common engineering material [1]. For many of these applications, it is important to know the thermal properties of the diamond film. However, the measured thermal properties of CVD-diamond film vary over more than an order of magnitude, depending, among other factors, on the amount of impurities and defects caused by the CVD-process [2 - 10]. Thus it becomes

Heidi Relyea, James V. Beck, John J. McGrath, Department of Mechanical Engineering, Michigan State University, East Lansing, Michigan 48824
Frank Breidenich, Rheinisch Westfälische Technische Hochscule Aachen, Germany

necessary to find a method to determine the thermal properties of CVD-diamond films.

The general objective of this research was the development of a new, quasi-steady state experiment utilizing non-contacting heating and measurement methods which would allow the determination of the thermal diffusivity parallel to the surface of a thin, circular specimen.  The primary objective of this research was to determine the thermal diffusivity of three CVD-diamond film samples manufactured by the Micro-Structures Laboratory at Michigan State University.  In order to accomplish this it was necessary to establish that the new experimental method and configuration represented a valid approach.  The validity of this approach was established using samples with  known thermal properties.

Several techniques which employ laser systems have be used to characterize the thermal properties of CVD-diamond films.  These techniques have based the calculation of the thermal diffusivity ($\alpha$) on relatively few temperature measurements taken at one particular time [2, 11-13].  The experimental configuration presented in this work used a system which took 360 temperature measurements at 120 locations and 3 different times.  The discipline of parameter estimation provided algorithms which utilized the data and the mathematical model to estimate the value of thermal diffusivity for the CVD-diamond film specimens.

**ANALYSIS/MODEL**

A mathematical model was developed for one dimensional, radial heat flow in a circular specimen, energized at the center by a concentrated, circular energy source (see Figure 1) for quasi-steady state time periods.

The basic law that describes the phenomenon of heat conduction in a solid body is the Fourier equation [14]:

$$\vec{q}(\vec{r},t) = -k(\vec{r},t)\nabla T(\vec{r},t) \tag{1}$$

Due to the high thermal diffusivity and thinness of the diamond film, the thin films are assumed to be isothermal in the axial direction (orthotropic) during heating. Therefore heat conduction is assumed to be restricted to the radial direction. Since the temperature levels in these experiments were at relatively small temperature

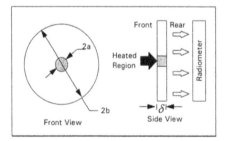

FIGURE 1: SCHEMATIC DIAGRAM OF SPECIMEN

differences (2-6°C) from room temperature and the specimens had high thermal conductivities, negligible convective and radiative heat losses from the sides of the specimen to the surroundings were assumed.  An investigation into the heat losses indicated convective and radiative heat losses to be on the order of 2-3% of the conduction heat transfer.  Thus, the one dimensional heat conduction equation used to model this process is equation (2) [15]:

$$\frac{1}{r}\frac{\partial}{\partial r}\left(r\frac{\partial T}{\partial r}\right)+\frac{q(r)}{k_r\delta}=\frac{1}{\alpha_r}\frac{\partial T}{\partial t} \qquad (2)$$

where q(r) is the input heat flux, $k_r$ is the radial thermal conductivity, $\alpha_r$ is the radial thermal diffusivity and $\delta$ is the thickness of the specimen.

The solution of equation (2) will consist of transient and quasi-steady state terms for this problem (see Appendix A).  The transient term is an infinite sum which decreases exponentially to zero for sufficiently large times.  After the transient term approaches zero, the spatial distribution of the temperature stops changing.  The temperature is said to be quasi-steady, because although the shape of the temperature distribution is fixed, the magnitude of the temperature distribution increases linearly with time [15].  As times become very large, the solution will approach steady-state and the disk will act like a fin (the energy input by the lamp will be equal to the energy convected and radiated away).  The time frame of interest for these experiments was in the quasi-steady state regime, during a period in which the convective and radiative losses were negligible.

The quasi-steady state solution is of main interest for this situation because the dimensionless time $(\alpha t / b^2)$ tends to be greater than 0.5 for all materials and heating times considered.  For this value of dimensionless time, even the first term of the transient solution sum is negligible compared with the total temperature.

The quasi-steady state solution for one dimensional radial heat flow in a circular disk with a natural boundary (temperature is finite) in the center and an insulated boundary condition at the edge (r = b) for the region a<r<b is equation (3):

$$T(r,t)=\frac{qa^2}{k_r\delta}\left[\frac{\alpha_r t}{b^2}+\frac{a^2}{4b^2}\left(\frac{r^2}{a^2}+\frac{1}{2}\right)+\frac{1}{2}\left(\ln\left(\frac{b}{r}\right)-\frac{3}{4}\right)\right]+T_i, \quad a<r<b \qquad (3)$$

where q is the input heat flux, $k_r$ is the radial thermal conductivity, $\alpha_r$ is the radial thermal diffusivity, $\delta$ is the thickness of the specimen, a is the radius of the heated area, b is the radius of specimen and $T_i$ is the initial temperature of the specimen.

## EXPERIMENTAL TECHNIQUE

Circular test pieces 50.8 mm in diameter of varying thickness (240 - 1200 μm) were utilized.  Copper and brass samples were obtained from the Goodfellow

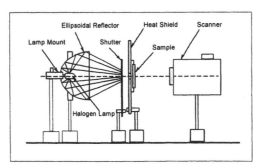

FIGURE 2: DIAGRAM OF EXPERIMENTAL CONFIGURATION

Corporation. The silicon and CVD diamond film samples were provided by the Micro-Structures Laboratory at Michigan State University.

Figure 2 shows the experimental set up. The circular specimens were heated over a 1.6 mm radius using a 250 W Quartz Tungsten Halogen Lamp equipped with an ellipsoidal reflector. A heat shield and shutter mechanism were utilized to ensure that the heating area was restricted to a 1.6 mm radius. Heating was also synchronized with a timer using the shutter. The temperature distribution along the back of the specimen was measured using an Inframetrics Model 600L Infrared Imaging Radiometer.

The thermal fields of the sample were captured on video tape while the specimen was energized. The thermal imaging processing system created temperature field data files for any given time from the video tape which were then used to create data files for the parameter estimation algorithms.

In order to prevent transmission of infrared radiation through the diamond samples and to increase the absorbtivity of the test pieces, the front and rear surfaces were coated with a thin layer (~35μm thick) of black, high temperature-resistant spray paint with a known emissivity of 0.975. A correction factor (see below) to account for the thermal diffusivity of the three-layer system was used to yield a value for the thermal diffusivity of the specimen material only.

## PARAMETER ESTIMATION

A tool which uses both the data given by the mathematical model and the experimental data to estimate the thermal diffusivity, $\alpha_r$, is found in the discipline of parameter estimation. The present problem is an inverse problem, as not all the constants are known in the model presented in Equation (3). The unknown constants include $\alpha_r$, $qa^2 / k_r \delta$, and $T_i$. Using parameter estimation, these unknowns (parameters) are obtained by measuring the temperature distribution inside the domain a<r<b. Here, a nonlinear, sequential parameter estimation computer program called NLIN [16], was utilized to estimate a value for the thermal diffusivity ($\alpha_r$) and values for the two other parameters which are the

group $qa^2 / k_r \delta$ and the initial temperature $T_i$. The thermal diffusivity also contains the thermal conductivity $(\alpha = k / \rho c_p)$, but only the thermal diffusivity can be estimated. This is analogous to the flash diffusivity experiment. The thermal conductivity could be estimated independently if the exact value of q were known. $T_i$ was estimated because there was a small uncertainty associated with the opening of the shutter mechanism.

NLIN uses the method of least squares which minimizes the sum S, with respect to the parameters [17]:

$$S = \sum_{j=1}^{N} \left[ Y_j - T_j \right]^2 \tag{4}$$

where N is the number of data points, $Y_j$ is the measured temperature and $T_j$ is the calculated temperature, both for the same particular location and time.

The program estimates the set of unknown parameters sequentially such that the error in S is minimized with respect to the parameters of interest $(B_n)$ as shown in equation (5).

$$\frac{\partial S}{\partial B_n} = -2 \sum_{j=1}^{N} \left( Y_j - T_j \right) \frac{\partial T_j}{\partial B_n} = 0, \quad n = 1, 2, 3 \tag{5}$$

Sequential estimation [17] means continually updating the parameter estimates as new observations are added. For good experimental data and an appropriate model, the parameter estimations will approach constants as the amount of data is increased.

## EXPERIMENTAL RESULTS

Figures 3 through 5 show the measured and calculated temperature distributions across the horizontal diameter of the CVD-diamond film specimens as a function of time and position for a<r<b . The calculated temperature curves were determined by the parameter estimation program NLIN using equation (3). The

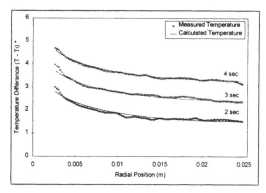

FIGURE 3: SAMPLE 1 (374μm THICK) TEMPERATURE DISTRIBUTION

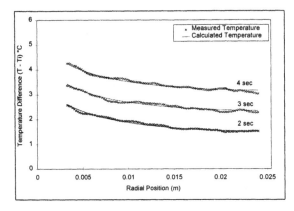

FIGURE 4:  SAMPLE 2 (240μm THICK) TEMPERATURE DISTRIBUTION

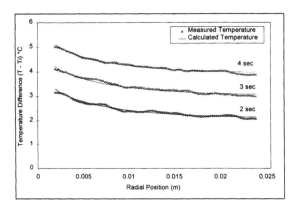

FIGURE 5:  SAMPLE 3 (370μm THICK) TEMPERATURE DISTRIBUTION

curves for both measured and calculated temperature data consist of 360 temperature values obtained for 120 different positions across the diameter of the disk at three different times. Utilizing assumed symmetry, the temperature values were averaged over the radius of the disk for a given time.

Heating times of 2.0, 3.0 and 4.0 seconds were selected for the CVD-diamond films in order to be within the quasi-steady state regime and to minimize heat losses due to convection while maximizing the relative temperature differences within the sample. The heat source had to be operated at maximum output power in order to obtain the highest possible relative temperature rise to increase the signal to noise ratio of the temperature measurements. In order to minimize convective and radiative heat losses, high temperature differences between the sample and the surroundings had to be avoided.

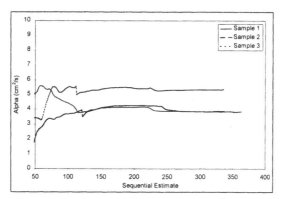

FIGURE 6: ESTIMATED THERMAL DIFFUSIVITIES FOR CVD-DIAMOND FILM

Even though the number of defects in the specimens are unknown, the assumption of an orthotropic specimen is validated by the good agreement of the model with the experimental data. The thermal field in the specimen is also axis-symmetric which again suggests orthotropy. The results show that the quasi-steady state assumption is valid because the measured temperature distribution increases linearly with time.

## PARAMETER ESTIMATION RESULTS

Figure 6 displays the sequential estimation of the thermal diffusivity of the CVD-diamond specimens using NLIN. The number of data points corresponds to the number of temperature measurements used for the estimation. The first set of 120 data points contains the temperatures measured at the first time during heating, the second set corresponds to the second time and data points from the last set of 120 data points describes the temperature distribution across the radius of the specimen at the third time during heating.

Using a sequential method, NLIN begins to estimate a value for $\alpha_r$ based on an initial estimate and just one data point. Then, another data point is added to the previous set of data and the estimation process is repeated before the next data point is added. The value obtained for alpha is "uncertain" when using few data points, which can be seen in Figure 6. For a larger number of data points, the estimated value of the thermal diffusivity converges indicating that the thermal model is good.

## RESIDUALS

Figure 7 shows the typical spatial distribution of the residuals for the three selected times. The residuals are the differences between the temperatures measured

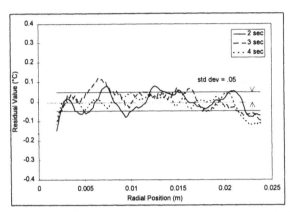

FIGURE 7:  TYPICAL RESIDUAL VALUES FOR CVD DIAMOND FILM SAMPLE
TEMPERATURE ESTIMATES

by the infrared radiometer and the calculated temperatures.   The standard deviation
for the residuals on all the experiments performed was less than 0.1°C.

CORRECTION OF THE ESTIMATED THERMAL DIFFUSIVITY

Since the mathematical model does not account for the paint on both sides
of the specimen, the value estimated for thermal diffusivity is not the value for the
film material itself but rather a value for the three-layer system.  In order to obtain a
value for the thermal diffusivity of the film material alone, a correction factor was
derived.

Due to the high thermal resistance of the paint in comparison with the lower
thermal resistance of the film, it is assumed that radial heat conduction only occurs
in the film layer and that no heat is transferred in the radial direction within the two
layers of paint.  An energy balance calculation using a high thermal contact
resistance ($< 1000$ W/m$^2$K) between the paint and film yields a small temperature
difference ($< .05$°C)  between the two at a given radial position (r).   This shows
that the effect of an interface boundary resistance is negligible. However, the
energy storage rate is now defined as a sum of two terms.  One term accounts for
the film material and another term for the layers of paint.  According to these
assumptions equation (6) is obtained based on equation (2) for one dimensional,
radial heat conduction within the non-heated region:

$$\frac{1}{r}\frac{\partial}{\partial r}\left(r\frac{\partial T}{\partial r}\right)=\frac{\left(\rho_f c_{pf}+\rho_p c_{pp}\,{\delta_p}/{\delta_f}\right)}{k_{rf}}\frac{\partial T}{\partial t} \tag{6}$$

where $\rho_f$ and $c_{pf}$ are the thickness and the specific heat capacity of the film, $\rho_p$ and
$c_{pp}$ are the total thickness and the specific heat capacity of the paint and $k_{rf}$ is the
thermal conductivity of the film material parallel to the surface.   The thermal

TABLE I: DETERMINATION OF CORRECTION FACTOR $F_{CORR}$.

| CVD Diamond Sample | $\delta_f$ [μm] | $\delta_{f+p}$ [μm] | $\delta_p$ [μm] | $\rho_f c_{pf}$ [kJ/Km$^3$] | $f_{corr}$ |
|---|---|---|---|---|---|
| 1 | 374 | 413 | 39 | 1790 | 1.087 |
| 2 | 240 | 276 | 36 | 1790 | 1.126 |
| 3 | 370 | 406 | 36 | 1790 | 1.082 |

diffusivity of the specimen $\alpha_f$ can be calculated by multiplying the estimated thermal diffusivity $\alpha_{NLIN}$ with the correction factor ($f_{corr}$):

$$\alpha_f = \alpha_{NLIN} f_{corr}; \qquad f_{corr} = \left[ 1 + \frac{\rho_p c_{pp} \delta_p}{\rho_f c_{pf} \delta_f} \right] \qquad (7)$$

The sample piece was measured using a micrometer before and after the paint was applied. The averaged thickness of the two layers of paint for three different locations was used (see Table I). A value for $\rho_p c_{pp}$ = 1500 kJ/Km$^3$ [14] was used and values of $\rho_f c_{pf}$ were taken from published literature [5]. The correction factors for the CVD-diamond samples are shown in Table I.

## SUMMARY OF ESTIMATED THERMAL DIFFUSIVITIES

As can be seen in Table II, the estimated thermal diffusivities for the copper, brass and silicon specimens match published values within 8% for the respective materials [14]. The estimated thermal diffusivities for the three CVD-diamond film samples also agree with the range of values published in other CVD-diamond film studies [2, 5, 11, 12].

Utilizing the value 1790 kJ/Km$^3$ [5] for the $\rho_f c_{pf}$ of the diamond films, the thermal diffusivity was converted into thermal conductivity. These values are

TABLE II: COMPARISON OF ESTIMATED AND PUBLISHED VALUES FOR THERMAL DIFFUSIVITY

| | Estimated $\alpha_E$ [cm$^2$/s] | Published $\alpha_P$ [cm$^2$/s] | Diff. (%) |
|---|---|---|---|
| Copper | $1.24 \pm 0.08$ | 1.17 [14] | 6 |
| Silicon | $0.82 \pm 0.05$ | 0.89 [14] | 8 |
| Brass | $0.36 \pm 0.03$ | 0.34 [14] | 6 |
| CVD-Diamond Sample 1 | $3.86 \pm 0.21$ | 2.91 - 7.46 [2, 5, 11-12] | |
| CVD-Diamond Sample 2 | $4.31 \pm 0.27$ | 2.91 - 7.46 [2, 5, 11-12] | |
| CVD-Diamond Sample 3 | $5.69 \pm 0.09$ | 2.91 - 7.46 [2, 5, 11-12] | |

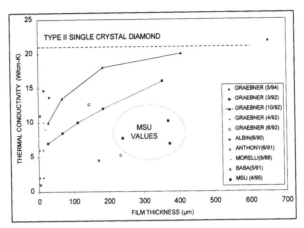

FIGURE 8: COMPARISON OF PUBLISHED THERMAL CONDUCTIVITIES FOR
CVD-DIAMOND FILMS

compared in Figure 8 with other published measurements for the thermal
conductivity of CVD- diamond films [3-10].

## CONCLUSIONS

The published values for the thermal diffusivities of copper and brass were
within one standard deviation of the estimated thermal diffusivities obtained
utilizing this new experimental technique. The value for silicon was within 1.5
standard deviations. This is taken to mean that it may be assumed the new
experimental method presented in this paper is capable of producing thermal
diffusivity values accurate to within 8% or better. This gives confidence in the
thermal diffusivity results obtained for the CVD-diamond film samples, which are
also within the range of values previously recorded for CVD-diamond film thermal
properties.

The sequential parameter estimation results for the thermal diffusivity,
indicate that the one-dimensional radial conduction model used for this process was
accurate.

There is still much to be learned about the thermal properties of CVD-
diamond films, as indicated by the large span of values for thermal conductivity in
Figure 8. Future research efforts are proposed which will concentrate on
determining the effects of the manufacturing parameters used to make the diamond
films (e.g. gas composition, temperature and deposition method) on the thermal
and material properties of the diamond films.

## REFERENCES

1. Zhu, W., B.R. Stoner, B.E. Williams and J. T. Glass, 1991. "Growth and Characterization of Diamond Films on Nondiamond Substrates for Electronic Applications." *Proceedings of the IEEE*, 79(5): 621-645.

2. Albin, S., W.P. Winfree and B. S. Crews, 1990. "Thermal Diffusivity of Diamond Films Using a Laser Pulse Technique." *J. Electrochem. Soc.*, 137(6): 1973-1976.

3. Anthony, T.R., J.L. Fleischer, J.R. Olson and D.G. Cahill, 1991. "The Thermal Conductivity of Isotopically Enriched Polycrystalline Diamond Films." *J. Appl. Phys.*, 69(12): 8122-8125.

4. Graebner, J.E. and T. M. Hartnett, 1994. "Improved Thermal Conductivity in Isotropically Enriched Chemical Vapor Deposited Diamond." *Appl. Phys. Lett.*, 64(19): 2549-2551.

5. Graebner, J.E., S. Jin, G.W. Kammlott, B. Bacon, and L. Seibles, 1992. "Anisotropic Thermal Conductivity in Chemical Vapor Deposition Diamond." *J. Appl. Phys.*, 71(11): 5353-5356.

6. Graebner, J.E., S. Jin, G.W. Kammlott, J. A. Herb, and C.F. Gardinier, 1992. "Unusually High Thermal Conductivity in Diamond Films. " *Appl. Phys. Lett.*, 60(13):1577-1578.

7. Graebner, J.E., S. Jin, G.W. Kammlott, J. A. Herb, and C.F. Gardinier, 1992. "Large Anisotropic Thermal Conductivity in Synthetic Diamond Films." *Letters to Nature*, 359: 401-404.

8. Graebner, J.E., J.A. Mucha, L. Seibles and G.W. Kammlott, 1992. "The Thermal Conductivity of Chemical-Vapor Deposited Diamond films on Silicon." *J. Appl. Phys.*, 71(7): 3143-3146.

9. Morelli, D.T., C.P. Beetz, and T.A. Perry, 1988. "Thermal Conductivity of Synthetic Diamond Films." *J. Appl. Phys.*, 64(6): 3063-3066.

10. Ono, A., T. Baba, H. Funamoto and A. Nishikawa, 1986. "Thermal Conductivity of Diamond Films Synthesized by Microwave Plasma CVD." *Jpn. J Appl. Phys.*, 25(10): L808-L810.

11. Lu, G., K.J. Gray, E. F. Borchelt, L. K. Bigelow and J.E. Graebner, 1993. "Free-Standing White Diamond for Thermal and Optical Applications." *Diamond and Related Materials*, 2:1064-1068.

12. Lu, G. and W.T. Swann, 1991. "Measurement of Thermal Diffusivity of Polycrystalline Diamond Film.." *Appl. Phys. Lett.*, 59(13): 1556-1558.

13. Petrovsky, N., A.O. Salnick, D.O. Muhkin, and B. V. Spitsyn, 1992. "Thermal Conductivity Measurements of Synthetic Diamond Films Using the Photothermal Beam Deflection Technique." *Materials Science & Engineering B: Solid-State Materials for Advanced Technology*, B11(1-4): 353-354.

14. Incropera, F.P. and D.P. De Witt, 1990. *Introduction to Heat Transfer*, 2nd. ed., New York: John Wiley & Sons.

15. Beck, J.V., K. Cole, A. Haji-Sheikh and B. Lithouhi, 1992. *Heat Conduction Using Greens Functions*, New York: Hemisphere Press.

16. Beck, J.V., 1994. "NLIN," FORTRAN Program, Department of Mechanical Engineering, Michigan State University, East Lansing, MI.

17. Beck, J.V. and K. J. Arnold, 1977. *Parameter Estimation in Engineering and Science*, New York: John Wiley & Sons.

## ACKNOWLEDGMENTS

The support of the State of Michigan Research Excellence Fund, Measurement of Properties of Diamond Films and J. Asmussen, Electrical Engineering Department, Michigan State University, East Lansing, MI are appreciated.

## APPENDIX A

SOLUTION OF EQUATION (2)

The solution for one dimensional radial heat flow in a circular disk with a natural boundary in the center and an insulated boundary condition at the edge (r = b) is:

$$T(r,t) = \frac{\alpha_r}{k_r} \int_{\tau=0}^{t} \int_{r'=0}^{b} \frac{q}{\delta} \left\{ \frac{1}{\pi b^2} \left[ 1 + \sum_{m=1}^{\infty} e^{-\frac{\beta_m^2 \alpha_r (t-\tau)}{b^2}} \frac{J_0\left(\frac{\beta_m r}{b}\right) J_0\left(\frac{\beta_m r'}{b}\right)}{J_0^2(\beta_m)} \right] \right\} 2\pi r' dr' d\tau + T_i \tag{8}$$

where $J_0(x)$ is the Bessel function of the first kind of order zero evaluated at x and the $\beta_m$'s are the eigenvalues of the eigencondition, $J_1(\beta_m) = 0$.

The solution of equation (8) consists of transient and quasi-steady state terms. The transient portion of the solution is:

$$T_{TR}(r,t) = -\frac{2qab}{k_r \delta} \sum_{m=1}^{\infty} e^{-\frac{\beta_m^2 \alpha_r t}{b^2}} \frac{J_0\left(\frac{\beta_m a}{b}\right) J_0\left(\frac{\beta_m r'}{b}\right)}{\beta_m^3 J_0^2(\beta_m)} \tag{9}$$

The first eigenvalue for this problem is $\beta_1 = 3.83$ [15] which indicates the transient term will be negligible for a Fo number of .5 or greater.

The quasi-steady state solution is given in equation (3). The complete derivation of the solution for a similar problem can be found in [15].

# Laser-Assisted AC Measurement of Thin-Film Thermal Diffusivity with Different Laser Beam Configurations

X. Y. YU, L. ZHANG and G. CHEN

## ABSTRACT

This work investigates and compares three laser-based ac heating methods for measuring thermal diffusivity of free-standing thin-film structures. These methods employ a modulated laser beam as the heating source and a miniature thermocouple as the temperature sensor. Three laser beam configurations including uniform illumination, a line, and a point are utilized as the heating sources. Different models and systems are developed for these beam configurations. Samples studied include a GaAs/AlGaAs two-layer thin-film structure, a periodic GaAs/AlAs thin-film structure, and a silicon film. Both the phase and the amplitude signals of the ac temperature rise of samples are used to derive their thermal diffusivities. It is found that the uniform illumination method is more susceptible to error than the other two configurations due to two-dimensional effects. Both the line and the point source configurations yield satisfactory results.

## INTRODUCTION

Thermal conductivity and diffusivity of thin films generally differ from their corresponding bulk values due to the microstructural and boundary effects. Various thin film measurement techniques have been developed to measure the thermal diffusivity or conductivity of thin film structures as summarized in several review articles [1-4]. An ac calorimetric method for measuring the thermal diffusivity of free-standing thin films in the parallel direction was developed by Hatta and co-workers in a series of publications [5-7]. This method employed a modulated radiation source to create a thermal wave in the sample and a temperature sensor to pick up its thermal response. The thermal diffusivity of the sample is obtained from the amplitude variation of the measured ac temperature as a function of the distance between the temperature sensor and the heating source. Based on this principle, several ac calorimetry measurement systems with different heating methods have been developed. Hatta et al. [5] used a uniform ac illumination obtained from a halogen lamp as the heating source and a thermocouple as the temperature sensor to

X.Y. Yu, Graduate Student; L. Zhang, Visiting Scientist, G. Chen, Assistant Professor, all with the Department of Mechanical Engineering and Materials Science, Duke university, Durham, NC 27708. Permanent Address of L. Zhang: Cryogenic Lab., Chinese Academy of Sciences, Beijing 100080, P. R. China.

measure the thermal diffusivity of thin films.  Kato et al. [7] developed the method of using a narrow radiation strip obtained from a modulated diode laser as the heating source.  These experiments are based on the amplitude signal of the ac temperature rise.  Yu et al. [8] employed a similar method to measure the thermal diffusivities of GaAs/AlAs thin-film structures and found that both the phase and the amplitude signals of the ac temperature rise can be used to derive the sample thermal diffusivity. Zhang et al. [9] focused a modulated laser beam to a point source to heat up sample and used an infrared detector to sense its ac temperature rise.  Other laser-based thermal-wave measurement systems for the thermal diffusivity of free-standing thin film are also developed [10,11].  No systematic study, however, has been reported on the effects of the laser beam configuration on the experimental results.

This paper presents a comparative study on the laser-based ac heating methods for measuring the thermal diffusivity of thin film structures using different laser beam configurations, including nearly uniform illumination and a line and a point heating source.  The thermal response of the sample is detected by a miniature thermocouple attached to the surface of the sample.  Samples studied include a GaAs/AlGaAs two-layer thin-film structure, a periodic GaAs/AlAs thin-film structure, and a silicon film. Different models and systems are developed for these laser beam configurations. Both the phase and the amplitude signals of the ac temperature rise of the sample are utilized to deduce the thermal diffusivity of the thin-film structures.  The results show that the uniform illumination method is more susceptible to error than the other two configurations due to two-dimensional effects.  Both the line and the point beam configurations yield satisfactory results.  The advantages and drawbacks of each experimental configuration are discussed.

## PRINCIPLES OF MEASUREMENT

Three different laser heating configurations are tested, as shown in Figs. 1(a)-1(c).  In Fig. 1(a), the laser beam is expanded to a nearly uniform illumination and a mask is used to control the distance between the heating source and the temperature sensor [5].  When the thermal diffusion length is much larger than the film thickness h, the sample temperature is uniform in the film thickness direction.  Assuming that the sample is infinitely large and the illumination is uniform, the sample ac temperature rise at the sensor can be expressed as [5]

$$\theta(x) = [Q_0 / 4\pi f\rho chA_1]\exp[-mx - i(mx + \pi / 2)] \qquad (1)$$

where $A_1$ is the illumination area, c the specific heat, f the modulation frequency, m $[=(\pi f/\alpha)^{1/2}]$ the inverse thermal diffusion length, $Q_0$ the absorbed power, x the distance between the sensor and the edge of the mask, $\alpha$ the thermal diffusivity of the film, and the $\rho$ the density of the film.  By moving the mask along the film, the relation $\theta(x)\sim x$ can be obtained.  The thermal diffusivity of the thin film can be derived from the amplitude or phase of the ac temperature signal as a function of the mask displacement,

$$\alpha_\theta = \frac{\pi f}{(d\ln|\theta| / dx)^2} \qquad (2)$$

$$\alpha_\varphi = \frac{\pi f}{(d\varphi / dx)^2} \qquad (3)$$

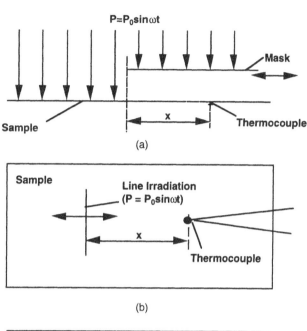

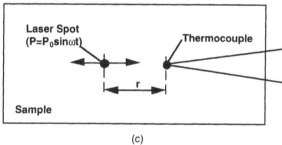

FIGURE 1     SCHEMATIC DRAWING OF THE PRINCIPLES OF THE
MEASUREMENT, (A) UNIFORM ILLUMINATION, (B) LINE SOURCE
IRRADIATION, AND (C) POINT SOURCE HEATING.

where $\varphi$ is the phase term in Eq. (1), and $\alpha_\theta$ and $\alpha_\varphi$ are the thermal diffusivity
values deducted from the amplitude and the phase signals, respectively.

In the second configuration, shown in Fig. 1(b), the laser beam is shaped into a
narrow strip and the distance between the heating strip and the temperature sensor is
altered.  By assuming a line heat source, the ac temperature rise along the sample in
the direction perpendicular to the heating strip can be solved as

$$\theta(x) = [Q_0 / 2\sqrt{2}mLh\kappa]\exp[-mx - i(mx + \pi / 4)] \tag{4}$$

where L is the length of line source and $\kappa$ is the thermal conductivity of the sample.
Again, the thermal diffusivity of the sample can be obtained from either the phase or

the amplitude signal. The expressions between the thermal diffusivity and the signals are identical to Eq. (2) and Eq. (3).

In the third laser beam configuration, Fig. 1(c), the laser beam is focused as a point source. The thermal wave generated by the modulated point source cylindrically propagates outward. By solving the one dimensional heat conduction equation in a cylindrical coordinate, the ac temperature rise is expressed in terms of the modified Bessel functions I and K as

$$\theta(r) = (\frac{1}{128\pi^2 m^2 r^2})^{1/4} \frac{Q_0 \alpha}{i\kappa f h A_3} \frac{I_1(\chi_0 \sqrt{i}) \cdot K_0(\chi \sqrt{i})}{I_1(\chi_0 \sqrt{i})K_0(\chi_0 \sqrt{i}) + I_0(\chi_0 \sqrt{i})K_1(\chi_0 \sqrt{i})} \qquad (5)$$

where r is the radial distance from the sensor to the center of the heating source, $r_0$ the radius of the laser spot, $A_3[=\pi r_0^2]$ the area of the laser beam, and $\chi$ [$=\sqrt{2}$ mr] a nondimensional parameter. When $\chi$ is larger than 8 and $r_0 \to 0$, Eq. (5) can be simplified to [12,13],

$$\theta(r) = (\frac{1}{128\pi m^2 r^2})^{1/4} \frac{Q_0}{\kappa h} \exp[-mr - imr + \varphi_0] \qquad (6)$$

within $10^{-7}$ accuracy, where $\varphi_0$ is a constant.

From Eq. (6), the thermal diffusivity of the sample can be derived as

$$\alpha_\theta = \frac{\pi f}{[d(\ln|\theta| + 0.5\ln r) / dr]^2} \qquad (7)$$

$$\alpha_\varphi = \frac{\pi f}{(d\varphi / dr)^2} \qquad (8)$$

When $\chi$ is less than 8, the above expressions for evaluating the thermal diffusivity will cause some error. Zhang et al. [9] employed a more accurate but nonlinear expression in deriving the properties from the measured amplitude and phase data. An error analysis can show that when $\chi > 0.9$, the relative error in diffusivity due to the use of the above approximation is less than four percent from the phase expression and less than 14% from the amplitude information. The error in using Eq. (7) drops to less than five percent as $\chi$ becomes larger than 1.5.

All the above equations are derived under the assumption that the convection loss is negligible compared to the heat conduction through the film. Hatta et al.[5] established than this is valid when $2\pi f\rho ch/U \gg 1$, where U is the heat transfer coefficient between the sample and its environment. For the current experiment under vacuum, it can be shown that this condition is readily satisfied.

The above discussions clearly show that the mathematical expression for deriving the thermal diffusivity of free-standing thin-film samples from the phase information remains the same for all three laser beam configurations. The only difference lies in the derivation based on the amplitude signals. For the focused laser beam configuration, determining the thermal diffusivity from the amplitude signal requires the knowledge of the exact location of the temperature sensor, while the other two laser beam configurations do not have this requirement. The temperature expressions for the three heating configurations also imply different laser power requirements in each method. It can be proven that for an equal ac temperature rise the power required for the uniform illumination is much higher than that for the other two methods. The power requirements for the line and the point source configurations are roughly comparable.

## EXPERIMENTAL SET-UP

Figures 2(a)-(c) illustrate the experimental set-ups. In Fig. 2(a), the beam from an argon laser, modulated by a mechanical chopper, is expanded and the center portion of the beam is utilized to heat up the sample. A mask installed on a micro-manipulator is employed to control the distance between the heating source and the temperature sensor. For the line source heating configuration, a diode laser is used because the beam from a semiconductor laser is close to elliptical and can be easily configured to the desired shape by adding an anarmorphic prism, as shown in Fig. 2(b). In this set-up, the distance between the heating strip and the thermocouple is controlled by a reflecting mirror on a micro-manipulator. The point source heating is realized by finely focusing an argon laser beam. The distance between the sensor and the heating spot is controlled by moving the sample installed on a micro-manipulator as illustrated in Fig 2 (c).

The samples tested include a free-standing GaAs/AlGaAs two-layer thin-film structure (MBE635), a free-standing multilayer GaAs/AlAs thin-film structure (MBE849) and a silicon single layer film. Samples MBE635 and MBE849 are grown by molecular beam epitaxy. Sample MBE635 consists of 10 $\mu$m GaAs and 1 $\mu$m $Al_{0.7}Ga_{0.3}As$. Sample MBE849 is made of periodic 700 Å/700 Å GaAs/AlAs layers to a total thickness of 10 $\mu$m. The substrates of sample MBE635 and MBE849 were removed by an selective chemical etching method. The silicon sample is a 4 $\mu$m single crystal thin-film and is coated with a 1000 Å gold film to increase its absorptivity. The free-standing area of the sample is about 4mm $\times$ 5mm.

A E-type thermocouple with a 35 $\mu$m junction diameter is attached onto the samples by small amount of vacuum grease.

## RESULTS AND DISCUSSION

Figures 3(a) and 3(b) show the normalized phase and amplitude of the ac temperature of sample MBE 635 as a function of the distance between the sensor and the heating source for the three different methods, respectively. The results demonstrate that the phase and the logarithm of the amplitude ($\ln|\theta|+0.5\ln r$ for the point source) are linearly proportional to the distance between the temperature sensor and the heating source, in agreement with the theoretical analysis. Table 1 presents the experimental results on MBE635 from the three methods at room temperature. For the multilayer structures, samples MBE635 and MBE849, the reported values are the effective thermal diffusivity of the composite structures. The measurement results by the line and the point source methods are close to each other, while the uniform illumination heating yields a considerably different value. The effective thermal diffusivity of MBE635 calculated from the bulk thermophysical properties of GaAs and $Al_{0.7}Ga_{0.3}As$ is 0.238 cm$^2$/s [11]. The line and point source methods yield results close to the bulk value. The uniform illumination method, however, gives a much lower value. It was estimated that size effects in the sample should be minimal and the measured value should be close to the bulk value [11]. This indicates that the uniform beam illumination method is more susceptible to error. It should be pointed out that Chen et al. [11] employed a similar uniform illumination method combined with a photolithographically patterned temperature sensor and obtained a sample diffusivity of 0.23 cm$^2$/s. The relatively large error in the current

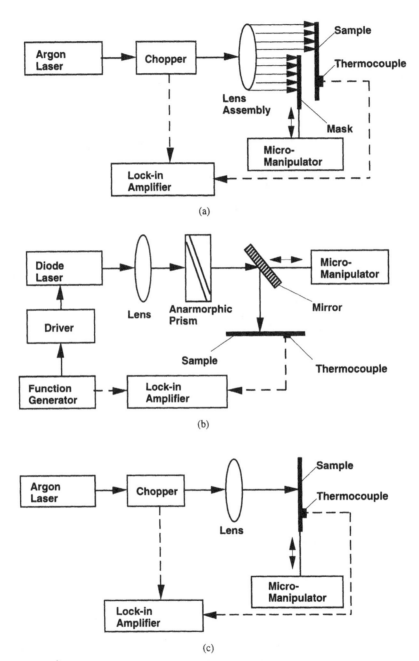

FIGURE 2 EXPERIMENTAL SET-UP, (A) UNIFORM ILLUMINATION,
(B) LINE SOURCE IRRADIATION, AND (C) POINT SOURCE HEATING.

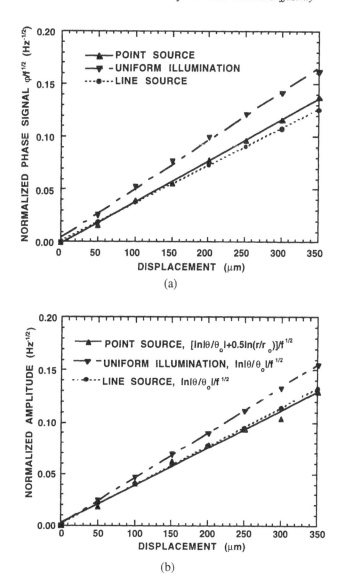

FIGURE 3    NORMALIZED (A) PHASE AND (B) AMPLITUDE SIGNALS FOR
SAMPLE MBE365.

uniform illumination method is suspected to stem from two-dimensional effects of
the Gauss intensity profile of the laser beam.   For the other two methods, the
Gaussian beam-profile does not affect the solution of the heat conduction equation
and thus the experimental results.   For each of the three methods, however, the $\alpha_\theta$
and $\alpha_\varphi$ are close.   Their deviation relative to $\alpha_{ave}$ is less than 10% and mainly

TABLE 1—EXPERIMENTAL RESULTS FOR SAMPLE MBE365 AT ROOM
TEMPERATURE

| Laser Beam Configuration | $\alpha_\varphi$ (cm²/s) | $\alpha_\theta$ (cm²/s) | $\alpha_{ave}$ (cm²/s) |
|---|---|---|---|
| Point | 0.203 | 0.242 | 0.223 |
| Line | 0.248 | 0.226 | 0.237 |
| Uniform | 0.148 | 0.164 | 0.156 |

depends on the quality of the system alignment. The difference between the thermal
diffusivity values derived from phase and the amplitude, which should be small, can
be used to adjust the system alignment. The following reported thermal diffusivity
values are the average of the values derived from the phase and the amplitude signals.

Another two-dimensional effect for the measurement is the finite size of the
sample. Equations (1-9) are derived under the assumption of infinite sample size.
The finite size of the samples causes reflections of thermal waves at the boundaries.
Hatta et al.[6] presented a one-dimensional model for the sample size effects on the
measured sample equivalent diffusivity. Chen and Yu [14] established a two
dimensional model for the finite sample effect, which yields the equivalent thermal
diffusivity as a function of the modulation frequency. Figure 4 shows the calculated
and measured equivalent thermal diffusivity as a function of the modulation
frequency for the line beam illumination method. Here the equivalent thermal
diffusivity is obtained by assuming that the one dimensional solutions, Eq. (2) and
Eq. (3) are valid for all the frequencies. Both the model and the experimental results
show that the thermal diffusivity approaches the infinitely-large thin-film solution
when the modulation frequency is larger than a certain frequency (50 Hz for sample
MBE635). Similar experimental results were obtained for the other two beam
illumination methods. In general, the larger the value of the thermal diffusivity, the
larger modulation frequency is required. A safe method in choosing the correct
modulation frequency is to compare the thermal diffusivity values obtained from the

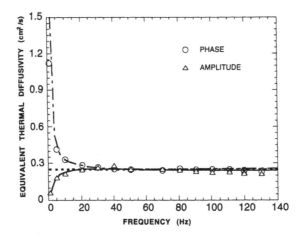

FIGURE 4    EQUIVALENT THERMAL DIFFUSIVITY AS A FUNCTION OF
THE MODULATION FREQUENCY FOR THE LINE SOURCE IRRADIATION
CONFIGURATION.

phase and the amplitude signals. The two signals should yield close results when the infinite sample size assumption is valid. It should be pointed out, however, that there exists an upper limit for the frequency that can be employed. This is imposed by the requirement that the temperature in the film must be uniform, which is satisfied if the thermal diffusion length is much smaller than the film thickness,

$$hm << 1 \qquad (9)$$

In the current experiment, the left hand side of the above inequility is of the order of $10^{-3}$ and the condition is always satisfied. For thicker films, cautioned must be exercised to satisfy both requirements.

For the point source method, data points with r>300 μm are used for the silicon sample. The corresponding $\chi$ value ranges from 0.9 to 3.5. The $\chi$ value for MBE635 falls between 1 and 3.1. In these ranges, the results obtained from the phase data are expected to be more accurate than those from the amplitude data.

Table 2 lists the experimental results of each sample at room temperature, by using the point (silicon sample) and the line source (MBE635 and MBE849) measurements. The thermal diffusivity value of MBE 849 is lower than its corresponding bulk value due to the interface effects [8, 16]. The measured thermal diffusity of the silicon sample at room temperature, 0.78 cm²/s, is slightly lower than the reported bulk value, 0.88 cm²/s [15]. The measured diffusivity value of sample MBE635 is close to the result of Chen et al. [11]. The narrow strip illumination method is also used to determine the temperature dependence of the two MBE samples. Figure 5 shows the experimental results.

Among the three heating methods, the point source offers several advantages. For any kind of laser, a focused point can be easily achieved and the laser energy is more concentrated. These features make the focused beam method applicable to a wide range of materials and temperatures. The choose of laser source depends largely on the material absorption properties. The disadvantage of this method is that it requires the exact distance between the sensor and the heating source and more complex fitting for higher accuracy. For the line source method, the data deduction is simple and the thermal diffusivity values from the phase and the amplitude measurement are close. This method is best suitable for diode laser-based system. Both of these methods are suitable for cryogenic experiments because of their lower power requirements. For the large area illumination method, the φ~x and the ln|θ|~x data have good linearity but the derived thermal diffusivity values are more susceptible to error due to the two-dimensional effects and the poor signal to noise ratio. The latter is because the large area illumination method requires a high laser power and thus creates a large dc temperature rise in the sample. When the ac temperature rise generated is not large enough, the system noise will surpass the ac temperature signal or make the signal unstable.

In general, the phase measurement of the ac temperature yields more satisfactory results because it is independent of the energy absorbed. The amplitude

TABLE 2—MEASURED THERMAL DIFFUSIVITIES OF THE SAMPLES AT ROOM TEMPERATURE.

| Sample | Thickness (μm) | $\alpha_{ave}$ (cm²/s) | $\alpha_{bulk}$ (cm²/s) |
|--------|----------------|------------------------|--------------------------|
| MBE635 | 11 | 0.23 | 0.24 |
| MBE849 | 10 | 0.25 | 0.39 |
| Silicon | 4 | 0.78 | 0.88 |

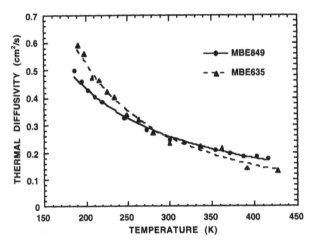

FIGURE 5  THERMAL DIFFUSIVITY AS A FUNCTION OF TEMPERATURE.

measurement, on the other hand, provides a comparison for the thermal diffusivity measurement. Such a comparison can be used to select the modulation frequency and to assist the system alignment.

## CONCLUSION

This work compares three different laser beam configurations, including a nearly uniform illumination and a line and a point heating source, for measuring the thermal diffusivity of free-standing thin-film structures. Different models and the experimental systems are developed for these three methods. The comparison of the experimental results shows that the point and the line source methods yield satisfactory results while the uniform illumination method is more susceptible to error. The phase measurement generally yields good results and the amplitude measurement can be employed to select the correct modulation frequency and to assist the system alignment.

## ACKNOWLEDGMENT

G.C. would like to thank Professor J. S. Smith's group at UC Berkeley for providing the MBE samples. He gratefully acknowledges the financial supports from the National Science Foundation Young Investigator Award, the Lord Foundation of North Carolina, and the School of Engineering at Duke University.

## REFERENCES

1. Cahill, D.G., Fischer, H.E., Klitsner, T., Swartz, E.T., and Pohl, R.O., 1989, "Thermal Conductivity of Thin Films: Measurement and Understanding," *Journal of Vacuum Science and Technology* A, Vol. 7, pp. 1259-1266.

2. Hatta, I, 1990, "Thermal Diffusivity Measurement of Thin Films and Multilayered Composites," *International Journal of Thermophysics*, Vol. 11, pp. 293-302.

3. Tien, C.L. and Chen, G., 1994, "Challenges in Microscale Radiative and Conductive Heat Transfer," ASME *Journal of Heat Transfer*, Vol. 116, pp. 799-807.

4. Goodson K.E. and Flik, M.I., 1994, "Solid Layer Thermal-Conductivity Measurement Techniques," *Applied Mechanics Review*, Vol. 47, pp. 101-112.

5. Hatta, I., Sasuga, Y., Kato, R., and Maesono, A., 1985, "Thermal Diffusivity Measurement of Thin Films by Means of an ac Calorimetric Method," *Review of Scientific Instruments*, Vol. 56, pp. 1643-1647.

6. Hatta, I., Kato, R., and Maesono, A., 1987, "Development of ac Calorimetric Method for Thermal Diffusivity Measurement. II. Sample Dimension Required for the Measurement," *Japanese Journal of Applied Physics*, Vol. 26, pp. 475-478.

7. Kato, R., Maesono, A., and Hatta, I., 1993, "Development of AC Calorimetric Method for Thermal Diffusivity Measurement V. Modulated Laser Beam Irradiation," *Japanese Journal of Applied Physics*, Vol. 32, pp. 36563658.

8. Yu, X.Y., Chen, G., Verma, A., and Smith, J.S., 1995, "Temperature Dependence of Thermophysical Properties of GaAs/AlAs Periodic Structure," *Applied Physics Letters*, Vol. 67, pp. 3554-3556.

9. Zhang, B., Imhof, R.E., and Hartree, W., 1994, "Thermal Surface Wave Technique for Thin-Film Thermal Diffusivity Measurement," *Journal De Physique* IV, Colloque C7, pp. 643646.

10. Kemp, T., Srinivas, T.A.S., Fettig, R., and Ruppel, W., 1995, "Measurement of Thermal Diffusivity of Thin Films and Foils Using A Laser Scanning Microscope," *Review of Scientific Instruments*, Vol. 66, pp. 176-181.

11. Chen, G., Tien, C.L., Wu, X., and Smith, J.S., 1994, "Measurement of Thermal Diffusivity of GaAs/AlGaAs Thin-Film Structures," ASME *Journal of Heat Transfer*, Vol. 116, pp. 325-331.

12. Abramowitz, M. and Stegun, I.A., 1964, *Handbook of Mathematical Functions*, Applied Mathematics Series, Vol 55, p. 385.

13. Spanier J. and Oldham, K., 1987, *An Atlas of Functions*, Hemisphere Publishing Corporation, Washington, p. 485-505.

14. Chen G. and Yu, X.Y., 1994, "Thermal Diffusivity of GaAs/AlAs Superlattices," AIAA paper No. 94-1964, presented at 6th AIAA/ASME Joint Thermophysics and Heat Transfer Conference, June 20-23, 1994, Colorado Spring, Colorado.

15. Touloukian, T.S., Powell, R.W., Ho, C.Y., and Nicolaou, M.C., 1973, *Thermophysical Properties of Matter*, IFI/Plenum, New York, Vol. 10, p. 160.

16. Chen, G., 1995, "Size and Interface Effects on Thermal Conductivity of Superlattices and Periodic Thin-Film Structures," submitted to ASME *Journal of Heat Transfer*.

# THEORY

SESSION CHAIRS
P. G. Klemens
J. V. Beck

# Thermal Conductivity of Zirconia

P. G. KLEMENS

## ABSTRACT

Stabilized cubic zirconia is used for thermal barrier coatings, and its thermal conductivity has been measured with conflicting results. To understand its behavior it is here calculated from thermal conductivity theory. The intrinsic thermal conductivity, which is not accessible to measurement, is estimated to be 1700/T W/m•K (T in Kelvin), in line with the measured value of 2800/T for titania. Point defect scattering of phonons is then estimated for additions of $Y_2O_3$, CaO, MgO and $CeO_2$, taking account of the vacancies which accompany the first three of these solute cations. The consequent thermal conductivity reductions are calculated. Grain boundary scattering and possible reductions by two-level tunnel states give further reductions. Theoretical values are compared with observations. There is rough agreement at low temperatures, but theory predicts larger decreases in conductivity than have been observed at high temperatures. This may be due to internal radiation.

## INTRODUCTION

Thermal barrier coatings of stabilized zirconia are used in gas-turbine engines to protect metal components from hot gases in the working chamber, allowing higher temperatures and greater efficiency. Increased solute content, small grain size and multilayer structures are ways of reducing the thermal conductivity. While mechanical and chemical stability are the most important constraints, there is a need to understand which factors are most significant in limiting the conductivity. This paper aims to provide a guide in terms of the theory of phonon interaction in dielectric crystals, and their effect on the thermal conductivity. One has to test the theory against existing observations before one can use it to suggest improvements.

P. G. Klemens, Dept. of Physics and Institute of Materials Science, University of Connecticut, Storrs, CT 06269.

In such a study one normally proceeds from measured thermal conductivities of a single crystal, or of fully dense and large-grained polycrystalline specimens, of the pure material, which yield the intrinsic thermal conductivity. Using the model of defects, one then calculates the reductions due to point defects, grain size or layer thickness, inclusions and other imperfections. There are however materials which are not available in the defect-free form. Thus tin-telluride always contains tin vacancies. Cubic zirconia requires solutes to stabilize the cubic phase, and its intrinsic conductivity is also not known. It must be estimated before one can discuss the influence of defects.

## INTRINSIC CONDUCTIVITY OF ZIRCONIA

We concentrate on the high temperature or classical equipartition regime, where the theory is simpler, and which is the regime of technical interest. The mean free path of phonons is limited by three-phonon interactions, owing to the cubic anharmonicities of the lattice forces. This was treated by Leibfried and Schlömann. Their theory exists in several variations, all agreeing in form, but differing in multiplicative constants, which depend on the way one sums over all interactions [1]. The present author obtained the following intrinsic phonon mean free path in a monatomic solid [2]:

$$1/\ell_i(\omega,T) = 2\gamma^2 \ (kT/\mu a^3)(\omega^2/v\omega_D) \tag{1}$$

Here $\omega$ is the phonon frequency (rad/sec), $\gamma$ the Grüneisen parameter, a measure of anharmonicity, $k$ is the Boltzmann constant, $T$ the absolute temperature. Also $\mu$ is the shear modulus, $a^3$ the atomic volume, so that $kT/\mu a^3$ is the mean square thermal strain due to one component of lattice vibrations, $v$ is the transverse phonon velocity, and $\omega_D$ the Debye frequency of the monatomic lattice. Thus

$$\omega_D^3 \ a^3/v^3 = 6\pi^2.$$

The thermal conductivity is then

$$\lambda = (1/3) \int C(\omega) \ v \ \ell(\omega) \ d\omega \tag{2}$$

where $C(\omega)$ is the spectral component of the vibrational specific heat per unit volume. In the equipartition limit

$$C(\omega) = 9ka^{-3} \ \omega_D^{-3} \ \omega^2 \tag{3}$$

From Equations (1) and (3), Equation (2) yields the intrinsic thermal conductivity

$$\lambda_i = (3/2\gamma^2) \ (\mu v^2/\omega_D) T^{-1}. \tag{4}$$

If thermal expansion is disregarded, $\lambda_i \propto T^{-1}$.

For solids which are not monatomic, but consist of molecular groups of N atoms, the integration in Equation (2) is not from 0 to $\omega_D$ but to $\omega_m = \omega_D N^{-1/3}$. Also in $\omega_D$ in Equation (1) should be replaced by $\omega_m$, so that $\lambda_i$ of Equation (4) should contain an additional factor $N^{-2/3}$. For zirconia, N equals 3.

Since $\gamma^2$ describing phonon scattering must always exceed that derived from thermal expansion [3], one may in the absence of better information, take $\gamma^2 = 4$, thus departing from Slack's analysis [4], who took $\gamma$ from thermal expansion data. Other parameters are: shear modulus $\mu = 5.3 \times 10^4$ J cm$^{-3}$, density $\rho = 5.83$ g/cc, average atom mass $6.82 \times 10^{-23}$ g, volume per atom $a^3 = 1.17 \times 10^{-23}$ cm. Thus the shear wave velocity is $v = (\mu/\rho)^{1/2} = 3.0 \times 10^5$ cm/s and $\mu a^3/k = 45,000$ K. From v and $a^3$, $\omega_D = 5.2 \times 10^{13}$ per sec. and $\omega_m = 3.6 \times 10^{13}$ per sec.

The intrinsic thermal conductivity, from Equation (4), reduced by a factor 2.08, becomes

$$\lambda_i = 1.7 \times 10^3 \ T^{-1} \qquad \text{(in W/m•K)} \tag{5}$$

a value which cannot be tested directly.

The intrinsic thermal conductivity of titanium dioxide is known [5]. Averaged over direction, $\lambda_i$ is $2.8 \times 10^3$ /T. Now the shear modulus does not vary substantially between similar materials, while v and $\rho$ differ mainly because of molar mass. Thus $\lambda_i$ should vary approximately as the inverse square root of the molar mass, which would make the conductivity of zirconia about 0.81 of that of titania, i.e. about 2300/T, or a factor 4/3 larger that the estimate of Equation (5).

## EFFECT OF POINT DEFECTS

Zirconia can be stabilized by additions of CaO, MgO, $Y_2O_3$ or $CeO_2$. All but the last also have oxygen vacancies to maintain charge neutrality. One expects vacancies to be stronger phonon scatterers than the solute metal ions. For reasons of chemical stability, CaO and MgO are not as suitable for thermal barrier coatings as the other two. However, existing thermal conductivity measurements lend interest to these cases [6, 7].

Point defects scatter phonons as the fourth power of frequency over most of the acoustic branch, so that they contribute to the attenuation as $1/\ell_p (\omega) = A \ \omega^4$. Adding this to the reciprocal mean free path $1/\ell_i (\omega, T)$, the combined mean free path becomes

$$\ell(\omega) = \ell_i(\omega,T) \ [\ 1 + (\omega/\omega_0)^2\ ]^{-1} \qquad (6)$$

where $\omega_0$ is the frequency for which $\ell_i\ (\omega\ ,\ T) = \ell_p\ (\omega_0)$. If one substitutes $\ell(\omega)$ into Equation 3 and integrates from 0 to $\omega_m$, the thermal conductivity in the presence of point defects becomes [8]:

$$\lambda_p = \lambda_i - \delta\ \lambda_p$$

where

$$\delta\lambda_p = \lambda_i\ [\ 1 - (\omega_0/\omega_m)\ \arctan(\omega_m/\omega_0)\ ] \qquad (7)$$

The stronger the point defect scattering, the smaller $\omega_0$. For very strong point defect scattering $\arctan \omega_m/\omega_0 \approx \pi/2$ and $\lambda_p \propto T^{-1/2}$ since $\lambda_i \propto T^{-1}$ and $\omega_0 \propto T^{1/2}$. In the opposite limit of weak point defect scattering, when

$$\delta\lambda_p/\lambda_i \sim (\omega_m/\omega_0)^2/3 \qquad (8)$$

this is equivalent to adding a temperature-independent resistivity $(\omega_m\ /\ \omega_0)^2\ /3\lambda_i$.

The reciprocal point defect mean free path due to substitutional atoms in a monatomic lattice, which differ from solvent atoms only in respect to mass, is given by

$$1/\ell_p = c\ (\Delta M/M)^2\ a^3\ (4\pi\ v^4)^{-1}\ \omega^4 \qquad (9)$$

where c is the concentration per atom, $a^3$ the atomic volume, M the mass of the solvent atom and $\Delta M$ the mass difference between solute and solvent [9]. In an oxide lattice, M becomes the average atomic mass and c the concentration per atom, which in the case of zirconia is 1/3 of the solute cation concentration per mole, while $a^3$ remains the volume per atom.

Except in the case of $CeO_2$ additions, there are also oxygen vacancies, assumed to scatter independently of the cation solutes, and contributing additively to $c(\Delta M/M)^2$. Each vacancy lacks the linkages of two atoms [10], that is $\Delta M/M = -2$, while the missing oxygen mass contributes -16/41, so that effectively $\Delta M/M = -2.39$, $(\Delta M/M)^2 = 5.7$.

There is also a contribution to phonon scattering by solute atoms because of size misfit. As shown elsewhere [11], the effective value of $\Delta M$ becomes

$$(\Delta M/M)_{eff} = \Delta M/M + 2\gamma\ \Delta V/V \qquad (10)$$

where $\Delta V/V$ is the fractional difference in ionic volume. However, this expression may not hold in the presence of vacancies. Therefore the strain correction has been used only in the case of $CeO_2$. In this case, strain enhances the effect of the mass difference.

## POINT DEFECTS - NUMERICAL RESULTS

Using Equations (1) and (9) and defining $\omega_o$ by $\ell_p(\omega_o) = \ell_i(\omega_o, T)$, one finds that

$$(\omega_0/\omega_m)^2 = (4\gamma^2/3) \ NkT/\mu a^3 \ [c(\Delta M/M)^2]^{-1}. \tag{11}$$

With the parameters previously chosen to calculate $\ell_i$,

$$(\omega_0/\omega_m)^2 = 5.66 \times 10^{-5} \ T \ [c(\Delta M/M)^2]^{-1} \tag{12}$$

(T in Kelvin). Table I gives values of $c(\Delta M/M)^2$ for 1 weight percent of the four solute oxides considered here. Thus one can calculate $\lambda$ for a dense large-grained zirconia specimen, where intrinsic scattering and point defect scattering should be the only important processes.

## MgO-STABILIZED ZIRCONIA

Flieger and Ginnings [6] measured the thermal diffusivity and derived the conductivity of one composition, 4.1 wt %. The diffusivity and conductivity decrease with increasing temperature, roughly as $T^{-1/2}$, from 400 to 1100 K, as expected from Equation (7) in the high concentration limit. This gives interest to this early measurement. Theoretical values are derived with $c(\Delta M/M)^2 = 0.35$. Values of

TABLE I SCATTERING PARAMETER OF FOUR SOLUTE OXIDES PER WEIGHT PERCENT

| Solute | $10^3 c$ | $\Delta M/M$ | $2\gamma \Delta V/V$ | $c(\Delta M/M)^2$ Solute | $c(\Delta M/M)^2$ Vacancy | $c(\Delta M/M)^2$ Total |
|--------|----------|--------------|----------------------|--------------------------|---------------------------|-------------------------|
| $YO_{1.5}$ | 3.62 | -0.05 | | $1 \times 10^{-5}$ | 0.0103 | 0.0103 |
| CaO | 7.33 | -1.24 | | 0.0113 | 0.0418 | 0.0645 |
| MgO | 10.19 | -1.63 | | 0.0271 | 0.0581 | 0.0852 |
| $CeO_2$ | 2.38 | +1.20 | +1.98 | 0.0241 | | 0.0241 |

TABLE II  THERMAL CONDUCTIVITY OF ZIRCONIA
with 4.1 wt% MgO, calculated with $c(\Delta M/M)^2 = 0.350$, and compared to measured values of Fleiger and Ginnings [6].

| T, K | $\omega_c/\omega_m$ | $\lambda_i$ W/m•K | $\lambda_{th}$, W/m•K | $\lambda_{obs}$, W/m•K |
|------|------|------|------|------|
| 373 | 0.246 | 4.50 | 1.47 | 1.54 |
| 573 | 0.304 | 2.93 | 1.14 | 1.35 |
| 773 | 0.354 | 2.17 | 0.95 | 1.22 |
| 973 | 0.397 | 1.73 | 0.82 | 1.18 |
| 1173 | 0.436 | 1.43 | 0.72 | 1.16 |

$\omega_c/\omega_m$ calculated values $\lambda_{th}$ and observed values of $\lambda$ are shown in Table II. Theory consistently underestimates the conductivity; the discrepancy increases with increase in T. No correction was made for porosity, which should have reduced the conductivity at low temperatures, and may have increased it at high temperatures by radiation within the pores. A higher value of $\lambda_i$, as deduced by scaling from $TiO_2$, should have increased $\lambda_{th}$ by 15% at 373 K, less at higher temperatures.


## CALCIA-STABILIZED ZIRCONIA

Mirkovich and Wheat [7] measured the thermal diffusivity of six compositions (5 to 22.5 mole %). In a separate apparatus, which required larger samples, they also measured the conductivity of four compositions these are shown in Table III at three temperatures. Significantly, they were unable to make the larger samples for 5 and for 22.5 mole %. Theoretical values of $c(\Delta M/M)^2$, $\omega_c/\omega_m$, and conductivity were calculated as in the previous case. One notes that the observed conductivities do not decrease with increasing temperature as rapidly as theory requires, and also that they do not decrease with solute content beyond 10%. The authors consider the 10% conductivities to be anomalously low; theory suggest that this decrease should be expected, and that the higher concentration samples may not be single phase.


## YTTRIA-STABILIZED ZIRCONIA

Because of its stability, this is the material used for thermal barrier coatings at high temperatures. The thermal diffusivity of 5.1 wt% material was measured, as function of porosity, by Mirkovich [11]. His starting material was very fine grained powder, so that the grains became large on sintering. He considered his results to be reliable up to 700°C. Table IV shows his results for 3% porosity, converted to conductivity assuming for the specific heat $C_p = 3.4$ J•cm$^{-3}$•K$^{-1}$.

### TABLE III THERMAL CONDUCTIVITY OF $ZrO_2$ WITH CaO
Calculated values compared to measurements of Mirkovich and Wheat [7].

| T, K | $\omega_0/\omega_m$ | $\lambda_i$, W/m•K | $\lambda_{th}$, W/m•K | $\lambda_{obs}$, W/m•K |
|---|---|---|---|---|
| 7.6 mole %: $c(\Delta M/M)^2 = 0.182$ | | | | |
| 400 | 0.353 | 4.2 | 1.83 | 2.25 |
| 600 | 0.432 | 2.8 | 1.41 | 2.01 |
| 800 | 0.499 | 2.1 | 1.16 | 1.80 |
| 10 mole %: $c(\Delta M/M)^2 = 0.240$ | | | | |
| 400 | 0.307 | 4.2 | 1.64 | 1.65 |
| 600 | 0.376 | 2.8 | 1.28 | 1.62 |
| 800 | 0.434 | 2.1 | 1.06 | 1.53 |
| 15 mole %: $c(\Delta M/M)^2 = 0.360$ | | | | |
| 400 | 0.251 | 4.2 | 1.40 | 1.74 |
| 600 | 0.307 | 2.8 | 1.09 | 1.76 |
| 800 | 0.355 | 2.1 | 0.92 | 1.62 |
| 20 mole %: $c(\Delta M/M)^2 = 0.480$ | | | | |
| 400 | 0.234 | 4.2 | 1.32 | 1.94 |
| 600 | 0.284 | 2.8 | 1.03 | 1.86 |
| 800 | 0.330 | 2.1 | 0.87 | 1.81 |

A similar composition ( 6 Wt %) was measured by Morell and Taylor [12], shown in Table IV. These authors also measured 8, 10, and 12 wt %, but for these compositions the conductivity depended on the shape of the powder particles before sintering and on the crack structure, more than on composition.

Anderson et al. measured 8 wt % material, shown in Table IV [13]. Eaton et al. measured 20 wt % material - see the Table [14]. It did not show significant temperature dependence, even a slight increase at high T.

In all these specimens, the expected decrease of conductivity with temperature is weaker than theory demands, even in Mirkovich's large-grained specimen, which comes closest to theoretical expectations. At the lowest temperatures there is rough agreement with theory.

### TABLE IV THERMAL CONDUCTIVITY OF $ZrO_2$ WITH $Y_2O_3$

| T, K | $\omega_o/\omega_m$ | $\lambda_i$, W/m•K | $\lambda_{th}$, W/m•K | $\lambda_{obs}$, W/m•K |
|---|---|---|---|---|
| 5.1 wt%, ref. [11]:  $c(\Delta M/M)^2 = 0.053$ | | | | |
| 473 | 0.71 | 3.55 | 2.4 | 2.3 |
| 900 | 0.98 | 1.87 | 1.5 | 1.8 |
| 6 wt%, ref. [12]:  $c(\Delta M/M)^2 = 0.062$ | | | | |
| 500 | 0.68 | 3.40 | 2.2 | 1.9 |
| 900 | 0.91 | 1.87 | 1.4 | 1.6 |
| 8 wt%, ref. [13]:  $c(\Delta M/M)^2 = 0.083$ | | | | |
| 300 | 0.65 | 5.67 | 3.6 | 1.5 |
| 700 | 0.69 | 2.43 | 1.6 | 1.3 |
| 900 | 0.78 | 1.83 | 1.3 | 1.1 |
| 20 wt%, ref. [14]:  $c(\Delta M/M)^2 = 0.181$ | | | | |
| 473 | 0.38 | 3.55 | 1.6 | 1.6 |
| 673 | 0.43 | 2.45 | 1.2 | 1.6 |
| 873 | 0.49 | 1.92 | 0.9 | 1.6 |
| 1,273 | 0.60 | 1.32 | 0.8 | 1.7 |
| 1,673 | 0.68 | 1.00 | 0.7 | 1.7 |

## CERIUM-OXIDE STABILIZED ZIRCONIA

Since oxygen vacancies are not needed to conserve charge, one may assume their absence. Phonon scattering is by the solute cations only, and as seen from Table I, distortion increases that scattering over mass difference scattering by a factor 7. Thermal diffusivities of 15 wt %, 20 wt % and 25 wt % have been measured from 450 K to 1400 K [15], showing very little decrease with temperature, but strong decrease with concentration, somewhat stronger than predicted. Table V compares theoretical versus observed values of conductivity for two concentrations and two temperatures each, using 2.4 J•cm$^{-3}$•K$^{-1}$ for the specific heat. The theoretical values are too high in all cases, with the biggest discrepancy at low temperatures and high concentrations.

TABLE V THERMAL CONDUCTIVITY OF $ZrO_2$ WITH $CeO_2$

| T, K | $\omega_q/\omega_m$ | $\lambda_i$, W/m•K | $\lambda_{th}$, W/m•K | $\lambda_{obs}$, W/m•K |
|------|------|------|------|------|
| 15 wt% $c(\Delta M/M)^2 = 0.362$ | | | | |
| 400 | 0.250 | 4.25 | 1.41 | 0.72 |
| 1000 | 0.395 | 1.70 | 0.80 | 0.67 |
| 25 wt% $c(\Delta M/M)^2 = 0.605$ | | | | |
| 400 | 0.193 | 4.25 | 1.13 | 0.36 |
| 1000 | 0.306 | 1.70 | 0.66 | 0.31 |

## GRAIN BOUNDARIES, FILM THICKNESS, INCLUSIONS

All these defects make a frequency-independent contribution to 1/L to the reciprocal of the mean free path. For grains and thin films, L is of the order of the mean grain diameter or of the film thickness. For small inclusions or voids, uniformly dispersed within each grain, and occupying a fractional volume p, L is given by

$$1/L = 3p/4R \tag{13}$$

where R is the radius of an inclusion.

This scattering mechanism is most effective at frequencies $<\omega_B$, where $\omega_B$ is defined by $\ell_i(\omega_B, T) = L$. Thus

$$(\omega_B/\omega_m)^2 = \mu a^3 (2\gamma^2 kT)^{-1} v/L\omega_m \tag{14}$$

and the reduction in thermal conductivity due to these scattering processes

$$\delta\lambda_B/\lambda_i = (\omega_B/\omega_m) \arctan(\omega_m/\omega_B)$$
$$\approx 1.57 \ \omega_B/\omega_m \tag{15}$$

In most cases $\omega_B \ll \omega_m$, and also $\omega_B$ is well below $\omega_o$, so that $\delta\lambda_B$ and $\delta\lambda_p$ are independent of each other. In the case of zirconia, with the parameters previously used,

$$\omega_B/\omega_D = 0.69 \ T^{-1/2} \ L^{-1/2} \tag{16}$$

where T is in Kelvin, L in micrometers. For L = 1 μm and T = 400 K, $\omega_B/\omega_m = 0.034$ and $\delta\lambda_B = 0.22$ W/m•K. Note that $\delta\lambda_B \propto T^{-3/2}$. At 900 K $\delta\lambda_B = 0.07$. Grain boundary scatting can thus reduce the conductivity at low temperatures and decrease its temperature dependence, but does not account for the fact that the observed conductivities do not decrease with increasing temperature at high temperature.

## OTHER PROCESSES

Single crystal yttria-stabilized zirconia has been measured at low temperatures by Walker and Anderson [17] and more recently by Morelli [18]. At very low temperatures $\lambda \propto T^2$, implying a scattering mechanism $1/\ell \propto \omega$. It has been suggested that an oxygen vacancy exists in two configurations, with a spacing between the energy levels which is not fixed, but varying randomly, depending on its surroundings. Extrapolating the observed scattering to higher frequencies is roughly equivalent to a value of L of about 0.5 μm. However, it is not certain whether this scattering mechanism should persist at higher temperatures, because the energy barrier between configurations may not be high enough.

At high temperatures, heat transfer by radiation may be significant. for a thin slab of transparent material of refractive index n and thickness L, the radiative component of thermal conductivity is

$$\lambda_R = 4\sigma T^3 n^2 L = 2.27 \times 10^{-7} T^3 n^2 L \tag{17}$$

(in W/m•K), where σ is the Stefan Boltzmann constant. If the material is transparent only at frequencies below $k\Delta/h$, and completely absorbent above,

$$\lambda_R = 8.73 \times 10^{-9} (T/\Delta)^3 J_4(\Delta/T) n^2 \Delta^3 L \tag{18}$$

where Δ is in K and $J_4(Z) = \int_0^z x^4 e^x (e^x - 1)^{-2} dx$. For $\Delta \to \infty$ Equation 18 becomes (17). For n = 2.7, L = 1 mm and T = 1000 K, $\lambda_r$ would be 1.65 W/m•K. However we know from the measurements of Brandt that the absorption edge Δ must be lower than 4700K, i.e. 3 μm [19]. With n = 2.7 and taking Δ = 2800 K (5 μm), the values given in Table VI were obtained.

It thus appears that radiative heat transfer could compensate a decrease in lattice conductivity at higher temperatures, leading to an overall weak temperature dependence. Since $Z^{-3} J_4(Z)$ saturates to a value of 1/3, so does $\lambda_r$ at highest temperatures. In the presence of an absorption edge, radiative heat transfer need not vary as $T^3$.

### TABLE VI RADIATIVE COMPONENT

| T (°K) | 600 | 800 | 1000 | 1500 | ∞ |
|---|---|---|---|---|---|
| $\lambda_r$ W/m•K | 0.19 | 0.27 | 0.33 | 0.39 | 0.47 |

## SUMMARY

The calculated conductivities of stabilized zirconia agree roughly with observations in magnitude, if one considers all the uncertainties which enter the theory. However the point defect scattering model is insufficient to explain the weak temperature dependence of the conductivity. At high temperatures this may be due to a radiative component. At low temperatures (below 400 K) it appears that at least one other scattering mechanism is active. Grain boundary scattering of phonons is not enough to account for this scattering. The effect of cracks and intermittent content between grains need to be better understood.

## ACKNOWLEDGEMENT

I am indebted to Dr. Maurice Gell, who introduced me to this problem and gave me much valuable information about thermal barrier coatings.

## REFERENCES

1. Leibfried, G. and E. Schlömann, 1954, Nachr. Akad. Wiss. Göttingen, IIa (4): 71.

2. Klemens, P. G., 1969, in Thermal Conductivity, R. P. Tye, ed., vol. 1, p.1, Academic Press, London.

3. Klemens, P. G., 1993, in Phonon Scattering in Condensed Matter VII, M. Meissner and R. O. Pohl, eds., p. 15, Springer, Berlin.

4. Slack, G. A., 1979, in Solid State Physics, H. Ehrenreich, F. Seitz and D. Turnbull, eds., vol. 34, p. 1, Academic Press, New York.

5. Touloukian, Y. S., R. W. Powell, C. Y. Ho and P. G. Klemens, 1970, Thermophysical Properties of Matter, Vol. 2 Thermal Conductivity, p. 208, Plenum, New York.

6. Flieger, H. W. and D. C. Ginnings, 1957, National Bureau of Standards Report No. 5642.

7. Mirkovich, V. V. and T. A. Wheat, 1985, High Temps. - High Pressures 17: 67.

8. Klemens, P. G., 1960, Phys. Rev., 119: 507.

9. Klemens, P. G., 1955, Proc. Phys. Soc. (London) A68: 1113.

10. Ratsifaritana, C. A. and P. G. Klemens, 1987, Int. J. Thermophysics 8: 737.

11. Mirkovich, V. V., 1976, <u>High Temps. - High Pressures</u> 8: 231.

12. Morrell, P. and R. Taylor, 1985, <u>High Temps. - High Pressures</u> 17: 79.

13. Andersson, C. A., S. K. Lau, R. J.. Bratton, S. Y. Lee, K. L. Rieke, J. Allen and K. E. Munson, 1982, <u>Advanced Ceramic Coating Development</u>, NASA Contractor Report CR 165619. Westinghouse Electric Corp.

14. Eaton, H. E., J. R. Linsey and R. B. Dinwiddie, 1994, in <u>Thermal Conductivity 22</u>, T. W. Tong, ed., Technomic, Lancaster, PA.

15. Brandon, J. R. and R. Taylor, 1989, <u>Surface and Coatings Technology</u> 39/40: 143.

16. Klemens, P. G., 1993, <u>Thermochimica Acta</u> 218: 247.

17. Walker, F. J. and A. C. Anderson, 1984, <u>Phys. Rev.</u> 29: 5881. See also Cahill, D. G., S. K. Watson and R. O. Pohl, 1992 <u>Phys. Rev.</u> B46: 6131.

18. Morelli, D. T., private communication..

19. Brandt, R., 1981, <u>High Temps. - High Pressures</u> 13: 79.

# A Rigid-Ion Lattice Dynamical Model for Calculating Group Velocities in Simple, Body-Centered, and Face-Centered Cubic Crystals

A. K. McCURDY

## ABSTRACT

A rigid-ion lattice dynamical model is developed for calculating phase and group velocities in dispersive sc, bcc, and fcc crystals. The model is easy to understand, allows non-central symmetric forces and requires no explicit use of group theory. Results are given for the sc lattice up to and including third nearest-neighbor interactions, and agree with Born-von Karman theory. This model allows, however, a maximum of 3 force constants per site whereas the general Born-von Karman model specifies, for example, 4 constants for fourth nearest-neighbors in bcc, and 4 constants each for third and fifth nearest-neighbors in fcc crystals. This model thus constrains one of the 4 force constants to be linearly dependent upon two of the other force constants for such neighbors. The model in its present form results in a symmetric force matrix and thus is not applicable for the diamond lattice. The model, however, provides a straightforward method for calculating phonon-focusing effects in sc, bcc and fcc crystals.

## INTRODUCTION

Phonon focusing in crystals in the long-wavelength limit depends upon the second-order elastic constants [1]. In elastically anisotropic crystals the phase velocity surfaces (i.e., the phonon constant-energy surfaces in wave-vector space) are nonspherical. The phonon phase velocity $s$ is parallel to the wave vector $k$, but the group velocity $v$ is normal to the constant energy surface in $k$ space. Phonon momentum is parallel to the wave vector, but energy flow is parallel to the group velocity. As a result the group velocity is in general, not parallel to the wave vector $k$. In fact an equidensity of wave vectors in solid angle gives a highly nonuniform density in group-velocity space [2]. This phonon focusing property is responsible for anisotropic heat conduction in cubic crystals in the boundary-scattering regime [3,4], and for energy-flow reversal (EFR) about symmetry axes for certain symmetric

Professor Emeritus Archie K. McCurdy, Electrical and Computer Engineering Department, Worcester Polytechnic Institute, 100 Institute Road, Worcester MA 01609

energy- and momentum-conserving 3-phonon and 4-phonon interactions [5-8].

For larger values of the wave vector dispersion becomes significant, and the phase and group velocities change with frequency. This paper develops a rigid-ion lattice dynamical model from which phase and group velocities can be easily calculated and effects of dispersion can be evaluated. The model gives the same results as the Born-von Karman general force model [9] except for fewer force constants for certain further-neighbor interactions. Results are given here in detail for the simple cubic lattice. Results for the bcc [10] and fcc [11] lattices are given in subsequent papers in this volume.

## THEORY AND ANALYSIS

Results for simple cubic crystals will be discussed in some detail. The analysis parallels the method discussed by Smith [12], but is generalized to allow for non-central forces up to and including 3rd nearest-neighbor (n-n), 5th n-n [10], and 6th n-n [11] interactions, respectively, in the sc, bcc, and fcc lattices.

Let the equilibrium positions of atoms (n) and (m) respectively be $\mathbf{r}_n$ and $\mathbf{r}_m$ and their displaced positions indicated by primes (see Fig. 1). Then $\mathbf{r}_m = \mathbf{r}_n + d\mathbf{i}$ where d

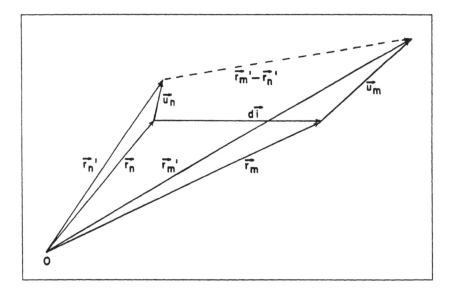

Fig. 1. Vector diagram showing the equilibrium and displaced positions of two interacting neighbors in a crystal lattice. Vectors $\mathbf{r}_n$ and $\mathbf{r}_m$ indicate the equilibrium positions of atoms (n) and (m) respectively, whereas the prime on these vectors indicate their displaced positions. Vectors $\mathbf{u}_n$ and $\mathbf{u}_m$ indicate the displacements of atoms (n) and (m) respectively.

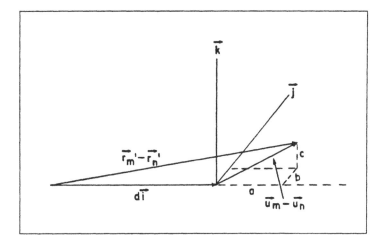

Fig. 2. Vector diagram showing two orthogonal components of the relative displacements $u_m$-$u_n$ of atoms (m) and (n). Vector di is directed from the equilibrium position of atom (n) toward the equilibrium position of atom (m). For first-nearest neighbors in the sc lattice d becomes the lattice constant.

is the equilibrium spacing between atoms (n) and (m) and **i** is a unit vector directed from atom (n) to atom (m). The displaced positions of these two atoms are:

$$\mathbf{r}'_n = \mathbf{r}_n + \mathbf{u}_n \text{ and } \mathbf{r}'_m = \mathbf{r}_m + \mathbf{u}_m$$

where $\mathbf{u}_n$ and $\mathbf{u}_m$ are their respective displacements. The force on atom (n) due to atom (m) is due to their relative displacements given by:

$$\mathbf{u}_m\text{-}\mathbf{u}_n = (\mathbf{r}'_m\text{-}\mathbf{r}'_n)\text{-}(\mathbf{r}_m\text{-}\mathbf{r}_n) = (\mathbf{r}'_m\text{-}\mathbf{r}'_n)\text{-d}\mathbf{i}.$$

Now $\mathbf{u}_m$-$\mathbf{u}_n$ will generally have components along **i** as well as two mutually orthogonal components along unit vectors **j** and **k** (see Fig. 2). Elementary treatments of lattice dynamics which make the central force approximation neglect these two orthogonal components of relative displacement.

Consider the force **F** on atom (n) due to relative displacements from equilibrium of atoms (n) and (m) and let the force constants along **i**, **j**, and **k** be $\alpha$, $\beta$, and $\gamma$ respectively. Then the relative displacement components along **i**, **j**, and **k** are $\mathbf{i}.(\mathbf{u}_m\text{-}\mathbf{u}_n)$, $\mathbf{j}.(\mathbf{u}_m\text{-}\mathbf{u}_n)$, and $\mathbf{k}.(\mathbf{u}_m\text{-}\mathbf{u}_n)$, respectively. The force on atom (n) is thus given by:

$$\mathbf{F} = \alpha\mathbf{i}[\mathbf{i}.(\mathbf{u}_m\text{-}\mathbf{u}_n)] + \beta\mathbf{j}[\mathbf{j}.(\mathbf{u}_m\text{-}\mathbf{u}_n)] + \gamma\mathbf{k}[\mathbf{k}.(\mathbf{u}_m\text{-}\mathbf{u}_n)].$$

For atom (n) at coordinates: (dr, ds, dt)

$$\mathbf{u}_n = \mathbf{a}_x u_{r,s,t} + \mathbf{a}_y v_{r,s,t} + \mathbf{a}_z w_{r,s,t}$$

and for a first-nearest neighbor (m) in the sc lattice at coordinates: (dr+d, ds, dt)

$$\mathbf{u}_m = \mathbf{a}_x u_{r+1,s,t} + \mathbf{a}_y v_{r+1,s,t} + \mathbf{a}_z w_{r+1,s,t}$$

where u, v, and w are the cartesian displacment components. For this particular first-nearest neighbor:

$$\mathbf{i} = \mathbf{a}_x, \ \mathbf{j} = \mathbf{a}_y, \text{ and } \mathbf{k} = \mathbf{i} \times \mathbf{j} = \mathbf{a}_z$$

and the three components of the force on atom (n) are:

$$F_x = \alpha_1(u_{r+1,s,t} - u_{r,s,t})$$

$$F_y = \beta_1(v_{r+1,s,t} - v_{r,s,t})$$

$$F_z = \gamma_1(w_{r+1,s,t} - w_{r,s,t}).$$

The force components for each of the remaining 6 first-nearest neighbors can be calculated in a similar manner, and summed to obtain the resulting force on atom (n).

Assuming plane wave solutions for u, v, and w of the form:

$$exp[i(\mathbf{k.r} - \omega t)] \text{ where } exp \text{ denotes the exponent of } e \text{ (i.e., 2.71828)}$$

so that, for example, $F_x = M\partial^2 u/\partial t^2 = -M\omega^2 u$, and

$$u_{r+1,s,t} = u_{r,s,t} \, exp(ik_x d) \text{ and } u_{r-1,s,t} = u_{r,s,t} \, exp(-ik_x d),$$

$$u_{r+1,s,t} + u_{r-1,s,t} = 2\cos(k_x d), \ u_{r+1,s,t} - u_{r-1,s,t} = 2i \sin(k_x d),$$

yields a set of three equations for the displacement components u, v, and w. These can be written for general neighbors in dynamical matrix form as:

$$\begin{vmatrix} M\omega^2 - \Gamma_{11} & -\Gamma_{12} & -\Gamma_{13} \\ -\Gamma_{12} & M\omega^2 - \Gamma_{22} & -\Gamma_{23} \\ -\Gamma_{13} & -\Gamma_{23} & M\omega^2 - \Gamma_{33} \end{vmatrix} \begin{vmatrix} u_{r,s,t} \\ v_{r,s,t} \\ w_{r,s,t} \end{vmatrix} = 0.$$

Note that M is the mass of the atom (n) and $\omega$ the radian vibrational frequency.

Considering atom (n) at (dr,ds,dt) to be at the origin of a simple cubic lattice, the 6 first-nearest neigbors are located at: ($\pm$d,0,0), (0,$\pm$d,0), (0,0,$\pm$d). Here because of symmetry the force constant $\gamma_1 = \beta_1$. The dynamical matrix elements for the first-nearest neighbors are:

$\Gamma_{11} = 2\alpha_1[1-\cos X] + 2\beta_1[2-\cos Y-\cos Z]$

$\Gamma_{22} = 2\alpha_1[1-\cos Y] + 2\beta_1[2-\cos Z-\cos X]$

$\Gamma_{33} = 2\alpha_1[1-\cos Z] + 2\beta_1[2-\cos X-\cos Y]$

$\Gamma_{12} = \Gamma_{23} = \Gamma_{13} = 0$

where $X = k_x d$, $Y = k_y d$, $Z = k_z d$.

The 12 second-nearest neigbors are located at: $(\pm d,\pm d,0)$, $(\pm d,0,\pm d)$, $(0,\pm d,\pm d)$. For the neighbor at $(d,d,0)$, for example, $\mathbf{i} = (\mathbf{a}_x+\mathbf{a}_y)/\sqrt{2}$ and one can choose $\mathbf{j}$ along the [-110] axis and $\mathbf{k} = \mathbf{i} \times \mathbf{j}$ so that $\mathbf{j} = (-\mathbf{a}_x+\mathbf{a}_y)/\sqrt{2}$, and $\mathbf{k} = \mathbf{a}_z$. This ultimately introduces a 1/2 factor in the $\mathbf{i}$ and $\mathbf{j}$ components of the force equation so it will be convenient to define the force constants for second-nearest neighbors as: $2\alpha_2$, $2\beta_2$, $2\gamma_2$. The dynamical matrix elements for the second-nearest neighbors are:

$\Gamma_{11} = 4(\alpha_2+\beta_2)[2-\cos X\cos Y-\cos Z\cos X] + 8\gamma_2[1-\cos Y\cos Z]$

$\Gamma_{22} = 4(\alpha_2+\beta_2)[2-\cos Y\cos Z-\cos X\cos Y] + 8\gamma_2[1-\cos Z\cos X]$

$\Gamma_{33} = 4(\alpha_2+\beta_2)[2-\cos Z\cos X-\cos Y\cos Z] + 8\gamma_2[1-\cos X\cos Y]$

$\Gamma_{12} = 4(\alpha_2-\beta_2)\sin X\sin Y$

$\Gamma_{23} = 4(\alpha_2-\beta_2)\sin Y\sin Z$

$\Gamma_{13} = 4(\alpha_2-\beta_2)\sin Z\sin X$

where $X = k_x d$, $Y = k_y d$, $Z = k_z d$.

Note that $\Gamma_{22}$ and $\Gamma_{33}$ can be obtained from $\Gamma_{11}$ by cyclic permutation of $X$, $Y$, and $Z$ (i.e., $X\rightarrow Y$, $Y\rightarrow Z$, $Z\rightarrow X$). Similarly, $\Gamma_{23}$ and $\Gamma_{13}$ are cyclic permutations of $\Gamma_{12}$.

The 8 third-nearest neigbors are located at $(\pm d, \pm d, \pm d)$. For the neighbor at $(d,d,d)$, for example, $\mathbf{i} = (\mathbf{a}_x+\mathbf{a}_y+\mathbf{a}_z)/\sqrt{3}$ and one can choose $\mathbf{j}$ along the [-110] axis and $\mathbf{k} = \mathbf{i} \times \mathbf{j}$ so that that $\mathbf{j} = (-\mathbf{a}_x+\mathbf{a}_y)/\sqrt{2}$, and $\mathbf{k} = (-\mathbf{a}_x-\mathbf{a}_y+2\mathbf{a}_z)/\sqrt{6}$. This ultimately introduces a 1/3 factor in the $\mathbf{i}$, $\mathbf{j}$ and $\mathbf{k}$ components of the force equation so it will be convenient to define the force constants for second-nearest neighbors as: $3\alpha_3$, $3\beta_3$, $3\gamma_3$. Symmetry requires: $\Gamma_{11} = \Gamma_{22} = \Gamma_{33}$ so that $3\gamma_3 = 3\beta_3$. The dynamical matrix elements for the third-nearest neighbors are:

$\Gamma_{11} = (8\alpha_3+16\beta_3)[1-\cos X\cos Y\cos Z] = \Gamma_{22} = \Gamma_{33}$

$\Gamma_{12} = 8(\alpha_3-\beta_3)\sin X\sin Y\cos Z$

TABLE I. Relation between these force coefficients for the sc lattice and those of the general force model in the Born-von Karman theory. The notation used for the Born-von Karman constants is consistent with Walker [11].

| Type of Neighbor | Force Coefficients (this paper) | Force Coefficients (Born-von Karman) |
|---|---|---|
| First | $2\alpha_1$ | $2\alpha_1$ |
| | $2\beta_1$ | $2\beta_1$ |
| Second | $4(\alpha_2+\beta_2)$ | $4\alpha_2$ |
| | $8\gamma_2$ | $4\beta_2$ |
| | $4(\alpha_2-\beta_2)$ | $4\gamma_2$ |
| Third | $8\alpha_3+16\beta_3$ | $8\alpha_3$ |
| | $8(\alpha_3-\beta_3)$ | $8\beta_3$ |

$\Gamma_{23} = 8(\alpha_3-\beta_3)\sin Y \sin Z \cos X$

$\Gamma_{13} = 8(\alpha_3-\beta_3)\sin Z \sin X \cos Y$

where $X = k_x d$, $Y = k_y d$ $Z = k_z d$.

Note that the dynamical matrix elements for all 3 types of neighbors is consistent with the general force model in Born-von Karman formulation (see Table I).

The determinant of the matrix equations can be factored along principal axes. For the [100] axis: $k_y = k_z = 0$ and

$\Gamma_{22} = \Gamma_{33}$ and $\Gamma_{12} = \Gamma_{23} = \Gamma_{13} = 0$

so there are two branches, $\omega_1$ and $\omega_2$ given by:

$M\omega_1^2 = \Gamma_{11}$ and $M\omega_2^2 = \Gamma_{33}$.

For the [110] axis: $k_x = k_y = k/\sqrt{2}$, $k_z = 0$ and

$\Gamma_{11} = \Gamma_{22}$ and $\Gamma_{13} = \Gamma_{23} = 0$

so there are three branches:

$M\omega_1^2 = \Gamma_{11}+\Gamma_{12}$, $M\omega_2^2 = \Gamma_{11}-\Gamma_{12}$, and $M\omega_3^2 = \Gamma_{33}$.

For the [111] axis: $k_x = k_y = k_z = k/\sqrt{3}$ and

TABLE II. Values of $\rho s^2$ in the long-wave elastic limit for the longitudinal ($L$) and the two transverse ($T$) modes for the principal axes of cubic crystals in terms of the second-order elastic constants.

| Axis | $\rho s^2_L$ | $\rho s^2_T$ | $\rho s^2_T$ |
|---|---|---|---|
| $\langle 100 \rangle$ | $C_{11}$ | $C_{44}$ | $C_{44}$ |
| $\langle 110 \rangle$ | $(C_{11}+C_{12})/2+C_{44}$ | $(C_{11}-C_{12})/2$ | $C_{44}$ |
| $\langle 111 \rangle$ | $(C_{11}+2C_{12}+4C_{44})/3$ | $(C_{11}-C_{12}+C_{44})/3$ | $(C_{11}-C_{12}+C_{44})/3$ |

$$\Gamma_{11} = \Gamma_{22} = \Gamma_{33} \text{ and } \Gamma_{12} = \Gamma_{23} = \Gamma_{13}$$

so there are only two branches:

$$M\omega_1^2 = \Gamma_{11}+2\Gamma_{12} \text{ and } M\omega_2^2 = \Gamma_{11}-\Gamma_{12}.$$

For very small values of wave vector k the phase velocity $s = \omega/k$ reduces to the value measured by ultrasonic methods. The value of $\rho s^2$ where $\rho$ is the density, is given in Table II for the 3 principal axes. This gives one 3 conditions so that only 4 zone points are needed to obtain the following set of 7 independent equations in the 7 unknown force constants.

$$
\begin{aligned}
dC_{11} &= \alpha_1 &&+4\alpha_2 +4\beta_2 &&+4\alpha_3 +8\beta_3 \\
dC_{44} &= \beta_1 &&+2\alpha_2 +2\beta_2 +4\gamma_2 &&+4\alpha_3 +8\beta_3 \\
dC_{12} &= -\beta_1 &&+2\alpha -6\beta_2 -4\gamma_2 &&+4\alpha_3 -16\beta_3 \\
dC_{ZE1} &= 4\alpha_1+8\beta_1 &&&&+16\alpha_3+32\beta_3 \\
dC_{ZE2} &= 4\alpha_1+4\beta_1 &&+8\alpha_2 +8\beta_2+16\gamma_2 \\
dC_{ZE3} &= 8\beta_1 &&+16\alpha_2+16\beta_2 \\
dC_{ZE4} &= 2\alpha_1+4\beta_1 &&+16\beta_2 +8\gamma_2+8\alpha_3+16\beta_3.
\end{aligned}
$$

A typical set of dispersion curves for the sc lattice are shown in Fig. 3 with labels to identify 5 important zone points. The *effective* elastic constants $C_{ZK}$ are defined in terms of these zone frequencies as: $dC_{ZK} = M\omega_{ZK}^2$. Only one known material, $\alpha$-polonium, is simple cubic, but no complete set of elastic constants or dispersion curves have been measured for this material so no reliable data are available from which to calculate the 7 force constants.

The matrix equations can also be factored in the (010) and (-110) symmetry planes. For the (010) plane $k_y = 0$ so that:

$$\Gamma_{12} = \Gamma_{23} = 0$$

and the determinant factors into a pure transverse mode:

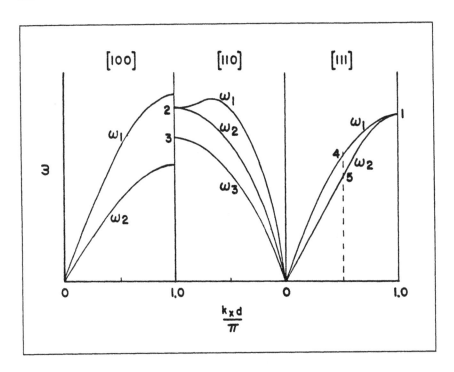

Fig. 3.  Schematic dispersion curves for a sc lattice defining the zone frequencies used to evaluate the force constants. Note that for the [100] axis the frequencies for the $\omega_1$ and $\omega_2$ branches at the zone edge where $k_x d = \pi$ are given by $4dC_{11} = M\omega_1^2$ and $4dC_{44} = M\omega_2^2$, respectively, so that these zone points are not independent of these second-order elastic constants. Also note that $d(C_{Z4}-C_{Z5}) = 12(\alpha_2-\beta_2)$.

$$M\omega_1^2 = \Gamma_{22}$$

and 2 mixed modes (i.e., a quasi-longitudinal (+), and (-) a quasi-transverse):

$$2M\omega^2 = \Gamma_{11}+\Gamma_{33} \pm [(\Gamma_{11}-\Gamma_{33})^2+4\Gamma_{13}^2]^{1/2}.$$

For the (-110) plane $k_x = k_y$ so that:

$$\Gamma_{11} = \Gamma_{22} \text{ and } \Gamma_{23} = \Gamma_{13}$$

and the determinant factors into a pure transverse mode:

$$M\omega^2 = \Gamma_{11}-\Gamma_{12}$$

and 2 mixed modes:

$$2M\omega^2 = \Gamma_{11}+\Gamma_{12}+\Gamma_{33} \pm [(\Gamma_{11}+\Gamma_{12}-\Gamma_{33})^2+8\Gamma_{13}^2]^{1/2}.$$

The phase velocity $s = \omega/k$ and is parallel to the wave vector $\mathbf{k}$, whereas the group velocity $\mathbf{v} = \partial\omega/\partial\mathbf{k}$. The group velocity components are:

$$v_x = \partial\omega/dk_x, \quad v_y = \partial\omega/dk_y, \quad v_z = \partial\omega/dk_z.$$

Note that the group velocity $\mathbf{v}$ is normal to the constant energy surface in $\mathbf{k}$ space. As a result the group velocity is not, in general, parallel to the phase velocity direction.

## DISCUSSION

The dynamical matrix elements for the simple cubic lattice are conceptually simple to understand but somewhat tedious to calculate. The method is attractive since it gives the same results as the Born-von Karman theory, but does not require the explicit use of group theory. Furthermore, it uses a maximum of 3 force constants per site and thus requires fewer force constants when certain further-neighbor interactions are included [10,11]. The method is its present form, however, gives a symmetric force tensor. It therefore would not be applicable when antisymmetric non-central forces are present, as for example, in the diamond lattice [13].

## REFERENCES

1. Maris, H.J., 1971. "Enhancement of Heat Pulses in Crystals due to Elastic Anisotropy." *J. Acoust. Soc. Am.* 50:812-818.

2. Taylor, B., H.J. Maris, and C. Elbaum, 1971. "Focusing of Phonons in Crystalline Solids due to Elastic Anisotropy." *Phys. Rev.* B3:1462-1472 .

3. McCurdy, A.K., H.J. Maris, and C. Elbaum, 1970. "Anisotropic Heat Conduction in Cubic Crystals in the Boundary Scattering Regime." *Phys. Rev.* B2:4077-4083.

4. McCurdy, A.K., 1982. "Phonon Conduction in Elastically Anisotropic Cubic Crystals." *Phys. Rev.* B26:6971-6986.

5. McCurdy, A.K., 1988. "Energy-Flow-Reversing *N* Processes in Elastically Anisotropic Crystals." *Phys. Rev.* B38:10335-10349.

6. Hasegawa, S., and A.K. McCurdy, 1990. "Four-Phonon Energy-Flow-Reversing *N* Processes About ⟨111⟩ Axes in Cubic Crystals," in *Thermal Conductivity* 21, H.A. Fine, and C.J. Cremers, eds. N.Y.:Plenum Press, pp. 401-411.

7.  Thomas, Jan, 1993. "Quantum Mechanical Coupling Coefficients for Energy-Flow-Reversing $N$ Processes in the Absence of Dispersion," M.S. Thesis, Worcester Polytechnic Institute, Worcester, MA.

8.  Thomas, Jan, and A.K. McCurdy, 1994. "Quantum Mechanical Coupling Coefficients for 3-Phonon Interactions," in *Thermal Conductivity* 22, T.W. Tong, ed. Lancaster, PA: Technomic Publishing Company Inc., pp. 374-382.

9.  Born, M., and K. Huang, 1954. *"Dynamical Theory of Crystal Lattices."*, New York: Oxford University Press, pp. 38-273.

10. Bell, M.G., and A.K. McCurdy, 1996. "Phonon-Focusing Efects in Dispersive Body-Centered Cubic Crystals." in *Thermal Conductivity* 23, K.E. Wilkes, R.B. Dinwiddie, and R.S. Graves,  eds. Lancaster, PA: Technomic Publishing Company, Inc., (this volume in Press).

11. Capriglione, G.S., and A.K. McCurdy, 1996. "Phonon-Focusing Efects in Dispersive Face-Centered Cubic Crystals", in *Thermal Conductivity* 23, K.E. Wilkes, R.B. Dinwiddie, and R.S. Graves, eds. Lancaster, PA: Technomic Publishing Company, Inc., (this colume in Press).

12. Smith, R.A., 1961. *Wave Mechanics of Crystalline Solids.* London: Chapman & Hall, pp. 204-209.

13. Lax, M., 1974. *Symmetry Principles in Solid State and Molecular Physics.* New York: Wiley & Sons, pp. 263-264.

# Phonon-Focusing Effects in Dispersive Body-Centered Cubic Crystals

M. G. BELL and A. K. McCURDY

## ABSTRACT

Phonon group velocities have been calculated for symmetry planes in bcc monotomic crystals using a rigid-ion lattice dynamical model utilizing 12 force constants and fifth nearest-neighbor interactions. Calculations are presented for two metals: sodium and tungsten for which experimental dispersion curves are available, and which reasonably agree with Born-von Karman predictions. Group velocities along all 3 principal axes are predicted to vanish at zone boundaries. Furthermore, group velocities along the $\langle 111 \rangle$ axes in both metals become negative for larger values of the wave vector within the first Brillouin zone. Polar plots of phase and group velocities for the (010) and (110) planes are given for several different magnitudes of the wave vector in both sodium and tungsten. The large cuspidal features in the group velocity of sodium undergo remarkable changes and spread to other modes as the magnitude of the wave vector is increased. Tungsten, although elastically isotropic for small values of the wave vector, exhibits pronounced cuspidal features in the group velocity for larger values of the wave vector. Since energy carried by phonons is in the direction of the group velocity such cuspidal features gives rise to strong focusing of phonons along cuspidal edges. These phonon-focusing effects arise because dispersion can be quite different along different crystallographic directions. Strong phonon focusing is predicted to dramatically change the phonon transport properties of these crystals and thus the phonon contribution to the thermal conductivity.

## INTRODUCTION

Phonon focusing in crystals depends upon the shape of the constant-energy surfaces in **k** space [1,2]. In the long-wavelength limit the constant-energy surfaces

Professor Emeritus Archie K. McCurdy, Electrical and Computer Engineering Department, Worcester Polytechnic Institute, 100 Institute Road, Worcester, MA 01609
Mark G. Bell, Electrical and Computer Engineering Department, Worcester Polytechnic Institute, 100 Institute Road, Worcester, MA 01609

in **k** space are nearly linear in **k** and thus can be determined entirely from the second-order elastic constants of the crystal [1]. For larger values of the wave vector, however, the presence of dispersion causes the constant energy surfaces to vary nonlinearly with wave vector **k** so that a lattice dynamical treatment is required for meaningful phase and group velocity calculations.

This paper explores the phonon-focusing properties of body-centered cubic crystals for larger values of the wave vector by calculating phase and group velocities using the rigid ion lattice dynamical model discussed in the previous paper [3]. Results for face-centered cubic crystals are given in a companion paper in this volume [4]. Results presented here show that dispersion radically effects the magnitude and direction of the group velocities, and thus dramatically effects the phonon-focusing and phonon transport properties of these crystals.

## THEORY AND ANALYSIS

Calculation of phase and group velocities have been discussed in some detail already [3] and will not be repeated here. Dynamical matrix elements are presented, however, for the bcc lattice up to and including fifth nearest-neighbor interactions. Note that for all neighbor interactions, except second-nearest neighbors, the **j** unit vector was chosen along the appropriate $\langle 110 \rangle$ direction.

The 8 first-nearest neigbors are located at $(\pm d/2, \pm d/2, \pm d/2)$. The force constants (see McCurdy [3]) are defined as: $3\alpha_1$, $3\beta_1$, $3\gamma_1$, with $3\gamma_1 = 3\beta_1$ required by symmetry. The dynamical matrix elements for the first-nearest neighbors are:

$$\Gamma_{11} = (8\alpha_1 + 16\beta_1)[1 - \cos X2 \cos Y2 \cos Z2] = \Gamma_{22} = \Gamma_{33}$$

$$\Gamma_{12} = 8(\alpha_1 - \beta_1)\sin X2 \sin Y2 \cos Z2$$

$$\Gamma_{23} = 8(\alpha_1 - \beta_1)\sin Y2 \sin Z2 \cos X2$$

$$\Gamma_{13} = 8(\alpha_1 - \beta_1)\sin Z2 \sin X2 \cos Y2$$

where $X2 = k_x d/2$, $Y2 = k_y d/2$, $Z2 = k_z d/2$.

The 6 second-nearest neigbors are located at: $(\pm d, 0, 0)$, $(0, \pm d, 0)$, $(0, 0, \pm d)$. The force constants are defined as: $\alpha_2$, $\beta_2$, $\gamma_2$, with $\gamma_2 = \beta_2$ required by symmetry. The dynamical matrix elements for the second-nearest neighbors are:

$$\Gamma_{11} = 2\alpha_2[1 - \cos X] + 2\beta_2[2 - \cos Y - \cos Z]$$

$$\Gamma_{22} = 2\alpha_2[1 - \cos Y] + 2\beta_2[2 - \cos Z - \cos X]$$

$$\Gamma_{33} = 2\alpha_2[1 - \cos Z] + 2\beta_2[2 - \cos X - \cos Y]$$

$$\Gamma_{12} = \Gamma_{23} = \Gamma_{13} = 0$$

where $X = k_xd$, $Y = k_yd$, $Z = k_zd$.

The 12 third-nearest neigbors are located at: $(\pm d, \pm d, 0)$, $(\pm d, 0, \pm d)$, $(0, \pm d, \pm d)$. The force constants are defined as: $2\alpha_3$, $2\beta_3$, $2\gamma_3$. The dynamical matrix elements for the third-nearest neighbors are:

$\Gamma_{11} = 4(\alpha_3+\beta_3)[2-\cos X\cos Y-\cos Z\cos X] + 8\gamma_3[1-\cos Y\cos Z]$

$\Gamma_{22} = 4(\alpha_3+\beta_3)[2-\cos Y\cos Z-\cos X\cos Y] + 8\gamma_3[1-\cos Z\cos X]$

$\Gamma_{33} = 4(\alpha_3+\beta_3)[2-\cos Z\cos X-\cos Y\cos Z] + 8\gamma_3[1-\cos X\cos Y]$

$\Gamma_{12} = 4(\alpha_3-\beta_3)\sin X\sin Y$

$\Gamma_{23} = 4(\alpha_3-\beta_3)\sin Y\sin Z$

$\Gamma_{13} = 4(\alpha_3-\beta_3)\sin Z\sin X$

where $X = k_xd$, $Y = k_yd$, $Z = k_zd$.

The 24 fourth-nearest neigbors are located at: $(\pm d/2, \pm d/2, \pm 3d/2)$, $(\pm d/2, \pm 3d/2, \pm d/2)$, $(\pm 3d/2, \pm d/2, \pm d/2)$. The force constants are defined as: $11\alpha_4$, $11\beta_4$, $11\gamma_4$. The dynamical matrix elements for the fourth-nearest neighbors are:

$\Gamma_{11} = A_4[2-\cos X2\cos Y2\cos 3Z2-\cos X2\cos 3Y2\cos Z2] + B_4[1-\cos 3X2\cos Y2\cos Z2]$

$\Gamma_{22} = A_4[2-\cos Y2\cos Z2\cos 3X2-\cos Y2\cos 3Z2\cos X2] + B_4[1-\cos 3Y2\cos Z2\cos X2]$

$\Gamma_{33} = A_4[2-\cos Z2\cos X2\cos 3Y2-\cos Z2\cos 3X2\cos Y2] + B_4[1-\cos 3Z2\cos X2\cos Y2]$

$\Gamma_{12} = C_4[\sin X2\sin Y2\cos 3Z2] + D_4[\sin X2\sin 3Y2\cos Z2+\sin 3X2\sin Y2\cos Z2]$

$\Gamma_{23} = C_4[\sin Y2\sin Z2\cos 3X2] + D_4[\sin Y2\sin 3Z2\cos X2+\sin 3Y2\sin Z2\cos X2]$

$\Gamma_{13} = C_4[\sin Z2\sin X2\cos 3Y2] + D_4[\sin Z2\sin 3X2\cos Y2+\sin 3Z2\sin X2\cos Y2]$

where $A_4 = 8\alpha_4+44\beta_4+36\gamma_4$, $B_4 = 72\alpha_4+16\gamma_4$

$C_4 = 8\alpha_4-44\beta_4+36\gamma_4$, $D_4 = 24(\alpha_4-\gamma_4)$

and $X2 = k_xd/2$, $Y2 = k_yd/2$, $Z2 = k_zd/2$,

$3X2 = k_x3d/2$, $3Y2 = k_y3d/2$, $3Z2 = k_z3d/2$.

The 8 fifth-nearest neigbors are located at $(\pm d, \pm d, \pm d)$. The force constants are

TABLE I.   Relation between these force coefficients for the bcc lattice and those of the general force model in the Born-von Karman theory [5]. The notation used for the Born-von Karman constants is consistent with Walker [6].

| Type of Neighbor | Force Coefficients (this paper) | Force Coefficients (Born-von Karman) |
|---|---|---|
| First | $8\alpha_1+16\beta_1$ | $8\alpha_1$ |
|  | $8(\alpha_1-\beta_1)$ | $8\beta_1$ |
| Second | $2\alpha_2$ | $2\alpha_2$ |
|  | $2\beta_2$ | $2\beta_2$ |
| Third | $4(\alpha_3+\beta_3)$ | $4\alpha_3$ |
|  | $8\gamma_3$ | $4\beta_3$ |
|  | $4(\alpha_3-\beta_3)$ | $4\gamma_3$ |
| Fourth | $72\alpha_4+16\beta_4$ | $8\alpha_4$ |
|  | $8\alpha_4+44\beta_4+36\gamma_4$ | $8\beta_4$ |
|  | $8\alpha_4-44\beta_4+36\gamma_4$ | $8\gamma_4$ |
|  | $24(\alpha_4-\gamma_4)$ | $8\delta_4$ |
| Fifth | $8\alpha_5+16\beta_5$ | $8\alpha_5$ |
|  | $8(\alpha_5-\beta_5)$ | $8\beta_5$ |

defined as: $3\alpha_5$, $3\beta_5$, $3\gamma_5$, with $3\gamma_5 = 3\beta_5$ required by symmetry. The dynamical matrix elements for the fifth-nearest neighbors are:

$$\Gamma_{11} = (8\alpha_5+16\beta_5)[1-\cos X\cos Y\cos Z] = \Gamma_{22} = \Gamma_{33}$$

$$\Gamma_{12} = 8(\alpha_5-\beta_5)\sin X\sin Y\cos Z$$

$$\Gamma_{23} = 8(\alpha_5-\beta_5)\sin Y\sin Z\cos X$$

$$\Gamma_{13} = 8(\alpha_5-\beta_5)\sin Z\sin X\cos Y$$

where   $X = k_x d$,  $Y = k_y d$,  $Z = k_z d$.

The relation between the above 12 force constants and those for the Born-von Karman theory are given in Table I. Calculation of these 12 force constants requires 12 independent equations. In the long wavelength limit where **k** is small the phase velocity approaches the value predicted by elastic theory. Expressions for $\rho s^2$, where s is the phase velocity and $\rho$ is the density, provide in this long-wavelength limit three of the required equations. As a result only 9 independent zone conditions are required to evaluate all 12 force constants. Solution for the force constants requires one to find 9 zone points which are mutually independent and which are independent

TABLE II. Values of the force constants in N/m for sodium and tungsten calculated from the zone frequencies labelled in Fig. 1. Frequencies for sodium were obtained from the 90°K data of Woods & Brockhouse *et al* [7]; frequencies for tungsten were obtained from the data of Chen & Brockhouse [8]. Elastic constants were taken from Landolt-Bornstein [9].

| Force Constant | Sodium | Tungsten |
|---|---|---|
| $3\alpha_1$ | 3.8315 | 60.1817 |
| $3\beta_1$ | -0.1034 | 4.4238 |
| | | |
| $\alpha_2$ | 0.9067 | 45.9782 |
| $\beta_2$ | 0.4919 | 0.9761 |
| | | |
| $2\alpha_3$ | -0.6189 | 8.1612 |
| $2\beta_3$ | 0.1540 | 0.5554 |
| $2\gamma_3$ | -0.2181 | -2.2232 |
| | | |
| $11\alpha_4$ | 0.0230 | -0.2476 |
| $11\beta_4$ | -0.0103 | -0.3483 |
| $11\gamma_4$ | -0.0147 | 0.3722 |
| | | |
| $3\alpha_5$ | 0.5135 | 1.0299 |
| $3\beta_5$ | -0.0610 | -1.3994 |

of the 3 second-order elastic constants. This can be realized by a number of different choices of zone points.

## RESULTS

Force constants for sodium and tungsten are given in Table II. Dispersion curves for sodium calculated from the force constants in Table II are shown in Fig. 1. Phase velocities for sodium are given for the {100} and {110} symmetry planes in Figs. 2 and 4, respectively, whereas group velocities are given for these same planes in Figs. 3 and 5. Group velocities for tungsten, which is elastically isotropic in the long-wavelength limit, are given in Figs. 6 and 7. Note that in Figs. 2-7 all wave vectors fall within the first Brillouin zone.

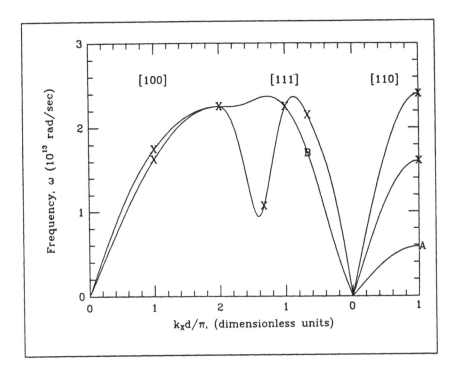

Fig. 1. Dispersion curves for a bcc lattice (sodium). Eight of the zone frequencies used to evaluate the 12 force constants are marked by an X. For sodium it was necessary to use point A instead of point B to obtain physically realizable dispersion curves. For tungsten point B was used instead of point A to obtain a better fit to the data.

## DISCUSSION

Dynamical matrix elements for all but fourth-nearest neighbors in the bcc lattice can be obtained using results from the sc lattice since the remaining neighbor types have similar coordinates. Note that this method uses a maximum of 3 force constants per site and thus requires one less force constant than the Born-von Karman theory for fourth-nearest neighbor interactions. As a result only 12 force constants are required to include fifth nearest-neighbor interactions. Note that in the Born-von Karman theory the four force constants for the fourth-nearest neighbors are considered independent, whereas in the present formulation the fourth force constant is a linear combination of two other force constants (see Table I). This method in its present form, however, gives a *symmetric* force tensor. It therefore would not be applicable when antisymmetric non-central forces are present, as for example, in the diamond lattice [10].

Dispersion curves for sodium are in good agreement with original experiments [7]. Since agreement depends somewhat upon the particular choice of zone points, zone points were chosen to closely match regions of negative group velocity. Plots of

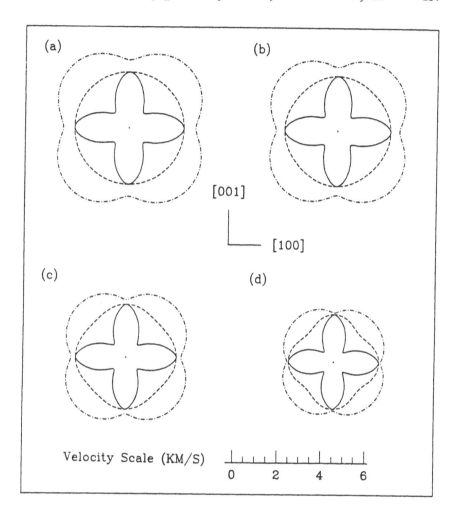

Fig. 2.  Polar plots of the phase velocity (km/sec) in the (010) plane for sodium using the force constants in Table II. Curves in (a) $kd/\pi = 0.01$; (b) $kd/\pi = 0.5$; (c) $kd/\pi = 0.9$; (d) $kd/\pi = 1.3$.

the group velocity in the elastic limit (i.e., $kd/\pi = 0.01$) show large cuspidal features about the $\langle 100 \rangle$ axes in $\{100\}$ planes, and about $\langle 100 \rangle$ and $\langle 110 \rangle$ axes in $\{110\}$ planes as a result of the large elastic anisotropy factor. These results contrast sharply with tungsten which is elastically isotropic, and exhibits constant phase and group velocities in the elastic limit.

For larger values of the wave vector, however, group velocities exhibit increasing anisotropy in both sodium and tungsten. At certain zone boundaries the

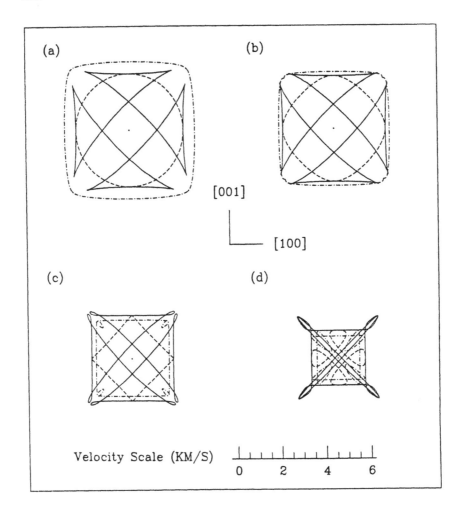

Fig. 3.   Polar plots of the group velocity (km/sec) in the (010) plane for sodium using the
force constants in Table II. Curves in (a) $kd/\pi = 0.01$; (b) $kd/\pi = 0.5$; (c) $kd/\pi = 0.9$;
(d) $kd/\pi = 1.3$.

group velocities in bcc crystals vanish, and in certain regions the group velocities
may be negative (i.e., reverse direction). As a result the group-velocity surfaces for
larger values of the wave vector can have rather bizarre shapes. Since energy flow is
parallel to the *group* velocity this dramatically affects the phonon-focusing and
phonon transport properties of these crystals. Note that in metals such as sodium and
tungsten most heat is carried by electrons, so phonon-focusing effects need to be very
large to make measurable differences in the thermal conductivity.

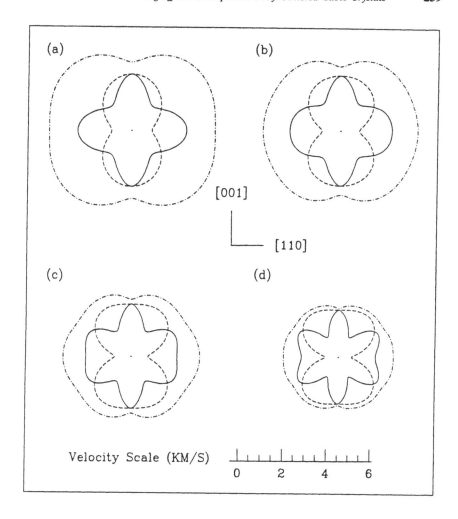

Fig. 4. Polar plots of the phase velocity (km/sec) in the (-110) plane for sodium using the force constants in Table II. Curves in (a) kd/π = 0.01; (b) kd/π = 0.5; (c) kd/π = 0.9; (d) kd/π = 1.3.

## ACKNOWLEDGEMENT

One of us (A.K.M.) acknowledges a useful discussion with P.G. Klemens.

## REFERENCES

1. Maris, H.J., 1971. "Enhancement of Heat Pulses in Crystals due to Elastic Anisotropy." *J. Acoust. Soc. Am.* 50:812-818.

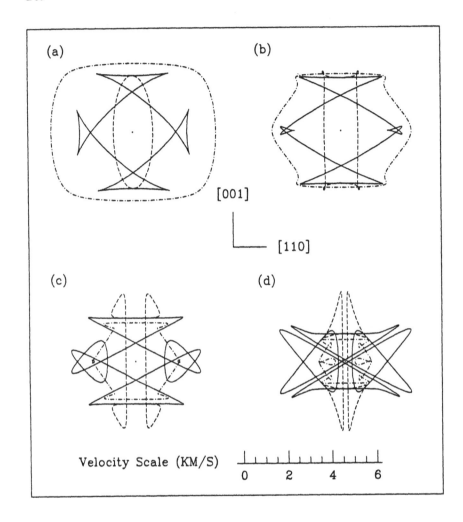

Fig. 5. Polar plots of the group velocity (km/sec) in the (-110) plane for sodium using the force constants in Table II. Curves in (a) kd/π = 0.01; (b) kd/π = 0.5; (c) kd/π = 0.9; (d) kd/π = 1.3.

2. Taylor, B., H.J. Maris, and C. Elbaum, 1971. "Focusing of Phonons in Crystalline Solids due to Elastic Anisotropy.". *Phys. Rev.* B3, 1462-1472.

3. McCurdy, A.K., 1996. "A Rigid-Ion Lattice Dynamical Model For Calculating Group Velocities in Simple, Body-Centered, and Face-Centered Crystals," in *Thermal Conductivity* 23, K.E. Wilkes, R.B. Dinwiddie, and R.S. Graves, eds. Lancaster, PA: Technomic Publishing Company, Inc., (this volume in Press).

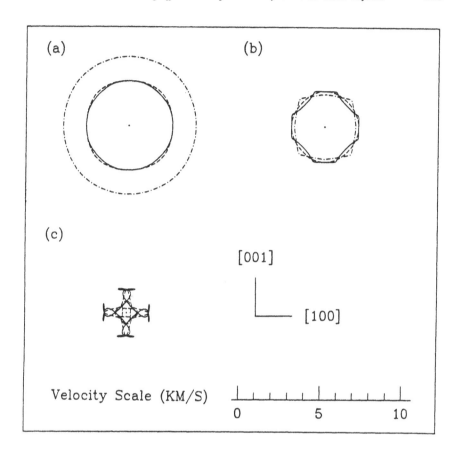

Fig. 6. Polar plots of the group velocity (km/sec) in the (010) plane for tungsten using the force constants in Table II. Curves in (a) kd/π = 0.5; (b) kd/π = 0.9, (c) kd/π = 1.3.

4.  Capriglione, G.S., and A.K. McCurdy, 1996. "Phonon-Focusing Efects in Dispersive Face-Centered Cubic Crystals," in *Thermal Conductivity* 23, K.E. Wilkes, R.B. Dinwiddie, and R.S. Graves, eds. Lancaster, PA: Technomic Publishing Company, Inc., (this volume in Press).

5.  Born, M., and K. Huang, 1954. *Dynamical Theory of Crystal Lattices.* N.Y., : Oxford University Press. pp. 38-273.

6.  Walker, C.B., 1956. "X-ray Study of Lattice Vibrations in Aluminum", *Phys. Rev.* 103, 547-557.

7.  Woods, A.D.B., B.N. Brockhouse, R.H. March, A.T. Stewart, and R. Bowers, 1962. "Crystal Dynamics of Sodium at 90°K." *Phys. Rev.* 128, 1112-1120.

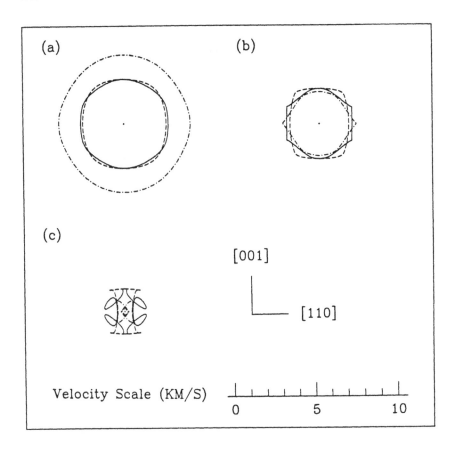

Fig. 7. Polar plots of the group velocity (km/sec) in the (-110) plane for tungsten using the force constants in Table II. Curves in (a) kd/π = 0.5; (b) kd/π = 0.9, (c) kd/π = 1.3.

8. Chen, S.H., and B.N. Brockhouse, 1964. "Lattice Vibrations of Tungsten", *Solid State Commun.* 2, 73-77.

9. Every, A.G., and A.K. McCurdy, 1992. *Landolt-Bornstein: Numerical Data and Functional Relationships in Science and Technology, New Series, Group III, Volume 29a: Second and Higher Order Elastic Constants*, O. Madelung, and D.F. Nelson, eds. Berlin: Springer-Verlag, pp. 11, 305.

10. Lax, M., 1974. "Symmetry Principles in Solid State and Molecular Physics." N.Y.: Wiley &, pp. 263-264.

# Phonon-Focusing Effects in Dispersive Face-Centered Cubic Crystals

G. S. CAPRIGLIONE and A. K. MCCURDY

## ABSTRACT

Phonon group velocities have been calculated for symmetry planes in fcc crystals using a lattice dynamical model utilizing 16 force constants and sixth nearest-neighbor interactions. Calculations are presented for one metal, copper, for which experimental dispersion curves can be matched with a Born-von Karman analysis. Group velocities are predicted to vanish at zone boundaries along $\langle 100 \rangle$ and $\langle 111 \rangle$ axes. Furthermore, the group velocity along $\langle 110 \rangle$ axes becomes negative inside the first Brillouin zone for larger values of the wave vector. Polar plots of phase and group velocities for the (010) and (110) planes are given for several different magnitudes of the wave vector. The pronounced cuspidal features in the group velocity of copper undergo remarkable changes and spread to other modes as the magnitude of the wave vector is increased. Since energy carried by phonons is parallel to the group velocity such cuspidal behavior gives rise to strong phonon-focusing effects. Such strong focusing is predicted to dramatically change the phonon transport properties of these crystals and thus the phonon contribution to the thermal conductivity.

## INTRODUCTION

Phase and group velocities have been calculated in simple and body-centered cubic crystals using a rigid-ion lattice dynamical model [1,2]. This paper explores the phonon-focusing properties of face-centerd cubic crystals for larger values of the wave vector by calculating phase and group velocities for this rigid-ion lattice dynamical model. Results presented here show that dispersion radically effects the magnitude and direction of the group velocities, and thus dramatically effects the phonon-focusing and phonon transport properties of these crystals.

Professor Emeritus Archie K. McCurdy, Electrical and Computer Engineering Department, Worcester Polytechnic Institute, 100 Institute Road, Worcester, MA 01609
Giovanni S. Capriglione, Physics Department, Worcester Polytechnic Institute, 100 Institute Road, Worcester, MA 01609

## THEORY

Since results for simple cubic crystals have been discussed in considerable detail [2] only salient results will be given here for the fcc lattice. The analysis is given for *symmetric* non-central forces up to and including 6th nearest-neighbor (n-n) interactions.

Consider the atom at the origin. Its 12 first-nearest neighbors are located at: $(\pm d/2, \pm d/2, 0)$, $(\pm d/2, 0, \pm d/2)$, $(0, \pm d/2, \pm d/2)$. The force constants (see McCurdy [2]) are defined as $2\alpha_1$, $2\beta_1$, and $2\gamma_1$. The relation between these force constants and those of the Born-von Karman model are given in Table I. Revelant dynamical matrix elements for these first-nearest neighbors are:

$$\Gamma_{11} = 4(\alpha_1 + \beta_1)[2 - \cos X2 \cos Y2 - \cos Z2 \cos X2] + 8\gamma_1[1 - \cos Y2 \cos Z2]$$

$$\Gamma_{22} = 4(\alpha_1 + \beta_1)[2 - \cos Y2 \cos Z2 - \cos X2 \cos Y2] + 8\gamma_1[1 - \cos Z2 \cos X2]$$

$$\Gamma_{33} = 4(\alpha_1 + \beta_1)[2 - \cos Z2 \cos X2 - \cos Y2 \cos Z2] + 8\gamma_1[1 - \cos X2 \cos Y2]$$

$$\Gamma_{12} = 4(\alpha_1 - \beta_1)\sin X2 \sin Y2$$

$$\Gamma_{23} = 4(\alpha_1 - \beta_1)\sin Y2 \sin Z2$$

$$\Gamma_{13} = 4(\alpha_1 - \beta_1)\sin Z2 \sin X2$$

where $X2 = k_x d/2$, $Y2 = k_y d/2$, $Z2 = k_z d/2$.

The 6 second-nearest neigbors are located at: $(\pm d, 0, 0)$, $(0, \pm d, 0)$, $(0, 0, \pm d)$. The force constants are defined as $\alpha_2$, $\beta_2$, and $\gamma_2$ with symmetry requiring $\gamma_2 = \beta_2$. The dynamical matrix elements for the second-nearest neighbors are:

$$\Gamma_{11} = 2\alpha_2[1 - \cos X] + 2\beta_2[2 - \cos Y - \cos Z]$$

$$\Gamma_{22} = 2\alpha_2[1 - \cos Y] + 2\beta_2[2 - \cos Z - \cos X]$$

$$\Gamma_{33} = 2\alpha_2[1 - \cos Z] + 2\beta_2[2 - \cos X - \cos Y]$$

$$\Gamma_{12} = \Gamma_{23} = \Gamma_{13} = 0$$

where $X = k_x d$, $Y = k_y d$, $Z = k_z d$.

The 24 third-nearest neigbors are located at: $(\pm d/2, \pm d/2, \pm d)$, $(\pm d/2, \pm d, \pm d/2)$, $(\pm d, \pm d/2, \pm d/2)$. The force constants are defined as $6\alpha_3$, $6\beta_3$, and $6\gamma_3$. The dynamical matrix elements for the third-nearest neighbors are:

$$\Gamma_{11} = A_3[1 - \cos X \cos Y2 \cos Z2] + B_3[2 - \cos X2 \cos Y2 \cos Z - \cos X2 \cos Y \cos Z2]$$

TABLE I. Relation between these force coefficients for the fcc lattice and those of the general force model in the Born-von Karman theory. The Born-von Karman constants [3] for first-, second- and third-nearest neighbors are those given by Walker [4]. An example calculation using the Born-von Karman formalism is given for first- and second-nearest neighbors in the fcc lattice in Ref. [5].

| Type of Neighbor | Force Coefficients (this paper) | Force Coefficients (Born-von Karman) |
|---|---|---|
| First | $4(\alpha_1+\beta_1)$ | $4\alpha_1$ |
| | $8\gamma_1$ | $4\beta_1$ |
| | $4(\alpha_1-\beta_1)$ | $4\gamma_1$ |
| Second | $2\alpha_2$ | $2\alpha_2$ |
| | $2\beta_2$ | $2\beta_2$ |
| Third | $32\alpha_3+16\gamma_3$ | $8\alpha_3$ |
| | $8\alpha_3+24\beta_3+16\gamma_3$ | $8\beta_3$ |
| | $8\alpha_3-24\beta_3+16\gamma_3$ | $8\gamma_3$ |
| | $16(\alpha_3-\gamma_3)$ | $8\delta_3$ |
| Fourth | $4(\alpha_4+\beta_4)$ | $4\alpha_4$ |
| | $8\gamma_4$ | $4\beta_4$ |
| | $4(\alpha_4-\beta_4)$ | $4\gamma_4$ |
| Fifth | $36\alpha_5+4\beta_5$ | $4\alpha_5$ |
| | $4\alpha_5+36\beta_5$ | $4\beta_5$ |
| | $40\gamma_5$ | $4\gamma_5$ |
| | $12(\alpha_5-\beta_5)$ | $4\delta_5$ |
| Sixth | $8\alpha_6+16\beta_6$ | $8\alpha_6$ |
| | $8(\alpha_6-\beta_6)$ | $8\beta_6$ |

$\Gamma_{22} = A_3[1-\cos Y \cos Z2 \cos X2] + B_3[2-\cos Y2 \cos Z2 \cos X-\cos Y2 \cos Z \cos X2]$

$\Gamma_{33} = A_3[1-\cos Z \cos X2 \cos Y2] + B_3[2-\cos Z2 \cos X2 \cos Y-\cos Z2 \cos X \cos Y2]$

$\Gamma_{12} = C_3[\sin X2 \sin Y2 \cos Z] + D_3[\sin X2 \sin Y \cos Z2+\sin X \sin Y2 \cos Z2]$

$\Gamma_{23} = C_3[\sin Y2 \sin Z2 \cos X] + D_3[\sin Y2 \sin Z \cos X2+\sin Y \sin Z2 \cos X2]$

$\Gamma_{13} = C_3[\sin Z2\sin X2\cos Y] + D_3[\sin Z2\sin X\cos Y2+\sin Z\sin X2\cos Y2]$

where $A_3 = 32\alpha_3+16\gamma_3$, $B_3 = 8\alpha_3+24\beta_3+16\gamma_3$,

$\qquad C_3 = 8\alpha_3-24\beta_3+16\gamma_3$, $D_3 = 16(\alpha_3-\gamma_3)$,

and $X2 = k_xd/2$, $Y2 = k_yd/2$, $Z2 = k_zd/2$,

$\qquad X = k_xd$, $Y = k_yd$, $Z = k_zd$.

The 12 fourth-nearest neigbors are located at: $(\pm d,\pm d,0)$, $(\pm d,0,\pm d)$, $(0,\pm d,\pm d)$. The force constants are defined as $2\alpha_4$, $2\beta_4$, and $2\gamma_4$. The dynamical matrix elements for the fourth-nearest neighbors are:

$\Gamma_{11} = 4(\alpha_4+\beta_4)[2-\cos X\cos Y-\cos Z\cos X] + 8\gamma_4[1-\cos Y\cos Z]$

$\Gamma_{22} = 4(\alpha_4+\beta_4)[2-\cos Y\cos Z-\cos X\cos Y] + 8\gamma_4[1-\cos Z\cos X]$

$\Gamma_{33} = 4(\alpha_4+\beta_4)[2-\cos Z\cos X-\cos Y\cos Z] + 8\gamma_4[1-\cos X\cos Y]$

$\Gamma_{12} = 4(\alpha_4-\beta_4)\sin X\sin Y$

$\Gamma_{23} = 4(\alpha_4-\beta_4)\sin Y\sin Z$

$\Gamma_{13} = 4(\alpha_4-\beta_4)\sin Z\sin X$

where $X = k_xd$, $Y = k_yd$, $Z = k_zd$.

The 24 fifth-nearest neigbors are located at: $(\pm 3d/2,\pm d/2,0)$, $(\pm 3d/2,0,\pm d/2)$, $(0,\pm 3d/2,\pm d/2)$, $(\pm d/2,\pm 3d/2,0)$, $(\pm d/2,0,\pm 3d/2)$, $(0, \pm d/2, \pm 3d/2)$. The force constants are defined as $10\alpha_5$, $10\beta_5$, and $10\gamma_5$. The dynamical matrix elements for the fifth-nearest neighbors are:

$\Gamma_{11} = A_5[2-\cos 3X2\cos Y2-\cos 3X2\cos Z2] + B_5[2-\cos X2\cos 3Y2-\cos X2\cos 3Z2]$

$\qquad + C_5[2-\cos 3Y2\cos Z2-\cos Y2\cos 3Z2]$

$\Gamma_{12} = 12(\alpha_5-\beta_5)[\sin 3X2\sin Y2+\sin X2\sin 3Y2]$

where $A_5 = 36\alpha_5+4\beta_5$, $B_5 = 4\alpha_5+36\beta_5$, $C_5 = 40\gamma_5$,

and $X2 = k_xd/2$, $Y2 = k_yd/2$, $Z2 = k_zd/2$,

$\qquad 3X2 = k_x3d/2$ $3Y2 = k_y3d/2$, $3Z2 = k_z3d/2$.

The remaining matrix elements can be found by cyclic permutation i.e., $X \to Y$, $Y \to Z$, $Z \to X$.

The 8 sixth-nearest neigbors are located at ($\pm$d, $\pm$d, $\pm$d). The force constants are defined as $3\alpha_6$, $3\beta_6$, and $3\gamma_6$ with symmetry requiring $3\gamma_6 = 3\beta_6$. The dynamical matrix elements for the sixth-nearest neighbors are:

$$\Gamma_{11} = (8\alpha_6 + 16\beta_6)[1 - \cos X \cos Y \cos Z] = \Gamma_{22} = \Gamma_{33}$$

$$\Gamma_{12} = 8(\alpha_6 - \beta_6)\sin X \sin Y \cos Z$$

$$\Gamma_{23} = 8(\alpha_6 - \beta_6)\sin Y \sin Z \cos X$$

$$\Gamma_{13} = 8(\alpha_6 - \beta_6)\sin Z \sin X \cos Y$$

where $X = k_x d$, $Y = k_y d$ $Z = k_z d$.

Note that this rigid-ion model requires 16 force constants to include sixth-nearest neighbor interactions whereas the Born-von Karman model [3,4] with general forces requires 4 force constants for third- and fifth-nearest neighbors for a total of 18 force constants.

## RESULTS

Force constants for copper are given in Table II. Dispersion curves for copper are shown in Fig. 1 with the zone points used to calculate the force constants clearly labelled with an X. For convenience these zone points have been chosen so that the defining equations have integral coefficients. Frequncies for copper were taken from the data of Svenson *et al* [6]. Elastic constant were taken from Landolt-Bornstein [7].

TABLE II. Values of the force constants in N/m for copper calculated from the 3 elastic constants and 13 zone frequencies defined in Fig. 1. Frequencies were obtained from the data of Svenson *et al* [6]. Elastic constants were taken from Landolt-Bornstein [7].

| Force Constant | | Force Constant | | Force Constant | |
|---|---|---|---|---|---|
| $2\alpha_1$ | 28.2099 | $6\alpha_3$ | 1.6909 | $10\alpha_5$ | -0.1586 |
| $2\beta_1$ | -1.2929 | $6\beta_3$ | -0.2222 | $10\beta_5$ | 0.5089 |
| $2\gamma_1$ | -0.6345 | $6\gamma_3$ | 1.0453 | $10\gamma_5$ | -0.4730 |
| | | | | | |
| $\alpha_2$ | -0.5868 | $2\alpha_4$ | -0.8946 | $3\alpha_6$ | 0.1257 |
| $\beta_2$ | 0.1075 | $2\beta_4$ | 0.7252 | $3\beta_6$ | 0.0744 |
| | | $2\gamma_4$ | 0.3103 | | |

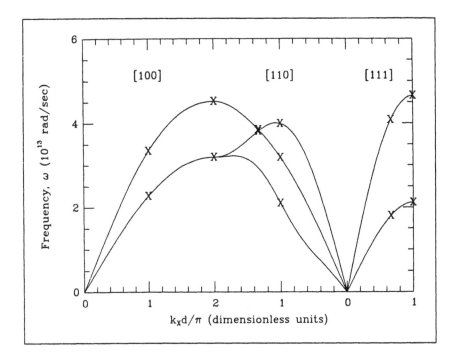

Fig. 1. Dispersion curves for a fcc lattice (copper) defining the 13 zone frequencies used to evaluate the force constants. Two of the frequencies (i.e., $k_x d = 4\pi/3$ for the [110] axis) are nearly equal. Frequency values were obtained from the data of Svenson *et al* [6].

Phase velocities for copper are given for the {100} and {110} symmetry planes in Figs. 2 and 4, respectively, whereas corresponding group velocities are given in Figs. 3 and 5. All wave vectors in these figures fall within the first Brillouin zone.

**DISCUSSION**

The force constants for the closest neighbors are in good agreement with those determined by Svenson *et al* [6] and Sinha [8]. Force constants for third- and fifth-nearest neighbors are much smaller than $2\alpha_1$ so that effects of only 3 force constants (i.e., $16(\alpha_3-\beta_3) = 8\delta_3$ and $12(\alpha_5-\beta_5) = 4\delta_5$) for these neighbors have little effect upon the dispersion curves or upon phase and group velocities.

Plots of the group velocity in the elastic limit (i.e., $kd/\pi = 0.01$) show striking cuspidal features about the $\langle 100 \rangle$ axes in {100} planes and about $\langle 100 \rangle$ and $\langle 110 \rangle$ axes in {110} planes as a result of the large elastic anisotropy factor. For larger values of the wave vector, however, the group velocities decrease and exhibit increased anisotropy. Near zone boundaries along $\langle 110 \rangle$ directions the group velocity

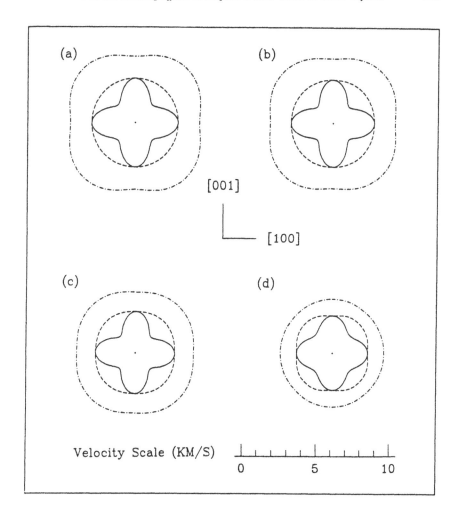

Fig. 2. Polar plots of the phase velocity (km/sec) in the (010) plane for copper using the force constants in Table II. Curves in (a) $kd/\pi = 0.01$; (b) $kd/\pi = 0.9$; (c) $kd/\pi = 1.3$; (d) $kd/\pi = 1.7$.

is negative. At zone boundaries along the $\langle 100 \rangle$ and $\langle 111 \rangle$ axes the group velocity vanishes. As a result the group-velocity surfaces for constant values of wave vector near Brillouin zone boundaries can exhibit rather bizarre behavior. Since energy flow is parallel to the group velocity such behavior dramatically affects the phonon-focusing and phonon transport properties of these crystals. Note, however, that in metals most heat is carried by electrons so that phonon-focusing effects need to be very strong to make measurable differences in the thermal conductivity. In dielectric crystals such as the alkali halides all heat is carried by phonons so that strong phonon-focusing effects can dramatically change the thermal conductivity. However,

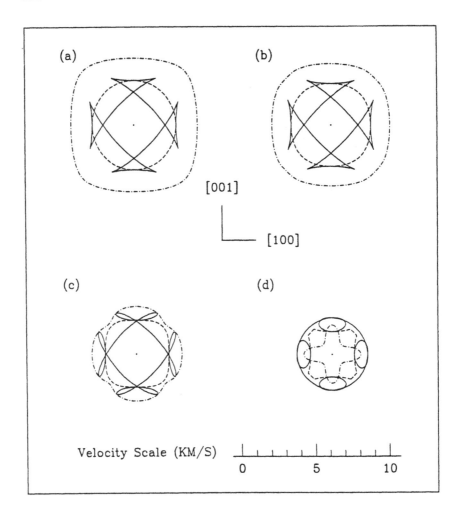

Fig. 3. Polar plots of the group velocity (km/sec) in the (010) plane for copper using the force constants in Table II. Curves in (a) kd/π = 0.01; (b) kd/π = 0.9; (c) kd/π = 1.3; (d) kd/π = 1.7.

because of significant electrostatic forces in the alkali halides neither this rigid-ion model nor the Born-von Karman model can accurately model the lattice dynamics of these crystals.

## REFERENCES

1.  Bell, M.G., and A.K. McCurdy, 1996. "Phonon-Focusing Effects in Dispersive Body-Centered Cubic Crystals," in *Thermal Conductivity* 23, K.E. Wilkes,

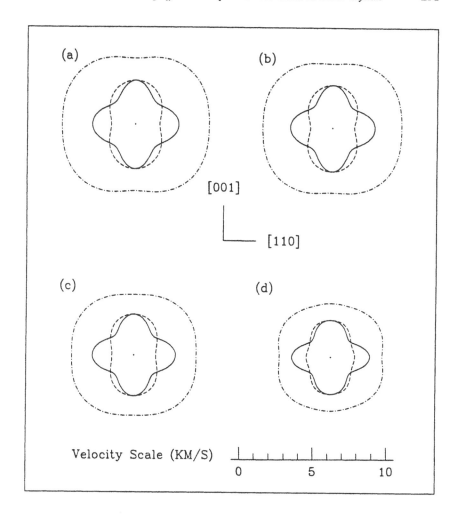

Fig. 4. Polar plots of the phase velocity (km/sec) in the (-110) plane for copper using the force constants in Table II. Curves in (a) kd/π = 0.01; (b) kd/π = 0.9; (c) kd/π = 1.3; (d) kd/π = 1.7.

R.B. Dinwiddie, and R.S. Graves, eds. Lancaster, PA: Technomic Publishing Inc., (this volume in Press).

2. McCurdy, A.K., 1996. "A Rigid-Ion Lattice Dynamical Model for Calculating the Group Velocities in Simple, Body-Centered, and Face-Centered Crystals," in *Thermal Conductivity* 23, K.E. Wilkes, R.B. Dinwiddie, and R.S. Graves, eds. Lancaster, PA: Technomic Pulishing Inc., (this volume in Press).

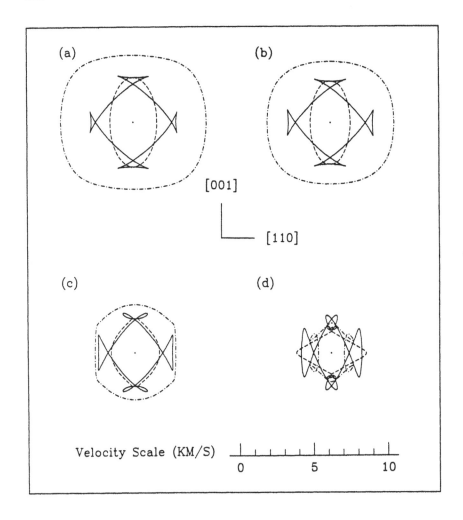

Fig. 5. Polar plots of the group velocity (km/sec) in the (-110) plane for copper using the force constants in Table II. Curves in (a) kd/π = 0.01; (b) kd/π = 0.9; (c) kd/π = 1.3; (d) kd/π = 1.7.

3.  Born, M., and K. Huang, 1954. *Dynamical Theory of Crystal Lattices*. Oxford, England: Clarendon Press, pp. 38-273.

4.  Walker, C.B., 1956. "X-ray Study of Lattice Vibrations in Aluminum." *Phys. Rev.* 103, 547-557.

5.  Venkataraman, G., L.A. Feldkamp, and V.C. Sahni, 1975. *Dynamics of Perfect Crystals*. Cambridge MA: MIT Press, pp. 106-111.

6. Svenson, E.C., B.N. Brockhouse, J.M. Rowe, 1967. "Crystal Dynamics of Copper." *Phys. Rev.* 155, 619-632.

7. Every, A.G., and A.K. McCurdy, 1992. in *Landolt-Bornstein: Numerical Data and Functional Relationships in Science and Technology, New Series, Group III, Volume 29a: Second and Higher Order Elastic Constants.* O. Madelung, and D.F. Nelson, eds. Berlin: Springer-Verlag, pp.12.

8. Sinha, S.K., 1966. "Lattice Dynamics of Copper." *Phys. Rev.* 143, 422-433.

# Energy-Flow-Reversing $N$ and $U$ Processes in Dispersive Crystals

A. K. McCURDY

## ABSTRACT

Energy-conserving interactions which reverse the energy flow can occur by way of a $N$ process or a $U$ process. Reversal of the energy flow in a 3-phonon $N$ process can occur in several different ways: (1) noncollinear processes exhibiting extreme elastic anisotropies, (2) dispersive noncollinear transitions to or from a region of an acoustic branch exhibiting negative group velocity, (3) collinear acoustic-to-optical transitions. Processes (1) and (2) will be numerically illustrated for copper by simple graphical methods applied to a lattice dynamical model. Process (3) will be illustrated graphically for a collinear interaction in an alkali halide. These and other examples presented represent the variety of energy-flow-reversing $N$ and $U$ processes that can be realized in dispersive single crystals.

## INTRODUCTION

Energy- and momentum-conserving phonon-phonon interactions can give rise to a reversal in the energy flow per unit area in crystals having sufficient elastic anisotropy [1]. This peculiar property results because the direction of the phonon momentum is, in general, not parallel to the energy-flow vector. In elastically anisotropic crystals the phonon constant energy surfaces are non-spherical. The phonon phase velocity $\mathbf{s}$ is parallel to the wave vector $\mathbf{k}$, but the group velocity is normal to the constant energy surface in $\mathbf{k}$ space. Phonon momentum is parallel to the wave vector $\mathbf{k}$, but energy flow is parallel to the group velocity $\mathbf{v}$.

In crystals exhibiting large elastic anisotropies the net energy-flow direction for the two decaying phonons can be in the *opposite* direction from the group velocity of the created phonon, and conversely. Conditions for such energy-flow reversal (*EFR*) have been calculated for 3-phonon processes in cubic crystals about 2-fold and 4-fold axes [1], and for 4-phonon processes about 3-fold axes [2]

Professor Emeritus Archie K. McCurdy, Electrical and Computer Engineering Department, Worcester Polytechnic Institute, 100 Institute Road, Worcester, MA 01609

in terms of second-order elastic constant ratios. In cubic crystals there is a total of eight different 3-phonon momentum- and energy conserving interactions that qualify for *EFR*. Some of these *EFR* interactions can be triggered by a phase transition driven by elastic instability [3].

Calculations of the quantum mechanical coupling coefficients for 3-phonon interactions show that phonons which have mutually orthogonal polarizations, or polarizations normal to the wave vector have zero coupling. As a result not all interactions allowed by energy and momentum conservation have a finite probability of forming a new quantum state in the long-wave, elastic limit [4,5]. Interactions, however, between mixed transverse waves and a pure longitudinal wave generally have nonzero coupling coefficients, and thus a finite probability of interacting [4,5]. Since *EFR* interactions in the elastic limit always involve mixed transverse waves interacting with a pure longitudinal wave, such interactions are generally allowed by quantum mechanics.

For larger values of the wave vector dispersion becomes important and the phonon frequency becomes nonlinearly dependent upon wave vector. Since this nonlinear dependence is very different along different crystallographic directions one needs a lattice dynamical model able to predict the phonon frequencies. Calculations for copper will be performed using the modified Born-von Karman rigid-ion model discussed in two of the previous papers in this volume [6,7]. Some interesting examples of *N* and *U* processes will be given and illustrated with graphical techniques. A number of these examples will exhibit *EFR*.

## THEORY AND RESULTS

Consider momentum- and energy-conserving interactions between two phonons *a* and *b* which combine to form a third phonon *c*. For energy conservation:

$$\hbar\omega_a + \hbar\omega_b = \hbar\omega_c. \tag{1}$$

For momentum conserving *N* processes:

$$\hbar\mathbf{k}_a + \hbar\mathbf{k}_b = \hbar\mathbf{k}_c. \tag{2}$$

For Umklapp or *U* processes, however:

$$\hbar\mathbf{k}_a + \hbar\mathbf{k}_b = \hbar\mathbf{k}_c + \hbar\mathbf{G} \tag{3}$$

where $\mathbf{G}$ is a vector of the reciprocal lattice in $\mathbf{k}$ space.

This paper will first consider noncollinear interactions in dispersive cubic crystals. Results will only be given for this case when phonons *a* and *b* arise from the *mixed* transverse branch of the phonon spectrum, and phonon *c* is a pure longitudinal mode. Furthermore, phonons *a* and *b* must have propagation vectors, $\mathbf{k}_a$ and $\mathbf{k}_b$ which are equal in magnitude, lie in the same (010) or (-110) symmetry

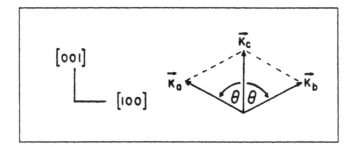

Fig. 1.  Vector diagram showing momentum conservation between two noncollinear phonons which can combine to from a third phonon for the special case in which $|k_a| = |k_b|$.

plane, and have the same angle $\theta$ with respect to a $\langle 100 \rangle$ or $\langle 110 \rangle$ axis, but differ in azimuthal angle by 180°. In this case the propagation vector, $k_c$, for phonon $c$ will be directed along a $\langle 100 \rangle$ or $\langle 110 \rangle$ axis.

Now the phase velocity is given by:

$$s = \omega/k \qquad (4)$$

and is parallel to the wave vector $k$. The energy conservation equation (Eq. 1) can therefore be expressed as:

$$k_c s_c = k_a s_a + k_b s_b. \qquad (5)$$

Now $k_c$ is aligned along an axis of 2-fold or 4-fold symmetry, both $k_a$ and $k_b$ have equal magnitude, and both are aligned at an angle $\theta$ to $k_c$ therefore from Fig. 1:

$$k_c = 2k_a \cos\theta \qquad (6)$$

and $$s_a = s_b. \qquad (7)$$

On substituting Eqs. 6 and 7 into Eq. 5 and dividing out common factors yields the simple but important result:

$$s_c \cos\theta = s_a. \qquad (8)$$

Note that $s_c$ is calculated using $k_c$ whereas $s_a$ is calculated using $k_a$. Both sides of Eq. 8 are plotted in Fig. 2 as a function of the angle $\theta$ for a constant value of $k_a$ for each of the propagating modes. Solutions exist where the left and right sides of this equation intersect.

Note that for $k_c$ along a $\langle 100 \rangle$ direction $s_c$ can propagate as a pure longitudinal or as a pure degenerate transverse wave. For a $\langle 110 \rangle$ direction,

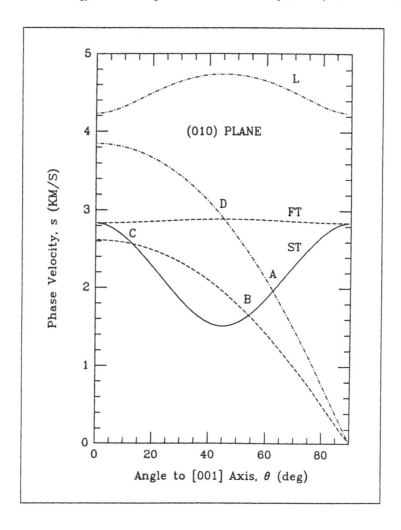

Fig. 2.  Loci for Eq. 8 showing multiple solutions, θ, for energy-and momentum
conserving interactions between noncollinear phonons in copper. Curves *L, ST*
and *FT* are the phonon phase velocities for the longitudinal, slower transverse,
and faster transverse modes, respectively, for wave vectors in the (010) plane as
a function of the angle θ to the [001] axis for $k_a d = 0.5\pi$. The *FT* mode is pure
transverse in copper, whereas *L* and *ST* are generally mixed or impure modes.
Intersections *A* through *D* are graphical solutions to Eq. 8. Only intersection *A*,
i.e., the interaction between a pure *L* and 2 mixed *T* modes, is listed in Table I.

however, $s_c$ can propagate as a pure longitudinal, or as one of 2 pure transverse
waves of different velocity. As a result $s_c$ can have 2 distinct values for a ⟨100⟩
direction, but have 3 distinct values for a ⟨110⟩ direction of wave vector. For

arbitrary values of $\theta$ in a symmetry plane $s_a$ can assume 3 different values corresponding to propagation as a mixed (or impure) longitudinal, a mixed transverse, or as pure transverse wave. As a result there can be multiple solutions to Eq. 8.

Quantum mechanical considerations in the long-wave elastic limit indicate that probabilities for interaction depend both upon the polarization of the interacting waves, and the third-order elastic constants [4,5]. In particular, waves with mutually orthogonal polarizations have zero probability for interaction. As a result Table I lists only the solutions to Eq. 8 where $s_c$ is a pure longitudinal mode and $s_a$ is a *mixed* transverse mode. No attempt is made, therefore, to calculate collinear interactions between a longitudinal, $s_c$, and transverse, $s_a$, modes since these interactions involve pure modes with mutually orthogonal polarizations. Collinear interactions between a longitudinal optic and longitudinal acoustic modes are discussed later in another paragraph.

Now the energy-flow density, $U_c$, for $N$ identical phonons per unit volume each with energy, $\hbar\omega_c$, and group velocity, $v_c$, is given by:

$$U_c = N\hbar\omega_c v_c \tag{9}$$

and is thus parallel to the group velocity, $v_c$. The resultant energy-flow density vector, $U_{a+b}$, for $N$ identical doublets of interacting phonons, $a$, and $b$ per unit volume is given by:

$$U_{a+b} = N\hbar\omega_a v_a + N\hbar\omega_b v_b. \tag{10}$$

For certain interactions $v_c$ can be antiparallel to $v_a + v_b$ leading to a condition known as energy-flow reversal (*EFR*). This is illustrated in Fig. 3 using the first solution listed in Table I.

For larger values of $k_a$ solutions to Eq. 8 give values of $k_c$ which extend beyond the edges of the first Brillouin zone. The values of $s_c$ obtained from $s_c = \omega_c/k_c$ are really fictitious values which need to be replaced by using the reduced wave vector, $k_c$(reduced), which has been brought back into the first Brillouin zone by the addition of a reciprocal lattice vector, $G$. The appropriate phase velocity under these conditions (i.e., a Umklapp process) is given by:

$$s'_c = \omega_c/|k_c(\text{reduced})| \tag{11}$$

where $k_c(\text{reduced}) = k_c - G$.

Note, however, that $\omega_c = s'_c k_c(\text{reduced}) = s_c k_c$ so that Eq. 8 is equivalent to:

$$s'_c k_c(\text{reduced}) = 2s_a k_a$$

or:
$$\omega_c = 2\omega_a \tag{12}$$

where $\omega_c$ is calculated at $k_c$ but $\omega_a$ is calculated at $k_a$. Solutions can be obtained to Eq. 12 by plotting $\omega_c$ and $2\omega_a$ as functions of the angle $\theta$ and determining their

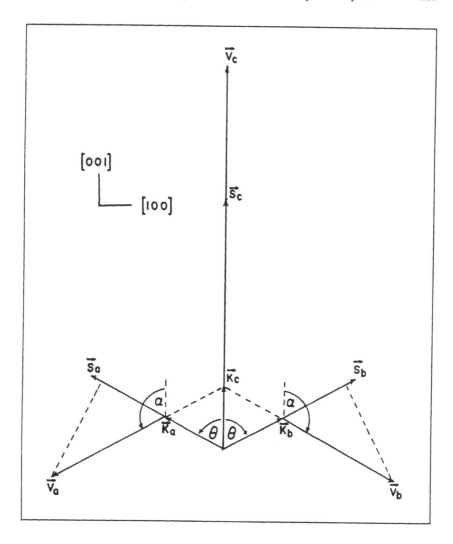

Fig. 3. A vector diagram showing the wave vectors, the phase velocity and the group velocity vectors for a energy- and momentum-conserving interaction between a pure $L$ and two mixed $T$ modes in copper for $k_a d = 0.5\pi$. Angle $\alpha$ is tabulated in Table I. Note that $\alpha$ is greater than 90 ° so that $v_a + v_b$ is antiarallel to $v_c$ leading to the condition of energy-flow reversal (*EFR*). Note also that because of dispersion $v.k/k$ is less than $s$ for phonons $a$ and $b$, and $v_c$ is less than $s_c$.

points of intersection. The results are equivalent to those determined from Eq. 8, but without the need to determine the fictitious phase velocity, $s_c$, when the value of $k_c$ extends beyond the zone boundary. Plotting Eq. 8, however, displays the phase velocity, and thus is particularly useful in the long-wavelength elastic limit.

Consider now collinear interactions between two phonons $a$ and $b$ which combine to form a third phonon $c$ that propagates along a symmetry axis. Along such an axis a wave propagates as a pure longitudinal or as a pure transverse wave. Quantum mechanical considerations indicate that waves with mutually orthogonal polarizations, and waves whose polarization is perpendicular to the propagation vector have zero probability for interaction. If wave $c$ propagates along a symmetry axis then the only energy- and momentum-conserving interaction allowed by quantum mechanics is:

$$L \leftrightarrow L + L. \tag{13}$$

For collinear interactions $\theta = 0$ and Eq. 8 becomes:

$$s_c = s_a = s_b \tag{14}$$

and Eq. 12 becomes:

$$\omega_c/2 = (\omega_a + \omega_b)/2 = \omega_a \tag{15}$$

where $\omega_c$ is calculated at $k_c = 2k_a$, but $\omega_a$ is calculated at $k_a$. In the presence of dispersion which lowers the phase velocity with increase in frequency, one cannot satisfy Eq. 14 if all three phonons arise from the same branch of the phonon spectrum. Eqs. 14 and 15 can be satisfied, however, if the higher frequency phonon arises from the longitudinal optical branch ($LO$), and the lower frequency phonons arise from the longitudinal acoustic branch ($LA$) of the phonon spectrum, that is: $LO \leftrightarrow LA + LA$. Fig. 4 illustrates a simple graphical solution for this interaction for one of the alkali halides, LiF. Note that although the *phase* velocity of the three phonons are equal, the group velocity of the $LO$ phonon is negative and differs in magnitude from the phase and group velocity of the $LA$ phonons. As a result, phonon $c$ travels in the opposite direction from phonon $a$ or $b$ yielding the peculiar condition of energy-flow reversal ($EFR$).

When the wave vector $k_c$ extends beyond the first zone boundary the value of the phase velocity, $s_c$, determined from $\omega_c/k_c$ is actually fictitious. The appropriate phase velocity is $s'_c$ where $s'_c = \omega_c/k_c(\text{reduced})$ and $k_c(\text{reduced})$ is the reduced wave vector resulting from the addition of the appropriate reciprocal lattice vector **G**. Fig. 5 illustrates a simple graphical solution for such a collinear $U$ process. The phase and group velocities for acoustic phonons $a$ and $b$ are determined at point $A$, whereas the phase and group velocities for optic phonon $c$ are determined at point $B'$ after the addition of a reciprocal lattice vector **G**. Note that the group velocity for both optic and acoustic phonons is positive (i.e., $\omega$ vs $k$ dispersion curve has a positive slope) and thus $EFR$ is absent in this interaction.

TABLE I. Solutions for momentum- and energy-conserving interactions between a pure longitudinal mode and two *mixed* transverse modes in copper for the special case in which $|k_a| = |k_b|$. For interactions numbered 1-4 $k_a d = 0.5\pi$ whereas for interactions 5-10 $k_a d = 1.3\pi$. Wave vectors $\mathbf{k_a}$ and $\mathbf{k_b}$ lie at an angle $\theta$ with respect to the wave vector $\mathbf{k_c}$, but differ in azimuthal angle by 180°. The corresponding angle for the group velocity is denoted by $\alpha$. Note that the wave vector $\mathbf{k}$ is always parallel to the corresponding phase velocity $\mathbf{s}$. Values of $\alpha$ denoted by a number sign # indicate energy-flow-reversing interactions. Negative values of $\alpha$ indicate that the positions of $s_a$ and $s_b$ are interchanged. Values of $k_c d$ denoted by an asterisk indicate an Umklapp process because these values extend beyond the first zone boundary. Values of $s_c$ for Umklapp processes listed here are given both by $s_c = \omega_c/k_c$ and (in the second entry) by $s'_c = \omega_c/k_c$(reduced) where $k_c$(reduced) has been brought back to the first Brillouin zone by the addition of a reciprocal lattice vector. Note that reducing $k_c$ back to the first zone changes the magnitude and reverses the sign of the phase velocity, but has no effect upon the frequency nor the group velocity. All vectors with negative values are directed vertically downward.

| Axis for $k_c$ | Plane of $s_a$ & $s_b$ | $k_c d$ | $s_c$ km/S | $v_c$ km/S | $s_a$ km/S | Angle $\theta$ deg | $v_a$ km/S | Angle $\alpha$ deg | No |
|---|---|---|---|---|---|---|---|---|---|
| [001] | (010) | 1.45 | 4.25 | 4.04 | 1.96 | 62.6 | 3.04 | 115.4# | 1 |
| [101] | (010) | 1.79 | 4.67 | 4.05 | 2.66 | 55.2 | 3.10 | 20.0 | 2 |
| [001] | ($\bar{1}$10) | 1.62 | 4.22 | 3.96 | 2.17 | 59.0 | 2.62 | 97.9# | 3 |
| [110] | ($\bar{1}$10) | 1.43 | 4.78 | 4.38 | 2.18 | 62.8 | 2.66 | 101.7# | 4 |
| [001] | (010) | 7.36* −5.21 | 2.16 −3.05 | −0.87 | 1.94 | 25.7 | 2.39 | −13.0# | 5 |
| [001] | (010) | 4.29 | 3.43 | 1.71 | 1.80 | 58.3 | 2.39 | 88.8 | 6 |
| [001] | ($\bar{1}$10) | 7.0* −5.57 | 2.31 −2.90 | 0.56 | 1.98 | 31.1 | 2.44 | −6.0 | 7 |
| [001] | ($\bar{1}$10) | 4.24 | 3.45 | 1.77 | 1.79 | 58.8 | 1.63 | 115.7# | 8 |
| [110] | ($\bar{1}$10) | 7.27* −6.49 | 2.14 −2.39 | 0.94 | 1.90 | 27.1 | 1.99 | −35.1 | 9 |
| [110] | ($\bar{1}$10) | 5.38 | 2.65 | −0.69 | 1.75 | 48.8 | 1.82 | 81.1# | 10 |

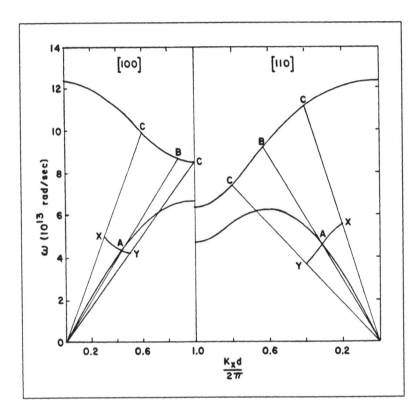

Fig. 4. Dispersion curves for longitudinal branches of LiF for the [100] and [110] axes from Dolling *et al* [8]. The upper curves are the optical branches (*LO*) whereas the lower curves are the acoustic (*LA*) branches. The figure illustrates collinear *N* processes between a phonon from the *LO* branch and two phonons from the *LA* branch. The short truncated curve *XY* is one-half the *LO* frequency plotted at one-half the corresponding wave vector $k_c$ and thus is a locus of the mid-points of line segment *OC* as a function of $k_c$. Point *A* is the intersection of this line segment with locus *XY* and yields the solution to Eq. 15, whereas point *B* is the corresponding solution for the optical branch. *EFR* results because the group velocities, $\partial\omega/\partial\mathbf{k}$, of the optical and acoustic phonons are in opposite directions. Note that for the [100] axis $k = k_x$, but for the [110] axis $k = \sqrt{2}\, k_x$.

## DISCUSSION

Note that 7 of the 10 noncollinear interactions listed in Table I are *N* processes; only 5, 7, and 9 are *U* processes. Five of the *N* processes, but only one of the *U* processes (i.e., 5) exhibit *EFR*. Note also that in Table I there are three types of energy-flow-reversing (*EFR*) interactions: (1) *N* processes in which *EFR* results from the large anisotropy between the phase velocities along different

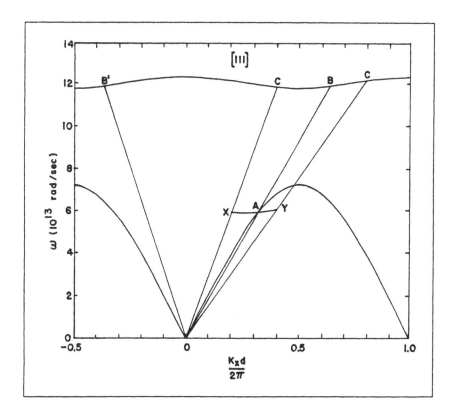

Fig. 5. Dispersion curves for longitudinal branches of LiF for the [111] axis from Dolling *et al* [8]. The upper curve is the optical branch (*LO*) whereas the lower curve is the acoustic (*LA*) branch. The figure illustrates a collinear *U* process between a phonon from the *LO* branch and two phonons from the *LA* branch. Note, in contrast to Fig. 4, the solution at point *A* yields a wave vector, $k_c$, which extends beyond the first zone boundary. Point *B'* is the equivalent zone point after adding a reciprocal lattice vector to reduce $k_c$ to be within the first zone. The appropriate phase velocity of the optical phonon is $s'_c = \omega_c/k_c$(reduced), whereas the phase velocity for each acoustic phonon is $s_a = s_b = \omega_a/k_a$. Note that the group velocities of both acoustic and optical phonons are positive so that *EFR* is absent in this interaction. Finally note that for the [111] axis $k = \sqrt{3}\, k_x$.

directions, e.g., 1, 3, 4 and 8; (2) *N* processes which yield a resulting wave vector $k_c$ falling in a region of the dispersion curves exhibiting negative slope, e.g., 10; and (3) Umklapp or *U* processes in which the reduced wave vector $k_c$(reduced) has reversed direction as a result of the addition of a reciprocal lattice vector, reversing the group velocity $v_c$, e.g., 5. Note, however, that *U* processes 7 and 9 do not exhibit *EFR* because the frequency of phonon *c* lies in a dispersion region having a positive slope. Points along the ⟨110⟩ directions beyond the first

Brillouin zone in fcc crystals, for example, are equivalent to points along the first Brillouin zone surface between the $\langle 110 \rangle$ and $\langle 100 \rangle$ directions. Here the dispersion curves might have a region of positive slope (as they do for this example), and thus a positive group velocity. Finally, note that for $k_a d = 1.3\pi$ there are no solutions for the [101] reference axis and (010) plane.

The collinear interactions in Fig. 4 illustrate two additional examples of *EFR*. Quantum mechanical considerations indicate that collinear interactions are more favorable than noncollinear interactions since the polarization vectors are mutually parallel and also parallel to the wave vector. Both interactions in Fig. 4 are examples of *N* processes since the resultant wave vector, $k_c$ falls within the first Brillouin zone. Since the group velocity $\mathbf{v} = \partial\omega/\partial\mathbf{k}$, it has a negative value at point *B* in the *LO* branch along the $\langle 100 \rangle$ and $\langle 110 \rangle$ axes within the first Brillouin zone. Since the group velocity of the *LA* branch is positive at point *A* along $\langle 100 \rangle$ and $\langle 110 \rangle$ axes both collinear *N* processes exhibit *EFR*. The collinear interaction in Fig. 5 illustrates a Umklapp process because the resultant wave vector $\mathbf{k}_c$ extends beyond the zone boundary. The reduced wave vector $\mathbf{k}_c$(reduced) is in the reverse direction from $\mathbf{k}_a$ with the actual phase velocity $\mathbf{s}_c$ parallel to $\mathbf{k}_c$, but of larger magnitude than $\mathbf{s}_a$. The group velocity at the equivalent zone points *B* and *B'* is positive, and thus has the same sign as the group velocity of the *LA* branch at point *A*. As a result this Umklapp process fails to exhibit *EFR*.

The graphical techniques discussed for noncollinear 3-phonon interactions require a model that can predict the dispersion relations in both symmetry planes of cubic crystals. The model used in this paper is a modified Born-von Karman rigid-ion model that has been discussed in one of the previous papers in this volume. Note, however, that these graphical techniques can be applied for other lattice dynamical models such as the dipole approximation model useful for alkali halides. The graphical technique discussed for collinear interactions along symmetry axes, however, only requires the knowledge of the *LO* and *LA* dispersion curves along the three principal axes, and thus can be used on raw data.

The peculiar condition of *EFR* is seen to be a direct result of the nonlinear dependence of phonon frequency on wave vector. Since this nonlinear dependence is always different along different crystallographic axes a large variety of phonon interactions are possible, with a significant fraction of these exhibiting *EFR*.

## REFERENCES

1. McCurdy, A.K., 1988. "Energy-Flow-Reversing *N* Processes in Elastically Anisotropic Crystals." *Phys. Rev.* B38, 10335-10349.

2. Hasegawa, S., and A.K. McCurdy, 1990. "Four-Phonon Energy-low-Reversing *N* Processes About <111> Axes in Cubic Crystals." in *Thermal Conductivity* 21, H.A. Fine and C.J. Cremers, eds. N.Y.: Plenum Press, pp. 401-411.

3. Thomas, Jan and A.K. McCurdy, 1994. "Energy-Flow-Reversing Three-Phonon Processes Near a Structural Phase Transition," in *Thermal Conductivity* 22, T.W. Tong , ed. Lancaster, PA: Technomic Publishing Inc., pp. 391-398.

4. Thomas, Jan, 1993. "Quantum Mechanical Coupling Coefficients for Energy-Flow-Reversing *N* Processes in the Absence of Dispersion," M.S. Thesis, Worcester Polytechnic Institute, Worcester, MA.

5. Thomas, Jan, and A.K. McCurdy, 1993. "Quantum Mechanical Coupling Coefficients for 3-Phonon Interactions," in *Thermal Conductivity* 22, T.W. Tong, ed. Lancaster, PA: Technomic Publishing Inc., pp. 374-382.

6. McCurdy, A.K., 1996. "A Rigid-Ion Lattice Dynamical Model For Calculating Group Velocities in Simple, Body-Centered, and Face-Centered Crystals," in *Thermal Conductivity* 23, K.E. Wilkes, R.B. Dinwiddie, and R.S. Graves, eds. Lancaster, PA: Technomic Publishing Inc., (this volume in Press).

7. Capriglione, G.S., and A.K. McCurdy, 1996. "Phonon-Focusing Efects in Dispersive Face-Centered Cubic Crystals," in *Thermal Conductivity* 23, K.E. Wilkes, R.B. Dinwiddie, and R.S. Graves, eds. Lancaster, PA: Technomic Publishing Inc., (this volume in Press).

8. Dolling, G., H.G. Smith, R.M. Nicklow, P.R. Vijayaraghavan, and M.K. Wilkinson, 1968. "Lattice Dynamics of Lithium Fluoride." *Phys. Rev.* 168, 970-979.

# Thermal Conductivity of Linear Chain Semiconductor $(NbSe_4)_3I$

A. SMONTARA, K. BILJAKOVIĆ, A. BILUŠIĆ, D. PAJIĆ,
D. STAREŠINIĆ, F. LÉVY and H. BERGER

## ABSTRACT

We present measurements and analysis of the thermal conductivity, K(T), of the linear chain semiconductor $(NbSe_4)_3I$, which undergoes a phase transition beween two semiconducting phases at $T_c$=274K. The obtained data cover the range 100-340 K. The behaviour of K(T) is similar to that expected for insulating crystals, where K(T) is mainly governed by the umklapp processes. Beyond that, the thermal conductivity anomaly appears at $T_c$ as a consequence of the structural phase transition that has been also seen in specific heat.

## INTRODUCTION

Quasi-one-dimensional halogened transition-metal tetrachalcogenides $(MSe_4)_xI$ (with M=Ta, Nb) have been extensively studied the last fifteen years [1]. These compounds consist of $MSe_4$ chains parallel to the c axis well separated from one another by iodine channels. In each $MSe_4$ chain the metal atom is located at the center of a rectangular antiprism of eight Se atoms. The $dz^2$ band filling of the transition metal M can be varied according to the composition [2]. This change in band filling leads to quite different structural and electrical properties. $(TaSe_4)_2I$ and $(NbSe_4)_{10}I_3$ undergo a Peierls transition at $T_p$=263 [3] and 280K [4] respectively [5], while $(NbSe_4)_3I$ exhibits a ferrodistortive structural phase transition at $T_c$=274K [5,6]. For this latter compound the $NbSe_4$ chain along c axis comprises six $NbSe_4$ units in the unit cell and is strongly distorted above $T_c$ with two long Nb-Nb distances of 3.25Å and a short Nb-Nb distance of 3.06Å [7]. Below $T_c$ the chains are less distorted with a Nb-Nb sequences of 3.31, 3.17 and 3.06Å [8]. This accounts for the semiconducting be-haviour with the activation energy corresponding to 2400K above $T_c$, and a

A. Smontara, K. Biljaković, A. Bilušić , D. Pajić, and D. Starešinić, Institute of Physics of the University, P.O.B. 304, HR-10000 Zagreb, Croatia
H. Berger, and F. Lévy, Institut de Physique Appliquée, Ecole Polytechnique Fédérale de Lausanne, CH-1015 Lausanne, Switzerland

much smaller gap of 900K below $T_c$ [5]. According to the activated behaviour at low temperatures two different types of crystals have been found. Crystals of type I keep the very small value for the gap measured below $T_c$. For type II the activation energy increases again to some intermediate value of 1700K [5,9]. Up to now it has not been possible to determine the physical parameters which distinguish these two types. Structural analysis [10] of two phases of (NbSe₄)₃I above and below $T_c$ shows that the phase transition is ferrodistortive but non-ferroelectric and non-ferroelastic. The space group change is P4/m$nc$→P42$_1c$ keeping the same number of formula units per unit cell ($Z=4$). The order parameter is of $B_{1u}$ symmetry and the corresponding soft mode was observed below $T_c$ in Raman spectra [11]. The infrared measurements show an additional strong positive contribution to the permitivity that is attributed to a hopping mechanism for the conductivity in (NbSe₄)₃I [12]. Softening of the $C_{44}$ elastic constant below $T_c$ has been studied by ultrasonic measurements [13] and the dispersion of acoustic branches have been measured by inelastic neutron scattering [14]. The phase transition has also been studied by x-ray diffraction [15], electron diffraction [16], NMR techniques[17] specific heat [18] and transport measurements [5,6,9,18]. Moreover it has been reported that another phase transition may take place near 100K with space group P4 ($Z=4$) [19,20,21]. The soft mode associated with this phase transition is a silent (IR- and Raman-inactive) mode of $A_2$ symmetry in the P42$_1c$ phase becoming again totally symmetric in the P4 phase. Raman measurements have reported the presence of a weak mode at 28cm$^{-1}$ showing a partial softening. Contrary to that the neutron scattering [13,22], infrared measurements [12] specific heat and thermal conductivity experiments [18,23] give no support to existence of this phase transition, and its existence is still questionable.

So far, only qualitative thermal conductivity measurements of (NbSe₄)₃I have been published previously [18] with some spurious effects due to the technological problems. In this paper, we report improved thermal conductivity measurements of (NbSe₄)₃I in the range 100-340K, with special attention to the behaviour at the phase transition.

## EXPERIMENTS

Single crystals of (NbSe₄)₃I were synthesised by the vapour-transport method. The procedure was described in detailed in ref. [7].

The low-electric field dc conductivity as a function of temperature was measured along the chain direction (the tetragonal c axis) using low frequency bridge. Four contacts were made with silver paint on the freshly cleaved surface of the sample. The quality of the electrical contacts was improved by coating the sample with a gold film at the places where the current and voltage leads were attached. The measurements have been carried out both in cooling and heating at a rate of 30 K/h; no evidence of thermal hysteresis has been observed.

Thermal conductivity measurements were performed with the comparative steady-state heat flow technique [24] relative to a constant foil. The heat flow was propagating along the c axis. The sample was attached with GE varnish to the reference sample which was also used as the thermal link between the sample and the heater. On the other side the sample was glued directly to a thermal sink. The sample chamber was maintained in vacuum better than $10^{-4}$Pa. In order to minimize the radiation heat losses, the sample holder was closely surrounded by the heat shield kept at almost the same temperature. The heat flow was estimated from the voltage difference between two chromel wires point-soldered to the constantan foil, so that the link was also used as the thermocouple element. The thermal gradients in the sample were measured by a chromel-constantan thermocouple glued to the sample with GE varnish. The temperature difference on the reference and on the sample was kept always smaller than 1K. The cooling (heating) rate was generally ~5K/h or less, even smaller than 1K/h in the transition region. The relative accuracy of the measurement (1-2%) was much better than that for absolute values (about 20% due to uncertainty in geometrical factors of the sample and the reference).

Specific heat measurements were made using the improved heat pulse technique previously applied to the calorimetric investigation of charge density-wave systems [25]. A single crystal was mounted on a small plate of copper foil. For a good thermal contact we used a thin layer of silicone high vacuum grease. Heating pulses of about 0.1-0.2K were applied to the sample using the small heater on the lower side of the copper plate. The relative temperature to the heat sink was measured by a cooper-constantan thermocouple atached by GE varnish on the upper side of the sample. The absolute temperature of the sink was determined with a platinum thermometer. The specific heat measurements were carried out in several experimental runs. The cooling (heating) rate was about 1K/h. Parasitic heat capacities were measured in a separate run. The resolution in the specific heat was better than 1%.

## EXPERIMENTAL RESULTS AND DISCUSSION

The electrical conductivity $\sigma(T)$ along the c axis and its logarithmic derivative ($d\ln(\sigma)/d(1000/T)$) with respect to the temperature are shown in Fig.1. The room temperature conductivity is about $85\Omega^{-1}m^{-1}$. The behaviour of $\sigma(T)$ is characteristic for the previously reported type II $(NbSe_4)_3I$ [1]. The conductivity obeys the activation law $\sigma(T)=\sigma_o e^{[E_g(T)/k_BT]}$ with temperature dependent activation energy $E_g(T)$. Its logarithmic derivative exhibits the sharp singularity at $T_c=$ 274K and a broad anomaly around 170K. The activation temperature, $E_g/k_B$, corresponding to the anomaly is found in the range 1500-1800K in agreement with the previously published data [1]. Electrical conductivity in the entire temperature range is appreciably lower than the submillimetre electronic conductivity [12], maybe owing to a hopping mechanism of the electron transport in these materials.

The total measured thermal conductivity $K_T$ on two samples (sample #1 (o), sample # 2 ($\square$)) of very different thickness (o d=1.62mm, $\square$ d=0.35mm) are shown in Fig.2. At room temperature (R.T.) we have found the value $K_T\approx17.5WK^{-1}m^{-1}$ and $K_T\approx14.0WK^{-1}m^{-1}$, respectively. $K_T$ increases upon cooling until a maximum is reached at ~10K [23]. There is the anomaly at $T_c=274K$. It can be seen from Fig. 2 that we observed the general features common to both samples of different sizes. However, the difference of about 25% in the abso-

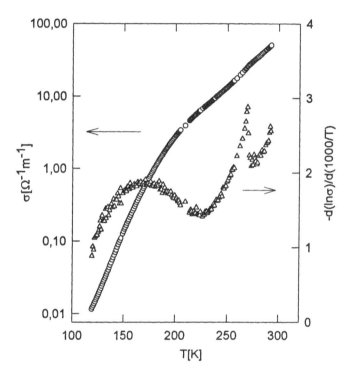

Figure 1: Low-field dc electrical conductivity $\sigma(T)$ along the c axis of sample #1(o) of (NbSe₄)₃I and its logarithimic derivative $(d(\ln\sigma(T)/d(100/T))$ ($\Delta$) with respect to the temperature. The conductivity anomaly is clearly observed as the logarithmic derivative exhibits a sharp singularity at $T_c =274K$.

lute values persists over the whole temperature range. It may come from the estimation of the geometrical factors. However to exclude completely the possible size effect at high temperatures in these quasi-one dimensional materials [26] some more detailed analysis (mainly on the same sample) should be done.

Due to the semiconducting nature of these solids, the total thermal conductivity $K_T$ can be given as the sum of the free carrier, $K_{fc}$, and the lattice $K_{ph}$, contributions [27]:

$$K_T = K_{fc} + K_{ph}. \tag{1}$$

The free carrier contribution can be estimated assuming the Wiedemann-Franz law, according to which $K_{fc}=L_{eff}\sigma T$, where $L_{eff}$ is the free carrier Lorenz number, and σ the electrical conductivity. The free carrier Lorenz number  is given by [28, 29]:

$$L_{eff} = L_o [ 1 + (3/\pi^2)[b/(1+b)^2][E_g/k_BT+4]^2 ] \qquad (2)$$

where $L_o$ is the Sommerfeld value of the Lorenz number ($L_o$ =2.45x10$^{-8}$WΩK$^{-2}$) , $E_g$ the energy gap, and b the electron-hole mobility ratio. Using the Eq. (2) we have estimated at room temperature (R.T.) the free carrier Lorenz number. $L_{eff}$ (R.T.)≈20$L_o$ where for the ratio of mobilities we estimated b=0.28 at T<$T_c$ from the thermopower measurements (ref. 23) and $E_g$ (T) from the temperature variation of the logarithmic derivative of the electrical conductivity. The free carrier contribution ($K_{fc}$(R.T.)≈0.01WK$^{-1}$m$^{-1}$) is less than 1% of the total thermal conductivity; i.e. it is in the range of the absolute accuracy of our measurement. So the total thermal conductivity  comes mainly from lattice vibrations.

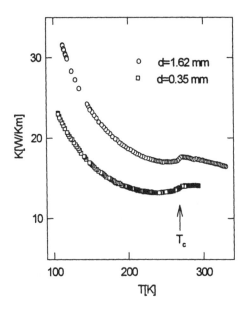

Figure 2: Thermal conductivity K(T) of (NbSe$_4$)$_3$I measured on two samples (sample #1 (o), sample #2 (□)) of very different thickness (o d=1.62mm, □ d=0.35mm). In both samples there is the anomaly at $T_c$=274K.

Thermal conductivity of (NbSe$_4$)$_3$I has temperature behaviour of crystalline materials. Fig. 3. shows  that K(T) obeys an exponential law such as K(T)~Ae$^{(\theta_D/\alpha T)}$ where α is a numerical factor and θ$_D$ Debye temperature. Debye temperature of these solids is relatively low [30]  and it is inside the temperature range of our measurement. The parameters obtained from the fit for the thermal conductivity are A=9.52W/Km and $\theta_D/\alpha$=67.6K (dashed line of Fig. 3). As $\alpha$ is the numerical factor of the order of 2 [31,32], for $\alpha$=2 with $\theta_D$=116K obtained from the low temperature specific heat measurements [30] we have good agreement.

The behaviour of thermal conductivity at the phase transition itself is distinctive. There is a broad anomaly, approximately 20K wide (Fig. 2) and the increase is slightly greater than 5% (Fig. 4 inset). The features of the anomaly were reproduced in two experimental runs on the same sample (sample #1). A second sample of (NbSe$_4$)$_3$I (sample #2) also showed anomaly, somewhat smea-

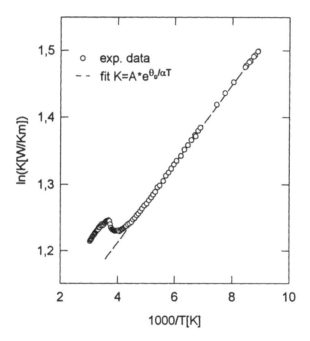

Figure 3: Themal conductivity K(T) with respect to the inverse temperature 1000/T in the sample of (NbSe$_4$)$_3$I (sample #1). The dashed line represents the phonon contribution fitted to the formula K(T)~Ae$^{(\theta_D/\alpha T)}$ with A= 9.52 [Wm$^{-1}$K$^{-1}$] and $\theta_D/\alpha$=57.6K.

red out as compared to the first one. The sample dependence at the phase transition seems to be correlated with the low-temperature behaviour. The sample with the larger low temperature conductivity maximum also has the narrow anomaly at T$_c$ [23]. Even though both of the samples were very different geometrically, it is likely that lattice defects affect these aspect of K$_T$.

The largest single crystal, which has mass 11.4mg, was used for the specific heat study. Fig. 4 shows the temperature variation of the total specific heat for (NbSe$_4$)$_3$I near the phase transition temperature T$_c$. The anomaly was found at 274K in correspondence with the structural transition and the anomalies of the electric and thermal conductivity (Fig.1 and Fig. 2) and thermoelectric power [23]. In order to analyze the features of the specific heat, the regular contribution from the background thermal excitation must be substracted from the total. Here, we have adopted a procedure for accomplishing the subtraction that we believe is reasonable for this case, and the conclusions that we draw do not depend on the details. A smooth polynominal fit to the background was made by forcing the error to be small at the temperature far from $T_c$ [23]. The subtraction of the fit from the measurements gives a crude estimate of the excess specific heat associated with the phase transition. Its value is aproximately 5% of the total specific heat. Moreover, the measurements of the specific heat were performed in cooling and heating runs: no apparent hysteresis with temperature variation was found. By using the procedure described in ref. [33], it was

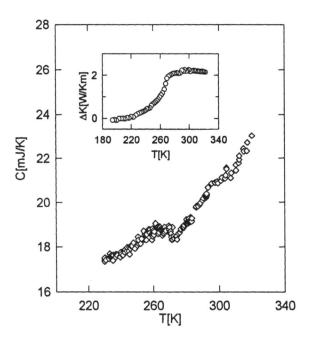

Figure 4: The temperature dependence of specific heat C(T) for $(NbSe_4)_3I$ near the transition temperature. Inset: Difference $\Delta K$ between the thermal conductivity data and the fit. The thermal conductivity anomaly at $T_c$ =274K, $\Delta K/K \geq 5\%$, is consistent with the specific heat jump, $\Delta C/C \geq 5\%$, associated with the structural transition.

found that no indication of a latent heat is present. These results confirm the assumption that the phase transition at 274K is of second order.

## CONCLUSION

We studied the phase transition by means of electrical and thermal conductivity and specific heat measurements. With respect to the phase transition at $T_c$=274K, we observed the increase of the band gap in the electrical conductivity. In addition, thermal conductivity shows the anomaly at 274K. This is consistent with the results of our specific heat measurements. A broad anomaly is observed and the jump associated with the phase transition is estimated.

In general, thermal conductivity has temperature behaviour of crystalline materials. The phonon thermal conductivity is dominant and decreases faster than 1/T even for temperatures $T>\theta_D$. We show that our results can be fitted with an exponential law, in agreement with the theory which takes into account the umklapp scattering for the phonon-phonon term [27, 32].

## ACKNOWLEDGEMENTS

One of us (A.S.) has benefited from enlightening discussions with P. G. Klemens. This work was financially supported by the Ministry of Science of the Republic of Croatia (1.03.055).

## REFERENCES

1. Monceau, P. 1985. "Charge-density wave transport in transition metal tri- and tetrachalcogenides" in *Electronic Properties of Inorganic Quasi One - Dimensional Compounds*, P. Monceau, ed. Dordrecth: D. Riedel Publishing Company, pp. 139-268.

2. Gressier, P., H. Whangbo, A. Meerschaut, and J. Rouxel, 1984. " Electronic Structures of Transition-Metal Tetrachalcogenides (MSe₄)ₙI (M=Nb,Ta)." *Inorganic Chemistry*, 23(9):1221-1228.

3. Wang, Z.Z., M. C. Saint-Lager, P. Monceau, M. Renar, P. Gressier, A. Meerschaut, L. Guemas, and J. Rouxel, 1983. "Charge Density Transport in a Novel Halogened Transition Metal Tetrachalcogenide (NbSe₄)₃.₃₃I." *Solid State Communications*, 46(4):325-328.

4. Wang, Z.Z., P. Monceau, M. Renar, P. Gressier, L. Guemas, and A. Meerschaut, 1983. "Charge density Transport in a novel Halogened Transition Metal Tetrachalcogenide (NbSe₄)₃.₃₃I." *Solid State Communications*, 47(6):439-443.

5. Gressier, P., A. Meerschaut, L. Guemas, J.Rouxel, and P. Monceau, 1984. "Characterization of the New Series of the Quasi One-Dimensional Compounds (MX₄)ₙY (M=Nb, Ta; X=S, Se; Y=Br, I)." *Journal of Solid State Chemistry*, 5:141-151.

6.  Izumi, M., T. Iwazumi, T. Seino, K. Uchinokura, R. Yoshizaki, and E. Matsuura, 1984. "Evidence of a Phase Transition in a New Type of Halogenated Niobium Tetraselenide $(NbSe_4)_xI$." *Solid State Communications*, 49(5):423-426.

7.  Meershaut, A., P. Palvadeau and J. Rouxel, 1977. "Preparation et Structure Cristalline de $I_{0.33}NbSe_4(I_4Nb_{12}Se_{48})$." *Journal of Solid State Chemistry*, 20 (1):21-27.

8.  Gressier, P., L. Guemas and A. Meerschaut, 1985. "$(NbSe_4)_3I$ Low Temperature Structure." *Materials Research Bulletin*, 20:539 -548.

9.  Tagushi, I., F. Lévy, and H. Berger, 1986. "Metal-Insulator Transitions in the Quasi-One-Dimensional Compound $(NbSe_4)_3I$." *Physica B*, 143:258-260.

10. Izumi, M., T. Toshiaki, K. Uchinokura, R. Yoshizaki, and E. Matsuura, 1984. " X-ray Diffraction Study of a Structural Phase Transition in $(NbSe_4)_3I$." *Solid State Communication*, 51(4):191- 194.

11. Sekine, T., K. Uchinokura, M. Izumi, and E. Matsuura, 1984. "Observation of Soft Phonon in Linear Chain Compound $(NbSe_4)_3I$ by Raman Scattering." *Solid State Communication.* 52(4):379-383 .

12. Zelezny, V, J. Petzelt, B. P. Gorshunov, A. A. Volkov, G. V. Kozlov, P. Monceau, and F. Lévy, 1989. "Far-Infrared Dielectric Response and Hopping Conductivity in Quasi-One-Dimensional $(NbSe_4)_3I$.", *Journal of Physics: Condensed Matter* 1(50):10585-10594.

13. Saint-Paul, M., P. Monceau, and F. Lévy, 1988. "Soft-Acoustic-Phonon Mode at the Phsase Transition in Quasi-One-Dimensional $(NbSe_4)_3I$.", *Rapid Comunnications, Physical Review B*, 37(2):1024-1027.

14. Monceau, P, L. Bernard, R. Currat, and F. Lévy, 1989. "Soft Mode Behaviour at the Structural Phase Transition in $(NbSe_4)_3I$.", *Physica B*, 156&157:20-22.

15. Iwazumi, T, M. Izumi, F. Sasaki , R. Yoshizak, and E. Matsura, 1986. "Formation of Long-Range Order near the Second-Order Phase Transition temperature in $(NbSe_4)_3I$." *Physica B*, 143:261-263.

16. Roucau, C, R. Ayrolles, P. Gressier, and A. Meerschaut, 1984. "Electron Microscopy of Ttransition-Metal Tetrachalcogenide $(MSe4)nI$ (M=Nb,Ta)." *Journal of Physics C : Solid State Physics*, 17:2993-2998.

17. Bataud, P., P. Segransan, C. Bertier, and A. Meerschaut, 1985. "[93]Nb NMR Study of CDW in $(NbSe_4)_{10/3}I$ Single Crystal." in *Lecture Notes in Physics 217, Charge Densi Waves in Solids* , Gy. Hutiray and J. Sólyom, eds. Springer-Verlag, Berlin Heidelberg New York Tokyo, pp. 121-124.

18. Smontara, A., K. Biljaković, L. Fórro, and F. Lévy, 1986. "Properties of the Phase Transition in (NbSe₄)₃I." *Physica B*, 14:264-266.

19. Izumi, M., T. Iwazumi, T. Seino, T. Sekine, E. Matsuura, K. Uchinokura, and R. Yoshizaki, 1987. " Crystal Structure and Far-Infrared Reflectance at 30K in Linear-Chain Semiconductor (NbSe₄)₃I. " *Synthetic Metals*, 19:863-868.

20. Sekine, T., M. Izumi, and E. Matsuura, 1987. "Raman Scattering in Linear-Chain Compounds (NbSe₄)₁₀/₃I and (NbSe₄)₃I." *Synthetic Metals*, 19:869-874.

21. Sekine, T, and M. Izumi, 1988. "Successive Phase Transitions in the Linear-Chain Semiconductor (NbSe₄)₃I Studied by Raman Scattering and Electrical Resistivity." *Physical Review B*, 38(3):2012-2020.

22. Lorenzo, J. E. 1992. "Etude par Diffusion de Neutrons des Transitions de Phase dans les Composes Quasi-Unidimensionnles (TaSe₄)₂I et (NbSe₄)₃I." *Ph.D. Thesis*, L'Universite Joseph Fourier-Grenoble, France, unpublished; Lorenzo J.E., et al. to be published.

23. Smontara A., K. Biljaković, L. Fórro, and F. Lévy, 1987. "Thermal Properties of (NbSe₄)₃I ", *Synthetic Metals*, 19:859-862.; Smontara, A., et al. to be published.

24. Smontara, A., Ž. Bihar, K. Biljaković, E. Tutiš, D. Šokčević, and S. N. Artemenko, 1994. "Heat Transport in Charge-Density-Wave Systems," in *Thermal Conductivity 22*, T. W. Tong, ed. Lancaster: Technomic Publishing Company, Inc., pp. 543-563.

25. Đurek, D., K. Franulović, M. Prester, S. Tomić, L. Giral, and J.M. Fabre, 1977. "New Phase Transition Observed at 46K in Tetrathiafulvalene-Tetracyanoquinodimethane (TTF-TCNQ)" *Physical Review Letters*, 38(13): 715-718.

26. There are some observation of the size dependence of the transport properties (f.e. for a germanium , Bergman R. 1976. "The Thermal Conductivity of Nearly Perfect Non-Metalic Crystals." in *Thermal Conduction in Solids*, Oxford, Claredon Press, pp. 45-72.; Geballe, T.H., and G. W. Hull, 1958. "Isotopic and Other Types of Thermal Resistance in Germanium." *Physical Review, Letters to the Editor*, 110 (2):773-775, and for a bismuth, Issi, J.P., and H. J. Mangez, 1972. "Size Dependence of the Transport Properties of Bismuth in the Phonon Drag Region."*Physical Review B*, 6(12):4429-4431; Issi., J.P. 1979. "Low Temperature Transport Properties of the Group V Semimetals." *Australian Journal of Physics*, 32:585-628.), and theoretical prediction of this effect, due to longitudinal phonons (Pomeranchuk, I.1941. "Thermal Conductivity of Dielectrics at Temperatures Higer than the Debye

Temperature", *Journal of Physics, U.S.S.R*, 4(3):259-268.; Pomeranchuk, I. 1941. "On the Thermal Conductivity of Dielectrics.", *Physical Review,* 60:820-821; Pomenranchuk, I. 1941."Thermal Conductivity of Paramagnetic Dielectrics at Low Temperatures, and Termal Conductivity of Dielectrics at Temperatures above the Debye Temperature. "*Journal of Experimental and Theoretical Physics USSR,* 11(2-3):226-254.). Therefore one has been claimed that thermal conductivity of $(NbSe_4)_3I$ with the low Debye frequen cy may also exibit a size effect at the temperatures higher than expected (Apostol, M. 1992. "On the Lattice Thermoconductivity of an Ideal Crystal", *Fizika A*, 1(2):175-180.

27. Berman, R. 1979. *Thermal Conduction in Solids*, Oxford, Claredon Press, pp. 169-178.

28. Smith, R.A. 1978. *Semiconductors*, Cambridge, Cambridge University Press, pp. 175-201.

29. Price, P.J. 1955. "Ambipolar Thermodiffusion Electrons and Holes in Semiconductors.", *Philosophical Magazine*, 46():1252 -1260.

30. Biljaković, K., J.C. Lasjaunias, F. Zougmore, P. Monceau, F. Lévy, L. Bernard, and R. Currat, 1986. "Contribution of Phasons and of Low-Energy Exictations to the Specific Heat of the Quasi One-Dinemsional Compound $(TaSe_4)_2I$.", *Phyical Review Letters*, 57(15):1907-1910.

31. Peierls, R.E. 1958. "Zur Kinetischen Theorie der Wärmeleitung in Kristallen.", *Annales de Physique*, 3:1055-1101.

32. Klemens, P.G. 1958. "Thermal Conductivity and Lattice Vibrational Modes.", *Solid State Physics*, 7:1-70.

33. Tomić, S., K. Biljaković, D. Đurek, and J. C. Cooper, 1981. "Calorimetric Study of the Phase Transitions in Niobium Triselenide $NbSe_3$." *Solid State Communications*, 38:109-112.

**SESSION 4**

# COMPOSITES

SESSION CHAIR
D. P. Hasselman

# Composition Effects on MMC's Thermal Properties and Behavior under Transient High Fluxes

J. J. SERRA,* E. MILCENT* and P. F. LOUVIGNÉ**

## ABSTRACT

The use of Metal Matrix Composites (MMC) is considered for applications in the hypervelocity vectors guidance domain. The guidance function have to be provided under extreme thermomechanical conditions but during a very short time. This paper deals with the determination of aluminium alloys based MMCs thermal properties and with their behavior under transient high fluxes.

Thermal diffusivity of Al-SiC composites for two matrix natures (X2080 and X7093) and different SiC/Al ratio have been determined. These measurements have been performed by using a method in which one face of a flat sample is exposed to a periodically modulated laser beam. The laser modulation shape was optimized so that multiple frequency measurement could be carried out during each test. Thermal diffusivity is identified by minimizing the differences between calculated and measured values for the phase shift between the incident flux and the resulting oscillations on the rear face temperature. Results show a diffusivity decrease when SiC fraction increases and an observable difference between the two composite families : diffusivity values being lower for X2080 matrix composites.

Transient heating behaviour of MMCs, involving ablation phenomena, have been studied using a flux controlled solar furnace. The fused metal was evacuated by a centrifugation process actuated by a pneumatic engine. Specimens weight loss and temperatures evolutions have been measured. It become apparent that X2080 matrix associated with a large amount of SiC particles delays the ablation phenomena.

## INTRODUCTION

Metal Matrix Composites (MMC) are commonly divided in two families defined  by the shape of the reinforcement. In one hand, long fibers addition in a metallic media gives Continuous Metal Matrix Composites (CMMC) in which

DGA/DRET/ETCA/Centre de Recherches et d'Etudes d'Arcueil
* Dpt Physique des Surfaces - Centre d'Essais d'Odeillo - BP6
66125 FONT ROMEU - FRANCE
** Dpt Matériaux en Conditions Sévères -16 bis, Ave Prieur de la Côte d'Or
94114 ARCUEIL - FRANCE

mechanical properties in the fibers direction are improved. On the other hand, Discontinuous Metal Matrix Composites (DMMC) can be presented as a dispersion of particulates, nodules or whiskers in a metallic phase which leads to quasi-isotropic thermomechanical characteristics. DMMC such as silicon carbide / aluminium (SiC/Al) are usually investigated as a structural material for near-room temperature applications in aeronautic, automobile and weapon industries because of their very good tribological and specific mechanical properties. Thermal properties of aluminium alloys reinforced by SiC particulates (SiCp/Al) have been also exploited in electronic packaging applications because of combination of a low coefficient of thermal expansion with a high thermal conductivity, and a quick heat dissipation. Both mechanical and thermal performances of SiCp/Al are taken into account in order to define the optimal composite for a missile fin application. Transient high temperature behaviour of the composites is also investigated using a third generation solar furnace facility by means of instrumented thermal shock testing.

## MATERIALS DESCRIPTION

NATURE

All of the composites have been provided by ALCOA and obtained by powder metallurgy process and extruded. In order to study separately the influence of the SiCp reinforcement and the influence of the constitutive aluminium alloy on the behaviour of the composite, eight different materials (see table I) have been tested.

TABLE I - Presentation of the materials studied

| Aluminium matrix | X2080 (1) | | | | X7093 (2) | | | |
|---|---|---|---|---|---|---|---|---|
| SiCp reinforcement (% vol) | 0 | 15 | 20 | 25 | 0 | 15 | 20 | 25 |

1= extruded and heat treated T4, 2=extruded and heat treated T7

The final heat treatment applied was T4 (498°C - 4 hours / quenching - 25° ageing) for X2080 based materials and T7E92 (487°C / 25°C - 4 days / 120°C - 24 hours / 150°C - 8 hours) for X7093 based materials.

Chemical composition of aluminium matrix has been determined by Electron Probe for Micro-Analysis (EPMA); the results are presented in table II.

TABLE II - Chemical composition (weight %) of 2080 and 7093 aluminium alloys

| Matrix \ Elements | Mg | Ni | Cu | Zn | Zr | Mn | Al |
|---|---|---|---|---|---|---|---|
| X2080 | 2,13 | <.01 | 3,36 | <.01 | 0,12 | 0,16 | bal. |
| X7093 | 2,44 | 0,11 | 1,56 | 9,06 | 0,12 | <.01 | bal. |

In all these materials, even in case of high volume fraction, the SiCp particulates are well distributed and there is no porosity. It can be observed a privileged particulate orientation in extrusion direction which leads to lightly non-isotropic properties.

## MECHANICAL CHARACTERISTICS

Tensile properties of alloys and composites have been determined in quasi-static conditions ($\dot{\varepsilon} = 10^{-3}s^{-1}$). The results are presented in table III. It can be seen that the addition of SiC particulates increases significantly the elastic moduli (E), the yield stress ($\sigma_y$) and the ultimate tensile strength (UTS). In the case of the X7093 based composites, the very high level of strength is associated with a dramatic loss in elongation (A%). On the other hand, the better ductility of the X2080 alloys leads to a quite satisfactory plastic behaviour of the reinforced matrix also for high volume fraction of particulates.

TABLE III - Tensile properties of pure matrix and SiC/Al composites in states 1 & 2

| Matrix | X2080 (1) | | | | X7093 (2) | | | |
|---|---|---|---|---|---|---|---|---|
| SiCp (% vol) | 0 | 15 | 20 | 25 | 0 | 15 | 20 | 25 |
| E (GPa) | 73±1 | 95±3 | 108±7 | 119±14 | 70±1 | 93±7 | 96±6 | 102±10 |
| $\sigma_y$ (MPa) | 345±20 | 335±30 | 410±23 | 413±30 | 615±18 | 618±30 | 636±23 | 658±15 |
| UTS (MPa) | 493±6 | 504±2 | 545±5 | 554±6 | 622±15 | 645±2 | 660±4 | 683±2 |
| A (%) | 14±5 | 8±1 | 5±1 | 4±1 | 10±1 | 3.1±1 | 2.8±1 | 2±1 |

## THERMAL DIFFUSIVITY MEASUREMENTS

Measurements of MMC samples thermal diffusivity have been performed by using a modulated heating method. A face of a plane sample is exposed to a periodically modulated radiant beam, as in [1]. In the Fourier symbolic domain, equations governing the system could be expressed as transfer function (attenuation and phase lag versus frequency). The temperature evolution (complex variable) at the sample rear face, $T_0(\omega)$ is given by :

$$\frac{T_0(\omega)}{i(\omega)} = H(Pv, \beta_i) \qquad (1)$$

where
- $i(\omega)$ the temporal boundary condition considered as thermal system input
- H is the transfer function, it depends on :
- the Predvoditeleev number : $Pv = e\sqrt{\omega/2a}$
(a is the thermal diffusivity and e the sample thickness)
- the Biot numbers : $\beta_i = eh_i/k$
($h_i$ are the heat loss coefficients for the surfaces located in x=0 and x=e, and k is the thermal conductivity. It is inferred subsequently that $\beta_0 = \beta_e = \beta$)

The phase shift $\delta_0$ between the incident flux and the resulting oscillations of the rear face temperature is then a function of two known parameters (thickness and frequency) and two unknown ones (diffusivity and Biot number). This phase shift is measured for different modulation frequencies in order to identify the unknown parameters [2,3].

The transfer function is determined by Fourier's transform of steady state periodic thermograms. Incident flux shape have to be chosen in order to combine high signal-to-noise ratio (SNR) and frequency richness. The two extremes are the Dirac (most frequency informations but low SNR) and the pure sine law (higher SNR but frequency unicity). The selected law is the repetitive pulse law which presents a good compromise between the two compulsions (fig 1.a). The control parameters are the fundamental frequency ($1/\tau_2$) and the pulse-to-period durations ratio ($\tau_1/\tau_2$).

The experimental setup is derived from the one described in [2] and adapted to monolayer geometry (fig 1.b). The temperature investigations are realized by setting the sample at the focus of a 1 m diameter solar furnace. This facility is able to deliver a flux density up to 90 W/cm² onto a 2.5 cm diameter target. It is equipped with a slow moving iris diaphragm which can reduce the incident flux, but it is unable to modulate this flux at high frequency. So, the modulated flux is delivered by a computer controlled 50 Watts $CO_2$ laser, for which beam is expanded to ensure an unidirectionnal thermal flow across the sample.

The rear face temperature is measured by optical pyrometry. The mean value is determined with a commercially available slow response pyrometer. The temperature oscillations are monitored with an arrangement involving a parabolic mirror and a fast response IR detector (HgCdTe). A desktop computer is used to control the laser waveform, level and frequency, and to record the temperature evolutions.

Experiments have been carried out on disc-shaped samples (50 mm diameter; 2 thicknesses ≈ 3 and ≈ 5 mm). The sample faces have been blackened using a high temperature black coating in order to increase the front face absorptivity and rear face emissivity.

A FFT analysis of thermograms allows the determination of phase lags at n frequencies, their number depending on the pulse-to-period durations ratio (1/4 or 1/8, in general). The error estimator is constructed with the square sum

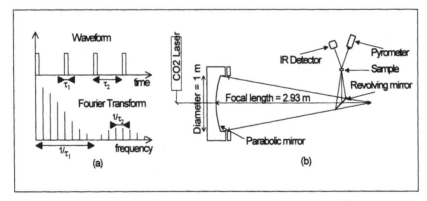

Figure 1 : a/ Thermal waveform and its Fourier Transform - b/ Experimental configuration for the thermal transfer function measurement.

of differences between measured (M) and calculated (C) values of the phase shift $\delta_0$ :

$$\sigma = \sum_{i=1}^{n} f_p \left[ \delta_{0i}^{C}(\omega) - \delta_{0i}^{M}(\omega) \right] \tag{2}$$

in which $f_p$ is a ponderation function taking into account random experimental errors and properties sensitivity to measured parameters.

The analytical formulation giving the theoretical phase lag [1] is :

$$\tan(\delta_0) = \frac{B(\tan Pv - \tanh Pv) + 2APv \tan Pv \tanh Pv + 2Pv^2(\tan Pv + \tanh Pv)}{B(\tan Pv + \tanh Pv) + 2APv - 2Pv^2(\tan Pv - \tanh Pv)} \tag{3}$$

where     $A = \beta_0 + \beta_e \approx 2\beta$ $\tag{4}$
and     $B = \beta_0 . \beta_e \approx \beta^2$ $\tag{5}$

The two unknown parameters (Pv and $\beta$) are determined by minimizing the estimator by using a Simplex minimization algorithm.

The figure 2 shows an example of such fitting, for a 5.18 mm thick, 2080 matrix plus 20% SiC composite.

RESULTS :

Measurements have been made at different temperature levels, but it appeared that thermal properties evolve in a non reversing way with thermal ageing. This is due to matrix recrystallisation phenomenon. So, the result presented in figure 3 are those obtained at the lowest test temperature ($\approx 150°C$).

As expected, the thermal diffusivity decreases when the SiC mass fraction increases. The X2080 matrix composites shows a lower diffusivity than X7093 ones.

## TRANSIENT THERMAL TESTS

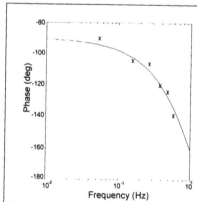

Figure 2 : Example of experimental data fitted for diffusivity identification

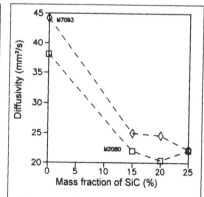

Figure 3 : Effective thermal diffusivity of Al/SiC composite versus mass percent of SiC particulates.

## EXPERIMENTAL APPARATUS, CONDITIONS AND MEASUREMENTS

The MMC samples have been submitted to very high thermal flux densities by means of a 45 kW solar furnace, on which a detailed description has already been published [4]. It is a two reflexions solar furnace as to keep the focal zone fixed. The solar rays first strike a 230 m² plane mirror (heliostat) which is automatically orientated in order to reflect the solar radiation in a horizontal north-south direction.

The solar beam being fixed that way is reflected towards a building which involves the concentrator, the modulator and the experiment chamber (figure 4a). The Davies-Cotton type concentrator have a 10.75 m focal length and covers a 10 x 10 m front area. This configuration gives a focal zone energy distribution that is quasi-constant (± 5%) over a diameter of 50 mm. The flux density obtainable at the focus can reach a 7 MW/m² level with a 1 kW/m² direct solar illumination.

The incident radiation is controlled by an attenuation system comprising a slow moving vertically sliding panel (whose position remains fixed for a test duration) and a group of twenty shutters moving round their vertical axis, acting like a venetian blind (whose minimum rising-falling time is 0.1 second). The position of both subsystems is controlled by a computer which real-time monitors the direct solar illumination, it allows calibrated heat fluxes to be delivered in a given timing sequence.

The apparatus generating the centrifugal stress is based on the use of a pneumatic engine whose maximum rotation speed is 20,000 rpm (figure 4.b).

The samples consist of 50 mm diameter discs, 5 mm thick, involving an axial fixing rod. They are set before a water-cooled diaphragm with an aperture hole having the same diameter as the sample, so that only the front face is exposed to the radiant flux.

Both faces of the samples have been coated with a high temperature black coating ($\alpha_s$ = 0.95). This paint is generally destroyed when the sample melts, but it homogeneizes the radiative exchanges of the different sample natures and furthers the high heating rate.

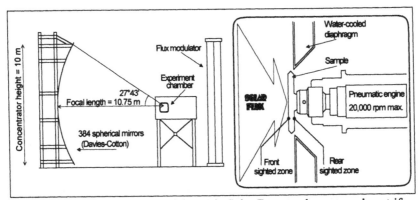

Figure 4 : Schematic drawing of the Main Solar Furnace elements and centrifugal stress arrangement.

Samples have been exposed to trapezoidal shape thermal pulses with variable plateau durations, rising and falling times being fixed at 0.2 s and illumination at 4.5 MW/m$^2$. The rotation speed remained constant at 2000 rpm.

The temperatures of both faces have been measured during the tests by using optical pyrometry techniques. The sighted zone was set approximatively at the middle of the sample radius (see fig.4b). The two pyrometers operated in the middle IR waveband because - due to the mirrors glass transmittance - the incident solar radiation is limited to the spectral range below 2.8 μm, and therefore does not interfere with the radiometric measurements. The use of a black coating allows the radiation analysis to be based on a unity value for the surface emissivity. In addition, samples mass measurements have been made before and after the exposition to the radiant flux.

## RESULTS

Experiments have been carried out for three test durations choosen - between 2.5 and 3.2 s - as to describe different degradation levels for the samples. An example of thermogram set obtained on basic 7093 alloy, is given in figure 5.

The front face heating shows different steps. First, the flux rising time (0.2 s), followed by a steady-state heating until the melting point be reached. This phenomenon causes a slope change (at t ≈ 1.3 s) followed by a constant temperature period which ends at the end of the flux plateau. The rear face temperature tends to a value slightly below the melting point. Similar behaviors have been observed, more or less distinctly, for other samples.

The degradation level has been characterized by the percentage of mass loss, defined as the ratio of mass variation on initial mass. The figure 6 presents the percentage of mass loss observed for the different samples as a function of the plateau duration. It appears that D and H samples (X2080 matrix with 20 and 25% SiC) are destroyed for tests duration larger (≈ 15%) than other ones. In spite of the small number of experiments, these results tend to demonstrate that X2080 matrix associated with a large amount of ceramic particles delays the ablation phenomenon.

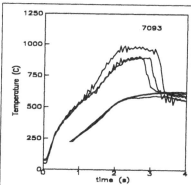

Figure 5 - Front and rear face thermograms on X7093 alloy, for 2.5, 2.7 and 2.9 seconds plateau.

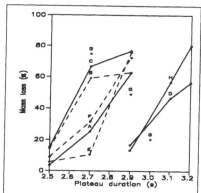

Figure 6 - Percentage of mass loss versus plateau duration.
(solid line :X 2080, dashed :X 7093)

## OBSERVATION OF IRRADIATED SAMPLES

Specimen exposed to the radiating flux have been observed. Figure 7 shows the final aspect of X2080 based samples exposed at the same plateau duration equal to 2.9 seconds. It can be seen clearly that degradation of the unreinforced X2080 matrix and the X2080+ 15%SiCp is mainly due to the flow of the liquid metal and, in a second step, fragmentation of the sample. In the case of high volume fraction MMCs, it can be observed that, at the same plateau duration, the mass loss is due to a controlled ablation of liquid metal in the front face of the specimen. When the degradation mechanism involved is the liquid ablation only, the sample integrity is conserved and the mass loss is limited to 15%. For the higher plateau duration, the second mechanism appears and leads to multi-fragmentation. In all cases, the observation of tested specimens reveals a recrystallization of the matrix.

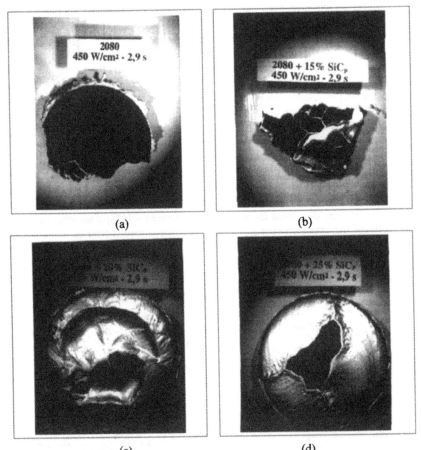

Figure 7 - Aspect of specimens after an exposure to a 450 W/cm² thermal flux during 2.9 seconds.   (a) alloy X2080,   (b) X2080 + 15%SiCp,   (c) X2080 + 20%SiCp, (d) X2080 + 25%SiCp

## CONCLUSION

The possible utilization of aluminium alloys matrix dispersed composites as materials for high velocity vector fin applications has been analyzed. Thermal and mechanical properties for two classes of matrix, and four SiC fraction for each of them, have been determined The whole of the compositions has been submitted to a severe thermomechanical test.

Preliminary results show that a high fraction of particulates decreases the thermal diffusivity and allows a better control of ablation phenomena. The matrix offering the lower diffusivity (and the higher ductility) shows the best behaviour towards the transient thermomechanical test.

Materials with an increased SiC fraction should be investigated in the near future, but this kind of composite are not off-the-shelf available because of their lower mechanical properties.

## REFERENCES

1. Cowan, R.D., 1961, "Proposed method of measuring thermal diffusivity at high temperature", J. Appl. Phys, 32, pp. 1363-1370

2. Chafik, E., Mayer, R and Pruschek, R., 1969, "Messung der Temperaturleitzahl fester Stoffe bei hohen Temperaturen", High Temp. High Press., 1, pp.21-26

3. Ranchy, E., Serra, J.J. and Hervé, Ph. 1994, "Determination of heat transfer modes in $Y_2O_3$-$ZrO_2$ plasma sprayed coatings", High Temp. Chem. Processes, 3, pp. 233-238

4.a Serra, J.J., 1991, "High Temperature Materials Characterization using a Flux and Temperature controlled Solar Furnace," in Metrology at High Temperature - ESA Workshop WPP-020 (Noordwijk NL - January 15-16, 1991)

4.b Suzanne, P. and Serra, J.J., 1994, "Thermal Flux Measurement Methodology and Flux Sensors Calibration at the ETCA Solar Furnace", Proc. 49th CNATI, Vol.3, pp. 2087-2095(Perugia I / September 26-30, 1994)

# Effect of Heat Treatment on the Thermal Conductivity of a Particulate SiC Reinforced 6061 Aluminum Matrix Composite

H. WANG and S. H. JASON LO

## ABSTRACT

The effects of aging on the thermal diffusivity/conductivity of a particulate-SiC-reinforced 6061 aluminum matrix composite were investigated. The experimental results have shown that the thermal diffusivity/conductivity of the particulate-SiC-reinforced 6061 aluminum matrix composite increases with the increase of aging time. This phenomenon is believed to be contributed by the increase in the thermal diffusivity/conductivity of the aluminum matrix alloy. However, it should be noted that the aging-time dependence of the thermal diffusivity/conductivity of the metal matrix composite (MMC) is different when compared with unreinforced alloy. This difference is due to the accelerated aging kinetics caused by the presence of reinforcement in the matrix. In summary, the increase in the thermal diffusivity/conductivity of the MMC could possibly be attributed to reduced solute concentration and decreased strain, before and after peak aging.

## INTRODUCTION

Lightweight materials with high thermal conductivity and low coefficient of thermal expansion are needed for electronic packaging applications. Silicon-carbide particulate-reinforced aluminum ($SiC_p/Al$) has been identified as an attractive material. A 6061 aluminum matrix reinforced with 40 to 50 vol% SiC exhibits a

Metals Technology Laboratories, CANMET, Natural Resources Canada, 568 Booth Street, Ottawa, Ontario, Canada K1A 0G1 coefficient of thermal expansion which is less than half of the corresponding value of the aluminum matrix [1]. Depending on the purity and degree of crystallinity, SiC exhibits high values of thermal conductivity [2] and can improve the thermal conductivity of the composites [3]. The above interesting properties have stimulated many research activities in studying the effects of reinforcement on the thermal conductivity of composites [4-7]. Other factors such as the type of Al alloy used as matrix alloy, as well as the heat-treatments, can significantly affect the thermal conductivity of composites. It is known that the thermal conductivity of 6061 Al alloy has a different value for different heat-treatment conditions [8], and a number of studies on the effect of heat-treatment on thermal conductivity of aluminum alloys were conducted [9]. However, up to the present few studies on the effects of heat-treatment on thermal conductivity of metal matrix composites (MMCs) have been reported [10].

Considering that the same heat-treatment condition could have different effects on MMCs and their unreinforced counterparts [11], this paper addresses the effects of heat-treatment on the thermal conductivity of SiC$_p$/6061 Al composite and 6061 Al alloy.


## EXPERIMENTAL

The MMC used in this work were commercial 6061 aluminum alloy reinforced with SiC particles. The materials were supplied by Advanced Composite Materials Corporation (ACMC) in a T6 condition. Using the image analysis technique, the volume fraction of SiC in the MMC was found to be 53 vol%. The size of the SiC particles is about 4 μm. For the purpose of comparison, a commercial 6061 aluminum alloy was also used.

To determine the microstructural condition of the matrix material after heat-treatments, microhardness measurements were conducted using a LECO DM-400 Hardness Tester equipped with a Vickers diamond pyramid indentor. Specimens were cut from the as-received stocks and were polished to a 1 μm finish. They were then solutionized[*] at 543°C for 2 hours in a purified argon atmosphere and quenched in cold water. Subsequently, the specimens were aged at 163 °C for

---

[*] Solutionization is a process of heating an alloy to a suitable temperature, holding at that temperature long enough to cause one or more constituents to enter into solid solution, and then cooling rapidly enough to hold these constituents in solution.

various lengths of time and quenched in cold water before hardness measurements were taken.

Thermal diffusivity was measured using a Holometrix Thermaflash 1100 Thermal Diffusivity Instrument, by means of the flash technique [12, 13] using a Nd:glass laser as a flash source. The specimens were cut from the solution heat-treated aluminum and composite materials in the form of ~12.7 mm diameter circular disks with a thickness of ~2mm, followed by aging. Both the solution and aging heat-treatments of the specimens were carried out under the same conditions as the microhardness specimens. Before measurements, specimens were sprayed with a thin layer of graphite to increase the absorptivity and emissivity of the front and rear surfaces of the specimen. The measurements were carried out at temperatures up to 400 °C in a protective atmosphere of high purity argon in order to avoid oxidation. No peeling-off of the graphite layer from the specimens was found by visual inspection after each run. Specific heat was measured using differential scanning calorimetry (DSC). Thermal conductivity was calculated as the product of thermal diffusivity, specific heat and density.

## EXPERIMENTAL RESULTS AND DISCUSSION

Figure 1 shows the microhardness data of 6061 Al and SiCp/6061 Al, aged at 163°C for different time intervals. Very similar results were also reported by others [11, 14]. Investigations using transmission electron microscopy have revealed that the process of aging of aluminum alloys involves the following stages [11, 15, 16]: (I) a supersaturated solid solution of alloying elements (e.g., Mg, Si); (II) formation of tiny precipitates and Guinier-Preston (GP) zones that form a dense network impeding the movement of dislocations during deformation and causing an increase in microhardness; (III) nucleation of coherent near-spherical GP-I zones. This is the stage of peak aging where maximum hardness is obtained. The formation of the GP-I zones is accompanied by the nucleation of needle-shaped $\beta''$ phase ( or GP-II zone); (IV) coarsening of the needles into the partially coherent $\beta'$ rods followed by a transformation into the equilibrium incoherent $\beta$ precipitates with a structure corresponding to a highly ordered $Mg_2Si$ [17]. As soon as only $\beta'$ rods are present, the stage of overaging is attained, characterized by a reduction in hardness.

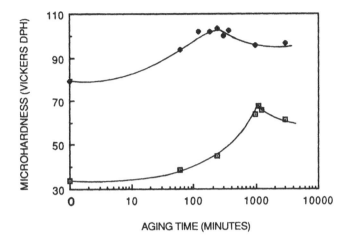

Figure 1        Microhardness data of 6061 aluminum (▣) and SiC$_p$/6061 Al (♦)
                aged at 163°C for different time intervals. (Reprinted with
                permission from Ref. [10].)

The mechanism of aging for an aluminum MMC is similar to
the Al alloy, except that the MMC requires less time to attain peak
hardness.  As seen from Fig. 1, the peak aging time of the 6061 Al
MMC is drastically reduced from 1080 to 240 minutes (or from 18 to
4 hours), compared to the Al alloy.  The accelerated aging in the Al
matrix of the composite is attributed to the relatively large difference
between coefficients of thermal expansion of the SiC particles and Al
matrix, which resulted in a substantial stress build-up leading to the
generation of dislocations at and near the interface between the SiCp
and Al matrix on quenching [18].  During aging, the high density of
dislocations would provide a short-circuit path for heterogeneous
nucleation and fast growth of precipitates resulting in accelerated
aging of the matrix.  This is further confirmed by the result that the
microhardnesses of the composite matrices are higher than the
unreinforced alloy under all heat-treatment conditions.

Figure 2 shows the experimental results for the thermal
diffusivity of the 6061 Al and the SiCp/6061 Al composite at 25°C, as
a function of aging time.  Generally, the thermal diffusivity

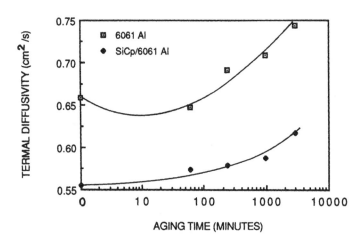

Figure 2        Thermal diffusivity of 6061 Al and SiC$_p$/6061 Al composite
                measured at 25 °C as a function of aging time. (Reprinted with
                permission from Ref.[10].)

increases with the increase of aging time. However, a decrease in the
thermal diffusivity of the 6061 Al alloy at the early stage of aging is
observed, which is very similar to the reported results on the decrease
in the electrical conductivity [19] or increase in electrical resistivity
[11, 20] at an initial stage of aging.

The decrease in thermal diffusivity at the first 60 minutes of
aging was not observed in the composite, indicating that the
precipitation process is faster in the MMC than in the alloy because
of the presence of the SiC particles which caused an accelerated
aging. Also as can be seen from Fig. 2, the thermal diffusivity of the
composite is lower than that of the Al alloy [10, pp.367-370].

Figure 3 shows the experimental results for the specific heat of the 6061 Al
and SiC$_p$/6061 Al composite, respectively. The specific heat of the MMC is lower
than that of the alloy. No clear aging-time dependence was found in the specific
heat.

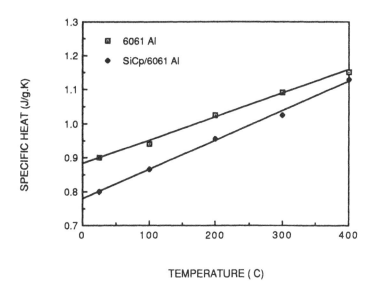

Figure 3    Experimental data for the specific heat of 6061 Al and
            SiCp/6061 Al.

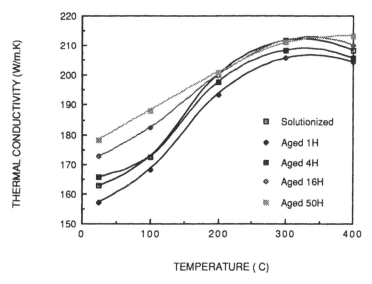

Figure 4    Thermal conductivity of 6061 Al at various temperatures.

Figure 4 shows the calculated values for the thermal conductivity of the 6061 Al at various temperatures. The positive temperature dependence of the thermal conductivity from 25 °C up to about 300 °C and the different temperature dependence for different heat-treatment conditions above 300 °C were observed. This phenomenon indicates certain changes in the material microstructure at around 300 °C for under-aged 6061 Al alloy. Similar results were reported by others on various aluminum alloys, although the transition temperature is alloy dependent. It occurred between 200 to 300°C [9].

Figure 5 shows the calculated values for the thermal conductivity of the SiC$_p$/6061 Al composite at various temperatures. At lower temperatures (< 100 °C), due to the presence of the 53 vol.% SiC particles, the temperature dependence of the thermal conductivity of the composite is more controlled by the negative temperature dependence of the lattice conductivity which obeys the 1/T law [21]. However, at high temperature (>100 °C), the temperature dependence of the thermal conductivity becomes positive and has a similar trend to the alloy at higher temperatures, suggesting that the electrons in the matrix are the dominant carrier of heat, despite the presence of the 53 vol.% SiC particles.

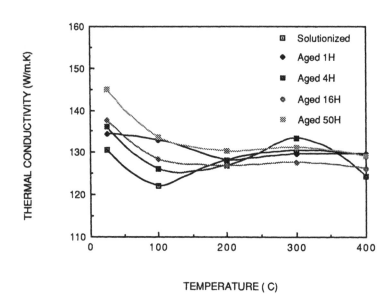

Figure 5          Thermal conductivity data for SiC$_p$/6061 Al composite at various temperatures.

To show the effect of aging on the thermal conductivity without the influence of elevated test temperature, Fig. 6 was plotted demonstrating the thermal conductivity at 25°C of the 6061 Al and SiC$_p$/6061 Al composite as a function of the aging time.

In the case of the alloy, at the early stages of aging, with the formation of coherent precipitates (GP and GP-I zones) and the transformation-induced strain field, the thermal conductivity temporarily decreases. With further aging, the diffusion of foreign atoms from the aluminum lattice tends to be dominant and increases the thermal conductivity. After the peak aging, the reduction of the point defects slows down, but at the same time, as the precipitates coarsen into the partially coherent β' and eventually the incoherent β, the strain decreases, as does the precipitate density [11]. The decrease in strain is the predominant cause of the increase in thermal conductivity after overaging.

In the case of the MMC, nucleation of precipitates in the matrices has been found to occur on two types of sites [11]: (i) dislocations generated as a result of the CTE mismatch between the matrix and the reinforcements and (ii) quenched-in vacancy loops. Only nucleation on the first type of sites was found to be accelerated in the case of composites. During subsequent aging, diffusion of magnesium atoms and hence precipitate growth were found to be significantly accelerated by the high density of dislocation in the matrix. This is supported by the rapid increase in thermal conductivity of composites after the first hour of aging. By the time of peak aging, the reduction in solute concentration slows down and the strain increases with the formation of the GP zones, resulting in less significant increases in the thermal conductivity from the 1st to 4th hour of aging. Dutta, *et al.* [11] have found that reinforcements appear to affect not only the kinetics of aging, but also the relative amounts of the various phases formed in the matrix alloy. The reinforcements seemed to stabilize the coherent GP zones thereby reducing the volume fractions of partially coherent β' and the incoherent β phase. The high strain is responsible for the less significant increase in the thermal conductivity after peak aging (from 4 to 16 h), because of the existence of a relatively large amount of the coherent precipitates. The strain reduction during overaging was not very dominant until an extended aging time (e.g., 50 hours) is applied, thereafter the thermal conductivity is largely increased.

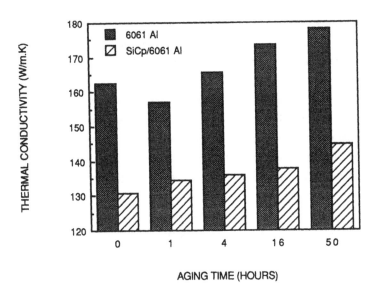

Figure 6        Thermal conductivity of 6061 Al and SiC$_p$/6061 Al composite at 25°C
                as a function of aging time.


## CONCLUSIONS

The results of this study indicate the very important role that heat-treatment
can play in the effective thermal conductivity of aluminum matrix composites. Aging
is found to increase the thermal diffusivity/conductivity of the particulate SiC
reinforced 6061 aluminum matrix composite, primarily due to the increase of the
thermal diffusivity/conductivity of the matrix alloy. The causes of the increases in
the thermal diffusivity/conductivity of both materials are the same: i.e., the reduced
electron scattering by defects due to reduced solute concentration before peak aging
and decreased strain after peak aging. However, with the presence of the
reinforcement, both the kinetics and the content of the microstructural change of the
composite matrix during aging are different from the unreinforced alloy, which
promotes the rate of increase in the thermal conductivity of the composite with the
increase of aging time. SiC particles also change the temperature dependence of the
thermal conductivity of the composite compared to its matrix alloy.

## ACKNOWLEDGMENTS

The authors would like to thank our colleagues at MTL/CANMET, Mr. B. Durocher for the microhardness measurements, Mr. R. Dos Santos for preparing the specimens and Mr. J. Barry for his assistance in heat-treatment.

## REFERENCES

1. German, R.M., K.F. Hens and J.L. Johnson, 1995. "Powder Metallurgy Processing of Thermal Management Materials for Microelectronicapplications", *Inter. J. Powder Metall.*, 30(2):205-215.

2. Hasselman, D.P.H. 1989. "Role of Structure and Composition in the Heat Conduction Behavior of SiC", in *Thermal Conductivity 20*, D.P.H. Hasselman and J.R. Thomas, Jr., eds. Plenum Press, New York, pp. 141-152.

3. Dolowy, J.F., B.A. Webb and M.R. van den Bergh., April 1988. "Metal Matrix Composites: Tailorable for Mechanical, Thermal, and Physical Characteristics", in *Proc. Composite-Technology: Fifth Annual Conference on Materials Technology*, Southern Illinois University at Carbondale, Materials Technology Center, Carbondale, Illinois, pp. 63-71.

4. Lai, Shy-Wen and D.D.L. Chung, 1994. "Aluminum-matrix Aluminum Nitride Particle Composite as a Low-Thermal-Expansion Thermal Conductor", in *Electronic Packaging Materials Science VII Materials Research Society Symposium Proceedings*, vol. 323, Materials Research Society, Pittsburgh, PA, pp. 207-212.

5. Geiger, A. L. and M. Jackson, July 1989. "Low-expansion MMCs Boost Avionics", *Advanced Materials & Processes*, pp. 23-30.

6. Schmidt, K., C. Zweben and R. Arsenault, 1990. "Mechanical and Thermal Properties of Silicon-Carbide Particle-Reinforced Aluminum", in *Thermal and Mechanical Behavior of Metal Matrix and Ceramic Matrix Composites*, J. M. Kennedy, H. H. Moeller and W. S. Johnson, eds. ASTM, pp. 155-164.

7. Hasselman, D. P. H, *et al.*, 1992. "Effect of Reinforcement Particle Size on the Thermal Conductivity of a Particulate-Silicon Carbide Aluminum Matrix Composite", *J. Am. Ceram. Soc.*, 75(11):3137-3140.

8. 1990, "Properties and Selection: Nonferrous Alloy and Special-Purpose Materials", *Metals Handbook,* 9th Edition, vol. 2, ASM International, pp. 102-103.

9.  Bogaard, R. H. and C .Y. Ho, 1989. "Thermal Conductivity of Selected Aluminum Alloys - A Critical Review", in *Thermal Conductivity 19*, D.W. Yarbrough, ed., Jr. Plenum Press, New York, pp. 551-560.

10. Wang, H. and S.H.J. Lo, 1996. "Effects of Aging on the Thermal Conductivity of a Silicon Carbide Particulate Reinforced 6061 Aluminum Composite", *J. Mater. Sci. Lett.,* 15:369-371.

11. Dutta, I., *et al.*, 1991. "Effect of Reinforcement on the Aging Response of Cast 6061 Al-Al$_2$O$_3$ Particulate Composites", *Metall. Trans.*, 22A:2553-2563.

12. Parker, W.J., *et al.*, 1961. "Flash Method for Determining Thermal Diffusivity, Heat Capacity, and Thermal Conductivity", *J. Appl. Phys.*, 32(9):1679-1684.

13. Heckman, R. C., 1973. "Finite Pulse-Time and Heat-Loss Effects in Pulse Thermal Diffusivity Measurements", *J. Appl. Phys.*, 44(4):455-1460.

14. Hickson, M. R., *et al.*, 1993. "The Influence of Aging on the Bauschinger Effect in Metal Matrix Composites", *Scripta Metall.*, 29:1023-1028.

15. Rack, H. J. and R. W. Krenzer, 1977. "Thermomechanical Treatment of High Purity 6061 Aluminum", *Metall. Trans.*, 8A:335-346.

16. Dutta, I. and S. M Allen, 1991. "A Calorimetric Study of Precipitation in Commercial Aluminum Alloy 6061", *J. Mater. Sci. Lett.,* 10:323-326.

17. Papazian, J. M., 1988. "Effects of SiC Whiskers and Particles on Precipitation in Aluminum Matrix Composites", *Metall. Trans.*, A19:2945-2953.

18. Chawla, K.K., *et al.*, 1991. "Effect of Homogeneous/Heterogeneous Precipitation on Aging Behavior of SiC$_p$/Al 2014 Composite", *Scripta Metall.*, 25:1315-1319.

19. Rosen, M. and E. Horowitz, 1982. "The Aging Process in Aluminum Alloy 2024 Studied by Means of Eddy Currents", *Mat. Sci. Eng.*, 53:191-198.

20. Dutta, I., *et al.*, 1994. "Role of Al$_2$O$_3$ Particulate Reinforcements on Precipitation in 2014 Al-Matrix Composites", *Metall. Mat. Trans.*, 25A:1591-1602.

21. Berman, R., 1976. *Thermal Conduction in Solids*, Oxford University Press, Oxford, Great Britain, pp. 145-148.

# Method for Measuring the Orthotropic Thermal Conductivity and Volumetric Heat Capacity in a Carbon-Carbon Composite

K. J. DOWDING, J. V. BECK and L. EILERS

## ABSTRACT

Composite materials are inherently anisotropic due to the nature of their non-homogeneous fiber-matrix construction. This directional dependence of the thermal properties is quite pronounced in carbon matrix carbon fiber composites. An experimental method to measure the thermal properties of a carbon-carbon material is discussed in this paper. The investigated material is characterized by an orthotropic thermal conductivity (two components) and isotropic volumetric heat capacity.

Thermal properties are estimated using parameter estimation techniques with experimentally measured surface heat flux and temperature histories for an experiment with two-dimensional heat flow. Particular attention is given in the paper to provide engineering insight into the experimental and analytical aspects of this method. The thermal conductivity measured parallel to the fibers is six times larger than the conductivity normal to the fiber. Experiments with one-dimensional heat flow show excellent agreement with the two-dimensional results.

## NOMENCLATURE

$C$  specific heat $[J/kg\,^\circ C]$
$e$  temperature residual $[^\circ C]$
$J$  number of temperature sensors
$k$  thermal conductivity $[W/(m-^\circ C)]$
$L$  length $[m]$
$N_t$  number of measurement times
$N_p$  number of parameters to estimate
$q''$  heat flux $[W/m^2]$
$S$  sum-of-squares function
$T$  temperature $[^\circ C]$

Department of Mechanical Engineering and Composite Materials and Structures Center, Michigan State University, A231 Engineering Building, East Lansing, MI, 48824

$\hat{T}$    calculated temperature [$°C$]

$t$    time [sec]

$\underline{U}$    weighting matrix for prior information

$\underline{W}$    weighting matrix for sensors

$Y$    measured temperature [$°C$]

$\underline{\beta}$    unknown parameter vector

$\rho$    density [$kg/m^3$]

$\underline{\mu}$    prior information vector

$\Omega$    resistance [ohms]

Subscripts

$cc$   carbon-carbon composite material

$Ins$   insulating material

$x$    direction parallel to the fiber direction

$y$    direction normal to the fiber direction

## 1. INTRODUCTION

Interest in the use of composite materials has broadened in the past twenty years. The increased use has been, at least partially, driven by reduced manufacturing cost and superior performance of these materials. Performance characterized by high strength to weight ratios, radar absorbing quality, or the capability of structural integrity at very high temperatures, attributes that make the materials well-suited for many diverse applications. Some applications are advanced aerospace applications, such as the National Aerospace Plane, cargo planes and fighter jets and bombers, automotive applications, including piston rings and engine blocks, and sporting applications, such as tennis rackets, baseball bats, and boats. The use of composite materials in most applications is superior to standard (metallic) materials.

The construction of the composite materials makes them more complex than ordinary materials. Composite materials are composed of a fiber and a matrix, which results in a non-homogeneous material. The resulting mechanical and thermophysical properties may, and often do, depend on the relative orientation of the fiber within the matrix. Hence, the properties normal to the fiber direction are quite different from the properties parallel to the fiber direction. The situation is further complicated for certain composite materials, such as some carbon-carbon composites, that require a conversion coating for protection against oxidation. The conversion coating, usually a silicon-carbide material, has considerably different thermal properties than the base carbon-carbon composite. Consequently, the resulting material properties are more complex.

The material investigated in this paper is a carbon matrix carbon fiber composite, without a conversion coating for oxidation protection. Characterization of the thermal properties (thermal conductivity and volumetric heat capacity) for this carbon-carbon material requires a multi-dimensional experiment, for the previous reasons stated. The carbon-carbon is modeled with two components (orthotropic) of thermal conductivity and an isotropic volumetric heat capacity. Simultaneously estimating two components of thermal conductivity requires a two-dimensional experiment. Effective properties are determined for the material, which assumes that the material is homogeneous.

A laboratory method to measure the thermal properties of a carbon-carbon composite is presented. An extensive experimental investigation of the thermophysical properties of the composite material is not proposed. The focus for this investigation is the methods used to measure the properties. Particular attention is given to provide engineering insight into the experimental aspects, as well as the analytical aspects of the method. Combining the aspects of experiments and theory (analysis), which have been traditionally considered separately, is part of a new research paradigm.

The method discussed in this paper is based on inverse analysis for estimating thermal properties. Other methods, such as the guarded hot plate, use steady-state analysis (Fourier's Law) to measure thermal conductivity. In such methods, the model equation for the process, in this case the heat conduction equation, is not necessarily satisfied or verified. Hence, the properties that are estimated may be inaccurate for the model equation. Furthermore, measuring the volumetric heat capacity requires a transient experiment. For these reasons, a transient experiment is used and analyzed with inverse (parameter estimation) techniques. This approach more closely combines the experiments and analysis, resulting in more accurate thermal properties. This work is quite similar to Dowding et al. [1]-[3], except that a different carbon-carbon material is studied.

An outline of the remainder of the paper is given. Experimental aspects are discussed in Section 2. The parameter estimation methods and sensitivity analysis are given in Section 3. Section 4 presents and discusses the results. The paper is concluded with a brief summary.

## 2. EXPERIMENTAL ASPECTS

A sketch of the experimental set-up used to estimate the thermal properties of the carbon-carbon material is shown in Figure 1. It consists of two nominally identical carbon-carbon composite specimens (5.08cm x 7.62cm x 2.54cm) and ceramic insulations (5.08cm x 7.62cm x 2.54cm, Zircar Products Inc., Florida, NY) with a mica heater assembly (Thermal Circuits, Inc., Salem, MA., $\Omega(T_{room}) = 33$ Ohms) located between the halves. The heater assembly consist of two (2.34cm x 7.62cm) independently controlled heaters that extend over the entire surface normal to the page for negligible temperature variation in this direction. For two-dimensional experiments only one of the two heaters is activated. Five sensors (Type E thermocouples, 0.254mm nom. wire diameter) are embedded on the surface of each carbon-carbon specimen at the heater/specimen interface. The sensors (insulation removed) are cemented into grooves (0.381mm wide x 0.457mm deep) that extend the length of the specimen (normal to the page). Two more sensors are at each interface of the carbon-carbon specimen and the ceramic insulation. Silicon grease is applied between the heater and specimen to improve thermal contact. The entire set-up is mounted between two 3.18mm thick aluminum plates that are connected with threaded rods and hold the layers firmly in place and placed in a furnace, which allows the initial temperature to be varied. A detailed discussion of the experimental aspects are

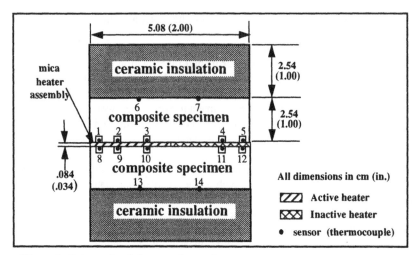

Figure 1. Schematic of the experimental set-up

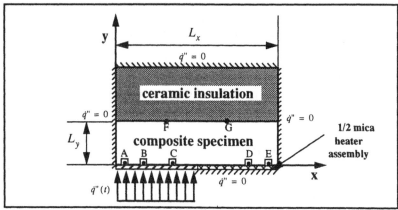

Figure 2. Heat transfer model for two-dimensional experiments

given in Ulbrich *et al.* [4].

The thermal experimental model is shown in Figure 2. All outer surfaces are assumed insulated, except for the surface where the energy is introduced by the heater. The energy to the heater is assumed to divide equally between the halves and emanate from the middle of the heater assembly ($y = -0.042\,\text{cm}$). The temperatures are averaged on opposite sides of the heater assembly to determine the temperature at each location. The location of the sensors are given in Table 1. The sensors that are embedded in the specimen are assumed to measure the temperature at the surface of the specimen. Since ($k_{y,cc} \gg k_{Ins}$) small temperature gradients will exist in the specimen near the specimen/insulation interface and the non-embedded sensors are assumed to measure the temperature at the backside of the specimen.

TABLE 1.     SENSOR LOCATIONS

| Senor ID | sensors | Location cm (in) | |
| --- | --- | --- | --- |
| | | x | y |
| A | 1,8 | 0.25 (0.1) | 0 |
| B | 2,9 | 0.76 (0.30) | 0 |
| C | 3,10 | 3.30 (1.3) | 0 |
| D | 4,11 | 4.38 (1.7) | 0 |
| E | 5,12 | 4.83 (1.9) | 0 |
| F | 6,13 | 1.69 (0.67) | 2.54 (1.00) |
| G | 7,14 | 3.39 (1.33) | 2.54 (1.00) |

## 3. ANALYSIS PROCEDURES

The techniques to estimate the thermal properties are detailed in a book by Beck and Arnold [5]. The basic process involves minimizing a sum of squares function,

$$S = (\underline{Y} - \hat{\underline{T}})^T \underline{W} (\underline{Y} - \hat{\underline{T}}) + (\underline{\mu} - \underline{\beta})^T \underline{U} (\underline{\mu} - \underline{\beta}) \tag{1}$$

where $\underline{Y}$ and $\hat{\underline{T}}$ are vectors of the measured and calculated temperatures and $\underline{w}$ is a weighting matrix for the sensors (typically the identity matrix). The last term in eq. (1) serves as regularization or allows for the inclusion of prior information about the thermal properties. It contains the difference between the prior information $\mu$ and present iteration estimates $\underline{\beta}$ with a symmetric weighting matrix $\underline{U}$. For this analysis no prior information was used. To determine the thermal properties the function $S$ is minimized with respect to the thermal properties, i.e. $\beta_1 = (\rho C)_{cc}$, $\beta_2 = k_{y,cc}$, and $\beta_3 = k_{x,cc}$. This is accomplished by setting the derivative equal to zero, and solving for the estimated parameters ($\hat{\beta}$). The resulting expression for the thermal properties (Beck and Arnold, 1977, Eq. 7.4.6) is

$$\hat{\underline{\beta}}^{(k+1)} = \hat{\underline{\beta}}^{(k)} + \underline{P}^{(k)} \left[ \underline{X}^T \underline{W} (\underline{Y} - \hat{\underline{T}}) + \underline{U} \left( \mu - \hat{\underline{\beta}}^{(k)} \right) \right] \tag{2}$$

$$\underline{P}^{(k)} = [\underline{X}^T \underline{W} \underline{X} + \underline{U}]^{-1} \tag{3}$$

The superscript ($k$) defines the iteration number, iteration is required even if the conduction problem is linear due to the non-linear nature of the sensitivity coefficients $\underline{X}$ (matrix). The sensitivity matrix is the first derivative of temperature with respect to the thermal property. It can provide considerable insight to the estimation problem and aid in the design of the experiment (Beck and Arnold, 1977, Chapter 8) for optimum accuracy in the estimates.

The sensitivity matrix and the temperatures are calculated with the properties from the previous iteration. Iteration continues until convergence of the estimated parameters is reached, as defined by, $\left| \hat{\underline{\beta}}^{(k+1)} - \hat{\underline{\beta}}^{(k)} \right| \leq \varepsilon \hat{\underline{\beta}}^{(k)}$, where $\varepsilon$ is a small number to quantify convergence, such as $\varepsilon = 0.001$.

The computer program **PROP2D** implements this inverse method to determine the two-dimensional properties. The program was developed at Michigan State University by taking the finite element code **TOPAZ2D** (Shapiro [6]) and combining this direct problem solver with these parameter estimation methods, to create a powerful algorithm. **PROP2D** allows for the estimation of the thermal properties for multiple materials, with irregular geometries, from transient temperature and heat flux histories. The thermal conductivity can be orthotropic and temperature dependent thermal properties are allowed.

As stated previously, observation of the sensitivity coefficients can provide insight to the estimation problem. An advantage of this method (parameter estimation), compared to other inverse methods, such as the adjoint or gradient method [7], is that the sensitivity coefficients are computed in the analysis. Hence, no additional computation are required to observe the coefficients. Observation of the sensitivity coefficients at this stage may be too late, since the experiment is essentially designed and moving sensors or changing the heated area is not easily done. However, some minor modifications may improve the accuracy, such as changing the heating duration or magnitude. When possible, an analysis of the sensitivity coefficients should be conducted prior to running the experiments.

Sensitivity coefficients provide insight to design experiments. In general, the sensitivity coefficients are desired to be large and uncorrelated (linearly independent). A sense of the magnitude of the sensitivity coefficients is gained through normalizing the sensitivity coefficients. Normalization is performed by multiplying by the parameters, resulting in units of temperature for the normalized sensitivity coefficients. The normalized sensitivity coefficient for parameter $\eta$ is

$$X_\eta = \eta \frac{\partial T}{\partial \eta} \qquad (4)$$

A comparison is then possible with the temperature rise of the experiment.

For the current experimental design, the normalized sensitivity coefficients are shown in Figure 3a-b. Figure 3a shows the sensitivity to $\rho C_{cc}$, $X_{\rho C}$, and Figure 3b the sensitivity to the thermal conductivities $k_{y,cc}$ and $k_{x,cc}$, $X_{k_y}$ and $X_{k_x}$. For clarity, every other sensor is shown. For the sensors not shown, sensor B would lie between the results for sensor A and C and sensors D and F would be within 5% of sensors E and G, respectively. Because the sensitivity coefficients are normalized, a direct comparison of their magnitudes is possible. Some observations are drawn from the sensitivity coefficient plots.

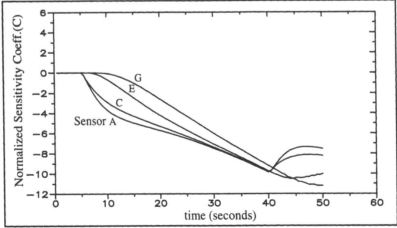

Figure 3a. Normalized sensitivity to volumetric heat capacity for experiment begun at $T = 25°C$

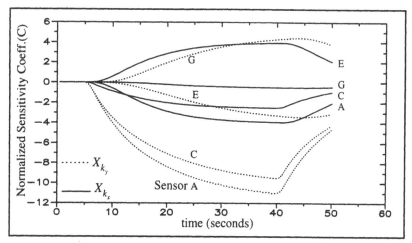

Figure 3b. Normalized sensitivity to thermal conductivity for experiment
begun at  $T = 25°C$

The most information is available on the active heater (sensors A, B and C in
Table 1). At these locations the sensitivity coefficients have the largest magnitudes
and are therefore most influential on the estimated properties. Notice that $x_{k_y}$ undergoes a sign change across the specimen (in the y-direction) and $x_{k_x}$ undergoes a sign
change along the specimen (in the x-direction). The implications of the these sign
changes are that there exist 1) a y-location within the body where the temperature is
insensitivity to $k_{y, cc}$ and more importantly 2) a x-location where the temperature is
insensitivity to $k_{x, cc}$. The latter result is more important for this case where surfaces
temperatures are measured, because seemingly logical locations along the measurement surfaces $y = 0, L_y$ may be insensitive to $k_{x, cc}$. Although it is desirable to avoid
locations where the sensitivity coefficients changes sign because the sensitivity is
quite small, sensor locations that have sensitivity coefficients with opposite signs
are a beneficial result. This situation produces contrasting effects which improves
the accuracy of the estimates. Hence, the choice of the extremes, which have opposite signs, for the measurement locations. Finally, the experiment was halted after
heating for 35 seconds to maintain that the magnitudes of the sensitivity coefficients
for all thermal properties were comparable. In Figure 3b $x_{k_x}$ has almost reached a
steady value, while the sensitivity to $x_{k_y}$ and $x_{\rho C}$ have not. If the experiment heated
for a longer duration $x_{k_y}$ and $x_{\rho C}$ would be much larger than $x_{k_x}$. Consequently, the
estimates for the two former properties would be more accurate than the latter property.

## 4. RESULTS AND DISCUSSION

A separate, independent set of one-dimensional experiments were performed to
determine the thermal properties of the mica heater (including the silicon grease)
and ceramic insulation in the model (Figure 2). Effective properties were determined to account for contact conductance between the layers. Therefore, only the

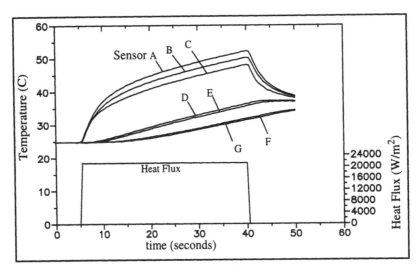

Figure 4. Experimental data for experiment begun at $T = 25°C$

TABLE 2. THERMAL PROPERTIES FROM ONE- AND TWO-DIMENSIONAL ANALYSES FOR THE CARBON-CARBON COMPOSITE

| Exp Type | Initial Temp | RMS | $k_{x,cc}$ | $k_{y,cc}$ | $\rho C_{cc} x10^{-6}$ | Heat Flux | |
|---|---|---|---|---|---|---|---|
| | (°C) | (°C) | W/m°C | W/m°C | J/m³°C | W/m² | seconds |
| 2D | 25 | 0.524 | 58.8 | 10.8 | 1.13 | 21,300 | 35 |
| 2D | 150 | 0.533 | 59.8 | 11.7 | 1.58 | 20,600 | 35 |
| 1D | 20 | 0.411 | | 10.6 +/- 0.5 | 1.16 +/- 0.03 | 13,000 | 40 |
| 1D | 150 | 0.405 | | 11.8 +/- 0.7 | 1.68 +/- 0.05 | 13,000 | 40 |

thermal properties of the carbon-carbon composite are unknown in the model (Figure 2). Furthermore, one-dimensional experiments were conducted to determine the thermal conductivity normal to the fibers ($k_{y,cc}$) and the volumetric heat capacity ($\rho C_{cc}$). The one-dimensional results provide initial estimates for the two-dimensional analysis and permit a comparison to demonstrate the accuracy and consistency of the methods.

The two-dimensional thermal properties of the carbon-carbon composite are discussed next. Experiments were conducted at two temperatures and analyzed assuming the thermal properties were constant for the duration of an experiment, but varied between experiments. The measured experimental data and details of the parameter estimation are presented and briefly discussed for one experiment.

Experimental data for a test at a temperature of $25°C$ are shown in Figure 4. A sampling interval of 0.5 seconds is used to acquired data for this experiment. The heating begins at approximately 5 seconds and ends at 40 seconds. Based on observation of the sensitivity coefficients and the criteria for optimal experiments, "D-optimality", (chapter 8, ref. [5]) the duration and magnitude of the heat flux were selected to be close to optimal experimental conditions for determining the two components of the thermal conductivity and the volumetric heat capacity.

For a two-dimensional analysis, as compared to a one-dimensional, the numerical issues are more important. The mesh size and time step selected for the finite element method used in the analysis can greatly influence the amount of time required to obtain a solution and the accuracy of this solution. The mesh used for this analysis contains 320 (quadrilateral) elements. Twenty elements are along the 5.08 cm surface (x-direction), for all materials. There is one element across the mica heater assembly, ten elements across each the carbon-carbon specimen, and five across the ceramic insulation (y-direction). The computational time step was 0.25 seconds, one-half the experimental time step. A typical two-dimensional analysis required 0.5-1 hours on a VAXstation 4000 Model 60 running VMS V5.5-2, depending on the number of iterations and accuracy of the initial parameter estimates; typically 4-5 iterations (16-20 direct finite element solutions) were required.

The two-dimensional thermal properties determined for the carbon-carbon composite are summarized in the first two rows of Table 2. The experiment type and the initial temperature are given in columns one and two. The next column presents the root-mean-square (*RMS*) of the difference between the calculated and measured temperatures. The next three columns are the two-dimensional thermal properties determined from the analysis. The final two columns give the duration and magnitude of the heat flux. Notice that the thermal conductivity in the direction of the fibers ($k_{x,cc}$) is approximately six times as large as that normal to the fibers ($k_{y,cc}$).

In addition to estimating the thermal properties, PROP2D provides some means to quantify the accuracy of the estimates. The *RMS*, which is given in Table 2, provides an indication of how well the calculated temperatures match the experimentally measured temperatures. The magnitude of the *RMS* can be compared to the temperature rise of the experiment, which is approximately 27°C and 18°C for the two experiments at initial temperatures of 25°C and 150°C, respectively. This gives a *RMS* that is within 2.0 and 3.0%, respectively, of the maximum temperature rise for the two experiments.

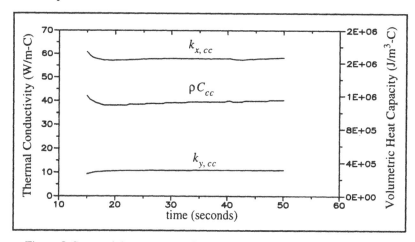

Figure 5. Sequential parameter estimates for experiment begun at $T = 25°C$

Other quantities can be observed to demonstrate the accuracy of the estimated properties. These include the sequential estimates of the properties and the residuals. Each is discussed below for the experiment with an initial temperature of $25°C$.

The sequential estimates demonstrate how the estimated properties vary as additional experimental measurements are considered. Figure 5 shows the sequential estimates for this experiment. The sequentially estimated property, at time $t_i$, represents the outcome if only data up to that time is used in the analysis. In other words, if the data is analyzed by adding one data set at each time, it shows how the estimated properties change as one more data set is added to the analysis. Initially the sequential estimates vary because there is not enough information (data) to accurately determine the properties; the information ($t < 10$ seconds) is excluded from the plot. However, as more data is considered, the property estimates approach constants values. If the experiment (or analysis) is ended at 30 seconds, the estimated

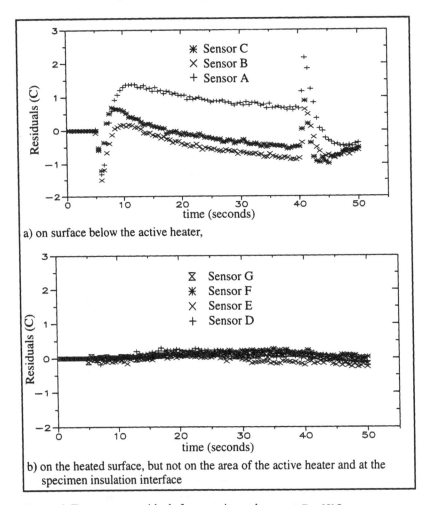

a) on surface below the active heater,

b) on the heated surface, but not on the area of the active heater and at the specimen insulation interface

Figure 6. Temperature residuals for experiment begun at $T = 25°C$

properties would not differ significantly from the properties at 45 seconds. In general, for a good estimation the sequential estimates converge to a constant and are steady with time. For this experiment the sequential estimates for $k_{x,cc}$, $k_{y,cc}$, and $\rho C_{cc}$ vary only 2.6, 0.9, and 0.6% over the final 20 seconds (30-50 seconds) of the experiment, which is excellent.

The temperature residuals ($e_{i,j} = Y_{i,j} - \hat{T}_{i,j}$) are related to the *RMS*. They represent the difference between the measured and calculated temperature for a particular time $(t_i)$ and sensor location $(x_j, y_j)$. The *RMS* gives an indication of the magnitude of the residuals for the analysis; the sign and magnitude of the individual residual can provide considerable insight. What is often done in the literature in similar investigations is that the measured and calculated values are plotted together to demonstrate the sufficiency of the results. Much more can be learned by plotting the difference of these quantities, the residuals.

The residuals for this experiment are shown in Figure 6a and b. Figure 6a presents the residuals for the sensors on the active heater and Figure 6b the residuals for the sensors on the heated surface but not on the active heater and for the sensors at the insulated surface. The residuals for the sensors on the heated surface, but not on the active heater and at the insulated surface (Fig 6b) are slightly correlated. The other residuals in Figure 6a are highly correlated and larger, 4-5% of the temperature rise on the heated surface. Although, it is not known exactly why the residuals are so highly correlated, a possible reason is the uncertainty in the location of the sensors with respect to the heater. There is some uncertainty in the alignment of the heater assembly and the specimen. The heating elements are contained within an opaque fiberglass with a mica layer on the outside. Since the fiberglass and mica extend beyond the heating elements, the alignment of the edge of the heater (element) with the specimen is difficult. The residuals for sensors away from the active heater are less sensitive to the location of the heat flux and therefore are less affected (within 2% of the temperature rise).

One-dimensional results are also given in Table 2. These results were obtained by heating over the entire surface instead of only one-half of the surface, as was done for the two-dimensional experiments. See Figure 1 and 2. A comparison of the values in Table 2 shows that there is essentially no difference in the thermal conductivity normal to the fiber. The difference between the one-dimensional results, which includes a confidence interval, and the two-dimensional results is inclusive in this interval for the thermal conductivity. The volumetric heat capacity at the lower temperature is within the confidence region, while the higher temperature value is just outside of this confidence region. Overall, excellent agreement is demonstrated between the one and two-dimensional results.

## 5. SUMMARY

A laboratory method to determine the thermal properties of carbon matrix-carbon fiber composite material was presented. Thermal properties were calculated for the carbon-carbon composite from two-dimensional experiments using measured temperature and heat flux histories.

The analysis methods and algorithms were very powerful; two components of thermal conductivity ($k_{x,cc}$, $k_{y,cc}$) and the volumetric heat capacity were simultaneously determined with two-dimensional experiments. The results of the two-dimensional analysis show excellent agreement with one-dimensional results.

It was demonstrated that considerable insight about the experimental model could be gained through sensitivity coefficients. Observation of the sensitivity coefficients can help maximize the information available from the experiment. When possible the sensitivity coefficients should be studied prior to conducting experiments.

## ACKNOWNLEDGMENTS

This research was partially supported by the Research Excellence Fund of the State of Michigan through the Composite Materials and Structures Center at Michigan State University.

## REFERENCES

1. Dowding, K., J. Beck, A. Ulbrich, B. Blackwell, and J. Hayes, 1995. "Estimation of Thermal Properties and Surface Heat Flux in a Carbon-Carbon Composite Material." *Journal of Thermophysics and Heat Transfer*, 9(2):345-351.

2. Dowding, K., J. Beck, and B. Blackwell, 1995. "Estimation of Directional-Dependent Thermal Properties in a Carbon-Carbon Composite." accepted for publication in International Journal of Heat and Mass Transfer.

3. Dowding, K. J. and J. V. Beck. February 1994. "Measurements of Transient Temperatures and Heat Fluxes in a Composite Material for the Estimation of Thermal Properties - Part II: Property Determination and Heat Flux Estimation," Technical report MSU-ENGR-004-94, Michigan State University, East Lansing, MI.

4. Ulbrich, A., J. V. Beck, and K. Dowding. December 1993. "Measurements of Transient Temperatures and Heat Fluxes in a Composite Material for the Estimation of Thermal Properties - Part I: Experimental Aspects," Technical report MSU-ENGR-009-93, Michigan State University, East Lansing, MI.

5. Beck, J. V. and K. J. Arnold, 1977. *Parameter Estimation in Engineering and Science*. New York, NY:John Wiley and Sons, Inc., Chapters 7 and 8.

6. Shapiro, A. B. 1986. "TOPAZ2D a Two-Dimensional Finite Element Code for Heat Transfer Analysis, Electrostatic, and Magnetostatic Problems," Lawrence Livermore National Laboratory, CA.

7. Jarny, Y., N. M. Ozisik, and J. P. Bardon, 1991. "A General Optimization Method Using Adjoint Equation for Solving Mulitdimensional Inverse Heat Conduction." *International Journal of Heat and Mass Transfer*, 34(11):2911-2919.

# Thermal Diffusivity Imaging of Continuous Fiber Ceramic Composite Materials and Components

S. AHUJA, W. A. ELLINGSON, J. S. STECKENRIDER and S. KING

## ABSTRACT

Continuous-fiber ceramic matrix composites (CFCCs) are currently being developed for various high-temperature applications, including use in advanced turbine engines. In such composites, the condition of the interfaces between the fibers and matrix or between laminae in a two-dimensional weave lay-up are critical to the mechanical and thermal behavior of the component. A nondestructive evaluation method that could be used to assess the interface condition and/or detect other "defects" has been developed at Argonne National Laboratory (ANL) and uses infrared thermal imaging to provide "single-shot" full-field quantitative measurement of the distribution of thermal diffusivity in large components. By applying digital filtering, interpolation, and least-squares-estimation techniques for noise reduction, shorter acquisition and analysis times have been achieved with submillimeter spatial resolution for materials with a wide range of "thermal thicknesses". The system at ANL has been used to examine the effects of thermal shock, oxidation treatment, density variations, and variations in fiber coating in a full array of test specimens. In addition, actual subscale CFCC components of nonplanar geometries have been inspected for manufacturing-induced variations in thermal properties.

## INTRODUCTION

Advanced ceramics that meet the requirements of tomorrow's technology are currently being introduced into the manufacturing community. For example, the titanium alloy with a silicon carbide (SiC)-reinforced ceramic-matrix, Timetal-21S (Ti-15Mo-3Nb-3Al-0.2Si) has reduced the weight of each of the engines of the Boeing 777 by 360 kg. These materials, in general, have high strength and stability at high temperatures due to the incorporation of continuous fibers in monolithic ceramics. CFCCs are being considered as replacements for traditional materials in numerous applications, including gas turbines, due to their relatively high strength and toughness at high temperatures (>1250°C), and lower density. Among the specific materials systems under consideration are $SiC_{(f)}/SiC$ and $Al_2O_{3(f)}/Al_2O_3$, which are desired specifically for their thermal properties (i.e., thermal diffusivity, thermal conductivity, etc.) and therefore any variations in those properties will significantly affect their ability to transfer heat properly. Such variations result from processing defects, thermal shock, or thermally induced degradation of the fiber/matrix interface and reduce the advantages for choosing these materials.

S. Ahuja, W. A. Ellingson, and S. King are with the Energy Technology Division, Argonne National Laboratory, Argonne, IL 60439; J. S. Steckenrider is with the Center for Quality Engineering and Failure Prevention, Northwestern University, Evanston IL 60208

Several nondestructive methods for the detection of thermal properties have been employed in the study of both ceramic and nonceramic materials systems [1]. These methods are primarily utilized for specimens whose thermal properties are assumed uniform [2-5]. The critical issue in most CFCC applications, however, is the distribution of thermal properties within a given component. In these situations, thermal infrared imaging methods, which are of greater utility, have been employed as a nondestructive testing method for CFCCs with the limitation that the test methods employed are either not portable or very expensive. CFCCs emit energy in different directions, parallel and perpendicular to the two dimensional weave. Transmissivity, the ratio of energy transmitted by the material to the incident energy, is used to determine the wavelength at which a body transmits infrared energy and is important in infrared detection. Nondestructive techniques such as ultrasonic testing have limitations due to the requirement that the test specimen be submerged in water, while CT scanning is effective but cannot be portable and inexpensive. Presently, acoustic emission and resonance are being researched, but more time is needed in order to determine their workability as test methods. Thermographic techniques are full-field and simultaneously and independently measure thermal properties over a 2-D distribution of locations. Since the detection and thermal excitation required is accomplished radiometrically, the method is inherently noncontact and can be performed at elevated temperatures.

This paper describes a modified thermographic imaging technique for the measurement of thermal diffusivity of a number of SiC/SiC CFCC specimens and components. Currently, the available commercial equipment for infrared analysis limits the size of components to 10 to 15 mm diameter and 6 mm thickness requiring high spatial resolution and long acquisition times [6]. The approach of the experiment is to use a commercially available infrared camera, a suitable thermal excitation device, and locally written software to automate the entire test setup. Complete thermal history is recorded for the specimen area of interest from the time of thermal excitation until the surface reaches its maximum temperature. From this temperature-time relationship, thermal diffusivity is calculated by two methods. By imaging at a higher resolution than required and spatially averaging the resulting images before computation of the diffusivity, sufficient noise reduction is achieved in a single heating cycle while maintaining the required spatial resolution.

## EXPERIMENTAL DETAILS

The thermal imaging system used for this work is illustrated in Fig. 1. This system uses the method of Parker et al. [7] to calculate thermal diffusivity requiring a thermal pulse of short duration to be incident upon the front surface of a specimen and the temperature of the back surface to be monitored as a function of time. This was accomplished by heating the front surface of the specimen using a 6 kJ photographic flash lamp with a pulse duration of less than 8 ms and monitoring the back-surface temperature using a commercially available scanning radiometer infrared camera. The camera was equipped with a 3-12 $\mu$m optical band-pass lens system and a liquid-nitrogen-cooled HgCdTe detector. Images were acquired using a Mac II with an on-board frame grabber capable of storing 128 images in real time at 512 x 512 pixel 8-bit resolution. The frame grabber board received standard RS-170 signals from the infrared camera and digitized each received image, which was then processed with locally developed software. The software extracted the average gray scale value representing the temperature of the specimen. Specimens were mounted in diatomaceous earth having low thermal conductivity and minimizing lateral heat flow. By determining the position of the scanning mirrors at the time of the flash (e.g., by determining which of the 512 image rows was excited by the flash) the exact flash time could be determined with sub-millisecond resolution.

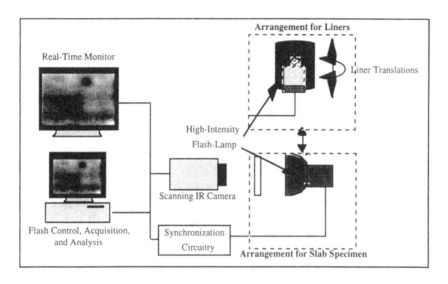

Figure 1:   Diagram of experimental arrangement used for thermal diffusivity
imaging.

Figure 2 shows the theoretically predicted back-surface temperature $T$ as a
function of time $t$ and specimen thickness $L$ according to Parker et al. [7] from the
relationship

$$T(L,t) = \frac{Q}{\rho CL}\left[1 + 2\sum_{n=1}^{\infty}(-1)^n \exp\left(\frac{-n^2\pi^2}{L^2}\alpha t\right)\right],\tag{1}$$

where $Q$ is the radiant energy incident on the front surface, $\rho$ is density, $C$ is specific
heat, and $\alpha$ is the thermal diffusivity.  Because the axes of Fig. 2 have been
normalized (where $T_M$ is the maximum back surface temperature), it is universal to
all specimens.  Two methods of determining $\alpha$ from the graph become evident.
First, the more common method notes that when $V = 0.5$ (the back-surface
temperature rise has reached half of its maximum), $\omega = 1.37$.  Thus, using the "half-
rise time" method,

$$\alpha = \left(\frac{1.37\,L^2}{\pi^2 t_{\frac{1}{2}}}\right).\tag{2}$$

As is evident from the plot of Fig. 2, any noise in the temperature signal can
significantly alter the computed thermal diffusivity value.  Because the noise level in
scanning radiometers is notoriously high (as much as 10% for temperature variations
on the order of a few degrees Celcius in the current system), some method of noise
reduction must be performed.  Previous researchers have performed a "cyclic
averaging" procedure in which the specimen is heated and allowed to cool a number
of times, and the time-temperature history for all such cycles is averaged [7].  They
have also employed a nonlinear least squares estimation of the entire time-temperature
relationship which requires significant computation time.  To minimize the time
necessary to measure thermal diffusivity, and because the curve is nearly linear in the
immediate vicinity, a linear least squares fit was used to interpolate the value of $t_{1/2}$ .

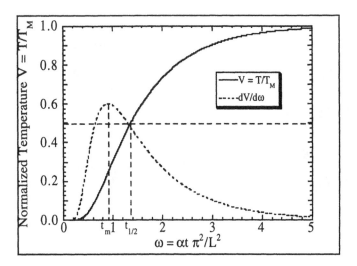

Figure 2. Theoretical plot of normalized back-surface temperature as a function of normalized time.

Furthermore, by imaging the components at higher resolution than required, spatial averaging, at a size which could be specified for the particular application, was permissible without loss of sensitivity to the minimum defect size. For the results presented here, a 10 x 10 pixel block was used. Thus a 512 x 413 pixel digitized image was transformed into a 51 x 41 pixel reduced image with a substantially better signal-to-noise ratio. As an additional means of noise reduction, the quarter-rise time ($t_{1/4}$) and three-quarter-rise time ($t_{3/4}$) were also determined for each block, where $\omega(t_{1/4}) = 0.92$ and $\omega(t_{3/4}) = 2.08$, according to the extension of Parker et al.'s procedure suggested in ASTM specification E 1461-92 [8]. The thermal diffusivity values obtained at these locations were averaged with that of the $t_{1/2}$ value to determine the effective thermal diffusivity. This permitted the determination of thermal diffusivity from a single thermal cycle, thereby reducing the total acquisition time.

The resulting thermal diffusivity values for all 10 x 10 pixel subsets were assembled into diffusivity images and were stored with 16-bit resolution. This avoided the need to scale each image independently (for better comparison between images) while retaining flexibility for image enhancement in display (which was performed at 8-bit resolution) [9]. The system was further modified by incorporating an autotrigger circuit, developed locally, to repeatedly fire the flash, giving a dependable time for the diffusivity calculations. This autotrigger circuit has eliminated the need to manually locate the frame of the flash thereby providing a more precise value of the flash-time. Another commercially available, fully integrated, self-contained infrared camera equipped with a 3-5 μm optical band-pass lens system and based on the second-generation focal plane technology was also utilized. The electronics of this camera allowed for user-selectable intensity transform algorithms for contrast enhancement of the output image. Nonuniformity correction was selected to eliminate the variations in pixels in the focal plane array. Automatic gain control dynamically optimized the video contrast and brightness on a frame-by-frame basis, providing additional resolution in the acquired images.

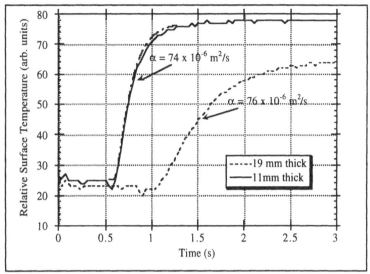

Figure 3.   Time-temperature plots for 11 mm and 19 mm NIST fine-grained isotropic graphite specimens obtained with the half-rise-time thermal diffusivity system.

## RESULTS

### CALIBRATION SPECIMEN

Samples from the National Institute of Science and Technology (NIST) were used to determine the accuracy of the infrared image analysis system.  The first specimens examined in this work were several thermal-property calibration standards (8426 graphite) with a stated thermal diffusivity value of $72 \times 10^{-6}$ $m^2/s$.  Three cylinders were cut from the 25.4-mm-diameter specimen provided with a diamond wafering saw in nominal thicknesses of 4, 11, and 19 mm.  Figure 3 shows typical time-temperature plots for the 11 and 19 mm specimens, from which thermal diffusivity values of 74 and $76 \times 10^{-6}$ $m^2/s$, respectively, were measured using the half-rise-time method and averaging over the center 1 $cm^2$.  The resulting thermal diffusivity image for the 19 mm specimen is shown in Fig. 4.  Despite the noise in the region surrounding the graphite disk, the thermal diffusivity clearly shows little variation across the face of the disk, indicating that the insulating mount was sufficient to ensure that lateral diffusion is indeed  negligible over the time  frame used to conduct these measurements. Furthermore, the standard deviation of thermal diffusivities measured in 1 $cm^2$ at the disk center was $\pm 2 \times 10^{-6}$ $m^2/s$.  Thus, the absolute value of the thermal diffusivity measured using this system deviated from the calibration value by $\approx 5\%$ or less (owing to thermal losses and approximations in the determination of $t_{1/2}$), and the variation of that measured diffusivity was less than 3%.

### SEEDED-DEFECTS SPECIMEN

The applications of the thermal diffusivity imaging system are primarily of interest in materials characterization.  While they use the full-field nature of the technique, they are interested only in a single measurement from a given  specimen, and  therefore do

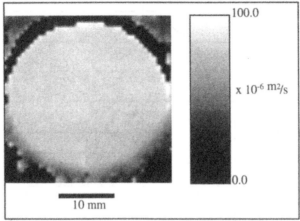

Figure 4. Thermal diffusivity image of 19 mm NIST graphite specimen, with absolute diffusivity scale shown at right.

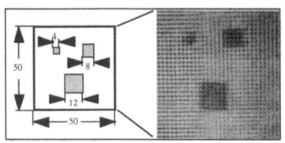

Figure 5. Schematic diagram of "seeded defect" $SiC_{(f)}/SiC$ composite panels used for defect detection (dimensions in mm) and X-ray radiograph.

not take full advantage of the system's capability to determine the distribution of thermal diffusivities. When examining actual components, the distribution of properties becomes critical, because what constitutes a potentially dangerous defect in one location might be an acceptable variation in another less hazardous location. To examine the system's capability to distinguish defects of various sizes in a $SiC_{(f)}/SiC$ CFCC, several "seeded-defect" $SiC_{(f)}/SiC$ composite panels were made. A total of 32 panels were created by varying several parameters (including graphoil section thicknesses, defects sizes, fiber coatings, defect depth, and infiltration density), and a diagram of one of these panels is shown in Fig. 5. The panels were made from 12 layers of 2-D plain weave Nicalon fabric. The simulated defects were created by cutting out small sections of two inner plies and replacing the fabric with graphoil sections. The weave was then infiltrated with SiC by CVI. Figure 5 also shows X-ray radiograph of the specimen indicating the defect locations. Figure 6 shows a comparison of the images generated for the $SiC_{(f)}/SiC$ panel. Figure 6a is the raw thermal image taken at the point of maximum thermal contrast between the defect and non-defect regions, and Fig. 6b is the same image after substantial image processing. While the defects are clearly evident in the processed image, the relative extent of "damage" is still unknown, and the image processing has required the interaction of a

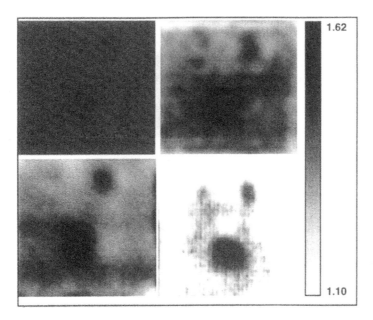

Figure 6.  Images of a SiC$_{(f)}$/SiC composite panels.  Image (a) is the raw thermal
image, (b) is the processed thermal image, (c) is the thermal diffusivity
image and (d) is the thermal diffusivity image with second generation
focal plane technology and user-selectable intensity transform algorithms
for contrast enhancement of the output image (absolute diffusivity scales
shown at right).

trained user.  Furthermore, direct quantitative comparison of the two images is
difficult at best.  Figure 6c is the thermal diffusivity image of the specimen.  Not only
is the contrast as good as or better than the processed thermal images, but the spatial
resolution is slightly better and no user interaction was required.  Since the thermal
diffusivity images are quantitative, direct comparison is possible not only between
the defect and nondefect regions, in which the defect region has a higher diffusivity,
but between the two porosities as well.  Figure 6d was acquired by user-selectable
intensity  transform algorithms for contrast enhancement of the output image; gain
control was dynamically adjusted to optimize video contrast and brightness on a
frame-by-frame basis, providing additional resolution in the acquired image.

## COMBUSTOR LINERS AND BURST RINGS

Infrared imaging techniques were applied to a combustor liner and two burst
rings.  Figure 7 shows photographs of the representative combustor liners and the
burst rings, respectively, that have been used for thermal diffusivity mapping.  A
typical burst ring had an approximate circumference of 64 cm, a diameter of 20 cm,
and  a thickness of 3 mm.  To reduce the effects of angular emissivity variation from
the cylindrical surface, 12 sections (each approximately 5.0 x 5.0 cm) of the burst
rings were imaged.  The combuster liner had an approximate circumference of 64
cm, a diameter of 20 cm, and a thickness of 3 mm.  Eleven sections per row of
approximately 5.0 x 6.5 cm were imaged, and the overall height of the map from top
to bottom was 15 cm.

Figure 7.   Photograph of combustor liner and burst ring.

The experimental configuration for the combustor liners and burst rings, as shown in Fig. 1, was modified in that the flash lamp was placed inside the liner while the IR camera monitored the temperature of the outside surface. Light from the flash lamp was concentrated by using aluminum foil inside the liner in an area ($\approx 100$ mm square) centered on but slightly larger than the segment of interest to prevent excessive lateral thermal diffusion from the region of interest. By containing the area exposed to the flash, the entire liner was not heated during each flash cycle. Thus, consecutive acquisitions of time-temperature profiles for nonadjacent segments were possible without allowing for the previously irradiated surface to cool completely, thereby minimizing overall acquisition time.

Figures 8 and 9 show the diffusivity maps for the burst rings along with the ultrasound testing maps which were available only for the burst rings. A direct correlation is observed in the maps in the identification of the voids, the point "A" being the orientation point of both the maps.

Thermal diffusivity images for the individual segments of the combustor liner were assembled into a single conformally mapped image as shown in Fig. 10a. The lines between segments are caused by tape used to mark their location. One large area of low thermal diffusivity is seen on the left side of the image, while other areas of higher diffusivity are seen as vertical bands across the image. The bands of higher diffusivity correlated well to the locations of seams in the weave lay-up. However,

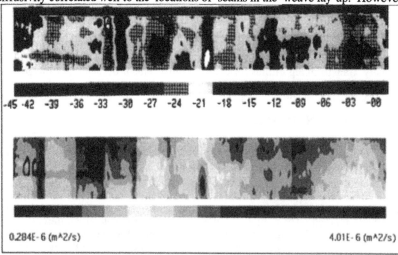

Figure 8:   Top: Ultrasonic testing map; Bottom: IR imaging map for burst ring.

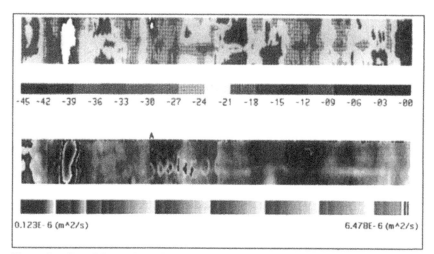

Figure 9:  Top: Ultrasonic testing map; Bottom: IR imaging map for burst ring.

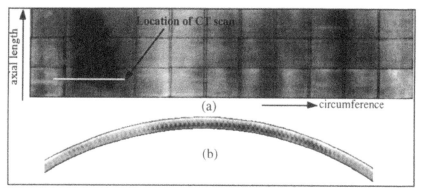

Figure 10.  (a) Conformally mapped thermal diffusivity image of subscale SiC$_{(f)}$/SiC combustor liner (where white = high diffusivity, black = low diffusivity) and (b) X-ray CT cross-section of same liner (where white = high density, black = low density) from location noted in (a).

the large region of low diffusivity was not related to any known geometrical condition of the liner. A density cross-section was therefore obtained for that region (as indicated in the figure) using a microfocus X-ray CT system, and this image is shown in Fig. 10b. Clearly, the region of low thermal diffusivity correlated directly to a region of low density (high porosity) in the liner, the result of poor matrix infiltration.

## SiC/SiC COMPOSITE PANEL

A diffusivity map was acquired of a 4 x 4 in. 3D composite panel of SiC/SiC with angle interlock defects in specific locations shown in Fig. 11. The through thicknesses and the weave row sketched in Fig. 12b are easily detected by the overall

Figure 11.   Digitized photograph of 4 x 4 in. SiC/SiC ceramic composite with special carbon fibers at selected locations.   Bottom half of the photograph shows the interweave.

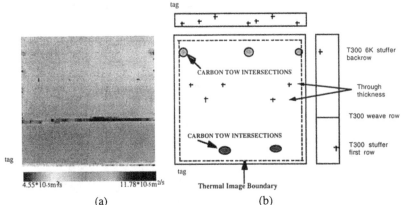

(a)                                      (b)

Figure 12.   (a) Thermal diffusivity image showing foreign fiber detection (b) schematic diagram of 3-D weave SiC/SiC specimen showing locations of carbon tows.

diffusivity map of the panel shown in Fig. 12a.  Data acquisition was conducted by dividing the panel in four sections, scanning, and obtaining four diffusivity maps.

## CONCLUSIONS

A full-field thermal-imaging-based nondestructive characterization method has been developed and is applicable to a wide variety of large continuous fiber ceramic components.  The data obtained by the thermal imaging nondestructive evaluation method have been correlated to delaminations, density variations, and other nondestructive evaluation methods. By applying digital filtering, interpolation, and least-square-estimation techniques for noise reduction, acquisition and analysis times of minutes or less have been achieved with submillimeter spatial resolution.  In the future, sensitivity of full-field thermal diffusivity imaging will be enhanced, fatigue damage will be correlated, and methods will be investigated to allow measurement of diffusivity at higher temperatures.

## ACKNOWLEDGMENTS

The authors would like to acknowledge the funding granted them by the U.S. Department of Energy, Energy Efficiency and Renewable Energy, Office of Industrial Technologies, under Contract W-31-109-ENG-88.

## REFERENCES

1. J. S. Steckenrider, W. A. Ellingson and S. A. Rothermel, "Full-Field Characterization of Thermal Diffusivity on Continuous-Fiber Ceramic Composite Materials and Components", *Thermosense XVII: An International Conference on Thermal Sensing and Imaging Diagnostic Applications*, SPIE Proceedings Vol. 2473, S. Semanovich, Ed., Bellingham, WA, 1994.

2. F. Enguehard, D. Boscher, A. Déom and D. Balageas, "Measurement of the thermal radial diffusivity of anisotropic materials by the converging thermal wave technique," *Mat. Sci. Eng.*, Vol. B5 (2), pp. 127-134 (1990).

3. A. Hafidi, M. Billy and J. P. Lecompte, "Influence of microstructural parameters on thermal diffusivity of aluminium nitride-based ceramics," *J. Mat. Sci.*, Vol. 27 (12), pp. 3405-3408 (1992).

4. J. Rantala, L. Wei, P. K. Kuo, J. Jaarinen, M. Luukkala and R. L. Thomas, "Determination of thermal diffusivity of low-diffusivity materials using the mirage method with multiparameter fitting," *J. Appl. Phys.*, Vol. 73 (6), pp. 2714-2723 (1993).

5. I. Hatta, "Thermal diffusivity measurements of thin films and multilayered composites," Int. J. Thermophysics, Vol. 9 (2), pp. 293-303 (1990).

6. K. E. Cramer, W. P. Winfree, E. R. Generazio, R. Bhatt and D. S. Fox, "The application of thermal diffusivity imaging to SiC fiber reinforced silicon nitride," in Review of Progress in Quantitative Nondestructive Evaluation, D. O. Thompson and D. E. Chimenti, eds., Vol. 12, p. 1305, Plenum, New York (1993).

7. W. J. Parker, R. J. Jenkins, C. P. Butler and G. L. Abbott, "Flash method of determining thermal diffusivity, heat capacity, and thermal conductivity," *J. Appl. Phys.*, Vol. 32 (9), pp. 1679-1684 (1961).

8. American Society for Testing and Materials, "Standard test method for thermal diffusivity of solids by the flash method," Annual Book of ASTM Standards, ASTM E 1461-92 (1992).

9. W. A. Ellingson, S. A. Rothermel, J. F. Simpson, "Nondestructive Characterization of Ceramic Composites used as Combustor Liners in Advanced Gas Turbines", *Trans. ASME*, 95-GT-404, American Society of Mechanical Engineers, New York, NY (1995).

10. A. Sala, Radiant Properties of Materials - Tables of Radiant Values for Black Body and Real Materials. Elsevier Science Publishing Co. Inc., New York, 1986.

11. W. R. Barron, "The Principles of Infrared Thermometry", Sensors. December, 1992.

# INSULATION

SESSION CHAIRS
A. O. Desjarlais
D. R. Flynn

# Effective Diffusion Coefficients for CFC-11 by Gravimetric Depletion from Thin Slices of PIR Foam

J. R. BOOTH,[1] R. S. GRAVES[2] and D. W. YARBROUGH[3]

## ABSTRACT

Chlorofluorocarbon (CFC) compounds have historically been used to achieve superior thermal performance in closed-cell polymeric foams. Changes in thermal performance which take place quite slowly occur as the gaseous components from the atmosphere and the insulating gases are rearranged. Thin-slicing of closed-cell core material and aging at room temperature has achieved meaningful reductions in the time required to evaluate the foam properties which define the time-scale for aging.

In a previous paper, changes in apparent thermal conductivity of stacks of thin-sliced foam were used to determine the effective diffusivity of the blowing agent (CFC-11) in the second region of polyisocyanurate (PIR) foam aging. There are two other methods for monitoring this time-dependent phenomena which are more directly connected to the diffusion process than thermal conductivity.

As specimens of PIR foam for the ORNL/Industry Aging Study were stored at 75 °F and 50 %RH, one set of thin-slices was intermittently subjected to thermal conductivity measurement while a second set of thin-slices were weighed. These two sets of data have been processed and each set was used to evaluate the effective diffusivity of CFC-11. Using 0.254 cm specimens, the effective diffusivity of CFC-11 in a closed-cell PIR foam of 29.2 kg/m$^3$ bulk density was found to be $0.9 \times 10^{-10}$ cm$^2$/s by gravimetric monitoring compared to $1.33 \times 10^{-11}$ cm$^2$/s determined by thermal conductivity.

## BACKGROUND

Foamed plastic insulations have been manufactured using high molecular weight volatile CFC compounds. These compounds, referred to as "blowing agents" were employed to reduce the bulk density and improve the insulation performance of the cellular plastics. These CFCs were found to destroy ozone in

[1] Department of Chemical Engineering, Tennessee Technological University, Cookeville, TN 38505
[2] R and D Services, Lenoir City, TN 37771
[3] The Oak Ridge National Laboratory, Oak Ridge, TN 37831

the upper atmosphere and their use for blowing agents has been eliminated by legislation. CFCs are currently being replaced with hydrochlorofluorocarbons (HCFC), hydrofluorocarbons (HFC) and other physical blowing agents.

The evaluation of replacement blowing agents and process modifications to produce marketable insulation products for the future require a new aging methodology. The evaluation of promising candidate blowing agents using appropriate component recipes and process modifications requires a reliable thin-slice methodology with a reduced testing interval. The development of prototype blowing agents requires an ultra-thin slice method to provide precise relative effective diffusion coefficients in specimens of "calibrated" generic foam structures having extremely compressed testing times.

## FULL-THICKNESS AGING

Moore [1] presented a series of full-thickness aging data for a series of polyurethane foams prepared with toluene-diisocyanurate (TDI), polymeric diisocynanurate (PMDI) and several polyols blown with CFC-11. The apparent thermal conductivity of these one-inch-thick foam boards was measured over a period of eight years. During this time air diffusion into the foam specimens appeared to achieve saturation but the testing interval was not sufficiently long to determine useful information about the loss of the CFC-11 blowing agent.

## THIN-SLICE AGING

Christian *et al.* [2] investigated a series of PIR boardstock specimens, prepared for a joint Industry-Government PIR alternative blowing agent aging investigation, using thin-slice aging. The boardstock was sliced to provide nominal 33, 19 and 10-mm-thick specimens. The apparent thermal conductivity of these specimens was repeatedly measured over a three-year period. When a scaling factor was applied to these specimens, the apparent thermal conductivity measurements formed a single curve for each blowing agent. In each case, the 33-mm-thick specimen was not completely saturated with air after three years while the 10-mm-thick specimen was saturated with air and was beginning to suggest a new slope representative of the loss of the replacement blowing agent.

Yarbrough *et al.* [3] investigated one of the joint Industry-Government PIR boardstock specimens blown with CFC-11. This specimen was sliced to provide sets of nominal 5.1, 2.5 and 1.3-mm-thick slices. The apparent conductivity of stacks, 13 to 15-mm thick, of these specimens were measured for a four month period. When a scaling factor was applied to these specimens, the apparent conductivity of the closed cell **core** of the specimens formed a single scaled aging curve. The scaled aging curve apparent thermal conductivity coincided with the results of Christian *et al.* [2] for the CFC-11 blown specimen.

The prediction of foam thermal insulation performance employing new prototype blowing agents and mixtures may be based on the physical properties, molecular weight, permeability or diffusivity and solubility in the polymer matrix.

Using these properties in a diffusion model one can estimate the average aged gas-phase compositions from which a gas-phase mixing rule can be used to estimate the aged average thermal conductivity of the gas-phase. The time-dependent thermal conductivity contribution of the gas-phase can be added to the solid matrix thermal conductivity and the radiation contribution to provide a predicted aging curve for the proposed foam. Once small scale samples of the proposed foam, containing the new prototype blowing agents, have been produced thin-slice aging can be used to rapidly evaluate the long-term thermal insulation performance of the proposed foam samples.

## DIFFUSION IN CELLULAR PLASTIC FOAM

Solutions of the transient diffusion equation have been developed using both continuum models [4,5,6,7] and an analysis of discrete networks [8,9]. Rapidly diffusing gases such as He and $CO_2$ have been suggested [10,11] as ideal materials to "calibrate" the structure of prototype foams.

It is preferable to evaluate the parameters in diffusion models using mass or concentration measurements; gas flow [5], pressure change [10,11] or weight change [12]. Alternate methods which use thermal measurements [13,14] monitor related phenomena (heat flux and temperature difference) and attempt to determine effective diffusion coefficients by working backwards from a complex multiple transport process [4,14] inside a distributed closed cellular heterogeneous medium. The actual diffusion distance (effective thickness) is reduced because of open (cut or damaged) cells at the cut surfaces. Effective thickness for diffusion can be computed from measured overall thickness and thickness of the destroyed surface layer (TDSL) [3].

$$X_{eff} = X_m - 2\ TDSL \qquad (1)$$

where:   $X_m$   = measured slice thickness, cm
$X_{eff}$  = effective diffusion thickness, cm
TDSL = thickness of destroyed surface layer, cm

The error function solution to the one-dimensional diffusion equation, provides a short-term approximation preferred [4,12] for evaluating the effective diffusion coefficient (diffusivity) using the effective thickness for the thin slices.

$$\frac{\Delta M_t}{\Delta M_\infty} \approx \frac{4}{\pi^{1/2}} \left( \frac{D_{eff}\ t}{X_{eff}^2} \right)^{1/2} \qquad (2)$$

where:   $M_o$   = initial mass of slice, gm
$M_t$   = instantaneous mass of slice, gm
$M_\infty$   = final mass of slice, gm
$\Delta M_t$   = instantaneous mass change ($M_o$ - $M_t$), gm
$\Delta M_\infty$   = equilibrium mass change ($M_o$ - $M_\infty$), gm
$D_{eff}$   = effective diffusivity, cm²/sec
t     = time, sec

The measured values of instantaneous mass of the specimen are plotted against the corresponding scaled age values. The slope of the initial linear portion of the data with scaled age is determined. This slope is divided by the equilibrium mass change, $\Delta M_\infty$. When the number of cells in the diffusion path of a slice allows the matrix to be viewed as homogeneous, and the Henry's law criteria are met, Equation 3 may be used to estimate effective diffusion coefficients for foam structures [4,5,6].

$$D_{eff} \approx \frac{\pi}{16}\left(\frac{d\dfrac{\Delta M_t}{\Delta M_\infty}}{d\dfrac{\sqrt{t}}{X_{eff}}}\right)^2 \tag{3}$$

## PURPOSE

The purpose of this project was to demonstrate a simple method for evaluating the parameters required for predicting the long-term thermal performance of gas-filled foam structures produced with replacement blowing agents and the reliable determination of effective diffusion coefficients of these blowing agents in closed-cell foams by gravimetric monitoring of thin and ultra-thin sliced foam specimens.

## EXPERIMENTAL

### MATERIALS

Experimental PIR foam slabstock insulation was produced for a Joint Industry/Government Aging Study [13]. The 38.1-mm-thick foams were produced as 1.2 x 2.4 m slabstock with three blowing agents, CFC-11, HCFC-123, HCFC-141a and a pair of mixtures of HCFC-123 and HCFC-141a, having permeable membranes on their surfaces. One of these specimens, blown with CFC-11 was used in this investigation.

### METHODS

Specimen Preparation and Physical Property Measurement

Specimen preparation, thickness, open-cell content and TDSL, bulk density and chord length were performed at the Granville Research Center of The Dow Chemical Company.

Specimens were prepared for measurement by cutting and trimming the foam samples to the desired final dimensions. The specimens were initially cut into 4.0 x 4.0 inch x 0.2 inch slices with a bandsaw. These rough-cut slices were then trimmed to 0.100, 0.050 and 0.026 inches thick using a precision foam slicer.

The thickness of the slices were measured using an ONO SOKKI Model GS-503 Linear Gauge and DG-325 Digital Display. The thickness at each of the corners and the center were used to compute an **average slice thickness**.

The **open-cell content** and **TDSL** of the foam specimens were determined by standard method ISO-4590 [15], using a sample volume of about 24 milliliters. A Beckman Model 930 Air Pycnometer was used to determine the air volumes. The slope of open-cell volume vs surface-to-volume ratio was interpreted as the TDSL.

The specimens prepared for determination of open-cell content were weighed on a Mettler Model H51AR analytical balance immediately after sample preparation. The apparent volume of the foam was computed from its orthogonal linear dimensions. The **bulk density** was computed by dividing the weight of the foam by its apparent volume as required by ASTM D 1622-88 [16].

The **chord length** in the vertical (rise) direction was determined by measuring the distance the traveling stage moved on a Unitron Model U-11 microscope to pass 50 cell walls under the cross-hairs of the 10X eye-piece. The length per cell wall was then computed by dividing the distance by the number of cell walls counted. This measurement was repeated 5 times in different regions of the foam and the average of these values was reported as the chord length.

Manual Gravimetric Monitoring of Blowing Agent Desorption

The diffusion of blowing agents from foam is described by a thickness-dependent diffusion scaling parameter, $\sqrt{t}/X_{eff}$ in Equation 2. The determination of effective diffusion coefficient by this parameter uses the diffusion time and the effective slice thickness. The diffusion time is the time since diffusion began for the thin sliced sample. Time of Day is the number of seconds since midnight.

$$t = 86400\,(JD_m - JD_i) + (TOD_m - TOD_i) \tag{4}$$

where:  t    = diffusion time, sec
$JD_m$ = Measurement Julian Date
$JD_i$ = Initial Julian Date
$TOD_m$ = Measurement Time of Day, sec
$TOD_i$ = Initial Time of Day, sec

The Mettler H51AR analytical balance was used for permanent blowing agent desorption. This balance was installed in the Constant Temperature/Constant Humidity Room, at the Granville Research Center to minimize static charges generated by handling the specimens. Sliced foam specimens were placed in a tray in contact with room air. Air temperature and humidity were controlled by a special humidification/air conditioning system and a strip-chart recorded these conditions.

The mass of each specimen was monitored on a schedule of increasing interval between measurements; daily for three days, every other day through the first week, twice weekly for 3 weeks and weekly until the mass of the specimen became constant and remained constant for a month. The mass data were entered into a computer data file manually.

Automated Gravimetric Monitoring of Blowing Agent Desorption

The relative rate of blowing agent desorption increases as specimen thickness decreases. For specimens less than 1-mm thick, a Mettler AT261 electronic analytical balance connected to a Compaq Deskpro 286e PC was used for the automated gravimetric monitoring of blowing agent desorption. The control program requested a measurement of the specimen, at 10 minute intervals. The date, time of day and the specimen mass were stored in a sequential data file. Between readings the computer would accept input from an operator. The operator was capable of manually terminating the collection of desorption data and closing the file.

## DISCUSSION OF RESULTS

### FOAM PROPERTIES

The volume and weight of six freshly cut regular rectangular prismatic specimens of ORNL PIR Boardstock Specimen T3B4 #3 were measured and the densities computed. The average bulk density of these specimens was 29.06 $\pm$0.06 kg/m$^3$.

The chord length of the rise (vertical) direction cells was measured on five precision sliced rise-transverse face specimens of T3B4 #3. The chord length averaged 0.20 $\pm$0.004 mm and ranged from 0.198 to 0.208 mm.

Standard method ISO-4590 was used to determine the intrinsic open-cell content and TDSL of the foam specimens. According to ISO-4590, a collection of specimens having approximately the same apparent volume but different surface areas are subjected to air pycnometry measurement. The surface value is the total surface area of all the pieces that make up the apparent volume.

The difference between the apparent volume of the specimen and the pycnometer volume is the open-cell volume. The data and results of the open cell fraction for the PIR foam blown with CFC-11 are reported in Table I.

The open-cell fraction was plotted against the surface to volume ratio of the specimens. Figure 1 shows the plotted data and the regressed linear least squares relationship. The intercept value, intrinsic open-cell content of the foam, was about 0.10 percent. The remainder of the open-cell content of the foam was attributed to surface damaged cells.

The increase in open cell volume with additional layers provides information about the TDSL of these additional surfaces. Hence, the slope of the regression, the TDSL of the foam, was 0.206 mm. The TDSL value was close to the vertical chord length (0.202 mm) for these specimens.

### MEASUREMENT OF SLOWLY DIFFUSING SOLUTES

CFC-11 desorption was monitored gravimetrically using specimens in the range of 0.0216 to 0.181 cm effective thickness. The specimens, conditioned at 50% relative humidity and 24°C, were intermittently weighed on the Mettler Model

**TABLE I. Determination of Open Cell Content and TDSL by ISO-4590**

| Specimen | Pcs | Length M$\times10^{-3}$ | Width M$\times10^{-3}$ | Height M$\times10^{-3}$ | Total Area M$^2\times10^{-3}$ | Apparent Volume M$^3\times10^{-6}$ | Pyc Volume M$^3\times10^{-6}$ | S/V M$^{-1}$ | Open[b] Cell fraction |
|---|---|---|---|---|---|---|---|---|---|
| 1 | 2 | 40.12 | 29.82 | 20.03 | 7.587 | 23.96 | 22.33 | 0.316 | 0.0680 |
| 2 | 2 | 40.07 | 29.78 | 19.76 | 7.534 | 23.58 | 21.99 | 0.319 | 0.0674 |
| 3 | 8 | 40.10 | 29.86 | 19.92 | 21.945 | 23.85 | 19.28 | 0.920 | 0.1916 |
| 4 | 8 | 40.10 | 29.84 | 19.95 | 21.936 | 23.87 | 19.26 | 0.919 | 0.1931 |
| 5 | 4 | 40.09 | 29.84 | 20.07 | 12.377 | 24.01 | 21.47 | 0.515 | 0.1058 |
| 6 | 4 | 40.12 | 29.82 | 20.09 | 12.381 | 24.04 | 21.43 | 0.515 | 0.1086 |
| 7 | 3 | 40.12 | 29.86 | 20.06 | 9.995 | 24.03 | 21.90 | 0.415 | 0.0886 |
| 8 | 3 | 40.09 | 29.81 | 19.78 | 9.936 | 23.64 | 21.60 | 0.420 | 0.0863 |
| 9 | 6 | 40.10 | 29.85 | 20.14 | 17.181 | 24.11 | 20.55 | 0.712 | 0.1477 |
| 10 | 6 | 40.10 | 29.85 | 19.84 | 17.139 | 23.75 | 20.25 | 0.721 | 0.1474 |

[b] Open-cell = (Apparent Volume - Pyc Volume)/Apparent Volume

H51AR analytical balance and the weights recorded. When the specimens were conditioned and weighed at ambient conditions less than 50% relative humidity, static charges produced by handling the foam specimens prevented obtaining accurate weights.

In addition to the specimens for a thermal aging investation [3] a pair of

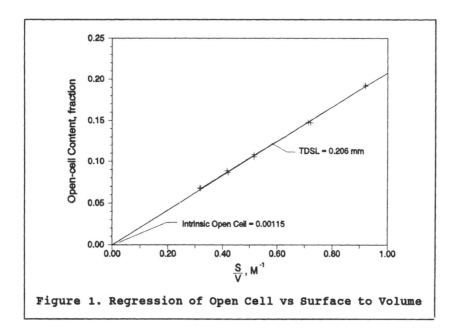

**Figure 1. Regression of Open Cell vs Surface to Volume**

specimens were prepared at each of several thicknesses for gravimetric desorption measurement. Two sets of specimens ORNL1 and ORNL2 were monitored by intermittent measurement on the Mettler H51AR. The data recorded in Table II bor ORNL1 represents a typical mass response for desorption. In addition, a single specimen ORNL3 was monitored by continuous monitoring on the Mettler AT261 at 10 minute intervals.

| TABLE II. Gravimetric Aging Data for Specimen ORNL1A | | | | | |
|---|---|---|---|---|---|
| Date | Time of Day | Julian Date | Time of Day sec | Scaled Age $sec^{1/2}$/cm | Weight gm |
| 09-29-1992 | 20:34:00 | 2448895 | 74040 | 330.4 | 0.37027 |
| 09-29-1992 | 21:15:00 | 2448895 | 76500 | 655.0 | 0.37170 |
| 09-29-1992 | 21:33:00 | 2448895 | 77580 | 754.6 | 0.37195 |
| 09-29-1992 | 22:02:00 | 2448895 | 79320 | 892.0 | 0.37237 |
| 09-29-1992 | 22:34:00 | 2448895 | 81240 | 1022.4 | 0.37280 |
| 09-29-1992 | 23:32:00 | 2448895 | 84720 | 1223.8 | 0.37330 |
| 09-30-1992 | 00:22:00 | 2448896 | 1320 | 1378.6 | 0.37370 |
| 09-30-1992 | 08:14:00 | 2448896 | 29640 | 2362.7 | 0.37559 |
| 09-30-1992 | 16:56:00 | 2448896 | 60960 | 3107.2 | 0.37600 |
| 10-02-1992 | 11:22:00 | 2448898 | 40920 | 5435.2 | 0.37595 |
| 10-03-1992 | 07:40:00 | 2448899 | 27600 | 6249.5 | 0.37538 |
| 10-04-1992 | 10:12:00 | 2448900 | 36720 | 7175.5 | 0.37530 |
| 10-13-1992 | 17:00:00 | 2448909 | 61200 | 12485.5 | 0.37285 |
| 10-17-1992 | 09:01:15 | 2448913 | 32475 | 14040.6 | 0.37226 |
| 10-24-1992 | 10:09:00 | 2448920 | 36540 | 16625.0 | 0.37119 |
| 11-01-1992 | 07:08:00 | 2448928 | 25680 | 19103.7 | 0.37035 |
| 11-12-1992 | 12:30:00 | 2448939 | 45000 | 22162.4 | 0.36905 |
| 12-18-1992 | 13:27:00 | 2448975 | 48420 | 29941.3 | 0.36547 |
| 01-05-1993 | 09:17:00 | 2448993 | 33420 | 33120.5 | 0.36444 |
| 03-07-1994 | 08:00:00 | 2449419 | 28800 | 76722.4 | 0.35403 |
| 04-12-1994 | 12:00:00 | 2449455 | 43200 | 79314.2 | 0.35282 |
| 03-06-1995 | 12:00:00 | 2449783 | 43200 | 99914.2 | 0.35123 |

Specimen ORNL1A had an average measured thickness of 0.1279 cm and a TDSL value of 0.0206 cm. Using Eq. 1, the effective thickness was 0.0887 cm.

Diffusion began for all five specimens of this investigation between 20:20:00 and 20:24:00 hours on 09-29-1992 (Julian Date 2448895) at the time of the precision slicing of the surfaces. The manual samples were first weighed from 20:34:00 to 20:38:00 hours. The automatic sample was installed and the monitoring program started at 20:40:00 hours.

The desorption response curves, specimen weight vs scaled age, are presented in Figures 2, 3 and 4. The linear region of the desorption curves following air saturation was used to extrapolate backward, to zero time, for the air-saturated initial weight. The final weight was **estimated** from the air-saturated initial weight, the blowing agent concentration in the feed and the fresh density. The slopes of linear region of the desorption curves following air saturation, initial weight and final weight were used to compute the dimensionless slope and hence to estimate the effective diffusion coefficient for each specimen.

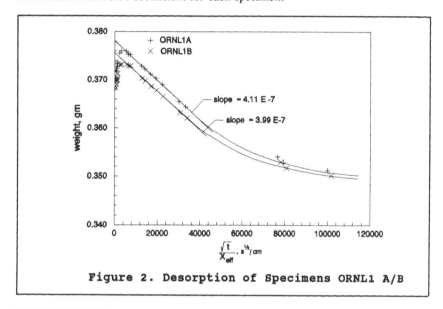

**Figure 2. Desorption of Specimens ORNL1 A/B**

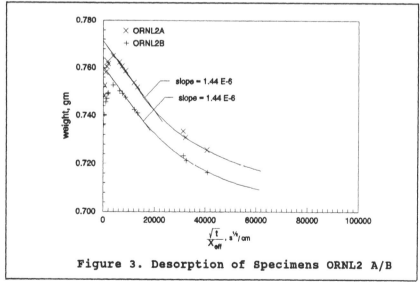

**Figure 3. Desorption of Specimens ORNL2 A/B**

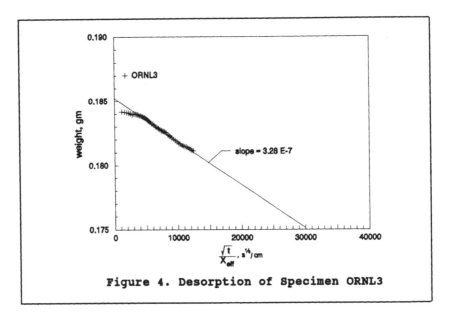

**Figure 4. Desorption of Specimen ORNL3**

## REPRODUCIBILITY OF GRAVIMETRIC MEASUREMENTS

The gravimetric desorption data from Table II and data collected for the other specimens were used to compute effective diffusion coefficients for CFC-11 from the PIR foam, which are recorded in Table III.

| Specimen | Effective Thickness cm | Initial Slope gm cm/sec$^{\frac{1}{2}}$ | Mass of CFC-11 gm | Effective Diffusivity cm$^2$/sec |
|---|---|---|---|---|
| ORNL3 | 0.0240 | $3.57*10^{-7}$ | 0.0076 | $4.3*10^{-10}$ |
| ORNL1A | 0.0877 | $4.11*10^{-7}$ | 0.028 | $0.42*10^{-10}$ |
| ORNL1B | 0.0861 | $3.99*10^{-7}$ | 0.028 | $0.39*10^{-10}$ |
| ORNL2A | 0.214 | $1.44*10^{-7}$ | 0.070 | $0.83*10^{-10}$ |
| ORNL2B | 0.215 | $1.44*10^{-7}$ | 0.062 | $1.05*10^{-10}$ |

**TABLE III.  Effective Diffusivity of CFC-11 in PIR with Thickness**

The results presented in Table III suggest that while there is agreement between samples of similar thickness, the agreement between specimens of different thickness is not good. Thermal measurements of this same PIR foam specimen were performed over a similar period of time in another in which the value of effective diffusivity was found to be $1.3*10^{-11}$ cm$^2$/sec [3]. Gravimetric desorption of CFC-11 for a single thickness of other specimens of PIR foam produced an effective diffusivity value of $4.6*10^{-11}$ cm$^2$/sec [17].

The effective diffusion coefficient does not appear to present a consistent trend with decreasing slice thickness. Variation of effective diffusion coefficient with sample thickness for ultra-thin specimens is disturbing, but not unexpected. De Nazelle *et al* [9] suggested that reducing the number of cells in the diffusion path would result in variability of the measured diffusion parameters. Perhaps sample ORNL3, being only one cell thick is not a representative sample.

As long as the number of cells in the diffusion path is known to be constant, relative measured effective diffusivies for prototype replacement blowing agents will represent real changes in their diffusion coefficients.

## PRACTICAL LIMITATIONS OF SLICE THICKNESS

A foam specimen thinner than twice the TDSL would not have a closed-cell core region to age. For the PIR foam tested in this investigation, such a slice would be 0.42 mm thick. Too few closed cells in the core region might produce conditions which would not conform to the continuum criteria generally assumed for the solution of the fundamental Fick's Law diffusion equation. However, a practical lower limit might result from the mechanical properties, the friability and method of slicing the foam. For the PIR tested in this investigation, specimens thinner than about 0.63 mm are too fragile to handle. In addition, a specimen 0.63 mm (0.024 inches) thick might have only a single closed cell core.

## CONCLUSIONS

The gravimetric measurement of ultra-thin specimens of PIR foam manufactured with CFC-11 provided the following conclusions:

1. The systematic bias in thermal conductivity analysis, caused by the air-filled TDSL of ultra-thin slices, does not create an error in gravimetric analysis of the effective diffusion coefficient, so long as the **effective slice thickness** is used to calculate the scaled age of the specimen.

2. The evaluation times reported in this study were sufficient to evaluate effective diffusion coefficients obtained by analysis of the slope of core weight change with scaled age for the secondary aging region (permanent blowing agent desorption) of CFC-11 blown foams.

3. The effective diffusivity of CFC-11 in a closed-cell PIR foam having a bulk density of 29.2 kg/m$^3$ was found to range from $0.39$-$4.3 \times 10^{-10}$ cm$^2$/sec by gravimetric analysis during the first 50 percent CFC-11 depletion compared to $1.33 \times 10^{-11}$ cm$^2$/sec by thermal conductivity analysis [3] during the same fractional depletion.

**ACKNOWLEDGEMENT**

The authors acknowledge the support of Government/Industry Task Group; (The Society of the Plastics Industry (SPI), The Polyisocyanurate Manufacturers Association (PIMA), National Roofing Contractors Association (NRCA) and the U.S. Department of Energy (DOE)). In addition, the support and interest of the Dow Chemical Company in pursuing this study is acknowledged.

**REFERENCES**

1. Moore, S. E., June 5-7, 1991, "Effect of Polymer Structure on the Long-term Aging of Rigid Polyurethane Foam, *2nd International Workshop on Long-term Thermal Performance of Cellular Plastics*, Niagara-on-the-Lake, Ontario, Canada.

2. Christian, J. E., R. S. Courville, R. L. Linkous, D. L. McElroy, R. S. Graves, F. J. Weaver and D. W. Yarbrough, 1991, "Thermal Measurement of in-Situ and Thin-Specimen Aging of Experimental Polyisocyanurate Roof Insulation Foamed with Alternative Blowing Agents", *Insulation Materials: Testing and Applications*, 2nd Volume, *ASTM STP 1116*, R. S. Graves and D. C. Wysocki, Eds, American Society for Testing and Materials, Philadelphia, pps. 142-166.

3. Yarbrough, D. W., R. S. Graves and J. R. Booth, March 23-25, 1993, " Aging of Thin-slices of PIR Foams Manufactured with Alternative Blowing Agents", Cellular Polymers II, Heriott-Watt University, Edinburgh.

4. Crank, J. and G. S. Park, 1968, "Methods of Measurement", in *Diffusion in Polymers*, J. Crank and G. S. Park, eds., Academic Press, London, pp. 1-39.

5. Shankland, I. R., 1986, "Gas Transport in Closed-cell Foams", in *Advances in Foam Aging*, Vol. 1, *A Topic in Conservation Series*, D. A. Brandreth, ed., Caissa Editions, pp. 60-88.

6. Bomberg, M. T., 1988, "A Model of Aging for Gas-Filled Cellular Plastics", *J. of Cellular Plastics*, 24: 327-347.

7. Sheffield, C. F., September 15-16, 1988, "Description and Application of a Diffusion Model for Rigid Closed-cell Foams", *Mathematical Modeling of Roof Systems*, The Oak Ridge National Laboratory, Oak Ridge, TN.

8. Reitz, D. W., M. A. Schuetz and L. R. Glicksman, 1984, " A Basic Study of Heat Transfer through Foam Insulation", *J. of Cellular Plastics*, 20: 332-340.

9. Du Cause de Nazelle, G. R. M., G. C. J. Bart, A. J. Dammers and A. Cunningham, September 28-29, 1989, "A Fundamental Characterization of the Aging of Polyurethane Rigid Foam", *International Workshop on Long-Term Thermal Performance of Cellular Plastics*, Huntsville, Ontario, Canada.

10. Schwartz, N. V., M. T. Bomberg and M. K. Kumaran, 1989, "Measurement of the Rate of Gas Diffusion in Rigid Cellular Plastics", *J. Thermal Insulation*, 13: 48-61.

11. Brehm, L. R. and L. R. Glicksman, September 28-29, 1989, "Implementation of a Constant-Volume Sorption Technique for Rapid Measurement of Gas Diffusion and Solubility in Closed-Cell Foam Insulation", *International Workshop on Long-Term Thermal Performance of Cellular Plastics*, Huntsville, Ontario, Canada.

12. Stern, S. A., U. M. Vakil and G. R. Mauze, 1989, "The Solution and Transport of Gases in Poly(n-butyl methacrylate) in the Glass Transition Region. I. Absorption-Desorption Measurements", *J. Poly. Sci.*, 27: 405-429.

13. McElroy, D. L., R. S. Graves, D. W. Yarbrough and F. J. Weaver, September 1991, "Laboratory Test Results on the Thermal Resistance of Polyisocyanurate Foamboard Insulation Blown with CFC-11 Substitutes-A Cooperative Industry/Government Project", ORNL/TM-11645, Oak Ridge National Laboratory, Oak Ridge, TN.

14. Booth, J. R., 1991, "Some Factors Affecting the Long-term Thermal Insulation Performance of Extruded Polystyrene Foams", *Insulation Materials: Testing and Applications*, 2nd Volume, *ASTM STP 1116*, R. S. Graves and D. C. Wysocki, Eds., American Society for Testing and Materials, Philadelphia.

15. ISO 4590-1981 (E), 1981, "Cellular plastics-Determination of volume percentage of open and closed cells of rigid materials", *International Organization for Standardization*, Switzerland.

16. ASTM D 1622-88, 1993, "Standard Test Method for Apparent Density of Rigid Cellular Plastics", *1993 Annual Book of ASTM Standards*, American Society for Testing and Materials, Philadelphia, Vol 08.01, pps. 491-493.

17. Bhattacharjee, D., P. W. Irwin, J. R. Booth and J. T. Grimes, 1994, "The Acceleration of Foam Aging by Thin Slicing-Some Interpretations and Limitations", *J. Thermal Insulation and Bldg. Envs.*, 17: 219-237.

# Influence of Moisture and Air Movement in Porous Materials on Thermal Insulation

H. STOPP, J. GRUNEWALD and P. HÄUPL

## ABSTRACT

The heat transmission through a moist porous material consists of the following components:
- the well-known moisture-dependent thermal conductivity $\lambda(w)$
- the enthalpy flow coupled with vapour and water mass-flow
- the latent enthalpy flow caused by phase changes
- the air enthalpy flow in case of air movement through the material

In this paper the model of the coupled heat, air and moisture transport in building structures and the results of the numerical solution are presented. Under typical meteorological climate conditions (test reference year: hourly values of temperature, relative humidity, direct and diffuse short wave radiation, longwave radiation balance, wind velocity, wind direction and rain) the moisture behaviour and the total heat losses for multilayered envelope components are analyzed. Moreover, the influence of moisture content and moisture movement on the thermal conductivity measurement (by means of a hot-plate-apparatus and a needle probe) and on the thermal resistance measurement (active u-values method ) is quantified.

## INTRODUCTION

There is no doubt that thermal insulation reduces energy consumption. Specifying how much to use requires knowing of the thermal quantities under service and natural climate conditions, because just the influence of moisture increases heat losses and causes damages of structures. [1]

Many standards and thermal measuring methods are available to determine and to control the thermal performance of buildings and building components, but in most cases they are only applicable for dry materials. In practice the conditions include driving rain, condensation within the structures and on their surfaces. To simulate such processes, a complete model [2] and a relevant software [3] for the nonlinear coupled heat- and mass transfer within the capillary porous materials were developed.

Horst Stopp, FH Lausitz, 01968 Senftenberg, John Grunewald and Peter Häupl, TU Dresden, 01062 Dresden, Germany

## MODELING OF COUPLED HEAT MOISTURE AND AIR TRANSFER

In the following, the heat conduction equation including transport functions [4] of the complete transport equation system is listed.

### TRANSPORT FUNCTIONS

$$\lambda = \lambda(w, w_i, T)$$
$$a_W = a_W(w, w_i, T)$$

$$a_i = 0$$
$$a_g = a_g(w, w_i, T)$$
$$\delta = \frac{D(T)}{\mu(w, w_i, T)R_V T}$$
$$l_A = l_A(w, w_i, T)$$

- thermal conductivity in W/(mK)
- capillary water diffusivity in m²/s
- no ice movement
- gravitative moisture diffusivity in s
- vapour conductivity in s
- permeability for air/vapour-mixture in m³s/kg

### HEAT CONDUCTIVITY EQUATION

$$\frac{\partial q_k}{\partial x_k} + \frac{\partial}{\partial t}\left[\frac{dU_{source}}{dV}\right] = \frac{\partial}{\partial t}\left[\frac{dU}{dV}\right]$$

- energy conservation theorem

$$\frac{\partial q_k}{\partial x_k} =$$

$$+ \frac{\partial}{\partial x_k}\left[\lambda \frac{\partial \mathbf{T}}{\partial x_k}\right]$$

$$+ \frac{\partial}{\partial x_k}\left[L_V \delta \frac{\partial \mathbf{p_V}}{\partial x_k}\right]$$

$$+ \frac{\partial}{\partial x_k}\left[c_{pV}\mathbf{T}\delta \frac{\partial \mathbf{p_V}}{\partial x_k}\right]$$

$$+ \frac{\partial}{\partial x_k}\left[c_W\mathbf{T}\rho_W a_W \frac{\partial \mathbf{w}}{\partial x_k}\right]$$

$$+ \frac{\partial}{\partial x_k}\left[c_W\mathbf{T}\rho_W a_g \mathbf{g_k}\right]$$

$$+ \frac{\partial}{\partial x_k}\left[c_{pA}\mathbf{T}\rho_A \cdot \dot{V}_{k,AV}\right]$$

$$+ \frac{\partial}{\partial x_k}\left[c_{pV}\mathbf{T}\rho_V \cdot \dot{V}_{k,AV}\right]$$

$$+ \frac{\partial}{\partial x_k}\left[L_V \rho_V \cdot \dot{V}_{k,AV}\right]$$

- local divergence of heat flow density

- classical heat flow caused by temperature gradient
- latent heat flow caused by vapour diffusion
- enthalpy flow caused by vapour diffusion

- enthalpy flow caused by capillary water transport
- enthalpy flow caused by gravitative water transport

- enthalpy flow caused by air-vapour-mixture-flow

$$\frac{\partial}{\partial t}\left[\frac{dU}{dV}\right] = \frac{\partial}{\partial t}\left[\rho_M c_M \mathbf{T}\right]$$

$$+ \frac{\partial}{\partial t}\left[\rho_w c_w \mathbf{Tw}\right]$$

$$+ \frac{\partial}{\partial t}\left[\rho_i c_i \mathbf{Tw}_i\right]$$

$$+ \frac{\partial}{\partial t}\left[\rho_v c_{pV}\mathbf{T}(w_s - \mathbf{w} - \mathbf{w}_i)\right]$$

$$+ \frac{\partial}{\partial t}\left[\rho_A c_{pA}\mathbf{T}(w_s - \mathbf{w} - \mathbf{w}_i)\right]$$

$$+ \frac{\partial}{\partial t}\left[\rho_v L_v\left(w_s - \mathbf{w} - \mathbf{w}_i\right)\right]$$

$$- \frac{\partial}{\partial t}\left[\rho_i L_i \mathbf{w}_i\right]$$

- storage terms of heat

$$w = \frac{V_{Water}}{V_{Material}} \qquad w_i = \frac{V_{Ice}}{V_{Material}}$$

$$w_h, w_s$$

$$p_V, p_s$$

$$p_V = p_s(T) \qquad\qquad w_h \le w < w_s$$

$$p_V = S_u(w,T)\, p_s(T) \qquad 0 \le w < w_h$$

$$\cdot \varphi = S_u(w,T)$$

$$w = S(\varphi, T) = S\!\left(\frac{p_D}{p_s}, T\right)$$

$$\frac{\partial p_V}{\partial x_k} = p_s(\mathbf{T})\left(\frac{\partial S_u}{\partial \mathbf{w}}\cdot\frac{\partial \mathbf{w}}{\partial x_k} + \frac{\partial S_u}{\partial \mathbf{T}}\cdot\frac{\partial \mathbf{T}}{\partial x_k}\right)$$

$$+ S_u(\mathbf{w},\mathbf{T})\frac{\partial p_s}{\partial \mathbf{T}}\frac{\partial \mathbf{T}}{\partial x_k}$$

- moisture and ice content in $m^3/m^3$ or Vol%
- hygroscopic and saturated moisture content
- partial and saturated pressure of vapour in Pa
- vapour pressure in the overhygroscopic area
- vapour pressure in the hygroscopic area
- humidity and inverse function of sorption isotherm
- sorption isotherm

- local gradient of the partial pressure

$$\dot{V}_{k,VA} = 1_{VA}\frac{\partial p_{VA}}{\partial x_k}$$

- air/vapour flow caused by total pressure gradient

## EQUATIONS FOR THE FREEZING POINT LOWERING

$w = w_{u,max}(T) \quad T = T_f$

$w_i = 0 \qquad T > T_0$

$w_{u,max}(T) = S(\varphi) = S\left[\exp\left(\dfrac{p_w(T)}{\rho_w R_D T}\right)\right] \qquad 0 \le w < w_h$

$w_{u,max}(T) = w(r) = w\left[-\dfrac{2\sigma(T)}{p_w(T)}\right] \qquad w_h \le w < w_s$

$$p_w(T) = \rho_w \left( \begin{aligned} &\dfrac{L_i}{T_0}(T - T_0) - (c_w - c_{i1})(T - T_0) \\ &+ (c_w - c_{i1} + c_{i2}T_0)T\ln\left(\dfrac{T}{T_0}\right) \\ &- \dfrac{c_{i2}}{2}(T^2 - T_0^2) \end{aligned} \right) \qquad T \le T_0$$

$c_i = c_{i1} + c_{i2}(T - T_0)$

- $T_f$    freezing temperature lowered by capillary suction

- $w_{u,max}(T)$    maximum unfrozen water content and dependence on temperature, sorption isotherm and capillary suction

- $p_w(T)$ additional suction in unfrozen water
- $L_i$    special melting heat for ice by 0°C
- $T_0 = 273.15$ °C

- $c_i$    specific thermal capacity of ice

## INFLUENCE OF THE MASS TRANSFER ON THE THERMAL TRANSMISSIVITY OF STRUCTURES

To characterize the heat losses of building envelopes, the overall heat transfer coefficient or u-value is used: $u = \dot{q}/\Delta\vartheta$. There are $\dot{q}$ the average heat flow density at the inner surface and $\Delta\vartheta$ the average temperature difference between indoor and outdoor. In the following constructions (fig. 1 .. 6) we show u-values for dry structures and for moist structures without and with moisture transport and with air transport, respectively, [6], [7].

**LIGHTWEIGHT WALL:**

Fig. 1   demonstrates the construction: 12mm gypsum board , 8mm plywood, 85mm mineral wool ,8mm plywood, 20mm PS-foam sheathing, 5mm outside plaster

Fig. 2   represents the moisture distribution in the tenth year (viewed from the inside). Basically condensation water has formed on the cold side of the mineral wool layer and penetrates into the plywood layer. The mineral wool itself is nearly dry. Driving rain ( peaks at the outside) wets the outside mortar but hardly penetrates into the polystyrene foam sheathing .

Fig. 3   shows the u-values during the heating period (October to April). By contrast with the dry structure, the whole heat loss (but without air moving) is more than 12% higher because the lightweight structure is vapour permeable.

**CELLULAR CONCRETE WALL:**

15 mm lime mortar , 300mm cellular concrete , 8mm outside mortar

Fig. 4   represents the moisture distribution of the west wall (viewed from the inside) in the tenth year. Although the outside mortar is relatively watertight ($\mu$=100, $a_{w,max}$=1 · $10^{-10}$m$^2$/s) some driving rain (peaks at the outside mortar) penetrates into the structure and   condensation water forms between the cellular concrete and outside mortar .

Fig. 5   indicates the  whole moisture content for the cellular concrete wall over the course  of ten years. The moisture values change between 7.2kg/m$^2$ (March) and 5.7kg/m$^2$ (Sept.).

Fig. 6   shows the u-values during the heating period for different cases. The highest value (balk 4) with all effects (moisture content, enthalpy flow caused by moisture and a small air flow through the structure) is 0.582W/m$^2$·K, the lowest value is 0.510W/m$^2$K (dry material),  a difference of 14%.

## DETERMINATION OF THE THERMAL RESISTANCE AND CONDUCTIVITY OF MOIST MATERIALS

To determine the thermal transmissivity and resistance of structures by means of measurement, a temperature gradient is necessary. If there is no gradient caused by natural climate conditions, a heating foil at the inner surface must be used (active u-value measurement method).

**WET CONCRETE WALL:**

Fig. 7   demonstrates the moisture field for a third multi-layer structure (compare fig. 9) during 1st/2nd year after its manufacture (construction moisture included, location Bannewitz).

Fig. 8   represents the change of the u-value over the course of 12 years. During the first years the u-value is about 30% higher than later, after   natural drying.

Fig. 9   shows the changing moisture profile due to the heating foil (heat power 70W) during the measurement. At the beginning of the measuring time (Feb. 1st, first heating period) the u-value is 0.42W/m$^2$·K (compare fig. 8 ) and at the end (March 2nd) it increases to 0.60W/m$^2$·K because the moisture has shifted partly into the insulation layer.

By means of the well-known guarded hot plate method in moist materials, a heat transfer coefficient is determined [9]. It is effected by the thermal conductivity itself, phase-change and moisture movement as long as moisture is shifted. The equilibrium is reached when the heat flow density is temporally constant. The duration of the field altering depends on the vapour resistance factor $\mu$ and, in the case of overhygroscopic moisture content, on the capillary diffusivity $a_w$.

**MINERAL WOOL SPECIMEN:**

Fig. 10   The time dependent moisture field is demonstrated for a specimen of mineral wool with a thickness of 20mm sandwiched between the hot and cold plate of a guarded hot plate apparatus (temperature difference 5K, initial moisture content 10vol%).

Fig. 11   The figure 11 shows the heat flow densities of the same sample over 16 days depending on the hygric transport coefficients. Although the moisture content for all three cases is 10 vol%, the heat flow densities and also the measured $\lambda$-values are very different.

For measuring of the thermal performance of erected buildings, short time measurement methods are necessary. In the past we developed a λ-needle probe using a nearly point-shaped and constant heated sensor fixed at the top of an about 120mm long needle [8].

**CALCIUMSILICATE SPECIMEN:**

Fig. 12   shows the shifted moisture in a sample of calciumsilicate (initial moisture content 10vol%) after a measuring time of 10min caused by a heating power of 0.15W for a needle with a diameter of 1.5mm. Directly at the surface of the needle the moisture content decreases to 9.5vol%, at a distance of 1mm from the needle the moisture changing is less than 0.2vol%. However, the temperature signal reaches a distance of about 10mm in the same time. Hence in this case, the heating influence is limited and the needle probe measures the λ-value for 10 vol%.

## CONCLUSIONS

The determination of the heat transfer in moist materials and structures requires the knowledge of the hygric transport coefficients. Given accurate coefficients, a numerical simulation is an appropriate method to quantify all effects on the thermal performance. Examples shown in this paper have the thermal transmissivity under usual climate conditions that are about 15% higher (4/5 caused by moisture content, 1/5 caused by mass transfer) than for dry structures.

## REFERENCES

[1]   Häupl, P.; Fechner, H.; Stopp, H.: Study of driving rain, T2-D-94/02. IEA-Annex 24 - Hamtie, Meeting Rom, Italy, October 1994

[2]   Häupl, P.; Stopp, H.; Strangfeld, P.; Fechner, H.: Vergleich gemessener und berechneter Feuchteverteilungen bei innerer Kondensatbildung in Baustoffproben. Bauphysik 16 (1994), H. 5, S. 138 -147

[3]   Fechner, H.; Grunewald, J.; Häupl, P.; Stopp, H.: Zur numerischen Simulation des gekoppelten Feuchte-, Luft- und Wärmetransportes in kapillarporösen Baustoffen. 9. Bauklimatische Symposium, Sept. 94, TU Dresden, Germany, Bd. 1, S. 125 - 141

[4]   Häupl, P.; Stopp, H.: Feuchtetransport in Baustoffen und Bauwerksteilen. Dissertation B, Bd. 1 - 5, TU Dresden, Germany, 1987

[5]   Stopp, H.; Häupl, P.; Fechner, H.; Neue, J.: Use of the thermal measurement methods within moist building materials. Int. symposium Non-Destructive Testing (NDT-CE), Sept. 95, Berlin, Germany, Vol. 1, p. 365 - 374

[6]   Häupl, P.; Stopp, H.; Fechner, H.; Grunewald, J.: Moisture conditions of non-ventilated wood-based membrane roof components, part 1,2, T1_D-94/02. IEA-Annex 24 - Hamtie, Meeting Trondheim, Norway, April 1994

[7]   Häupl, P.; Grunewald, J.; Xu, Y.: Different parts of heat losses within the moist structures. 4th congress, Building Physics of China, Xi-an, May 1995

[8]   Häupl, P.; Stopp, H.: Application of double-wires heated thermocouples for the measurement of thermal conductivity in solid and bulk materials. IMEKO TC 12, Temperature measurement in industry and science, 2nd symposium, October 84, Suhl, Germany, Vol. 2, p. 413 - 422

[9]   Sandberg, P. J.: Determination of the thermal conductivity of moist masonry materials. Building Physics in the nordic countries, Sept. 93, Copenhagen, Denmark, Vol. 1, p. 401 - 409

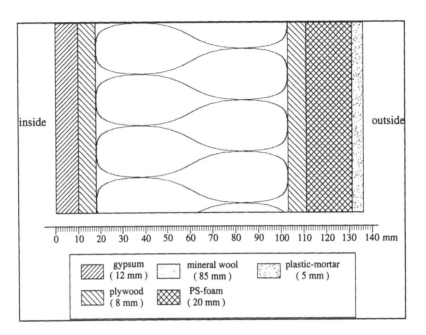

Figure 1    Construction of the lightweight wall

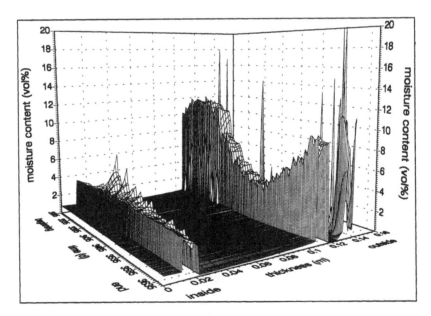

Figure 2    Moisture field of the lightweight-construction (acc. Fig. 1) in the
10th year, Climate: TRY Essen westwall

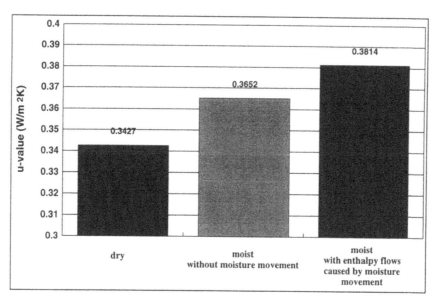

Figure 3    U-value of the lightweight-construction (acc. Fig. 1) in the heating
            period, Climate: TRY Essen westwall

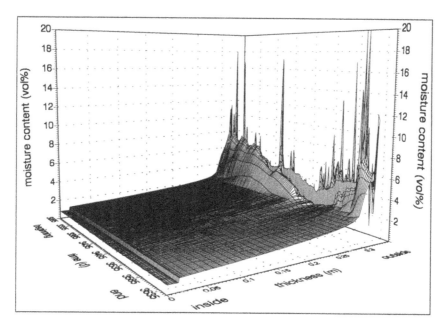

Figure 4    Moisture field of a cellular concrete wall in the 10th year
            Materials: inside mortar (15mm) - cellular concrete (300mm) - outside
            mortar (8mm), Climate: TRY Essen westwall

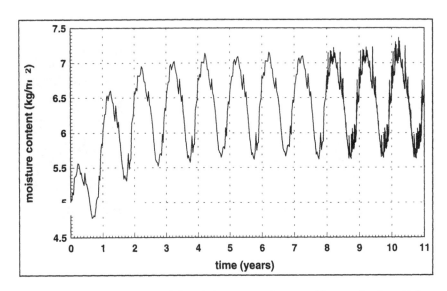

Figure 5    Moisture content of the whole construction over 11 years for the cellular concrete wall (acc. Fig. 4), Climate: TRY Essen westwall

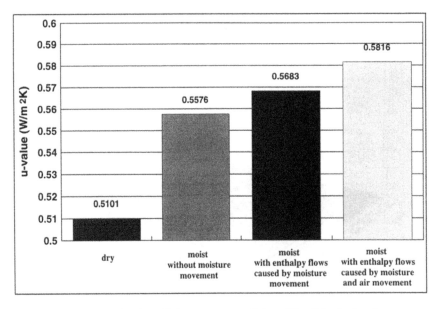

Figure 6    U-values for the cellular concrete wall (acc. Fig. 4) in the heating period, Climate: TRY Essen westwall

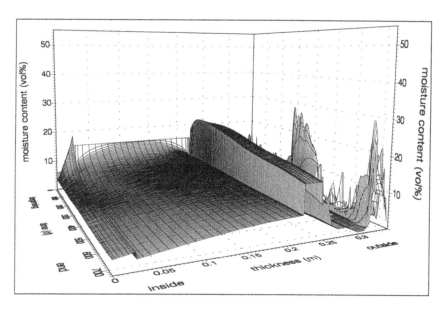

Figure 7    Moisture field in the 1st and 2nd year for the location Bannewitz
Materials: chip board (25mm) - concrete (200mm) - chip board (25mm)
- PUR-foam (80mm) - outside mortar (20mm)
Climate: TRY Essen westwall

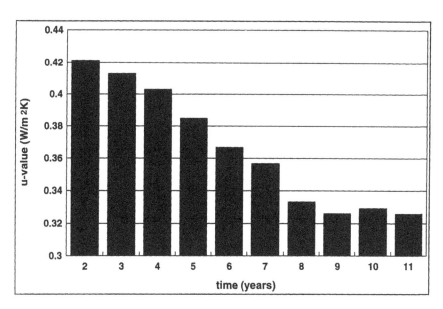

Figure 8    U-values during the first 11 heat periods (october-april) for the wall-
construction on the location Bannewitz (acc. Fig. 7)

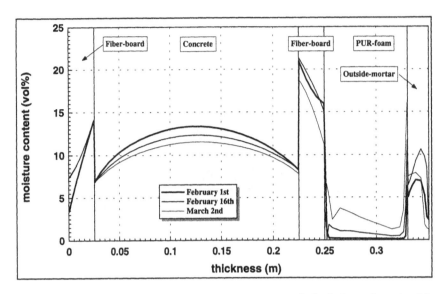

Figure 9    Moisture movement in the wet concrete wall (built-in moisture) with outside insulation (location Bannewitz acc. Fig. 7) caused by active u-value measurement method during the first heating period

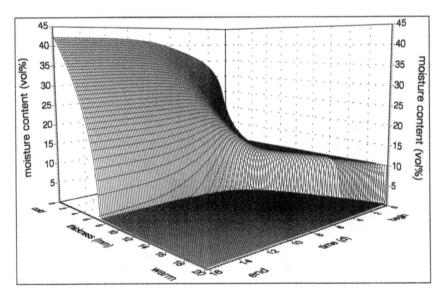

Figure 10   Moisture distribution within a mineral wool specimen during the λ-measurement in a guarded hot-plate-apparatus, $\Delta T = 5K$

(w(0)=10vol%, $\lambda_{dry}$=0.04W/mK, $a_{w,max}$=4.0E-11m$^2$/s, $\mu$=1.5)

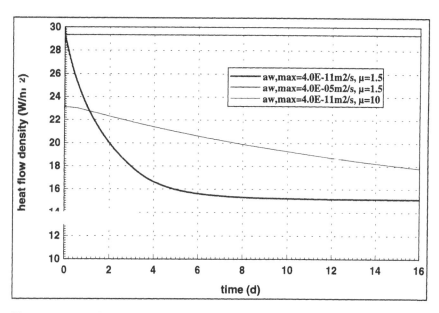

Figure 11    Heat flow densities of a mineral wool specimen, acc. fig. 10
the capillary conductivity and the vapour conductivity have been varied

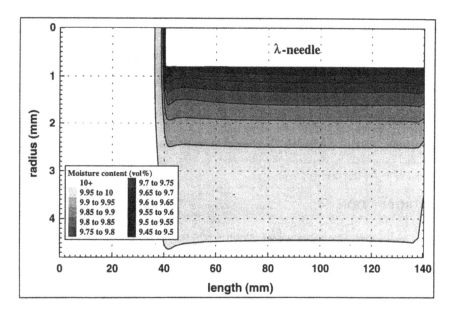

Figure 12    Moisture field in surrounding of the λ-needle-probe (l=100mm) after
10min heating time, (P=0.15W, w(0)=10vol%, μ=6, $a_{w,max}=1.0 \cdot 10^{-7} m^2/s$,
$\rho=230kg/m^3$)

# Heat and Moisture Transfer in Fiberglass Insulation under Air Leakage and Frosting Conditions

D. R. MITCHELL,[1] Y.-X. TAO[2] and R. W. BESANT[1]

## ABSTRACT

This paper summarizes a study of combined heat and moisture transfer in sample fiberglass insulation boards using both experimental measurements and numerical modeling. Special attention is given to the conditions where moist air passes through the insulation, simulating air leakage, with a sub-freezing temperature boundary condition at one side of the insulation board. Measurements include temperature, moisture content, air flow rate through the insulation and heat flux with the temperature range from 20 °C to -20 °C. The effects of air exfiltration and infiltration on the heat and moisture transport characteristics within a medium density fiberglass insulation material are investigated. The experimental results, which are typical of cold climate building envelope applications, indicated that for all of the air exfiltration tests carried out, the majority of the moisture and frost accumulation was within the insulation slab adjacent to the cold surface. For air infiltration, it was discovered that the drying rate was substantially higher for lower airflow rates. The numerical results agree reasonably well with the experimental results for exfiltration cases. For infiltration, a relatively large discrepancy between prediction and test data exists. It is expected that thermal instability, because the thermal gradient is opposite in direction to the moisture gradient, is partly responsible for this discrepancy.

## INTRODUCTION

In the past few decades, thick layers of fiberglass insulation have been used to conserve energy within building and refrigerated space envelopes. Failure to eliminate the natural convection of air and moisture diffusion within typical fiberglass insulated roof structures in a cold climate has been shown to reduce the apparent thermal resistance of the insulation by up to 50% [1]. The accumulation of moisture not only increases the energy transfer across the insulation, it can also lead to mold growth, degradation of any organic materials such as wood, rust damage to any metallic components, freezing damage in the condensation zone at low temperatures,

---

[1]Department of Mechanical Engineering, University of Saskatchewan, Saskatoon, Canada
[2]Department of Mechanical Engineering, Tennessee State University, Nashville, TN, USA

as well as an overall reduction in the quality of the insulation [2]. To inhibit the flow of air and moisture through the wall structure a vapor retarder is a required building construction detail in cold climates, especially where the insulation systems operate below freezing conditions [3]. However, in practice, the consequences of the frequently observed failure of the vapor retarder in air-tight structures due to installation problems (among other factors) are well known in the design and consulting communities [4]. Moisture transport within building wall and ceiling insulation generally takes place either through the process of vapor diffusion, or as a result of the flow of moist air through the insulation. In addition, adsorption and desorption can occur [5]. Although there is a large accumulation of literature on combined heat and moisture transfer in fiberglass insulation, the laboratory tests that quantify the effect of air movement on moisture and frost accumulation for cold climate applications, are still limited. There is also a need to provide experimental data for validation of numerical models in this area.

We have performed a series of tests and modeling to investigate the effects of both air exfiltration (which results in the accumulation of moisture and frost within the porous insulation inside wall cavities), and air infiltration (which may tend to remove moisture / frost from some regions of an insulation slab and deposit this moisture as condensation or frost elsewhere). In the following, we briefly summarize the experimental and numerical methods and concentrate on the discussion of the measured and predicted results.

## EXPERIMENTS

A macroscopically uniform, one-dimensional flow of air through a porous fiberglass insulation slab was established in an apparatus. The air flow was to simulate exfiltration and infiltration conditions through a house wall or ceiling during cold weather (e.g., internal environment: moist air at +20°C; external environment: dry air at -20°C). Although typical indoor relative humidity conditions during cold winter weather rarely exceed 50% RH in most buildings, the exfiltration tests were conducted with warm, room temperature air with a high relative humidity (60-90% RH). These high relative humidities were used in order to obtain experimental data with small relative errors and to magnify the effects that varying the relative humidity might have on the heat, moisture and air transport characteristics within the porous insulation slab. It should be noted, however, that for certain buildings and rooms (some houses, animal barns, indoor swimming pools, hospital operating rooms, computer rooms, etc.) the internal relative humidity can reach 50% or higher during cold weather.

A schematic drawing of the experimental test section is provided in Fig. 1. The test insulation slab is placed horizontally and cooled from below. This configuration allows us to exclude buoyancy flow from the model. For the air exfiltration tests warm air was supplied to the air inlet ports at the top of the test cell (eight are provided in the top cover plate) and flowed uniformly down toward the cold plate on the bottom. For the air infiltration tests cold air was supplied to the air gap through eight other inlet ports located in the bottom air gap region and the air flowed uniformly up toward the top cover plate which was at room temperature.

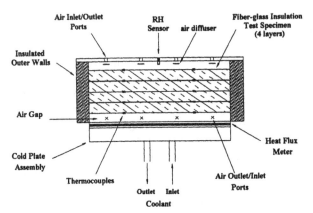

Figure 1 Schematic Drawing of Experiment Test Section

A four-layer, medium-density ($50 \ kg/m^3$), glass-fiber slab with a total dimension of 275 x 600 x 90 mm was placed in the test section for each test. A cold plate, which could be cooled well below the triple point temperature of water through a heat exchanger, was located at the bottom of the test section. By placing nine small plastic disc spacers on top of the cold plate, an air gap was created between the bottom surface of the insulation slab and the cold plate (approximately 12 mm in height). This permitted the exfiltration or infiltration air to flow uniformly through the insulation slab. To start a test, the volumetric airflow rate was set using a mass flow controller, and the ethylene glycol - water solution, which was used as the heat exchanger coolant for the cold plate, was pumped from a storage tank which is placed in an environmental chamber.

The range of air flow velocity used in this study was established based on a field investigation by Shaw [6]. This investigation found that a typical detached Canadian residential house experiences an air exfiltration rate of 0.98 L/(m²·s). For a typical house size this means an average air speed through the building envelope filled with insulation of about 1.0 mm/s. According to Ogniewicz and Tien [7], air velocities encountered in typical porous insulation applications, due to free convection and air infiltration, are also of the order of 1.0 mm/s. In this study, tests were carried out with the average air flow velocity through the insulation at 0.5 mm/s, 1.0 mm/s, and 1.5 mm/s. The detailed information about the apparatus can be found elsewhere [8].

## NUMERICAL MODEL

A schematic drawing detailing the geometry for the numerical model is provided in Fig. 2. The phenomena occurring within the insulation are assumed to be one-dimensional; i.e., the vertical side walls are treated as adiabatic and impermeable and the air flow is only in the z direction. Since the experimental work has been conducted to minimize or eliminate any two-dimensional free convection effects occurring within the insulation slab, a one-dimensional numerical model is taken to be sufficient.

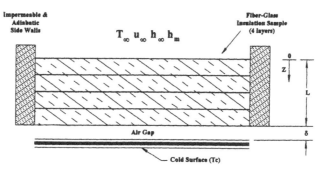

Figure 2  Geometry Used in the Numerical Modeling: $T_\infty$=ambient temperature, $u_\infty$=air velocity, $h_\infty$=heat transfer coefficient, and $h_m$ =mass transfer coefficient.

The local volume averaging technique [9] is applied to the basic equations of conservation of mass, momentum, and energy transfer in order to establish governing equations which describe the processes occurring within the insulation on a macroscopic level. Using this method, the local properties in the equations are averaged over a representative elementary volume; i.e., the smallest volume that represents the local average properties. For many practical porous media problems, the solution of the volume averaged transport equations is considered adequate [10]. Furthermore, the assumption of local thermal equilibrium is a common simplification procedure to tackle problems involving heat and mass transport in porous material [11]. Detailed information pertaining to the theory involved in the local volume averaging technique can be found in Kaviany [10] and Whitaker [11].

For the numerical simulations, the liquid capillary velocity is assumed to be zero. According to Vafai and Tien [12] the liquid contained within a porous insulation is essentially immobile, due to a lack of interpore connections, if the level of saturation is below 0.10; i.e., the water condensate tends to exist in a discontinuous pendular state. For the work presented here, the saturation never exceeds this level, therefore this assumption is considered valid. Furthermore, since the glass fibers within the insulation has a low volume fraction and the flow velocity through the insulation is low (0.5 to 1.5 mm/s), thermal dispersion and mass dispersion are neglected. It is also assumed that in the frosted or ice crystal region the frost does not significantly change the homogeneity of the porous matrix. Finally, the analysis is restricted to a rigid porous matrix with constant properties for each constituent and no chemical reactions taking place except those related to hygroscopicity.

The numerical model is intended to model transient one-dimensional processes of simultaneous heat and mass transfer within a porous media (insulation material) subject to pressure driven gas flow and phase changes (condensation / evaporation, sublimation / ablimation, and adsorption / desorption). The model is based, in part, on one presented by Tao et al. [13,14]. The model was modified to include airflow through the porous insulation [9]. The details regarding the governing equations, initial and boundary conditions can be found in Mitchell [8].

Heat transfer taking place within the fiberglass porous media occurs mainly through conduction and gaseous phase convection. Some radiation effects may have occurred within the insulation, especially in the gap regions near the bottom or top surfaces of the slab (since the insulation is very porous). However, these effects are neglected in this study because they are expected to be less than 10% with respect to conduction and convection [15]. Furthermore, an attempt was made to reduce any radiation effects near the top surface of the sample by coating the inner surface of the test apparatus with aluminum foil.

## RESULTS AND DISCUSSION

### AIR EXFILTRATION

Validation of the numerical models for both air exfiltration and air infiltration was obtained by comparing the numerically predicted temperature and moisture profiles to those obtained experimentally. The experimental temperature measurements were obtained for five positions across the insulation sample ($z/L$ = 0.00, 0.25, 0.50, 0.75, 1.00). Since the moisture accumulation distribution was measured at the end of each 1, 2, or 3 hour test, continuous moisture accumulation versus time profile comparisons were not possible. Comparisons between the numerical and experimental moisture accumulation profiles for each layer within the insulation sample are performed based on the results obtained at the end of each test period.

Figures 3 and 4 are comparisons of the simulated and measured data for temperature and moisture accumulation for one typical case, 75% inlet air relative humidity ($\phi$), and 1.0 mm/s air speed ($u_g$). The air exfiltration numerical temperatures from the model are well within the experimental uncertainty of ±0.2K for temperature and ±1.0K in set point temperatures at the two boundaries. The exfiltration of warm air causes warmer temperatures throughout most of the insulation except near the bottom surface where a very large temperature gradient exists. The maximum temperature discrepancy for all of the air exfiltration experiments was less than 1.0K for a temperature step decrease of 30K at $z = L$.

Each experimental data point in Fig. 4 represents a ratio of the mass of the accumulated moisture within each insulation layer divided by the dry mass of each layer of insulation. Figure 4 indicates that the majority of the moisture accumulation occurred within the insulation layer closest to the cold surface. With the temperatures within the bottom layer dropping below the saturation temperature, based on the vapor density at such locations, condensation or ablimation takes place. Although moisture adsorption is evident by the temperature rise within the upper regions of the slab immediately after the start of the experiment, the amount of moisture accumulation was much less than within the condensation region (since adsorption deals with the deposition of moisture onto the fiber surfaces in molecular layers).

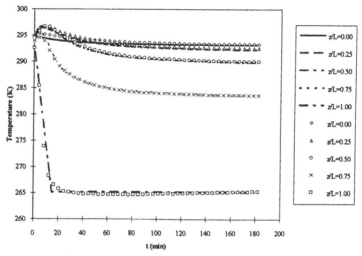

Figure 3   Air Exfiltration Temperature Profile : $\phi$ (z=0) = 75%, $u_g$ = 1.0 mm/s.

For these test conditions, the numerical model underpredicts the moisture accumulation within the bottom layer of insulation (the condensation and frosting region). This is partially explained by noting that the numerical model only dealt with the insulation slab itself (z/L =0.00 to z/L =1.00). It does not take into account any moisture/frost which might have accumulated on the bottom surface of the slab (towards the cold plate). Although the temperature comparisons agreed at the five specified locations as shown in Fig. 3, we would not know if the predicted temperature profile within the bottom layer of the insulation slab differed from experimental data at other points without additional temperature sensors.

Radiative heat flux between the cold plate surface and the bottom surface of the insulation slab could have contributed to the additional moisture and frost accumulation measured experimentally. Calculations were carried out, but are not presented here, to estimate the expected radiation flux within the air gap and the effect that this radiation might have on the frost accumulation within this gap. Radiative heat flux is expected to be a primary mode of heat transfer in the gap and this heat flux is expected to be mostly balanced by frost growth on both the cold plate surface and the bottom surface of the insulation slab.

This trend of having the maximum amount of moisture accumulation near the cold surface is similar to the test results obtained with no exfiltration airflow (vapor diffusion only) by Tao et al. [14]. In another study by Tao et al. [13], it was also found that during the initial period, more moisture accumulated near the warm side of the insulation slab while later more moisture and/or frost was found near the cold side of the slab. It should be noted that for operating temperatures above the freezing point, moisture accumulation is not significant even with air infiltration/exfiltration. Vafai and Tien [12] reported, from their numerical results, a liquid volume fraction of the order of $10^{-4}$ with the operating temperatures above

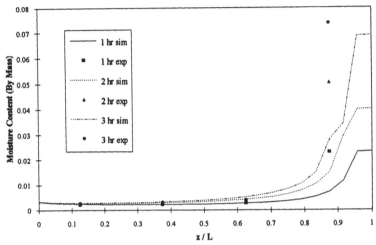

Figure 4 Moisture Profile: Air Flow Rate - 10 l/min, Air Relative Humidity - 75 %

the freezing point. With frosting, however, the air exfiltration effect caused about 45 times as much moisture accumulation; the moisture content reached 0.09 by mass or about $4.5 \cdot 10^{-3}$ liquid volume fraction for RH $\phi = 90\%$ and $u_g = 1.0$ mm/s case.

Calculations were made to determine the heat flux leaving the bottom of the insulation slab based on the experimental results. However, due to radiative heat flux within the air gap between the cold plate and the insulation slab and frost growth on the upper and lower surfaces of the gap, the uncertainty in the heat flux at $z/L = 1.0$ computed from the measured data was nearly 100%. Therefore, the heat flux values were obtained instead from the numerical simulation. The conductive heat flux ($q''_{cond}$) leaving the bottom surface of the insulation slab ($z / L = 1.0$) was determined using the following equation:

$$q''_{cond}(L) \quad = \quad -k_{eff} \frac{\partial T}{\partial x}\bigg|_{Z=L} \tag{1}$$

The effective thermal conductivity ($k_{eff}$) was determined using values for the porosity of the insulation ($\varepsilon$), the degree of saturation at $z = L$ (which is a function of the liquid or frost volume fraction, $s = \varepsilon_\beta / \varepsilon$), and the thermal conductivity of the three individual phases; liquid or frost, gas, and solid (fiberglass) ($k_\beta, k_g, k_s$) [13]. The temperature gradient at the bottom surface of the insulation was calculated using the backward differencing approximation.

The numerical results were also used to obtain the heat flux results in relation to the convection process over the entire insulation slab. The following equation is used to evaluate the sensible convective heat transfer rate per unit flow area ($q''_{conv}$) for the air not involved in the phase transfer process:

$$q''_{conv} = \frac{\dot{m}\, c_p\, \Delta T}{A} \tag{2}$$

where $\Delta T$ is the temperature difference between the top of the insulation slab and the bottom, $T(z = 0) - T(z = L)$, m is the flow rate, $c_p$ is the specific heat of air and A is the cross-sectional area of the fiberglass slab.

For air exfiltration through the envelope of a building, the magnitude of the convective heat flux term represents the amount of energy which would be required to heat the makeup ambient air required to compensate for the amount of air lost through the wall structure during air exfiltration. A heat flux calculation was also performed for a dry insulation sample with no air exfiltration or vapor diffusion (based on the same boundary conditions). This value was calculated using equation (1), with the effective thermal conductivity based on a dry insulation slab, and the temperature gradient taken across the entire insulation slab. A summary of these results is presented in Table I.

Table I  Numerical Heat Flux Results (2 hr data) - Air Exfiltration

| Flow Rate (L/min) | RH (%) | $q''_{cond}$ (L) $(W/m^2)$ | $q''_{conv}$ $(W/m^2)$ | $q''_{total}$ $(W/m^2)$ | $q''_{dry-noflow}$ $(W/m^2)$ | $q''_{total}$ / $q''_{dry-noflow}$ |
|---|---|---|---|---|---|---|
| 5 | 75 | 32.8 | 20.8 | 53.6 | 10.4 | 5.2 |
| 10 | 60 | 40.0 | 33.4 | 73.4 | 9.17 | 8.0 |
| | 75 | 44.0 | 35.7 | 79.7 | 9.03 | 8.8 |
| | 90 | 49.4 | 35.2 | 84.6 | 8.95 | 9.5 |
| 15 | 60 | 49.1 | 49.7 | 98.8 | 8.43 | 11.7 |
| | 75 | 55.6 | 47.7 | 103.3 | 8.11 | 12.7 |
| | 90 | 60.4 | 46.0 | 106.4 | 7.85 | 13.6 |

For air exfiltration it is evident that increasing the airflow rate or the inlet air relative humidity results in an increase in the conductive heat flux through the bottom surface of the insulation slab. It is also apparent that increasing the air flow rate had a more significant effect on the total heat flux than increasing the inlet air RH. For example, increasing the inlet air RH from 60 to 90%, resulted in a 23.0% increase in the total heat flux for a flow rate of 10 L/min ($u_g$= 1.0 mm/s), and a 23.5% increase in the total heat flux for a flow rate of 15 L/min ($u_g$= 1.5 mm/s). Increasing the air flow rate from 5 to 15 L/min ($u_g$= 0.5 to 1.5 mm/s) for an inlet air RH of 75% resulted in a 70% increase in the total heat flux. The results also indicate that, in comparison to the convection heat flux values, the conductive heat flux through the bottom of the insulation was of similar magnitude. Furthermore, Table I shows that the total heat loss due to air exfiltration through the insulation was more than 13 times that for a sample of insulation under similar boundary conditions, which had no moisture accumulation and experienced neither vapor diffusion nor airflow.

It should be noted that for these simulations, the modeling parameters (properties of all phases involved, the grid spacing, and all of empirical constants) were kept constant, however, as discussed above, there is uncertainty in these heat flux results and it is possible that some of these values might not have been completely accurate for the system under study. Nonetheless, the trends indicated in Table I are believed to be quite relevant to typical building envelope applications.

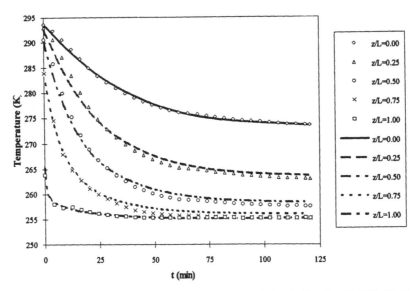

Figure 5 Temperature Profile: Air Flow Rate - 10 L/min, Exfiltration Air RH - 75 %

## AIR INFILTRATION

The graphs shown in Figs. 5 and 6 compare the experimental and numerical temperature and moisture accumulation results for air infiltration with the flow velocity at $u_g$ = 1.0 mm/s (10 L/min) and the exfiltration inlet air RH at 75%. Comparisons for all of the test conditions studied indicated a maximum temperature discrepancy between the experimental and numerical results of less than 3K for a temperature difference of 33K across the slab of insulation.

Figure 6 indicates that, in the upper layer of the insulation slab (z/L=0 to z/L=0.25) the moisture accumulation was over predicted. In the remaining layers of insulation the predicted moisture accumulation values agreed well to those obtained experimentally. The discrepancy between the data and the simulation in the top layers of insulation could be reduced and possibly eliminated by varying the diffusion coefficient across the insulation slab, decreasing the diffusion for small values of z. During the development of this model, it was discovered that for air infiltration, the diffusion coefficient had a significant influence on the resulting temperature and moisture accumulation profiles. Although a value of $0.26 \times 10^{-4}$ $m^2/s$ (based on a binary diffusivity between air and water vapor) performed quite well for the air exfiltration simulations, it did not work well for air infiltration where sublimation occurred in the bottom layer and ablimation or condensation occurred in the upper layers. Instead, a constant value of $0.52 \times 10^{-4}$ $m^2/s$ was used for all of the air infiltration simulations. However, examination of the measured and simulated results suggests that perhaps the diffusion coefficient should not remain constant across the

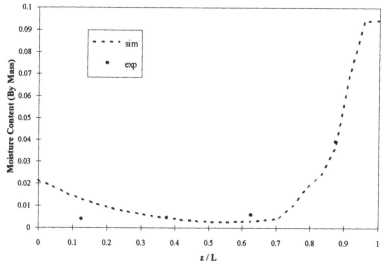

Figure 6  Moisture Profile: Air Flow Rate - 10 L/min, Exfiltration Air RH - 75 %

insulation slab. It also suggests that the assumption of local thermal equilibrium may not be satisfied for air infiltration, where the flow direction was the same as the temperature gradient. Furthermore, a local thermal instability in the flow could exist due to the heat release during condensation or ablimation which is similar to a fluid being heated from below (Benard Problem, [16] ). A flow instability could have occurred if, due to the energy released during the re-deposition of moisture in the upper layers of insulation, the local temperature gradient became reversed relative to the temperature gradient across the entire slab of insulation. These refinements are beyond the scope of this paper and need further study.

The calculation for the conductive heat flux through the bottom surface of the insulation sample, based on the numerically predicted air infiltration results, was carried out using the same equations as for air exfiltration (4,17). This also applied to the convective heat flux component, which was calculated using equation (18). For air infiltration this represents the amount of sensible heating which would be required to heat cold air, which enters the building envelope due to air infiltration, to a temperature of 20°C. The results obtained for the five test conditions for air infiltration are presented in Table II.

Table II indicates that heat conduction through the bottom surface was small compared to convective heat transfer, except for low air speeds. Furthermore, relative to a dry insulation sample which experiences neither vapor diffusion nor forced airflow, infiltration increased the total heat loss by more than a factor of 6. This ratio is independent of the temperature to which the air is heated during the infiltration process, but is strongly dependent on the air infiltration rate.

Table II  Numerical Heat Flux Results - Air Infiltration

| Test Conditions (exf = exfiltration) | $q''_{cond}$ (W/m$^2$) | $q''_{conv}$ (W/m$^2$) | $q''_{total}$ (W/m$^2$) | $q''_{dry-noflow}$ (W/m$^2$) | $q''_{total}$ / $q''_{dry-noflow}$ |
|---|---|---|---|---|---|
| 5 L/min, 75% exf RH | 4.51 | 15.2 | 19.7 | 7.39 | 2.67 |
| 10 L/min, 60% exf RH | 0.766 | 25.2 | 26.0 | 6.06 | 4.29 |
| 10 L/min, 75% exf RH | 0.691 | 24.6 | 25.3 | 5.90 | 4.29 |
| 10 L/min, 90% exf RH | 0.743 | 24.5 | 25.2 | 5.86 | 4.3 |
| 15 L/min, 75% exf RH | 0.017 | 28.9 | 28.9 | 4.54 | 6.36 |

## SUMMARY

This work sought to model the processes taking place within wall or roof fiberglass insulation with frosting effects and show the influence of airflow through the insulation (exfiltration / infiltration) on the heat loss to the exterior of a building. The development and validation of the numerical model to simulate such processes will aid in studying various other test conditions. Some discrepancies exist between the numerical predictions and the experimental results near the cold surface gap for frost accumulation related to air exfiltration and near the warm surface for moisture removal and deposition during air infiltration. Possible explanations for these differences have been made; however, more experimental and numerical research work will be required to entirely resolve the discrepancies. Based on the results presented, the physical importance of reducing or eliminating any air flow through insulated structures with the use of a properly installed vapor retarder has been made clear.

## REFERENCES

1. C. P. Hedlin, Effect of moisture on thermal resistances of some insulations in a flat roof system under field-type conditions, ASTM STP 789, 602-625 (1983).

2. J. Lstiburek and J. Carmody, Moisture Control Handbook, Oak Ridge National Laboratory, ORNL/Sub/89-SD 350/1 (1990).

3. NBC, National Building Code of Canada, National Research Council, Canada (1990).

4. ASTM, Water in Exterior Building Walls: Problems and Solutions, (Edited by T. A. Schwartz), ASTM STP 1107, Philadelphia (1991).

5. C. J. Simonson, Y.-X. Tao, and R. W. Besant, Thermal Hysteresis in Porous Insulation, *Int. J. Heat Mass Transfer* 36, 4433-4441 (1993).

6. C. Y. Shaw, Methods for Estimating Air Exchange Rates and Sizing Mechanical Ventilation Systems for Houses, *ASHRAE Trans.* 93, 2284-2302 (1987).

7. Y. Ogniewicz and C. L. Tien, Analysis of Condensation in Porous Insulation, *Int. J. Heat Mass Transfer* 24, 421-429 (1981).

8. D. R. Mitchell, The effects of air exfiltration/infiltration on the transient thermal response of a glassfiber insulation slab, M. Sc. Thesis, University of Saskatchewan (1994).

9. D. R. Mitchell, Y.-X. Tao, and R. W. Besant, Air filtration with moisture and frosting phase changes in fiberglass insulation: II model validation, *Int. J. Heat Mass Transfer*, (1995).

10. M. Kaviany, *Principles of Heat Transfer in Porous Media*, Springer-Verlag, New York (1991).

11. S. Whitaker, Simultaneous heat, mass and momentum transfer in porous media: a theory of drying, In: *Advances in Heat Transfer*, (Edited by J. P. Hartnett and T. F. Irvine, Jr.) Vol. 13, Academic Press, New York (1977).

12. K. Vafai and H. C. Tien, A numerical investigation of phase change effects in porous materials, *Int. J. Heat Mass Transfer* 32, 1261-1277 (1989).

13. Y.-X. Tao, R. W. Besant and K. S. Rezkallah, The transient thermal response of a glass-fiber insulation slab with hygroscopicity effects, *Int. J. Heat Mass Transfer* 35, 1155-1167 (1992).

14. Y.-X. Tao, R. W. Besant and K. S. Rezkallah, Unsteady heat and mass transfer with phase changes in an insulation slab: Frosting effects, *Int. J. Heat Mass Transfer* 34, 1593-1603 (1991).

15. P. Boulet, G. Jeandel and G. Morlot, Model of radiative transfer in fibrous media - Matrix method, *Int. J. Heat Mass Transfer* 36, 4287-4297 (1993).

16. S. Chandrasekhar, *Hydrodynamic and Hydromagnetic Stability*, Oxford University Press, 9-75 (1961).

# Low-Pressure Thermophysical Properties of EPB—Expanded Perlite Board

G. MILANO, F. SCARPA and G. TIMMERMANS

## ABSTRACT

Experimental results concerning the temperature dependent thermophysical properties of EPB-Expanded Perlite Board below atmospheric pressure are presented. Thermal conductivity and specific heat are simultaneously estimated as a function of temperature by processing the thermal response of the material subjected to a transient heat conduction experiment. The Kalman filtering technique is adopted as the identification algorithm in the iterative linearized version. Tests are performed with dried air as gas filler from atmospheric pressure to 5 μbar. The temperature interval investigated in the experiments ranges from 290 K to 400 K. Preliminary results show that under vacuum condition the apparent thermal conductivity of the expanded perlite board, which is an open pore material, is reduced by approximately a factor four with respect to the corresponding values at atmospheric pressure.

## INTRODUCTION

The thermal performance of open-cell insulating materials under evacuated conditions has been the subject of many experimental and theoretical studies. The increase of thermal resistivity obtainable at reduced pressure is of great interest in cryogenic applications such as the storage and transportation of liquefied gas. Recently, evacuated thermal insulation has been shown to have the potential for significant energy conservation also in residential and commercial refrigeration systems [1] and in highly insulated building envelope [2]. Among the porous insulators, expanded perlite and perlite-based evacuated materials have been intensively tested [3,4,5]. Experimental results are available [8] on the thermal conductivity of EPB-Expanded Perlite Boards at atmospheric pressure,

Guido Milano and Federico Scarpa, DITEC Dipartimento di Termoenergetica e Condizionamento Ambientale, Università di Genova, Via All'Opera Pia 15 A, I 16145, Genova - ITALY.
Georges Timmermans, Permalite Europe Ottergemsesteenweg 473, 9000 Gent - BELGIUM

but not under vacuum.   This kind of insulator has a thermal conductivity normally greater than that of light fibrous or foamed materials but it is available in rigid panel with high compressive strength (200 kPa with 2% deformation). This characteristic can be considered advantageous for potential applications in evacuated condition. In fact the evacuated EPB panels could be covered by simple deformable plastic envelopes rather than by rigid  plastic or metallic containers.

One of the aims of this work is the experimental determination of the apparent thermal conductivity of EPB in evacuated condition, from 290 K to 400 K.

A second purpose consists in the development of a transient measurement technique which allows the simultaneous identification of the temperature dependent thermal conductivity and specific heat. This last parameter is useful in transient conductive calculation but it is rarely measured in case of non homogeneous insulating materials.  Specific heat is often evaluated by weighing the value of each component of the material and usually it is assumed constant with temperature.

## HEAT TRANSFER MODEL AND ESTIMATION TECHNIQUE

The specimen, having a plane geometry, is subjected to a transient heating from an initial temperature up to a prescribed maximum temperature.  Then the thermophysical properties are reconstructed in form of cubic B-splines approximation by solving the inverse conduction equation in which the above properties are the temperature dependent unknown functions and the time temperature and heat flux histories on the boundaries and inside the specimen, at known positions, are the measured variables.

The 1-D, non linear, heat conduction equation for opaque, homogeneous and isotropic materials is assumed as heat transfer model :

$$\frac{\partial}{\partial z}\left[k(T)\frac{\partial T}{\partial z}\right] = C(T)\frac{\partial T}{\partial t} \tag{1}$$

$$0 < z < b; \quad 0 < t \le t_m$$

The initial  and  boundary conditions are:

$$T(z,0) = T_0(z) \; ; \quad 0 \le z \le b \tag{2a}$$

$$T(0,t) = u_1(t) \; ; \qquad T(b,t) = u_2(t) \tag{2b}$$

where: $z$ is the spatial co-ordinate, $T$ temperature, $t$ time; $k(T)$ and $C(T)$ are the temperature dependent thermal conductivity and volumetric heat capacity; $b$ is the specimen thickness and $t_m$ the total test duration.

The measured quantities are temperatures and heat fluxes at several inner locations $z_i$ inside the specimen as follows:

$$\hat{T}_i(t) = T(z_i,t); \qquad \hat{q}_i(t) = q(z_i,t) = -k[T(z_i,t)]\frac{\partial T}{\partial z}\bigg|_{z=z_i} \qquad (3)$$

$$0 < t \le t_m$$

Thermal conductivity and specific heat are parameterized with cubic B-splines approximations as follows:

$$k(T) = \sum_p k_p B'_p(T); \qquad C(T) = \sum_q C_q B''_q(T) \qquad (4)$$

where: $B'_p(T)$ and $B''_q(T)$ are the well known basis functions for uniform cubic B-splines; $k_p$ and $C_q$ are the set of control vertices (the unknown parameters to be determined); $T$ is temperature defined in the interval $[T_{min},T_{max}]$ where $T_{min}$ and $T_{max}$ are respectively the minimum and maximum temperature value reached during the experiment.

The conductive process is represented in a discretized form by means of an explicit formulation and the related inverse problem is solved using the Kalman algorithm. The Kalman filter is a stochastic algorithm based on a sequence of recursive predictor-corrector calculations by which the current estimate is continuously improved as a new set of measurements becomes available from the experiment. The sequential equations of the linearized Kalman filter used in this work are similar to those reported in [6,7,8,9] and are not repeated here. In respect to previous experimental work, in which the estimated thermophysical properties were thermal diffusivity and normalized volumetric capacity, the measure of heat flux history allows us to identify simultaneously the temperature dependent thermal conductivity and specific heat provided that density is also measured. Temperature and heat flux measurements are optimally weighted by the Kalman *gain* matrix according to the quality of the measurements and to the sensitivity coefficients of the unknown parameters.

## EXPERIMENTAL APPARATUS

Tests are carried out with an apparatus already described [8,10,11]. Here only additional devices and new measurement techniques are reported in detail. The heated plate, the specimen and the cold plate are placed inside an airtight thermostatted bell-shaped vessel ( Fig.1).

A system of vacuum pump and of air drier is connected to the bell jar in order to control pressure and the micro-climate of air surrounding the specimen. Pressure has varied from the atmospheric value down to 5 µbar. During the transient test pressure can be controlled by an automatic on-off valve connected to the vacuum pump circuit or by regulating a continuous leakage of dried air through a precision pin valve.

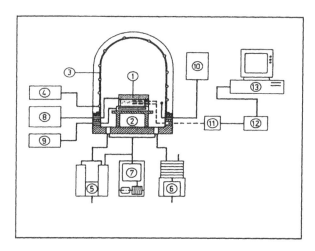

**Figure 1** - Schematic view of the experimental set-up: (1) test section; (2) base plate; (3) bell-jar; (4) thermostatted coil; (5) air drier; (6) vacuum pump; (7) air compressor; (8) heater; (9) cold plate; (10) hygro-thermometer; (11) multiplexer; (12) A/D converter; (13) data acquisition system.

The humidity of the air surrounding the specimen and therefore the moisture content that establishes inside the material is an important parameter to control in this kind of experiment. In fact the moisture contained in the material, even if in small percentage, is subjected during the transient heating to a process of evaporation and mass diffusion. This may cause large distortions of temperature profiles in respect to the assumed conductive model. To minimize this effect, before the test, the specimen is kept inside the apparatus at uniform temperature of about 420 K and at a pressure of few μbar for a period of 5-6 days. Then, the experiments at the minimum pressure value are performed. Afterwards the bell jar is filled with dried atmospheric air, at a nominal relative humidity less then 1.0 % and all other experiments are realized.

Owing to the importance of the moisture effect a theoretical and experimental research program is planned on the problem of heat and mass diffusion within open-cell porous materials during thermal transient.

The dynamic experiment is simply realized by supplying constant electric power to the ribbon resistance allocated inside the heating plate. The temperature increase of the heater is approximately exponential (reaching an asymptotic value), while the cold plate is kept at the initial constant temperature. The value of electric power is arranged in each test in such a way to operate at constant Fourier number. At atmospheric pressure the temperature increase is of about 100 °C in one hour while at the lowest pressure, when thermal diffusivity reduces to approximately one fourth, the same temperature increase is reached in about four hours.

The temperature variation of the hot and cold plate are assumed as control functions in the inverse conduction problem. The response of the material is measured by inserting temperature and heat flux sensors between adjacent slices. Several sheathed thermocouples are inserted at different locations while the transient heat flux is measured by three micro-foil heat flux-meter (2.5 cm x 2.0 cm x 0.015 cm) placed between the first and the second slice. It is known [12,13] that the optimum theoretical position for heat flux sensors is on the surface of material in contact with the heater where the sensitivity coefficients are higher. However, some preliminary tests have shown that, in our experimental set-up, the heat flux sensors, when placed in that location, give inaccurate heat flux measurements . This is probably due to an imperfect thermal contact between the micro-foil and the rigid heater.

The micro-foil heat flux sensors are calibrated by the maker: however, a new accurate calibration has been performed with the use of the Guarded Hot Plate (GHP) in steady state [14]. The heat flux-meters are introduced between slices of EPB material similar to that used for transient experiments and subjected to a stationary heat flux in the GHP apparatus. Several values of specific heat flux and average temperatures are tested. Moreover, the possible influence of the loading pressure of the material on the flux-meter signal has been also investigated. As the influence of load was negligible within the pressure range 1-4 kPa, the transient experiments are realized with a loading pressure of 2.3 kPa that is a common value for GHP measurements.

For each kind of sensor, a transient conductive model is introduced in the Kalman filter in order to take into account the delay in the time response. At low pressure, in fact, the thermal contact resistance between sensors and material plays a relevant role in the signal dynamic. After the arrangement of the sensors inside the material their nominal position is normally known within ±0.1 cm. Uncertainty of this order of magnitude can produce strongly biased estimates values. To greatly reduce this kind of error the position of each internal sensor is considered as further parameter to be determined as described in [9,15]. To correct the typical step-wise disturbances produced by commercial A/D converter, the signals are re-sampled by a simulated A/D converter having the same error of the electronic circuitry but opposite sign [8]. In each transient test a large number of single temperature and heat flux signals are analyzed. In this way the variance of the white noise affecting the signal is reduced along with the variance of the estimated parameters. This fits well with the Kalman filter which is a sequential estimator. Processing the set of measurements in a sequential way avoids storage of large quantities of data.

## RESULTS

The EPB material tested in this study has an average density, in dry condition, $\rho = 148$ kg/m$^3$ and is made in Europe. It consists mainly (2/3) of expanded perlite grains held together by a network of a mixture of cellulose and glass fibers. Some bituminous compound is also present as a binder.

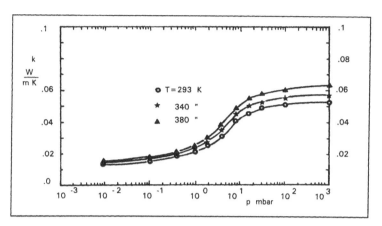

**Figure 2** - Apparent thermal conductivity of EPB-expanded perlite board as a function of pressure at different temperature values.

In Fig. 2 the apparent thermal conductivity is reported as a function of pressure for three different values of temperature. At a reference room temperature T=293 K, if the pressure is reduced from the atmospheric value below 0.1 mbar, thermal conductivity reduces to about one fourth, i.e. from 0.0515 to about 0.012 W/mK. In Figs. 3 and 4 the apparent thermal conductivity and specific heat are reported as a function of temperature for decreasing pressure values. While the curves of thermal conductivity show a reduction in their values and slopes when pressure decreases, the specific heat seems to be independent on the pressure variation and increases linearly with temperature with a slope similar to that found for EPB materials in [8]. The relatively high increase of specific heat with temperature is probably due to the presence of long molecular chain organic compound (cellulose, bitumen).

In Tables I and II a comparison is given between the present results (EPB-2, $\rho$=148 kg/m³) and those reported in [8] ( EPB-1, $\rho$=140 kg/m³ and EPB-3, $\rho$=157 kg/m³). We note, however, that the materials EPB-1 and EPB-2 are made in Europe, while EPB-3 is made in USA. In USA the network fibers are made by cellulose while in Europe a mixture of cellulose and glass-fiber is utilized. For the three tested materials the linear increase with temperature of thermal conductivity, at atmospheric pressure, and of specific heat are summarized by the following equations:

$$k_1 = 0.0488 + 1.317 \cdot 10^{-4} \cdot ( T\text{-}290 ) \ \text{W/m K} \tag{5}$$
$$k_2 = 0.0512 + 1.270 \cdot 10^{-4} \cdot ( T\text{-}290 ) \qquad " \tag{6}$$
$$k_3 = 0.0527 + 1.450 \cdot 10^{-4} \cdot ( T\text{-}290 ) \qquad " \tag{7}$$

$$c_1 = 890 + 3.45 \cdot ( T\text{-}290 ) \qquad \text{J/kgK} \tag{8}$$
$$c_2 = 880 + 3.61 \cdot ( T\text{-}290 ) \qquad " \tag{9}$$
$$c_3 = 900 + 4.39 \cdot ( T\text{-}290 ) \qquad " \tag{10}$$

As one can see, both thermal conductivity and specific heat values obtained in this study ( $k_2$, $c_2$ ) are consistent with those ( $k_1$, $c_1$ and $k_3$, $c_3$ ) reported in [8] and measured at atmospheric pressure respectively with guarded hot plate and differential scanning calorimeter.

A detailed theoretical characterization of the thermal performance of EPB as a function of pressure and temperature is beyond the scope of this work which is of preliminary nature. As well known, many accurate predictive models of thermal performance are available in literature for open-cell non homogeneous insulators. However they refer to fibrous materials [16] or foam insulation [17,18] or powders and spherical particles [19,20,21]. To the authors' knowledge no combined model has been investigated for open pore insulators in which the solid phase is composed both by spherical and fibrous materials like EPB and this is left for future works. Here the experimental data are analyzed with a pure conductive model and this presumes that the possible radiative contribution could be included in a generalized radiative-conductive law following the Rosseland approximation. The convective heat transfer may be excluded since the average cell size is too small (15 μm) and the direct radiative transmission may be excluded too because the material is opaque and relatively compact.

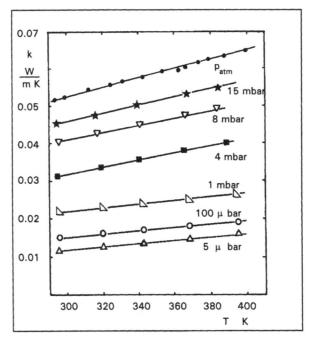

**Figure 3** - Apparent thermal conductivity of EPB-expanded perlite board as a function of temperature at different pressure values

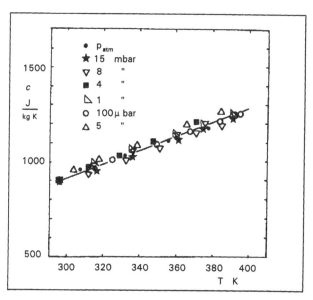

**Figure 4** - Specific heat of EPB-expanded perlite board as a function of temperature at different pressure values.

TABLE I - THERMAL CONDUCTIVITY (W/mK) AT Patm

| | 290 K | 310 K | 330 K | 350 K |
|---|---|---|---|---|
| EPB-1 (*) ( $\rho$ = 140 kg/m³ ) | 0.0488 | 0.0514 | 0.0541 | 0.0567 |
| EPB-2 (**) ( 148 kg/m³ ) | 0.0512 | 0.0537 | 0.0563 | 0.0588 |
| EPB-3 (*) ( 157 kg/m³ ) | 0.0527 | 0.0556 | 0.0585 | 0.0614 |

TABLE II - SPECIFIC HEAT (J/kgK)

| | 290 K | 320 K | 350K | 380 K |
|---|---|---|---|---|
| EPB-1 (*) ( $\rho$ = 140 kg/m³ ) | 890 | 995 | 1105 | 1200 |
| EPB-2 (**) ( 148 kg/m³ ) | 880 | 985 | 1095 | 1205 |
| EPB-3 (*) ( 157 kg/m³) | 900 | 1030 | 1165 | 1295 |

(*) Ref. [8];   (**) present study

As it can be seen in Fig. 5 the standard deviation of temperature residuals are very low (less than 0.01 K) both for tests at atmospheric pressure and in vacuum condition. Moreover no relevant biases are detectable from the residuals and this can reasonably support the validity of the simplified model assumed in this study.

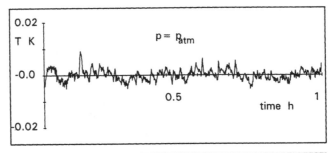

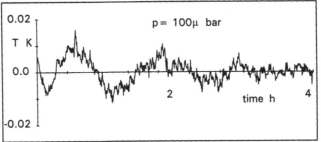

**Figure 5** - Temperature residuals at the distance x = 10.4 [mm] from the heater, for two pressure values: p = $p_{atm}$ and p = 100 μ bar

## CONCLUSION

The apparent temperature dependent thermal conductivity and specific heat of EPB Expanded Perlite Board ($\rho = 148$ kg/m$^3$) have been simultaneously identified with the use of a transient technique and the Kalman filter as parameter estimator. Pressure has been varied from the atmospheric value down to about 5 μbar and temperature from 290 to about 400 K. In the limit of the temperature range investigated, a non linear conductive model appears suitable for treating the heat transfer process in EPB material even in evacuated conditions. The preliminary results obtained in this study show that at room temperature (293 K) and at pressure below than 0.1 mbar the apparent thermal conductivity reduces to about one fourth of the atmospheric value. Specific heat is practically not influenced by the pressure variation and increases linearly with temperature.

The moisture content of the material and the delay in the temperature response of the sensors are important parameters that affect the accuracy of the thermal conductivity reconstruction, chiefly at low pressure. For this reason more investigation at low pressure is advisable in order to asses the reliability of the results.

## ACKNOWLEDGEMENT

This work has been supported by Permalite Europe, Belgium and by the Italian MURST (Ministero per l'Università e la Ricerca Scientifica)

## REFERENCES

1. Fine, H., A., 1989. "Advanced Evacuated Thermal Insulation: The State of the Art", *Journal of Thermal Insulation*, **12** : 183-208.

2. Svendsen, S., 1995. "Highly Insulated Building Envelope- Energy Saving Potential and Problems" Eurotherm Seminar N°44- Advances in Thermal Insulation, Espinho, Portugal, pp. 1-7.

3. Adams, L., 1965. "Thermal Conductivity of Evacuated Perlite", *Cryogenic Technology*, **1**,6 : 249-251.

4. Yarbrough, D. W., R. S. Graves, F. J. Weaver and D. L. McElroy, 1986. "The Thermal Resistance of Perlite-Based Evacuated Insulations for Refrigerators", ORNL/CON-215, Oak Ridge National Laboratory, pp. 1-15.

5. Fay, R. M. and M. Albers, 1993. "Fiber Glass for Insulating Cryogenic Tanker Systems", *Journal of Thermal Insulation*, **17** : 20-33.

6. Scarpa, F., and G. Milano, 1993. "Identification of Temperature Dependent Thermophysical Properties from Transient Data: a Kalman Filtering Approach", Internal Report EGR/21, Dipartimento di Termoenergetica e Condizionamento Ambientale, Università di Genova, Genova, pp. 1-16.

7. Scarpa, F., G. Milano and D. Pescetti, 1993. "Thermophysical Properties Estimation from Transient Data: Kalman Versus Gauss Approach", *ASME Book N°100357*, pp. 109-116.

8. Milano, G., F. Scarpa and G. Timmermans, 1994. "Thermophysical Properties of EPB-Expanded Perlite Board: a Transient Measurements Technique", *Thermal Conductivity 22*, T.W.Tong Ed., Lancaster, Penn., USA, Technomic Publishing Co., pp. 263-274.

9. Scarpa, F., R. Bartolini and G. Milano, 1991. "State-Space (Kalman) Filter in the Reconstruction of Thermal Diffusivity from Noisy Temperature Measurements", *High Temperature-High Pressure* , Pion Limited, London, **23** : 663-642.

10. Milano, G., F. Scarpa and R. Bartolini, 1994. "Determination of the Thermophysical Properties of Insulating Materials from Transient Data", *High Temperatures-High Pressures*, Pion Limited, London, **26** : 331-338.

11. Bartolini, R., F. Scarpa and G. Milano, 1991. "Determination of Thermal Diffusivity of Fibrous Insulating Materials", *High Temperatures-High Pressures*, Pion Limited, London, **23** : 659-673.

12. Garnier, B., D. Delaunay and J. V. Beck, 1994. "Improved Measurement of the Surface Temperature of Composite Materials for the Optimal Estimation of their Thermal Properties", *High Temperatures-High Pressures*, Pion Limited, London, **26** : 15-23.

13. Scarpa, F. and G. Milano, 1994. "Influence of the Boundary Conditions on the Estimation of Thermophysical Properties from Transient Experiments", *High Temperatures-High Pressures*, Pion Limited, London, **26** : 163-175.

14. Cartesegna, M., F. Scarpa and G. Milano, 1995. "On the Use of Micro Foil Heat Flux Meter for Estimation of Thermophysical Properties of Insulating Materials in Transient Tests", Internal Report DITEC, Dipartimento di Termoenergetica e Condizionamento Ambientale, Università di Genova.

15. Milano, G. and F. Scarpa, 1991. "Numerical Experiments on Thermophysical Properties Identification from Transient Temperature Data", Proceedings of the Fourth Annual Inverse Problems in Engineering Seminar, Michigan State University, J.V. Beck editor.

16. Stark, C. and J. Fricke, 1993. "Improved Heat-Transfer Models for Fibrous Insulations", *Int.J.Heat Mass Transfer*, **36** ,3 : 617-625.

17. Doermann, D. and J. F. Sacadura, 1995. "Prediction of Thermal Performance of Open Cell Foam Insulations", Eurotherm Seminar N°44- Advances in Thermal Insulation, Espinho, Portugal, pp. 1-8.

18. Campanale, M., F. De Ponte and L. Moro, 1995. "Theoretical Characterization of Non-Homogeneous Cellular Plastic Materials, Eurotherm Seminar N°44- Advances in Thermal Insulation, Espinho, Portugal, pp. 1-12.

19. Luikov, A.V., A.G. Shashkov, L.L. Vasiliev, E. Fraiman, 1968. "Thermal Conductivity of Porous Systems", *Int.J.Heat Mass Transfer*, **11** : 117-140.

20. Tien, C. L. and G. R. Cunnington, 1973. "Cryogenic Insulation Heat Transfer", *Advances in Heat Transfer*, **9** :349-417.

21. Brodt, K. H., 1995. Ph.D. Thesis TU Delft University, Netherlands.

# Thermal Conductivity of Spun Glass Fibers as Filler Material for Vacuum Insulations

R. CAPS, J. HETFLEISCH, TH. RETTELBACH and J. FRICKE

## ABSTRACT

The thermal conductivity of spun glass fibers has been measured as a function of gas pressure and temperature at an external load of 0.1 MPa. At room temperature the thermal conductivity of the evacuated fibers was as low as $1.6 \cdot 10^{-3}$ W/(m·K). An analysis of the heat transfer modes revealed that solid conduction contributes with $0.6 \cdot 10^{-3}$ W/(m·K) and radiative conduction with $1.0 \cdot 10^{-3}$ W/(m·K) to the thermal conductivity at room temperature.

## INTRODUCTION

Thermal super insulations are employed for cryogenic tasks, as transport and storage of cold liquid gases. The envelope of these super insulations has to be of cylindrically or spherically shaped thick metal cases in order to sustain the external pressure load. The insulation material (often stacks of foils) generally is not loaded in these systems. Large, evacuated flat panels on the other hand have to contain load bearing filling materials which sustain the external pressure load.

In the 1980s a highly efficient load bearing super insulation was developed in order to reduce the thermal losses of the sodium-sulphur battery which is operated at temperatures between 300 and 350°C [1]. Microglass fiber boards were used as filling material. At room temperature thermal conductivities between 2 and $3 \cdot 10^{-3}$ W/(m·K) and at a mean temperature of 170°C conductivities between 4 and $5 \cdot 10^{-3}$ W/(m·K) were measured.

Meanwhile many activities have started to develop evacuated insulation panels for applications at room temperature, e.g. for the efficient thermal insulation of refrigerators and cold stores. Degussa, Germany, has commercialized a powder filled vacuum panel with a multilayer plastic envelope [2]. It has a thermal conductivity of around $7 \cdot 10^{-3}$ W/(m·K). Due to the fine-grained precipitated silica powder an

R. Caps, J. Hetfleisch, Th. Rettelbach, J. Fricke, Bayerisches Zentrum für Angewandte Energieforschung e.V., -ZAE Bayern-, Am Hubland, D-97074 Würzburg, Germany

internal gas pressure of up to 10 mbar can be tolerated without surpassing $8 \cdot 10^{-3}$ W/(m·K). Pressure loaded powders based on aerogels, which have been opacified with carbon black, have conductivities between 3 and $4 \cdot 10^{-3}$ W/(m·K) for gas pressures below 0.1 mbar [3]. These powders are now being used as filler material for the vacuum insulation of latent heat storage devices used in cars [4]. Microglass fiber boards as used in the vacuum insulation of the Na/S-battery have still lower conductivities. They are, however, rather expensive to produce and their content of fibers with diameters below 1 µm may pose a health hazard.

Therefore we investigated the insulation potentials of glass fiber materials with considerably larger fiber diameters. Especially spun glass fiber materials with a typical fiber diameter of 10 µm have been selected as they can be produced in thin layers which can be stacked easily. Thus an almost ideal orientation of the fibers in a plane perpendicular to the heat flux is achieved.

From a simple geometrical model [5] it is expected that the solid conductivity here is as low as in the microfiber boards because the solid conductivity is strongly influenced by the contact resistances between the fibers and thus should only weakly depend on the fiber diameter.

In order to be able to separate radiative and solid conductivity we performed the thermal conductivity measurements on the spun glass fibers as a function of temperature. Additionally, the dependence of the thermal conductivity on the internal gas pressure was measured at room temperature.

## THERMAL CONDUCTIVITY MEASUREMENTS

### APPARATUS

The thermal conductivity measurements were performed using the evacuable guarded hot plate apparatus LOLA 4 [6]. The guarded section of the hot plate has a diameter of 140 mm. It is surrounded by two guard rings of width 35 mm each, thus the total diameter of the heating and cooling plates amounts to 280 mm. The electrical power P fed into the central section is converted into a heat flux which is symmetrically transmitted through both samples and absorbed by the heat sinks. The apparent thermal conductivity then is

$$\lambda = \frac{P \cdot D}{2 \, A \cdot \Delta T} \tag{1}$$

with      D: sample thickness,
           A: area of central hot plate (A = 0.0156 $m^2$ at room temperature
                 including half of gap between central hot plate and the inner
                 guard ring) and
           $\Delta T$: temperature difference between hot and cold plates.

The expansion of the area A as function of temperature is accounted for. Temperature and voltage data are registered by a computer within one measuring cycle of about 60 s duration. After the end of each single measuring cycle the power supplies are reset by analog signals to the new power ratings according to a proportional-integral control algorithm. Temperature measurements are performed with Pt-100 platinum resistors in 4-wire technique. The measurement accuracy for the temperatures is about 0.1 K at room temperature.

The measurement of a temperature difference between the central hot plate and the inner guard ring is most critical to the accuracy of the thermal conductivity measurement. A small temperature offset of e.g. 0.1 K results in an additional heat flux from the central hot plate to the guard ring. If, however, one can assume that this unknown heat loss $\delta P$ is independent of the temperature difference $\Delta T$ between the hot plate and the cold plate (e.g. $\Delta T = 20$ K or 40 K) a correction procedure [6] can be established as follows: For the same mean temperature $T_m$ both a small temperature difference $\Delta T_1$ and a larger temperature difference $\Delta T_2$ across the specimen is chosen yielding two corresponding electrical power ratings $P_1$ and $P_2$ for the central plate.

With help of the two equations relating the thermal conductivity to the power ratings the unknown heat loss $\delta P$ can be eliminated and a corrected thermal conductivity $\lambda_{corr}$ then is calculated according to:

$$\lambda_{corr} = \frac{(P_1 - P_2)\, D}{2\, A \cdot (\Delta T_1 - \Delta T_2)} \tag{2}$$

If the measurements are performed with more than two temperature differences $\Delta T$, the corrected conductivity can also be extracted from an extrapolation of the conductivity in a $\lambda$ versus $1/\Delta T$- plot (which should yield a straight line) to $1/\Delta T = 0$.

The overall measuring uncertainty including errors of electric power, radial heat losses and the sample thickness is about 5%, the latter contributing the main part.

## SAMPLE PREPARATION

The sample of spun glass fiber fleece [7] shows a relatively narrow fiber diameter distribution around 10 μm. Thin layers of glass fibers are stacked on top of each other; they form a loose fleece material with a low density. The fibers are mainly oriented in-plane. Compression with a load of 1 bar yields a density around 300 kg/m$^3$.

Prior to the measurements the fiber material has been heated in an oven for 5 h at a temperature of 250°C in order to remove organic residues. Two samples with a diameter of 280 mm and masses of 194 g and 185 g, respectively, have been cut out of the fleece material. Due to the compression the sample diameter slightly increased to 290 mm within the apparatus. The mean mass per area m" of the sample stack during measurement thus was 2.86 kg/m$^2$. Within the apparatus the fiber board was loaded with 1 bar and baked at 160°C in a vacuum for 50 hours. The mean sample thickness D of both samples during the measurement was 9.3 mm and the density $\rho$ = 308 kg/m$^3$.

## RESULTS

At a mean temperature of 293.5 K (20$^\circ$C) the apparent thermal conductivity of the samples was measured as function of nitrogen gas pressures. A temperature difference $\Delta T = 40$ K was chosen.

The results are depicted in fig.1. The thermal conductivity varies between $1.6 \cdot 10^{-3}$ W/(m·K) at $p_g < 10^{-3}$ mbar and $34.1 \cdot 10^{-3}$ W/(m·K) at ambient gas pressure. Thus by evacuation the thermal conductivity can be reduced by a factor of 20 compared to conventional, air-filled insulations. A significant rise of the thermal conductivity can be recognized according to fig.2 for gas pressures above 0.01 mbar. At 0.04 mbar gas pressure the conductivity exceeds $2.0 \cdot 10^{-3}$ W/(m·K).

In a second stage the thermal conductivity of the evacuated sample was measured as function of temperature. The temperature difference $\Delta T$ between the hot and cold plates was either set to $\Delta T_1 = 20$ K or $\Delta T_2 = 40$ K. In table 1 the resulting power ratings $P_1$ and $P_2$ and the corresponding calculated thermal conductivities $\lambda_1$ and $\lambda_2$ according to eq.(1) and the corrected thermal conductivities $\lambda_{corr}$ according to eq.(2) are given.

As can be seen from table 1 the absolute correction is less than $0.2 \cdot 10^{-3}$ W(m·K) and the relative correction less than 5% (T = 433 K) even if the conductivity $\lambda_1$ at the smallest temperature difference $\Delta T_1$ is considered for comparison. At room temperature the correction is negligable.

*Table 1:   Thermal conductivities $\lambda_1$ and $\lambda_2$ as function of mean temperature T measured with different temperature differences $\Delta T_1$ and $\Delta T_2$ and the corresponding corrected conductivity $\lambda_{corr}$ according to eq.(2)*

| mean temperature T K | temperature difference $\Delta T_1$ K | power rating $P_1$ W | thermal conductivity $\lambda_1$ $10^{-3}$ W/mK | temperature difference $\Delta T_2$ K | power rating $P_2$ W | thermal conductivity $\lambda_2$ $10^{-3}$ W/mK | thermal conductivity $\lambda_{corr}$ $10^{-3}$ W/mK |
|---|---|---|---|---|---|---|---|
| 293.5 | 9.75 | 0.05094 | 1.56 | 29.85 | 0.15540 | 1.55 | 1.55 |
| 313.5 | 19.78 | 0.12317 | 1.85 | 40.07 | 0.24616 | 1.83 | 1.80 |
| 333.5 | 20.0 | 0.14214 | 2.11 | 40.0 | 0.28075 | 2.08 | 2.06 |
| 363.5 | 20.0 | 0.17481 | 2.59 | 40.0 | 0.34363 | 2.55 | 2.50 |
| 393.5 | 20.0 | 0.21185 | 3.14 | 40.0 | 0.41560 | 3.08 | 3.01 |
| 413.5 | 20.0 | 0.24096 | 3.57 | 40.0 | 0.47237 | 3.49 | 3.42 |
| 433.5 | 20.0 | 0.27631 | 4.05 | 40.0 | 0.53631 | 3.96 | 3.88 |

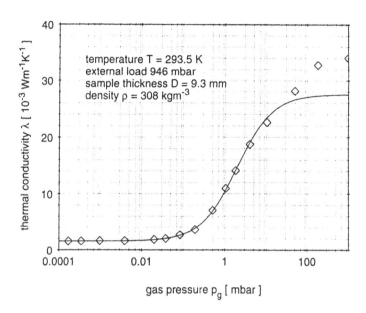

Figure 1:    Thermal conductivity $\lambda$ of spun glass fiber fleece as function of gas pressure $p_g$; $\diamond$ measurements, —— fit according to eq.(4)

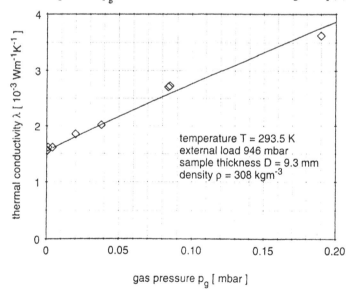

Figure 2:    Dependence of thermal conductivity $\lambda$ on gas pressure for gas pressures $p_g < 0.2$ mbar

The corrected thermal conductivity of the evacuated sample $\lambda_{corr}$ is depicted in fig. 3 for mean temperatures between $20\,^{\circ}C$ and $160\,^{\circ}C$. For the upper temperature the conductivity is still below $4 \cdot 10^{-3}$ W/(m·K).

Additionally at room temperature (T = 20 $^{\circ}C$) the thermal conductivity of the evacuated sample has been measured with a reduced external load of 64 mbar. The sample thickness increased to 14.2 mm corresponding to a density of 200 kg/m$^3$. The thermal heat transfer coefficient decreased from 0.166 W/m$^2$K to 0.087 W/m$^2$K and the corresponding thermal conductivity $\lambda$ from 1.6 to $1.2 \cdot 10^{-3}$ W/(m·K).

## DISCUSSION OF RESULTS

Heat transfer in fiber boards occurs via solid conduction along the fibers and across the contact points between the fibers, by radiative heat transfer and by conduction of residual gas.

For small residual gas pressures $p_g$ where the interaction of solid and gaseous conduction can be neglected, the three heat transfer modes are the sum of solid conductivity $\lambda_s$ of the evacuated sample, the radiative conductivity $\lambda_r$ and residual gaseous conductivity $\lambda_g$:

$$\lambda(T,p_g) = \lambda_s(T,p_g=0) + \lambda_r(T) + \lambda_g(T,p_g) \qquad \textbf{(3)}$$

The gaseous conductivity $\lambda_g$ as function of gas pressure $p_g$ can be described by

$$\lambda_g(T,p_g) = \frac{\lambda_g^{o}(T)}{1 + \dfrac{p_{1/2}(T)}{p_g}} \qquad \textbf{(4)}$$

where $\lambda_g^{0}(T)$ represents the themal conductivity of the free gas. The characteristic pressure $p_{1/2}$ can be fitted to the experimantal data for $p_g$ below 10 mbar which yields $p_{1/2} \approx 2$ mbar at T = 20°C (see fig. 1).

As the interfiber contact resistances are thermally shorted by the gas molecules at gas pressures above 100 mbar the measured total conductivity is higher than predicted by eq.(3) [8]. This causes an additional conductivity contribution of $\lambda(ambient) - \lambda_g^{0}(N_2) - \lambda_r = 7 \cdot 10^{-3}$ W/(m·K) at ambient gas pressure.

If the optical thickness of the sample is sufficiently high for the calculation of the radiative conductivity $\lambda_r$ the diffusion model can be applied [9]:

$$\lambda_r = \frac{16}{3} \; \frac{n^2 \, \sigma \, T^3}{E^*} \qquad \textbf{(5)}$$

with $\qquad$ $n^2$: squared effective index of refraction of insulation medium

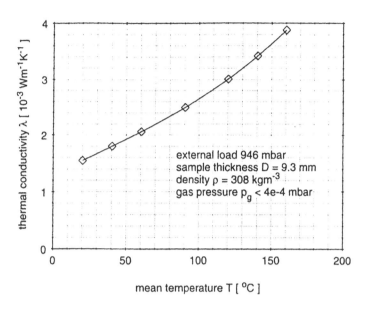

Figure 3: Thermal conductivity λ of spun glass fiber fleece as function of mean temperature T; ◇ measurements, —— fit

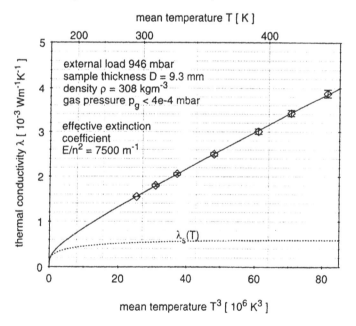

Figure 4: Thermal conductivity λ versus third power of mean temperature T; ◇ measurements, —— fit of total conductivity λ(T), --- derived solid conductivity $\lambda_s(T)$

E*: mean effective extinction coefficient
$\sigma$: Stefan-Boltzmann constant $= 5.67 \cdot 10^{-8}$ W/(m$^2$K$^4$).

The effective extinction coefficient E$^*$ includes the influence of forward scattering on radiative transfer [9]. In general both the effective extinction coefficient and effective index of refraction depend on wavelength. At small wavelengths $\Lambda$ far away from absorption bands the square of the effective index of refraction n$^2$ can be approximated by:

$$n^2 \approx 1 + 2 \cdot (n_0 - 1) \cdot \rho/\rho_0 \approx 1.12 \qquad (6)$$

with $n_0$: index of refraction of the glass material ($\approx 1.5$),
  $\rho_0$: density of the glass fiber material ($\approx 2500$ kg/m$^3$),
  $\rho$: density of fiber board ($\approx 300$ kg/m$^3$).

Near absorption bands, however, the index of refraction becomes complex and this approximation is not valid any more. Estimations show that at densities $\rho$ of 300 kg/m$^3$ the radiative energy transfer within the absorption bands may be considerably higher than predicted by eq.(5) with n$^2 = 1.12$.

Thus as both the n$^2$-value and the extinction coefficient depend on wavelength only a mean value over the thermal spectrum (Rosseland mean [10])

$$E^*/n^2 = 1 / \langle n^2(\Lambda) / E^*(\Lambda) \rangle \qquad (7)$$

can be obtained by thermal conductivity measurements.

In general the thermal conductivity may depend on the thickness D if the optical thickness of the sample is low. Then only an apparent thermal conductivity can be defined. In cases where the boundary emissivity of the sample is not too low ($\varepsilon \geq 0.5$) the apparent thermal conductivity can be described by a slightly modified form of eq.(5) [11]

$$\lambda_{r,app} = \frac{16}{3} \frac{n^2 \, \sigma \, T^3}{E^* + \frac{4}{3 \cdot D}} \qquad (8)$$

The thermal conductivity versus the third power of absolute temperature T in fig. 4 shows an almost linear dependence. If we assume as a first approximation that the solid conductivity $\lambda_s$ is temperature independent and if we apply eq.(5), then the slope of the line yields a temperature independent extinction coefficient E$^*/n^2 = 7500$ m$^{-1}$. If eq.(8) instead is used for the line fit, the influence of the insulation thickness D = 9.4 mm yields a slightly reduced, true extinction coefficient of E*/n$^2 = 7500$ m$^{-1}$ - 4/(3$\cdot$D$\cdot$n$^2$) = 7370 m$^{-1}$. Thus in our case the "thickness effect", i.e. the change of apparent thermal conductivity with thickness is less than 2% compared to an infinite thick insulation (where eq.(8) reduces to eq.(5)) and the results can be stated as "thermal conductivities" instead of "apparent thermal conductivities".

At room temperature the radiative conductivity $\lambda_r$ according to fig. 4 amounts to $1 \cdot 10^{-3}$ W/(mK) and rises to $3 \cdot 10^{-3}$ W/(mK) at a mean temperature of 150 °C. A more accurate analysis may also consider the temperature dependence of the thermal conductivity of solid glass (which is proportional to $\lambda_s(T)$, see fig.4) but yields about the same extinction coefficient and radiative conductivity as above.

According to fig. 4 the fitted solid conductivity $\lambda_s$ at room temperature is $0.6 \cdot 10^{-3}$ W/(m·K), which is an extremely low value if compared with other evacuated load bearing insulations (where usually $\lambda_s > 1 \cdot 10^{-3}$ W/(m·K) [12].

In comparison to fibrous insulations with fiber diameters in the range of 1 µm to 5 µm the radiation extinction of the spun glass fibers is about half as large and thus the radiative conductivity here is increased by a factor of two. Due to the optimal arrangement of the spun glass fibers in the plane perpendicular to the heat flux, however, essentially only the extremely low heat flux via the fiber contacts is responsible for the solid conductivity. The combined small solid and the somewhat higher radiative conductivity of the spun glass fibers thus yields an overall conductivity $\lambda$ which at room temperature is as low as or even lower than the thermal conductivity of evacuated melt blown fibers.

For applications as vacuum super insulation panel usually a rigid board is required, which retains its thickness if the pressure is released. From experience with other fibrous samples [13] it is expected that a temper process under load and at temperatures around 400 and 500°C can be conducted in such a way that the additional increase in solid conductivity can be limited to $1.0 \cdot 10^{-3}$ W/(m·K) at the most.

## CONCLUSIONS

It has been shown that loaded spun glass fibers have conductivities well below $2 \cdot 10^{-3}$ W/(mK) at room temperature. The residual gas pressure should be kept below $10^{-2}$ mbar in order to suppress gaseous conductivity. Despite the relative large fiber diameters the radiative conductivity remains in the range of $1 \cdot 10^{-3}$ W/(m·K) at room temperature.

A vacuum insulation panel made of spun glass fibers as filler material needs a metal barrier case which keeps the residual gas pressure below 0.01 mbar within lifetime. Due to edge losses through the metal rim the overall thermal conductivity of a vacuum panel may be considerably higher than the thermal conductivity of the pure evacuated fibers (usually measured as "center value"). Main effort of future developments has to concentrate on vacuum panel designs with edge losses well below the total heat losses through the fiber filling itself.

## ACKNOWLEDGEMENTS

We thank Schuller GmbH, Wertheim for generously providing glass fiber samples. This work was supported by the BStMWVT/Munich.

## REFERENCES

[1]     R. Knödler, H. Reiss, 1983. "Heat Loss Measurements on an Enclosure for High Temperature Batteries", *J. Power Sources* 9: 11-17

[2]     German Patent No. DE 4029405, Degussa AG, Frankfurt/Main

[3]     E. Hümmer, X. Lu, Th. Rettelbach, J. Fricke, 1992. "Heat Transfer in Opacified Aerogel Powders", *J. Non-Cryst. Solids*, 145: 211-216

[4]     W. Zobel, R. Strähle, 1995. "Heat Storage Battery for Car Applications", *IMechE* 379-385

[5]     M. G. Kaganer, 1969. *Thermal Insulation in Cryogenic Engineering*, Israel Program for Scientific Translation, Jerusalem

[6]     U. Heinemann, J. Hetfleisch, R. Caps, J. Kuhn, J. Fricke, 1995. "Evacuable Guarded Hot Plate for Thermal Conductivity Measurements between -200°C and 800°C", Proc. Eurotherm Seminar No.44, Lisboa

[7]     Schuller GmbH, D-97865 Wertheim, Germany

[8]     C. Stark, J. Fricke, 1993. "Improved heat-transfer models for fibrous insulations", *Int. J. Heat Mass Transfer* 36:617-625

[9]     H. C. Hottel, A. F. Sarofin, 1967. *Radiative Transfer*, McGraw-Hill, New York

[10]    R. O. Buckius, 1982. "The Effect of Anisotropic Scattering on Molecular Gas Radiation", Proc. 7th Int. Heat Transfer Conf. Munich

[11]    R. Caps, J. Fricke, 1986. "Infrared Radiative Heat Transfer in Highly Transparent Silica Aerogel", *Solar Energy* 36: 361-364

[12]    J. Fricke, D. Büttner, R. Caps, J. Gross, O. Nilsson, 1990. "Solid Conductivity of Loaded Fibrous Insulations", in: Insulation Materials, Testing and Applications, ASTM STP 1030, D.L. McElroy and J.F. Kimpflen, Eds., ASTM Philadelphia, 66-78

[13]    J. Trappmann, 1990. "Wärmetransport in getemperten Glasfaserwärmedämmungen", Diplom thesis, University of Würzburg, Germany

# Thermal Conductivity of Evacuated Insulating Powders for Temperatures from 10 K to 275 K

TH. RETTELBACH, D. SATOR, S. KORDER, J. FRICKE

## ABSTRACT

The thermal conductivity of thermally insulating powders like silica aerogel (opacified with carbon black), precipitated silica, perlite and diatomite was measured by a stationary method in an evacuated guarded flat plate apparatus. The mean temperature was varied between 10 and 275 $K$, while the uniaxial external pressure load was varied from 0.15 up to $1 bar$.

The measured thermal conductivities are in the range of $0.1 \cdot 10^{-3} W/(m \cdot K)$ at $T = 20K$ up to $3 \cdot 10^{-3} W/(m \cdot K)$ at $T = 275K$ for silica aerogel powders, $0.2 \cdot 10^{-3} W/(m \cdot K)$ and $3.5 \cdot 10^{-3} W/(m \cdot K)$ for precipitated silica, $0.3 \cdot 10^{-3} W/(m \cdot K)$ and $5 \cdot 10^{-3} W/(m \cdot K)$ for perlite and $0.1 \cdot 10^{-3} W/(m \cdot K)$ and $7 \cdot 10^{-3} W/(m \cdot K)$ for diatomite, respectively.

The results are interpreted using two models which describe the thermal conductivity of a powder fill in terms of density, external load, elastic properties and the thermal conductivity of the monolithic material.

For better comprehension geometric factors $g' = \lambda_{mat}/\lambda_{pow}$ (ratio of the thermal conductivity of monolithic material and powder) and $g = \lambda_{vitreous\,silica}/\lambda_{pow}$ (ratio of the thermal conductivity of vitreous silica to the one of powder) are introduced. They describe the 'reduction' of thermal conductivity caused by the 'dilution' of the material and the intergranular contact resistances.

The experimentally derived values of $g$ are in the range of 100 to more than 1000. The values of $g'$ are in the range of 4 to 20. We tried to reproduce these values by using suitable models and realistic parameters.

Th. Rettelbach, D. Sator, S. Korder, J. Fricke
Bavarian Center for Applied Energy Research e.V. (ZAE Bayern)
Am Hubland, D - 97074 Würzburg, Germany

## INTRODUCTION

Evacuated powders are efficient low temperature thermal insulations (see, for example, [1]). They are not as optimal as non-load-bearing evacuated multilayer insulations though on the other hand they tolerate background gas pressures in the 0.1 mbar range and are load bearing.

The reason for the smaller insulating capability of powders is the higher solid thermal conductivity. The optimization of low-temperature powder insulations requires both theoretical investigations of the heat transfer modes within a powder fill as well as thermal conductivity measurements. The description of the thermal transport in an inhomogenous system of irregularly shaped powder grains touching each other is a non-trivial problem. The models published up to now always contain simplifications or are only semiempirical correlations. In this paper experimental results of thermal conductivity measurements are presented and compared with predictions of resistance models of Chan and Tien [2] and of Leyers [3].

## THEORY OF HEAT TRANSFER IN POWDEROUS MATERIALS

### PARTICIPATING HEAT TRANSFER MODES

Powders like the investigated ones can be treated as fine granular systems. The powder grains are touching each other, and there are voids between the grains. As the investigated powder fills are evacuated down to gas pressures of $\leq 10^{-5} mbar$, gaseous thermal conductivity is negligible. The remaining heat transfer modes are solid thermal conduction within the grains and via grain to grain contact areas as well as radiative heat transfer.

### RADIATIVE THERMAL CONDUCTIVITY

The radiative heat transfer through a powder fill can be treated as a diffusion through a quasihomogenous optically thick medium, if the geometrical thickness of the layer is large compared to the mean free path $\ell = 1/E$ and the grains and voids have sizes small compared to $\ell$. $E$ is the extinction coefficient. The radiative thermal conductivity is given by:

$$\lambda_{\text{rad}} = \frac{16}{3} \frac{\sigma n^2 T^3}{E(T)} \tag{1}$$

with: $\sigma$ being the Stefan-Boltzmann number, $n$ the refractive index and $T$ the absolute temperature.

The relevant parameters are the optical thickness $\tau_0$ of the fill or the mass specific extinction $e = E/\rho$ of the powder, with $\rho$ being the density of the powder fill. They can be determined by infrared optical measurements.

On the other side, if the grain diameter is considerably larger than the mean free path of the photons the radiation transport parameters are dominated by the grain diameter and the size of the intergranular voids. The overall radiative transfer would have to be described by a complex diffusive/ballistic process.

## MODELS FOR THE THERMAL CONDUCTIVITY OF THE POWDER GRAIN SYSTEM

Both analytical and numerical methods to describe the thermal conductivity of a granular system have been tried. One possibility is to consider an elementary cell of the powder and describe it as a resistance network of thermal conductors of different conductivities. Maxwell [4] calculated the temperature field in and around a sphere of one material embedded in a medium of another material. So he could derive an effective thermal conductivity of the system. An essential condition was that the spheres must be well separated. Other authors like Tsao [5] or Crane and Vachon [6] considered an elementary cell with touching particles (not necessarily spheres) in a nonconvecting fluid and altered the matter-void system at a given density to simplify the analogous resistance system. Among other authors Kaganer [1] studied the effect of the external load and elastic properties of the grain material on the effective thermal conductivity of the powder. Two neighboring spheres are assumed to touch each other in a small contact circle. Its radius $r_{ct}$ is given by the Hertz formula:

$$ r_{ct} = r_o \cdot \left( \frac{1 - \mu_o^2}{Y} \cdot F r_o \right)^{1/3}, \tag{2} $$

with $r_o$: radius of sphere, $Y$: Young's modulus, $\mu_o$: Poisson ratio, $F$: force onto the contact.

In Kaganer's model the effective thermal conductivity is predicted to be dependent on the third root of the external load. The assumptions concerning the geometry and the heat transfer direction inside the elementary cell are not realistic however. The absolute values of the model predictions cannot be verified by experiments.

A better description of thermal resistances of an array of touching spheres is given by Chan and Tien [2]. They derived equations for the thermal resistance given by a sphere in a regular packing (simple cubic, body centered cubic, face centered cubic) in dependence of the contact radius of the area between two touching spheres. The resulting formulas for the effective solid thermal conductivity $\lambda_s$ of the powder are:

$$ \text{(sc)} \quad \frac{\lambda_s}{\lambda_{\text{mat}}} = \frac{\pi}{4} \cdot \frac{r_{ct}^2}{r_0^2} / \sum_{n=1}^{\infty} \frac{1}{2n-1} \frac{1}{4n-1} \cdot \left( P_{2n-2}(x) - P_{2n}(x) \right)^2 \tag{3} $$

$$ \text{(bcc)} \quad \frac{\lambda_s}{\lambda_{\text{mat}}} = \frac{\sqrt{3}\pi}{4} \cdot \frac{r_{ct}^2}{r_0^2} / \sum_{n=1}^{\infty} \frac{1}{2n-1} \cdot \left( P_{2n-2}(x) - P_{2n}(x) \right) \cdot \left( 1 + 2P_{2n-1}(\tfrac{1}{3}) \right) \tag{4} $$

$$ \text{(fcc)} \quad \frac{\lambda_s}{\lambda_{\text{mat}}} = \frac{\pi}{\sqrt{2}} \cdot \frac{r_{ct}^2}{r_0^2} / \sum_{n=1}^{\infty} \frac{1}{2n-1} \cdot \left( P_{2n-2}(x) - P_{2n}(x) \right) \cdot \left( 1 + 2P_{2n-1}(\tfrac{1}{2}) \right), \tag{5} $$

where $\quad x = \sqrt{1 - \dfrac{r_{ct}^2}{r_o^2}}\quad$ and

$P_n(x)\quad$ denotes the Legendre polynomial of order n and argument $x$.

As the authors supposed the intergranular voids to be empty and nonconducting, their model should be applicable to evacuated powders at low temperatures. However, such analytical relations are only available for regular packing patterns like the ones given above. An interpolation or extrapolation to other porosities is not possible.

Leyers [3] describes the heat transfer in a system of packed spheres in a conducting but nonconvecting fluid. The derivation is done for three packing patterns: simple cubic, orthorhombic and face centered cubic. As the resulting effective thermal conductivity of the powder is dependent on the ratio $L = \lambda_{mat}/\lambda_{gas}$ (which compares the thermal conductivity of the grain material to the one of the intergranular gas) and the porosity (or matter density), a phenomenological interpolation between the three lattice configurations is given. Also corrections concerning the contact area between the grains are considered. The overall resulting formula for the combined contributions of solid and gaseous thermal conduction to the thermal conductivity of the powder is given by:

$$\lambda_{s,g} = \lambda_{mat} \cdot \left(1 - \frac{\pi}{4} + \frac{\pi}{2}\frac{L}{L-1}\left(\frac{L}{L-1}\ln L - 1\right)\right)$$

$$\cdot \left(1 + A\left(\frac{L-1}{L}\right)^2 \ln L\right) \cdot \left(1 + 2\left(1 - \frac{1}{L^{0.3}}\right)\left(\frac{p}{p_o} - 1\right)\right) \cdot K\,, \qquad (6)$$

where $\quad A = A_o \cdot \left(8.4476\dfrac{p}{p_o} - 2.9522\dfrac{p^2}{p_o^2} - 4.4954\right)\quad$ is an empirical value,

$A_o = p_o^3/\sqrt{3} = 0.082877,$

$p\quad$ the matter density of the actual fill, and

$p_o = 0.523599$ the mass density of the simple cubic packing pattern.

The factor $K$ denotes the correcture to be made by thermal transport via the contact areas between the grains:

$$K = 1 + \left[\frac{\pi}{4}L\frac{r_{ct}^2}{r_o^2} \;+\; \frac{\pi}{2}\frac{L}{L-1}\cdot\left(1 - \sqrt{1 - \frac{r_{ct}^2}{r_o^2}}\right) - \frac{\pi}{2}\frac{L^2}{(L-1)^2}\right.$$

$$\left.\cdot\ln\left(L-(L-1)\sqrt{1 - \frac{r_{ct}^2}{r_o^2}}\right)\right] \;\div\; \left[\left(1-\frac{\pi}{4}+\frac{\pi}{2}\frac{L}{L-1}\left(\ln L(\frac{L}{L-1})-1\right)\right)\right] \qquad (7)$$

Both in the model of Chan and Tien as well as in the one by Leyers the contact radius $r_{ct}$ is calculated via Hertz' equation (2).

The effective thermal conductivity in Leyers' model is calculated under assumption that the thermal conductivity $\lambda_{gas}$ of the intergranular matter is non-zero and the ratio $L = \lambda_{mat}/\lambda_{gas}$ is not too large. In the present experiment

L was in the order of $10^5$ which is not in the range considered by the original paper [3]. However, calculations of $\lambda_{s,g}$ as a function of $L$ showed no divergence for $L \to \infty$.

## GEOMETRIC FACTOR $g'$

There is one common property in the introduced models: The effective thermal conductivity $\lambda_{pow}$ is given in terms of the thermal conductivity of the grain material $\lambda_{mat}$. Therefore it is useful to calculate a geometric factor

$$g' := \lambda_{mat}/\lambda_{pow} \tag{8}$$

from the experimental data. In this relation all the information about the reduction mechanisms of the thermal conductivity by the dilution of matter is contained.

## EXPERIMENTAL SETUP

### APPARTATUS FOR THERMAL CONDUCTIVITY MEASUREMENTS

Thermal conductivity measurements were performed in an evacuable guarded hot plate apparatus which has been described in detail in [7] and [8]. The two identical disk-like specimens are placed between one "hot" (being at temperature $T_1$) and two cold plates (temperature $T_2$). The heat transfer coefficient $k$ can be derived from:

$$k \quad = \quad \frac{P}{2\,A\,(T_1 - T_2)}, \tag{9}$$

where P is the electrical power fed into the central metering section (area $A$) in order to keep the hot plate at temperature $T_1$. The sample thickness is measured in situ by a slide wire bridge. The mean thermal conductivity of the two samples is $\lambda = k \cdot d$, with $d =$ sample thickness. The external load onto the samples is applied by a $He$-gas filled bellow above the sample-plate-stack. The $He$-pressure can be varied. To prepare the sample disks, the powders were filled into an evacuable powder container with variable height.

### INVESTIGATED POWDERS

The powders being investigated are known to be efficient thermal insulators at ambient temperatures. All are based on silica, but are different in structure: Three silica aerogels, made by a sol-gel-process followed by supercritical drying, one subcritically dried precipitated silica, and two 'natural' materials: expanded perlite and diatomite. The silica aerogel powders were derived from granules by milling and contained carbon black powder. All relevant data of the powders are given in table I.

| sample name | delivered by | type and constitution | density / kg/m$^3$ | | |
|---|---|---|---|---|---|
| | | | loose | at 150 mbar | at 1000 external load |
| AP01 | BASF | 90% silica aerogel 10% carbon black physical mixture | 95 | 118 | 123 |
| B292/77 | BASF | 90% silica aerogel 10% carbon black physical mixture | 94 | 120 | 125 |
| G 6555 | BASF | 90% silica aerogel 10% carbon black chemical mixture | 107 | 120 | 135 |
| FK 500 LS | Degussa | 98.5 % silica precipitated silica | 150 | 193 | 210 |
| diatomite | | natural mineral | 310 | 350 | 366 |
| Otaperl | Otavi | expanded perlite | 44 | 58 | 64 |

TABLE I: Relevant sample data.

## MEASUREMENT PARAMETERS AND RESULTS

### THERMAL CONDUCTIVITY MEASUREMENTS

The total thermal conductivity of all powders was measured as a function of temperature $T$ at an external load of $1.5 \cdot 10^4 Pa$ (fig. 1). Additionally the thermal conductivity was determined as a function of external load at a given temperature of $T = 35K$ (fig. 2).

### DERIVATION OF $\lambda_{rad}$ AND $\lambda_s$

The spectral mass specific extinction $e(\Lambda)$ of all powders was determined by directional-hemispheric transmission and reflection measurements for wavelengths between 2.3 and 300$\mu m$ [9], followed by Rosseland averaging to obtain $e(T)$ [10].

The total thermal conductivity $\lambda$ can be regarded as the sum of the solid thermal conductivity $\lambda_s$ through the packed grain system and a radiative thermal conductivity $\lambda_{rad}$ which can be calculated from the ir-optical properties via eq. (1). The resulting solid thermal conductivities $\lambda_s$ are shown in fig. 3. Below $100K$ the difference between the total and the solid thermal conductivity is less than the measurement error. Radiative contributions to the total thermal conductivity are significant only above $100K$.

### MEASUREMENT ERRORS

Errors in the measured total thermal conductivity result from errors in the electrical heating power, a non-negligible temperature difference betweeen cen-

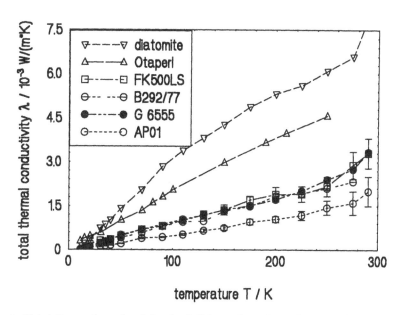

Fig. 1: Total thermal conductivity $\lambda$ of all investigated powders as a function of temperature $T$. The external load $p_{\text{Ext}}$ onto the fill was $150\,mbar$.

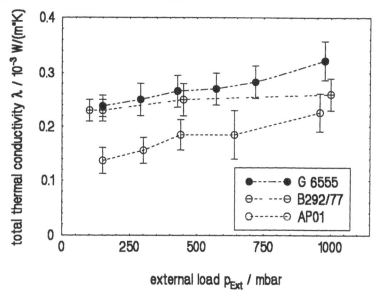

Fig. 2: Total thermal conductivity $\lambda$ of silica aerogel powders as a function of external load $p_{\text{Ext}}$. The temperature $T$ was $35\,K$.

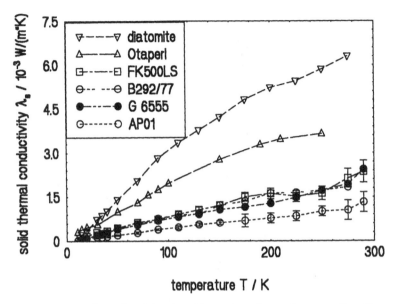

Fig. 3: Solid thermal conductivity $\lambda_s$ of investigated powders as a function of temperature $T$ at an external load $p_{Ext}$ of $150\, mbar$.

tral metering section and guard rings in the flat plate apparatus and from the uncertainty in sample thickness. Errors in heating power are below 2%; errors due to a radial heat flow can be eliminated by measuring the thermal conductivity at the same mean temperature but different temperature differences [11]. Corrections due to this procedure remained below 5%. The sample thickness during measurement was determined to within ±10%. Thus the total error in total thermal conductivity is about 10%.

Errors in the derived solid thermal conductivity $\lambda_s$ are influenced by the uncertainties in the ir-optical determination of $e(T)$. The accuracy can be given within ±20%.

## INTERPRETATION

GEOMETRIC FACTOR $g = \lambda_{vs}/\lambda_{pow}$ COMPARED
TO VITREOUS SILICA

Scheuerpflug [7] showed that above $50\,K$ the temperature dependence of the solid thermal conductivity $\lambda_s$ of monolithic silica aerogel is the same as for vitreous silica. The reduction factor $g$ only depends on the aerogel structure and density, not on temperature. Reduction factors $g$ for silica aerogel powders were also reported by the authors of this paper [8], [12]. A comparison is given here for the powders diatomite, perlite and precipitated silica (fig. 4). The thermal conductivity of vitreous silica was taken from [13]. The curves show clearly that above $50K$ no significant difference in the temperature dependence of the heat transfer mechanisms in vitreous silica and the investigated powders is detectable. The reduction factor $g$ is 1000 to 3000 for the aerogel powders and 100 to 1000 for the other powders.

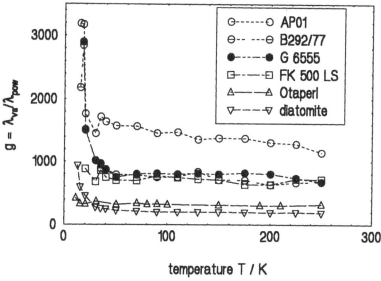

Fig. 4: Reduction factor $g = \lambda_{vs}/\lambda_{pow}$ (comparing the thermal conductivities of the silica based powders to the one of vitreous silica) as a function of temperature $T$ at an external load $p_{Ext}$ of $150\,mbar$.

## GEOMETRIC FACTOR $g'$, WHICH COMPARES MONOLITHIC MATERIAL TO POWDER

The material the aerogel powder grains are made of is monolithic silica aerogel, for which thermal conductivities have been measured by Scheuerpflug et. al. [7] at different densities. The data from that paper have been interpolated in temperature and density to calculate the geometric factors $g'$ for the silica aerogel powders. They are shown in fig. 5 and 6 for the different measurement parameters ($T$ or $p_{Ext}$, respectively).

### EVALUATION OF MODEL PREDICTIONS

The geometric factors $g'$ for the silica aerogel powders can be calculated from the models in chapter 2. This requires the knowledge of parameters like the elastic properties or the different densities both of the powders as well as the monolithic silica materials. Several investigations have been made concerning the elastic properties of monolithic silica aerogels with different densities [14] (see table II). To calculate $g'$ by Leyers' model the apparent thermal conductivity of $He$ at a pressure of $10^{-5}mbar$ and a mean free path of $10^{-5}m$ (the typical

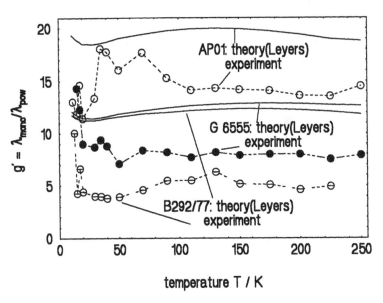

Fig. 5: Reduction factor $g' = \lambda_{\mathrm{mono}}/\lambda_{\mathrm{pow}}$ (comparing the thermal conductivities of monolithic silica aerogels to the one of aerogel powders) as a function of temperature $T$ at an external load $p_{\mathrm{Ext}}$ of $150\,mbar$.

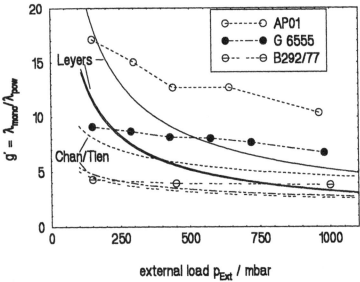

Fig. 6: Reduction factor $g' = \lambda_{\mathrm{mono}}/\lambda_{\mathrm{pow}}$ as a function of external load $p_{\mathrm{Ext}}$ at $T = 35\,K$. The model predictions are shown as solid and dotted lines.

integranular pore size) was chosen. This gives a conductivity ratio $L$ of about $10^5$. The predicted geometric factors $g'$ are shown as the solid(Leyers) and dotted(Chan/Tien) lines in fig. 6. Experiment and theories agree to within one order of magnitude and in the qualitative load dependence. This should be judged as a success, as the relevant parameters of silica aerogels are strongly dependent on the density and not well known for the investigated powders. The theoretically predicted increase of $g'$ for low external loads was not observed experimentally. This may be due to a non-elastic behaviour of the powder fills with non-reversible changes in the grain arrangement.

Up to now it is impossible to apply the two models to the other silica powders like diatomite, perlite or precipitated silica as the inner structure, mass density and elastic properties of the material forming the powders grains is not known. The idea to describe these systems as highly porous powders of particles made from massive vitreous silica is not very realistic.

| sample name | powder density | grain density | Young's modulus [14] |
|---|---|---|---|
| | $kg/m^3$ | $kg/m^3$ | $Pa$ |
| AP01 | 118 - 123 | 210 | $1.8 \cdot 10^7$ |
| B292/77 | 120 - 125 | 150 | $7 \cdot 10^6$ |
| G 6555 | 120 - 135 | 186 | $9 \cdot 10^6$ |

TABLE II: Material and sample properties for the evaluation of the theoretical geometric factors $g'$. The grain densities (column 3) were provided by the manufacturer.

## CONCLUSIONS

The temperature dependence of the thermal conductivities of all powders are the same as for vitreous silica in a wide temperature range. This holds for nanostructured aerogels as well as for microstructured diatomite, perlite and precipitated silica. The reduction of thermal conductivity compared to the grain material may be explained by a 'geometrical model'. The heat transfer modes within the grains are basically the same as in vitreous silica. To predict the order of magnitude of the thermal conductivity of a powder the model of Leyers as well as the one of Chan and Tien can be used.

Acknowledgements

This work was supported by Bayerisches Staatsministerium für Wirtschaft, Verkehr und Technologie, Munich, Germany. The authors like to thank the following german companies: BASF AG Ludwigshafen, Degussa AG Hanau, Deutsche Perlite Gesellschaft mbH Dortmund and Otavi Minen Dorfprozelten am Main for generously providing powder samples.

## REFERENCES

1. M.G. Kaganer 1969. *Thermal insulation in cryogenic engineering.* Jerusalem: Israel program for scientific translations.

2. C.K. Chan, C.L. Tien, 1973. "Conductance of Packed Spheres in Vacuum." *Journal of Heat Transfer - Transactions of the ASME* C **95**: 302-308.

3. H.J. Leyers, 1972. "Wärmeleitung von losen Kugelschüttungen in stagnierenden Medien." *Chemie-Ing.-Techn.* **44**: 1109-1115.

4. J.C. Maxwell, 1892. *A Treatise on Electricity and Magnetism.* London: Oxford University Press, Vol.1, 34.ed.

5. G.T. Tsao, 1961. "Thermal Conductivity of Two-Phase Materials." *Industrial and Engineering Chemistry* **53** (5): 395-397.

6. R.A.Crane, R.I. Vachon, 1972. "Effective thermal conductivity of granular materials," in *Proceedings of XII International Conference on Thermal Conductivity,* Birmingham, pp. 99.

7. P. Scheuerpflug, H.J. Morper, G. Neubert, J. Fricke, 1991. "Low Temperature Thermal Transport in Silica Aerogels." *J. Phys. D* **24**: 1395.

8. Th. Rettelbach, J. Säuberlich, S. Korder, J. Fricke, 1995. "Thermal conductivity of IR-opacified silica aerogel powders between 10 K and 275 K." *J. Phys. D* **28**: 581-587.

9. J. Kuhn, S. Korder, M.C. Arduini-Schuster, R. Caps, J. Fricke, 1993. "Infrared-optical transmission and reflection measurements on loose powders." *Rev. Sci. Instrum.,* **64** (9): 2523-2530.

10. R. Siegel, J.R. Howell, 1972. *Thermal Radiation Heat Transfer,* Tokyo: McGraw-Hill Kogakusha.

11. J. Hetfleisch, U. Heinemann, R. Caps, J. Kuhn, J. Fricke, 1995. "Evacuable Guarded Hot Plate For Thermal Conductivity Measurements Between $-200^\circ C$ and $800^\circ C$." *Proceedings of the Eurotherm Seminar 44,* Porto, Portugal

12. Th. Rettelbach, J. Säuberlich, S. Korder, J. Fricke, 1995. "Thermal conductivity of silica aerogel powders at temperatures from 10 K to 275 K." *J. Non-Cryst. Solids* **186**: 278-284.

13. Y.S. Touloukian, R.W. Powell, C.Y. Ho and P.G. Klemens, 1970. *Thermophysical Properties of Matter, Vol. 2, Thermal Conductivity - Nonmetallic Solids.* New York: IFI Plenum, pp. 923-932.

14. J. Fricke, J. Groß, 1994. "Aerogels - Their Manufacture, Structure, Properties and Applications," in *Chemical Processing of Ceramics,* B.I. Lee and E.J.A. Pope eds. New York: Marcel Dekker, pp. 311-336.

# Precision and Bias of the Large Scale Climate Simulator in the Guarded Hot Box and Cold Box Modes

K. E. WILKES, T. W. PETRIE and P. W. CHILDS

## ABSTRACT

Tests to establish the precision and bias of R-values for the central 2.44-m square part of 4.1-m square test panels were done using the large scale climate simulator (LSCS) at a U.S. national laboratory. Two different panels provided geometric variations. Air temperature in the climate chamber was varied from test to test, but was maintained near room temperature and above in the guard and metering chambers.

For most of the tests, the metering chamber required net heating to achieve steady temperatures. This is termed guarded hot box mode. For conditions where net cooling is required in the metering chamber, the LSCS is said to be in guarded cold box mode. A cooling loop was installed to provide slight overcooling of the metering chamber and allow control of temperature by the same heaters as used in guarded hot box mode.

In guarded hot box mode, intervals for 95% confidence ranged from ±1.8 to ±2.3% about linear regressions of the measured R-values with panel temperature. They increased slightly away from the respective average temperatures of all tests with each panel. Bias for the panel from layers of expanded polystyrene (EPS) over gypsum board and between and over wood joists varied from -1.3 to +2.4% of the R-value from a model using the panel's geometry and independently determined thermal conductivities of its components. The other panel was a uniform slab of EPS. Unlike for the first panel, thermocouples were imbedded into the top and bottom surfaces and a layer of paint was applied to both surfaces. Bias for this panel ranged from -6.2 to -5.2% of the R-values from the full thickness divided by the measured thermal conductivity of the EPS. The bias is attributed to the imbedded thermocouples and the paint.

For the slab of EPS, R-values from a fit of its conductances, which are expected to be linear with temperature, were extrapolated slightly to the higher mean EPS temperatures of the guarded cold box tests. With optimum cooling, the LSCS produced R-values in guarded cold box mode within the 95% confidence interval for the guarded hot box tests.

Kenneth E. Wilkes, Metals and Ceramics Division, Oak Ridge National Laboratory, Oak Ridge, TN 37831-6092

Thomas W. Petrie and Phillip W. Childs, Energy Division, Oak Ridge National Laboratory, Oak Ridge, TN 37831-6070

## INTRODUCTION

A large scale climate simulator (LSCS) was built at a U.S. national laboratory that provides a user facility for buildings technology. The LSCS allows studies of the thermal behavior of all types of commercial and residential roof systems, under climatic conditions representative of any location in the world. The LSCS was put into operation in 1987, and has been used for several projects, including studies of commercial roofing systems [5], a system with large reflective air spaces [10], moisture movement in low-slope roofing systems [9], and residential attic insulation systems [13, 14, 15].

Some of these studies were conducted using heat-flux transducers to measure the heat flow through small specimens; others, in particular those dealing with residential attic insulation, have used the guarded hot box mode (wherein net heating is required to maintain steady temperatures in the metering chamber) to measure the heat flow for large non-uniform specimens. A recent study of superinsulation in the attic space of a manufactured home [11] used both the guarded hot box and guarded cold box modes (wherein net cooling is required to maintain steady temperatures in the metering chamber). In studies involving large specimens, the behavior of the R-value of the assembly over the full range of expected climatic conditions is of interest.

The purpose of this paper is to establish the precision and bias of R-value measurements for large specimens in the LSCS and, if heat flows and test conditions are similar, in other apparatus similarly designed and operated. Precision and bias are established through measurements on two panels of moderate thermal resistance over a wide range of climatic conditions. Two panels provide variety in geometry and construction features. ASTM definitions for precision and bias are used [2]. "Precision is the closeness of agreement between randomly selected individual measurements or test results." Precision is presented as the limits of an interval about the linear regression with specimen temperature of R-values for a specimen. The interval conveys to 95% confidence the R-value of a specimen at a given temperature. "Bias is a systematic error that contributes to the difference between a population mean of the measurements or test results and an accepted reference or true value." The population mean is given herein by least-squares fits of the R-values or their inverse (termed conductances) with temperature. The true value is obtained from the geometry of the large specimen and independently determined thermal conductivities of its components.

## EXPERIMENTAL FACILITY

The large scale climate simulator, shown in Figure 1, consists of climate, metering and guard chambers, with the 4.1-m square roof test specimen situated between the climate chamber and the metering and guard chambers [8]. The climate chamber simulates the expected U.S. range of climatic conditions in either a steady-state or dynamic mode, the latter including diurnal conditions. Air temperatures in the climate chamber can range from 66°C to -40°C and the top surface of a test panel can be heated to 93°C using a bank of infrared lamps. In these tests, the infrared lamps were not used. Air temperatures went from 52°C to -32°C. The guard and metering chambers can be controlled independently between 66°C and 7°C. A range from 38°C

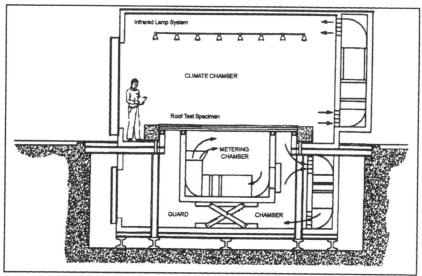

**Figure 1. Cross Section View of the Large Scale Climate Simulator.**

to 20°C was used here. Temperatures in all three chambers can be held steady to ±0.06°C except in the climate chamber when the infrared lamps are operating.

With multi-panel test specimens, the metering chamber is lowered and local heat flowrates through the multiple test panels can be measured using heat-flux transducers. In the guarded hot and cold box modes of interest for studies with single large panel specimens, the metering chamber is sealed against the panel and measures the heat flow through the central portion of the panel. The metering chamber has inside dimensions of 2.44 m by 2.44 m, with a wall thickness of 11.1 cm, of which 10.2 cm is polyiso-cyanurate foam insulation. Heat flowrates measured for the metering chamber include:

- $Q_{fan}$, input from the fans, measured by direct current (DC) voltage and current
- $Q_{heat}$, input from resistance heaters, measured by DC voltage and current
- $Q_{cool}$, removed by a chilled water coil, measured as $mc\Delta T$ where m is mass flowrate (with a Coriolis meter), c for water is assumed to 4.19 kJ/(kg·K) and $\Delta T$ is temperature rise (with a differential platinum resistance temperature device)
- $Q_{wall}$, gains through the metering chamber walls, measured with 32 differential thermocouples, known surface area, and independently measured thermal resistance of the polyisocyanurate foam.

The quantity of interest, $Q_{specimen}$, is defined as the heat gain through the open top of the metering chamber, which is sealed against the specimen. It is what yields an energy balance of zero: $Q_{specimen} + Q_{fan} + Q_{heat} - Q_{cool} + Q_{wall} = 0$. If heat loss through the top of the metering chamber ($Q_{specimen} < 0$) exceeds heat into the metering chamber through the walls of the box, and from the fans, input from the resistance heaters is needed to hold steady metering chamber temperatures. This is termed guarded hot box mode of operation. If the heat loss is small or there is a heat gain through the top, cooling is required and this is termed guarded cold box mode of operation.

Guarded cold box operation is more difficult to do than guarded hot box operation because of the thermal inertia inherent to a loop of circulating water compared to an electric resistance heater. The cooling loop originally designed for the metering chamber of the LSCS attempted to combine the cooling and control functions. Intervals for 95% confidence were ten times larger with it than with control by heating only. A new cooling loop was designed and is shown in Figure 2. It features a Coriolis flowmeter, which allows mass flowrates through the finned tube heat exchanger to be accurately measured when the flowrates are small enough to keep the temperature difference across the finned tube heat exchanger above 5°C. This combination allows for accurate measurement of the cooling. Equally as important, with the new cooling loop the control function for the metering chamber is separate from the cooling, yielding only slightly less precision in guarded cold box operation than in guarded hot box operation. For example, in five tests with $Q_{specimen}$ about 900 W, obtained from readings every four minutes over more than eight hours of steady-state, the standard deviation in $Q_{specimen}$ was ± 17 W in two tests with heating only compared to ± 44 W in three tests with fixed overcooling and control by heating.

A set amount of cooling is achieved by manual positioning the micrometer head valves in the flow control manifold so that a small amount of cool water is allowed through the Coriolis flowmeter and the metering chamber heat exchanger. If the water is cool enough, the heat exchanger provides a steady rate of slight overcooling as long as a guarded cold box experiment is ongoing. The electric resistance heater that achieves good control of the metering chamber temperature in guarded hot box operation does the same in guarded cold box operation. It holds temperature constant by the amount of steady or variable heating that is required. Variable heating is needed when dynamic conditions are imposed on the test specimen during a series.

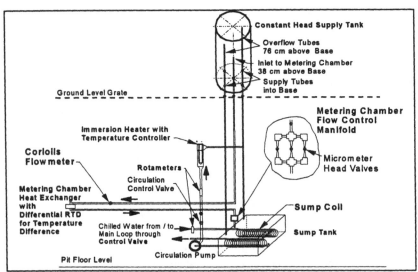

**Figure 2.  Schematic of the Cooling Loop to Achieve Steady Cooling in the Metering Chamber.**

## TEST PANELS

Two different panels of known thermal resistance but differing construction were used in the experiments. Panel A was the part of an attic test section over the metering chamber. Shown in Figure 3, Panel A consisted of a gypsum ceiling fixed to wood joists. Two layers of expanded polystyrene (EPS) foam insulation cut from 10.2 cm thick slabs completed the configuration. One layer was planed to fit between the wood joists; the other was planed to yield a total foam thickness of 12.7 cm. The thermal conductivity of the EPS was measured using a thin-heater apparatus and fit the following expression, with a probable error of ±1.2% [6,7]:

$$k_{EPS} = 0.0336 + 1.303 \times 10^{-4} \, T \tag{1}$$

where k is in W/(m·K) and T is the mean temperature in °C. Thermocouples were surface mounted on the exposed EPS surface using masking tape. Existing thermocouples on the bottom surface of the gypsum board were used for the other temperatures in calculating thermal resistance. The expected thermal resistance of this panel was calculated by the finite difference code HEATING [4], using Equation (1) for the thermal conductivity of the foam, a thermal conductivity of 0.160 W/(m·K) for gypsum board [1], and the following expression for the thermal conductivity of the wood joists [1,13]:

$$k_{wood} = 0.1105 + 3.263 \times 10^{-4} \, T \tag{2}$$

where the units are the same as for Equation (1). This analysis gave the following expression for the thermal resistance of the panel, including the effects of the foam, gypsum board, and wood joists:

$$R = 3.618 - 0.01325 \, T_m \tag{3}$$

where R is in m²·K/W and $T_m$, in °C, is the average of the temperatures on the top of the foam and the bottom of the gypsum board. The accuracy of Equation (3) is dominated by the accuracy of Equation (1).

Panel B was 10.2 cm thick, comprising slabs of EPS from the same lot as used in Panel A but painted on both surfaces like the EPS used for an ASTM/DOE hot box round robin [3]. Thermocouple beads for surface temperature measurements were pressed into the surfaces of the insulation so that the beads were flush with the outer surfaces. Panel B's R-value is its effective thickness divided by $k_{EPS}$ from Equation (1).

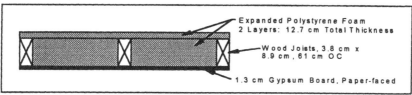

**Figure 3. Cross Section of Panel A.**

Typical temperature instrumentation of the panels consisted of 25 thermocouples on each surface of the panel in the central 2.44 m square metered area. Additional arrays of thermocouples were used to measure air temperatures but the air temperatures were not used for the R-values discussed herein.

## DATA COLLECTION AND REDUCTION

Tests were run for about 24 hours, and the most stable eight hour period, usually near the end of the test, was selected for use in data analysis. The heat flowrate through the metered area of the panels and the average of all temperatures over the array for a particular surface were obtained every four minutes. Using quantities averaged over the eight hour period, the surface-to-surface thermal resistance, $R_{ss}$, was calculated from the relationship:

$$R_{ss} = \frac{A\ (T_b - T_t)}{Q_{specimen}} \tag{4}$$

where A is the effective area of a panel exposed to the metering chamber, $T_b$ is the temperature at the bottom of a panel, $T_t$ is the temperature at the top of a panel, and $Q_{specimen}$ is the heat flowrate through the open top of the metering chamber.

The temperatures and heat flowrate in Equation (4) are determined from measurements. The effective area of a panel must be estimated because the 11.1-cm thick walls of the metering chamber are in contact with a test panel. The lower bound on this area is 5.95 m², the open area of the metering chamber. The upper bound (assuming symmetry due to balanced temperatures in the metering and guard chambers) is 6.50 m², the area measured to the centerlines of the metering chamber walls. The lower bound should apply to very thin test panels; the upper bound to very thick panels with the effective area for intermediate thickness panels between these two bounds. The effective area chosen for data reduction, 6.45 m², was slightly smaller than the upper bound and equalled the effective area of Panel A from a two-dimensional analysis using HEATING. The effective area for Panel B from HEATING was about 1% less.

## RESULTS AND DISCUSSION

### PANEL A

Six tests in guarded hot box mode were run with Panel A. For these tests, the metering chamber air temperature was maintained near 21°C, while the climate chamber air temperature was varied from -28°C to 7°C. Figure 4 shows a plot of thermal resistance versus mean temperature for Panel A. The solid line represents the linear regression and the dashed ones the limits of the 95% confidence interval about the regression. In equation form,

$$R_A = 3.690 - 0.0211\,T_{mean} \pm 0.064\,[\,1 + 0.0036(T_{mean} - 3)^2\,]^{1/2} \tag{5}$$

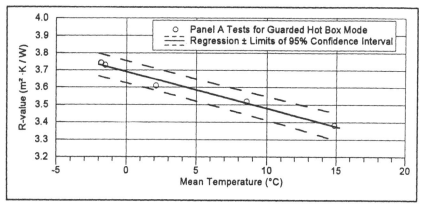

**Figure 4. Thermal Resistance versus Mean Temperature for Panel A, 12.7 cm of Expanded Polystyrene in the Attic Test Module.**

where $R_A$ is in $m^2 \cdot K/W$, $T_{mean}$ is in °C, and the first two terms represent the regression, for which the coefficient of determination ($r^2$) is 0.984. For four tests, the guard and metering chamber temperatures were both 21°C. In one of these, the climate chamber was at -28°C. The other two had climate chamber temperatures of -28°C, metering chamber temperatures of 21°C, but guard chamber temperatures of 22°C and 20°C, respectively. The resulting thermal resistances for the mean temperatures near -2°C are within 0.3%. Assigning the average of temperature differences from five or six differential thermocouples on each metering chamber face to the entire area of the face produces accurate estimates of heat flows between the metering and guard chambers.

The precision and bias for Panel A are shown in Table I for the average data at the four mean temperatures of the tests. Known resistances to calculate the bias are from Equation (3). The limits of the 95% confidence interval range from ±1.8 to ±2.3%, and the bias ranges from +2.4 to -1.3%, changing sign due to the different temperature dependencies shown by Equations (3) and (5). Overall, for Panel A, the bias is relatively small, falling generally within the precision range. It is thought that this is due to accurate knowledge of the effective thickness of the panel, since the thermocouples were not pressed into the surfaces and the foam surface was not painted.

TABLE I.    PRECISION AND BIAS FOR TESTS ON PANEL A

| Mean Temperature, °C | Regression Resistance, m²K/W | 95% Confidence Interval, m²K/W | 95% Confidence Interval, % | Known Resistance, m²K/W | Bias, % |
|---|---|---|---|---|---|
| -1.84 | 3.73 | ±0.067 | ±1.8 | 3.64 | +2.4 |
| 2.13 | 3.65 | ±0.063 | ±1.8 | 3.59 | +1.6 |
| 8.58 | 3.51 | ±0.067 | ±1.9 | 3.50 | +0.2 |
| 14.86 | 3.38 | ±0.077 | ±2.3 | 3.42 | -1.3 |

### PANEL B

Thirty-seven tests in guarded hot box mode were run with Panel B over a period of nearly two years. For these tests, climate chamber temperatures ranged from -28 to 9°C. The metering chamber was maintained near 21°C except for the last three tests wherein the metering chamber temperature was maintained at 38°C. For most of the tests, the guard and metering chamber temperatures were the same. Several tests were performed with the guard chamber temperature intentionally set either 1.1°C higher or 1.1°C lower than the metering chamber temperature. Thermal resistances for these tests varied by less than 1%, indicating again that corrections for heat flows through the walls of the metering chamber are being estimated adequately.

Nine tests in guarded cold box mode were done with the old metering chamber cooling loop during the original test period. The resulting R-values had 95% confidence interval limits of ±0.6 m²·K/W, ten times larger than in guarded hot box mode. Consequently, a new cooling loop was installed and three checks were made in guarded hot box mode followed by another series of nine runs in guarded cold box mode.

Figure 5 shows a plot of thermal resistance versus mean temperature for Panel B. The solid line is R-values from the linear regression of the conductances (the reciprocal of the R-values) with temperature. Conductance is directly proportional to the thermal conductivity of the EPS in Panel B and is expected to be a linear function of temperature. This form was chosen to allow moderate extrapolation, shown by the long-dashed line, of the R-values to the mean temperatures for guarded cold box mode. The regressed data are only those for the guarded hot box mode, not including the three checks after installation of the new cooling loop. The short-dashed lines are the limits of the 95% confidence interval about the linear regression of the R-values themselves. This regression and the confidence interval limits are given by

$$R_B = 2.861 - 0.0113\,T_{mean} \pm 0.054[1 + 0.00036(T_{mean} - 7)^2]^{1/2} \qquad (6)$$

where $R_B$ is in m²·K/W and $T_{mean}$ is in °C. The first two terms are the regression, for which the coefficient of determination ($r^2$) is 0.936. The residuals showed no systematic variations with mean temperature or with the order in which tests were performed.

Table II shows precision (as 95% confidence intervals) and bias for the R-values of Panel B at low, mid-range and high mean temperatures experienced in these tests in guarded hot box mode. The precision ranges from ±1.9 to ±2.2%, which is essentially constant. Two sets of numbers are given for the bias: the first uses the full 10.16 cm thickness of the foam, and the second uses a thickness of 9.96 cm. The latter thickness takes into account that the thermocouple beads were about 0.10 cm in diameter and were pressed into the surfaces of the foam. Known R-values are obtained by dividing thickness by thermal conductivity from Equation (1). With 10.16 cm, the measured resistances are 5.2 to 6.2% lower than the "known" values; with 9.96 cm, the measured values are 3.3 to 4.3% lower.

Accounting for penetration of the thermocouple beads is a step in the right direction. Possible causes for the remaining bias include: bias in measuring temperature difference or heat flowrate; error in the effective heat flow area; and, error in the thermal resistance used as the known value. The bias is relatively independent of mean

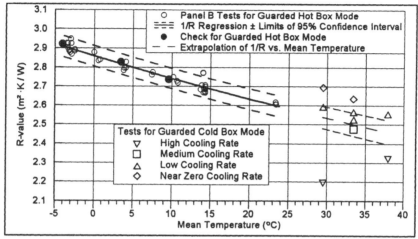

**Figure 5. Thermal Resistance versus Mean Temperature for Panel B, a 10.7 cm Layer of Expanded Polystyrene Foam.**

temperature, but temperature differences and heat flowrate are more accurate at low mean temperatures. The heat flow area needed to account for the remaining bias exceeds the upper bound area for thick specimens discussed above. Therefore, the thermal resistance used as the known value, found from the thickness divided by the thermal conductivity, is thought to be the most likely cause of the remaining bias.

Regarding the thermal conductivity, R.S. Graves used a thin-heater apparatus to measure the thermal conductivity of one specimen of the expanded polystyrene foam in 1988 and again in 1991, with results that agreed to within 0.5%. With two other specimens from the same lot of material, he conducted a round robin that involved four laboratories and three different types of apparatus (thin-heater, heat flow meter, and guarded hot plate apparatus). Thermal conductivities from this round robin ranged from 7.1% lower to 0.6% higher than that given by Equation (1) [6]. Since use of these conductivities would generally make the bias larger, it is concluded that the bias is not likely due entirely to an uncertainty in the thermal conductivity of the foam.

Error in assigning the thickness of the panel also contributes. If, on both sides of the foam, the paint penetrated 2.85 mm, which is roughly three-fourths the diameter of a bead of expanded polystyrene, the effective thickness of the foam would be 9.59 cm,

TABLE II.   PRECISION AND BIAS FOR TESTS ON PANEL B

| Mean Tempera-ture, °C | 1/R Regression Resistance, m²·K/W | 95% Confidence Interval, m²·K/W | 95% Confidence Interval, % | Thickness =10.16 cm | | Thickness=9.96 cm | |
|---|---|---|---|---|---|---|---|
| | | | | Known Resistance, m²·K/W | Bias, % | Known Resistance, m²·K/W | Bias, % |
| -3.46 | 2.90 | ±0.055 | ±1.9 | 3.06 | -5.2 | 3.00 | -3.3 |
| 6.91 | 2.78 | ±0.053 | ±1.9 | 2.94 | -5.4 | 2.88 | -3.5 |
| 23.36 | 2.60 | ±0.056 | ±2.2 | 2.77 | -6.2 | 2.71 | -4.3 |

not the 9.96 cm between thermocouple beads. The average bias would be zero. The measurements on Panel A partially confirm the truth of this hypothesis. Its EPS was from the same lot, but the thermocouple beads were taped to the unpainted surface. Table I showed that the bias was small and bi-directional for Panel A.

With small bias for both Panel A and Panel B, the LSCS shows the ability to produce accurate R-values for both complex constructions and thick specimens and, as Figure 5 shows, over a large temperature range for guarded hot box mode. Figure 5 also shows data from Panel B after the new metering chamber cooling loop was installed. Results of three checks in guarded hot box mode (without use of the cooling loop) are shown by the solid circles. They prove that reproducibility is excellent, relative to the earlier results in guarded hot box mode, despite the change to heating only for the temperature control scheme in the metering chamber.

With the new cooling loop, heating alone can control temperature as long as the cooling exceeds the sum of the heat gains through the walls and top of the box and from the circulating fans inside it. The rate of cooling is another parameter that must be set to operate in guarded cold box mode. The purpose of the series of tests in this mode was to establish the effect of cooling rate on the measured R-value for Panel B at high mean temperatures. Different amounts of cooling were imposed. Results were compared to the accepted values for this panel represented by R-values from the fit of conductances versus panel temperature extrapolated slightly to higher temperatures.

The first test with the new cooling loop was at a mean temperature of 29.5°C and high cooling rate. The tests at a mean temperature of 38°C, two more tests at a mean of 29.5°C, and the tests at a mean of 34°C followed. The cooling rates corresponding to the labels in Figure 5 are over 620 W for 'high', about 440 W for 'medium', about 160 to 270 W for 'low' and about 90 to 110 W for 'near zero.'

From the very first test, which was the shakedown run for operation of the new cooling loop, there was a systematic effect of cooling rate on the bias of the R-value. Despite this systematic effect, the final run at 'medium' cooling rate and its immediate predecessor at 'low' cooling rate show that accurate results can be obtained in a controlled fashion. The results of these guarded cold box tests except the shakedown run were reanalyzed to predict the amount of cooling that would have been required for each one's R-value to exactly equal the expected R-value for its mean temperature. A linear regression of these ideal cooling rates against the actual cooling rates shows that 330 W is the cooling rate at which this system yields the most accurate results. This cooling rate will exceed heat gains for moderately to well insulated test sections even at extreme summer conditions. Achieving this rate within ±60 W yields R-values with Panel B within the 95% confidence interval from guarded hot box operation. We can control the metering chamber cooling rate within ±15 W of the optimum value. Not enough tests were done with the optimum cooling rate to directly observe the 95% confidence interval for the guarded cold box mode of operation.

## CONCLUSIONS

A series of experiments has been performed to establish the precision and bias of a large scale climate simulator when operated in the guarded hot box and cold box modes. Tests for the guarded hot box mode were performed with two panels: A) expanded polystyrene foam, 12.7-cm thick and installed in an attic test module, and B) foam from the same lot, but in the form of a single 10.16-cm thick slab with thermocouples imbedded in its top and bottom surfaces and both surfaces painted. The limits of the 95% confidence intervals for both panels ranged from ±1.8 to ±2.3%. The bias for Panel A was both positive and negative, ranging from -1.3 to +2.4%. The bias for Panel B was consistently negative, ranging from -4.3 to -3.3% if corrections are made for penetration of the thermocouple beads into the surfaces . It is concluded that the precision (i.e., interval for 95% confidence) is smaller than ±2.5%, and the bias is less than ±5% in guarded hot box mode. Further measurements using a slab with neither painted surfaces nor thermocouple beads pressed into the surfaces might lower the limit on bias to about ±3%.

To improve precision in guarded cold box mode, a new cooling loop and temperature control scheme for the metering chamber were required. The optimum amount of cooling was determined empirically. It allowed the system in guarded cold box mode to produce R-values that were within the 95% confidence interval from the guarded hot box mode. Too few tests were run with the optimum amount of cooling to establish the precision of the guarded cold box mode directly.

## REFERENCES

1. ASHRAE, 1993. *1993 ASHRAE Handbook–Fundamentals*. Atlanta, GA: American Society of Heating, Refrigerating, and Air-Conditioning Engineers, pp. 22.6-9.

2. ASTM, 1986. *Form and Style for ASTM Standards, 7th Edition*. Philadelphia, PA: American Society for Testing and Materials, pp. 8-9.

3. Bales, E. 1988. "ASTM/DOE Hot Box Round Robin," Oak Ridge National Laboratory Report No. ORNL/Sub/84-97333/2, Oak Ridge, TN.

4. Childs, K.W. 1993. "Heating 7.2 Users' Manual," Oak Ridge National Laboratory Report No. ORNL/TM-12262, Oak Ridge, TN.

5. Courville, G.E., P.H. Shipp, T.W. Petrie, and P.W. Childs, 1989. "Comparison of the Dynamic Thermal Performance of Insulated Roof Systems," in *Proceedings, 9th Conference on Roofing Technology*. Rosemount, IL: National Roofing Contractors Association, pp. 50-55.

6. Graves, R.S. 1988 and 1992. Private Communication. Oak Ridge, TN: Oak Ridge National Laboratory.

7. McElroy, D.L., R.S. Graves, D.W. Yarbrough, and J.P. Moore, 1985. "A Flat Insulation Tester That Uses an Unguarded Nichrome Screen Wire Heater," in *Guarded Hot Plate and Heat Flow Meter Methodology*, ASTM STP 879. Philadelphia, PA: American Society for Testing and Materials, pp. 121-139.

8. Huntley, W.R. 1989. "Design Description of the Large Scale Climate Simulator," Oak Ridge National Laboratory Report No. ORNL/TM-10675, Oak Ridge, TN.

9.  Pedersen, C.Rode, T.W. Petrie, G.E. Courville, A.O. Desjarlais, P.W. Childs, and
    K.E. Wilkes, 1992. "Moisture Effects in Low-slope Roofs: Drying Rates after
    Water Addition with Various Vapor Retarders," Oak Ridge National Laboratory
    Report No. ORNL/CON-308, Oak Ridge, TN.

10. Petrie, T.W., G.E. Courville, P.H. Shipp, and P.W. Childs, 1989. "Measured R-
    values for Two Horizontal Reflective Cavities in Series," in *Proceedings, Thermal
    Performance of the Exterior Envelopes of Buildings IV*. Atlanta, GA: American
    Society of Heating, Refrigerating, and Air-Conditioning Engineers, pp. 250-257.

11. Petrie, T.W., J. Kośny, and P.W. Childs, 1995. "The Evaluation of Superinsulation
    to Improve the Energy Efficiency of Manufactured Housing: Tests in the Large
    Scale Climate Simulator," Oak Ridge National Laboratory Report No.
    ORNL/CON-407, Oak Ridge, TN.

12. Wilkes, K.E. 1979. "Thermophysical Properties Data Base Activities at Owens-
    Corning Fiberglas," in *Proceedings, Thermal Performance of the Exterior
    Envelopes of Buildings IV*. Atlanta, GA: American Society of Heating,
    Refrigerating, and Air-Conditioning Engineers, pp. 662-677.

13. Wilkes, K.E., R.L. Wendt, A. Delmas, and P.W. Childs. 1991. "Attic Testing at
    the Roof Research Center - Initial Results," in *Proceedings of the Third
    International Symposium on Roofing Technology*. Rosemont, IL: National
    Roofing Contractors Association, pp. 391-400.

14. Wilkes, K.E.. R.L. Wendt, A. Delmas, and P.W. Childs. 1991. "Thermal
    Performance of One Loose-fill Fiberglass Attic Insulation," in *Insulation
    Materials: Testing and Applications*, 2nd Volume, ASTM STP 1116.
    Philadelphia, PA: American Society for Testing and Materials, pp. 275-291.

15. Wilkes, K.E., and P.W. Childs. 1992. "Thermal Performance of Fiberglass and
    Cellulose Attic Insulations," in *Proceedings of the ASHRAE/DOE-ORNL
    Conference, Thermal Performance of the Exterior Envelopes of Buildings V*,
    Atlanta, GA: American Society of Heating, Refrigerating, and Air-Conditioning
    Engineers. pp. 357-367.

**ACKNOWLEDGEMENT**

Research sponsored by the Office of Buildings Energy Research, U.S. Department of
Energy under Contract No. DE-AC05-84OR21400 with Lockheed Martin Energy
Research Corporation as part of the National Program for Building Thermal Envelope
Systems and Materials. W.R. Huntley designed and supervised the construction of the
LSCS. R.S. Graves measured the thermal conductivity of the expanded polystyrene
foam used for the test panels and conducted a round robin to check the measurements.

# Thermal Conductivity of Resorcinol-Formaldehyde Aerogels

TH. RETTELBACH,* H.-P. EBERT,* R. CAPS,* J. FRICKE,*
C. T. ALVISO** and R. W. PEKALA**

## ABSTRACT

We measured the total thermal conductivity $\lambda$ of four resorcinol-formaldehyde (RF-) aerogel tiles in a guarded hot plate apparatus and a hot-wire device. The temperature was varied between 20 and $80°C$, the gas pressure (air) from $1000\ mbar$ down to $1 \cdot 10^{-4}\ mbar$.

All samples with bulk densities of $\varrho = 158, 180, 205$ and $236\ kg/m^3$ have been prepared using the same molar ratio of resorcinol to catalyst ($R/C = 200$).

The measured thermal conductivities are between 5 and $8 \cdot 10^{-3}\ Wm^{-1}K^{-1}$ in the evacuated state and in the range of 11 to $13 \cdot 10^{-3}\ Wm^{-1}K^{-1}$ in air at room temperature. The thermal conductivity data derived from the hot plate and the hot-wire device agree within $10^{-3}\ Wm^{-1}K^{-1}$ or 10 to 20 % of the absolute value, respectively.

From the thermal conductivity measurements as a function of air pressure a typical pore size in the aerogels between $20\ nm$ for $\varrho = 236\ kg/m^3$ and $30\ nm$ for $\varrho = 158\ kg/m^3$ was derived.

From additional infrared-optical transmission and reflection measurements the mass specific extinction $e$ and the temperature dependent radiative conductivity could be calculated. The solid conductivity was separated by subtracting the infrared-optically derived radiative conductivity from the total conductivity of the evacuated samples. The solid thermal conductivity was found to increase by about 20 % in the temperature range from $20°C$ to $80°C$.

The total thermal conductivity in air as a function of the aerogel density shows a broad minimum. The optimal density for minimized thermal conductivity at room temperature was found to be about $180\,kg/m^3$. To our knowledge the measured conductivities of the resorcinol-formaldehyde aerogels are within the lowest thermal conductivities ever measured for any solid body in air.

*  Bavarian Center for Applied Energy Research e.V. (ZAE Bayern)
   Am Hubland, D - 97074 Würzburg, Germany
** Lawrence Livermore National Laboratory, POB 808, Livermore, CA 94550, USA

## INTRODUCTION

Organic aerogels have been investigated with numerous methods since their first production 7 years ago [1] [2]. Among other properties the low thermal conductivity is very interesting. Due to the intricate network and the high porosity in the nanometer-range the gaseous thermal conductivity is strongly reduced compared to free, nonconvecting air. Also the solid conductivity of aerogels is much smaller than the conductivity of non-porous polymers. Organic aerogels from melamine-formaldehyde (MF), resorcinol-formaldehyde (RF) as well as carbon (C) aerogels have been produced and investigated. Recent thermal conductivity measurements by Lu et al. [3] with the hot-wire method yielded very low thermal conductivities down to $12 \cdot 10^{-3} W/(m \cdot K)$ in air for a RF-aerogel (made at a resorcinol to catalyst ratio of 200) at room temperature. The authors predicted a minimum in the thermal conductivity for densities between 150 and 200 $kg/m^3$. The aim of this work is to verify the former measurements using another especially accurate measurement technique: Stationary guarded hot plate conductivity measurements on four large monolithic RF-aerogel discs are presented, the densities of which are chosen close to the supposed minimum of the thermal conductivity in air at room temperature.

## THERMAL TRANSPORT IN RF-AEROGELS

In porous materials heat can be transferred by conduction via the solid structure, by gaseous conduction within the pores and by radiation. The total thermal conductivity $\lambda_{\text{tot}}$ of RF-aerogels can be treated as the sum of solid ($\lambda_{\text{sol}}$), gaseous ($\lambda_{\text{gas}}$) and radiative thermal conductivity($\lambda_{\text{rad}}$):

$$\lambda_{\text{tot}} = \lambda_{\text{sol}} + \lambda_{\text{gas}} + \lambda_{\text{rad}}. \tag{1}$$

The solid thermal conductivity strongly depends on the aerogel density $\varrho$. The dependence can be expressed with a power law: [1]

$$\lambda_{\text{sol}} \propto \varrho^{\alpha} \qquad \text{with } \alpha \approx 1.5. \tag{2}$$

The gaseous thermal conductivity $\lambda_{\text{gas}}$ depends primarily on the gas pressure $p_{\text{gas}}$ and a typical distance $\Phi$ which can be interpreted as a pore size in porous sytems like aerogels [1]:

$$\lambda_{\text{gas}} = \frac{\lambda_{\text{gas},0}}{1 + p_{1/2}/p_{\text{gas}}} \tag{3}$$

with $\lambda_{\mathrm{gas},0}$ being the thermal conductivity of free, nonconvecting air and $p_{1/2}$ a characteristic gas pressure which is related to the pore size $\Phi$ according to:

$$p_{1/2}/mbar = \frac{110}{(\Phi/\mu m)}. \tag{4}$$

This equation holds for air at room temperature.

The radiative thermal conductivity $\lambda_{\mathrm{rad}}$ can be calculated using the formula for optically thick media:

$$\lambda_{\mathrm{rad}} = \frac{16n^2\sigma T^3}{3E(T)}, \tag{5}$$

with $E = e \cdot \varrho$ being the extinction coefficient, $e$ the mass specific extinction, $\varrho$ the density and $n$ the index of refraction.

## EXPERIMENTAL SETUP

### SAMPLE PREPARATION

The samples were produced via the aqueous polycondensation of resorcinol and formaldehyde [4]. Sodium carbonate was used as a base catalyst. The molar ratio of resorcinol to catalyst was 200. After gelation in flat cylindrical molds the reaction solution was exchanged by acetone and subsequently by liquid $CO_2$. The supercritical drying was performed with respect to $CO_2$ ($T_c = 31°C$, $p_c = 74 bar$). Discs of 200 $mm$ in diameter, thicknesses between 13 and 15 $mm$ and densities of 158, 180, 205 and 236 $kg/m^3$ resulted.

### GUARDED HOT PLATE APPARATUS

The evacuable guarded hot plate apparatus LOLA 2 (see fig. 1) was used for the thermal conductivity measurements. Two disc-shaped samples were placed between three temperature-guarded plates. The plates were electrically heated and the heating power was PI-controlled in order to keep each plane at the required temperature. The temperatures were measured by Pt-100 resistance thermometers. The middle plane was divided into a central metering section and two guard rings. The latter were at the same temperature as the center in order to avoid radial heat flow. Under stationary conditions the thermal conductivity was derived from:

$$\lambda = \frac{P \cdot d}{2 A \cdot (T_1 - T_2)}, \tag{6}$$

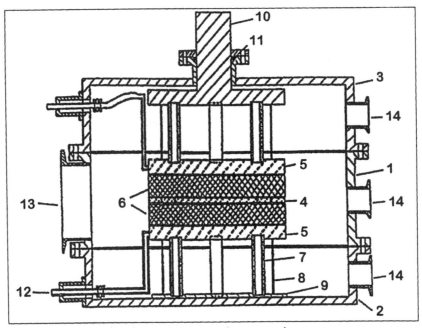

| 1-3: | vacuum chamber | 10: | piston |
|------|----------------|-----|--------|
| 4: | central metering plate with two guard rings | | |
| 5: | cold plates | 11: | sealing |
| 6: | samples | 12: | fluid pipes |
| 7-9: | mechanical support | 13-14: | flanges |

Figure 1: computer controlled evacuable guarded hot plate apparatus LOLA 2

where $P$ was the electrical power fed into the central metering section kept at the elevated temperature $T_1$, $A$ the area of the central metering section, $d$ the sample thickness and $T_2$ the temperature of the cold plates.

For two-sided measurements the upper and the lower sample must be identical. As only one specimen of a given density was available, one - sided measurement were performed. In this case the temperature of one of the cold plates was held at the same temperature as the middle plate and served as a guard. Two of the RF-aerogel discs (with different densities) at a time were placed between the three plates, the more rigid sample with the higher density was placed onto the lower plate. The one-sided measurement procedures were applied twice in order to measure the thermal conductivity of both samples.

## HOT-WIRE MEASUREMENTS

For the hot-wire measurements, RF-aerogel discs used in the guarded hot plate measurements, were cut into rectangular pieces of 12 x 3 x 1.5 $cm^3$. A platinum wire with a length of about 8 cm was squeezed between two identical rectangular pieces of RF-aerogels. Pressure was applied via four loading screws in the sample holder (see fig. 2). The pressure was increased until no air slit was visible between the aerogel tiles.

For the evaluation of the measured $\Delta T(t)$-curves we used a new theoretical solution which considers a contact resistance between wire and specimen as well as heat losses caused by the wire connections [6]. Fit parameters were the thermal conductivity, the thermal contact resistance and the initial temperature. In addition heat capacity data were obtained from measurements with a differential scanning calorimeter (DSC- 400, Netzsch Gerätebau) [7].

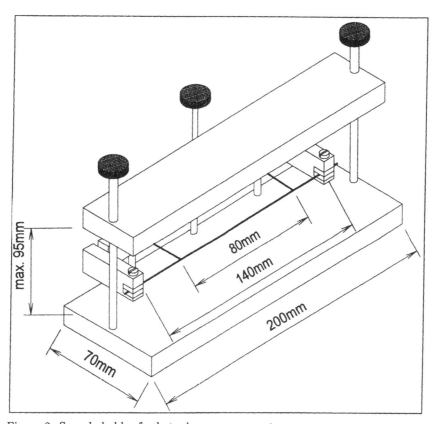

Figure 2: Sample holder for hot-wire measurements.

## MEASUREMENT PARAMETERS AND RESULTS

### MASS SPECIFIC EXTINCTION

The mass specific extinction of RF-aerogel was determined by infrared optical hemispheric transmission and reflection measurements. For this purpose, thin slices of about $200\ \mu m$ thickness were cut from the RF-aerogels with a circular saw. The experimental setup and the calculation parameters are given in [8].

The resulting mass specific extinction is plotted in fig. 3 as a function of wavelength $\Lambda$. The spectrum is dominated by a strong $OH$-absorption peak at $\Lambda = 3\mu m$ and absorption bands at wavelengths between 6 and $10\mu m$ which are due to $C - C$ valence oscillations.

In fig. 4 the resulting Rosseland-averaged temperature dependent mass specific extinction is shown. Within the broad maximum at $T = 250K$ $e$ is about $40\ m^2/kg$. For temperatures between 0 and $100°C$ the specific extinction decreases from 40 to $30\ m^2/kg$ monotonously. The uncertainty in the specific extinction is in the range of 10 to 20 %.

### THERMAL CONDUCTIVITY

After being evacuated and heated up to $80°C$ for 6 hours the thermal conductivity of each sample was measured in the guarded hot plate apparatus at a mean temperature of $20°C$ as a function of gas pressure (air). Measurements were also performed in the evacuated state (gas pressure below $10^{-3}mbar$ at three different temperatures ($20°C$, $50°C$ and $80°C$). The external load onto the samples remained less than $10^2 mbar$ during the measurements.

With the hot-wire technique the thermal conductivity of two samples ($\varrho = 180\ kg/m^3$ and $205\ kg/m^3$) was measured as a function of gas pressure at a temperature of $20°C$.

The results of the gas pressure dependence are depicted in figs. 5a and b. Under evacuation the thermal conductivity is 5.5 to $8 \cdot 10^{-3} W/(m \cdot K)$, depending on density. Only above a gas pressure of about $100\ mbar$ does the thermal conductivity vary in a significant way: it increases by 4 to $6 \cdot 10^{-3}\ W/(m \cdot K)$ yielding thermal conductivities in air between 11 and $13 \cdot 10^{-3} W/(m \cdot K)$. There is a good agreement between stationary and dynamically derived data.

The results of the temperature dependent measurements with the guarded hot plate are depicted in fig. 6. The samples were evacuated down to $10^{-4}\ mbar$. At $20°C$ the thermal conductivity ranges from 5 to $8 \cdot 10^{-3} W/(m \cdot K)$. A distinct increase with temperature is observed.

### MEASUREMENT ERRORS

In order to quantify and to eliminate errors due to radial heat flow within the middle plane, the thermal conductivity was measured applying two different temperature differences (10 K, 20 K) and extrapolating the plot of the thermal heat transfer coefficient as a function of $1/\Delta T$ towards $1/\Delta T \rightarrow 0$. This

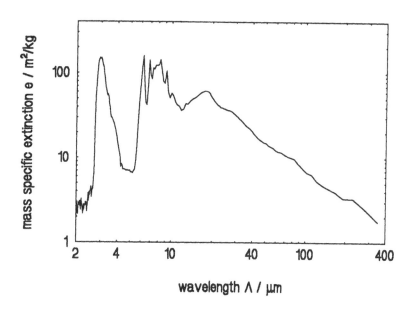

Figure 3: Spectral mass specific extinction $e$ of RF-aerogel in the wavelength range $2.3..350\mu m$.

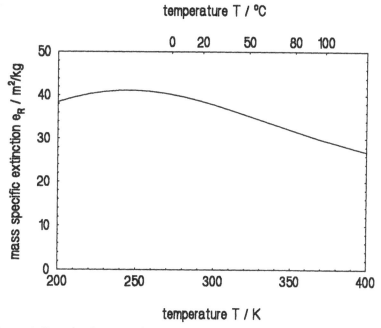

Figure 4: Rosseland averaged mass specific extiction $e(T)$.

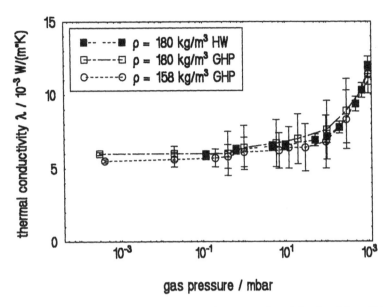

Figure 5a: Thermal conductivity of RF-aerogels with densities 158 and 180 $kg/m^3$ as a function of gas pressure $p_{gas}$, $T = 20°C$. The empty symbols represent the data from guarded hot plate measurements, the full symbols from hot-wire measurements.

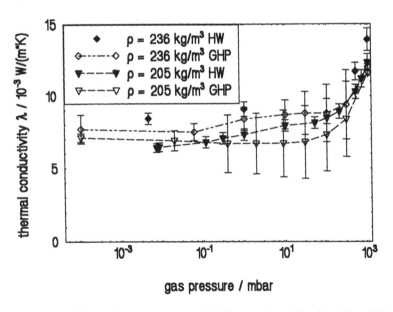

Figure 5b: Thermal conductivity of RF-aerogels with densities 205 and 236 $kg/m^3$ as a function of gas pressure $p_{gas}$, $T = 20°C$. The empty symbols represent the data from guarded hot plate measurements, the full symbols from hot-wire measurements.

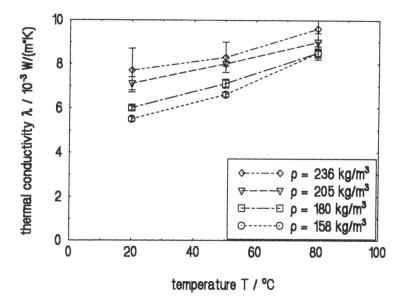

Figure 6: Thermal conductivity of evacuated RF-aerogels as a function of temperature $T$, $p_{\text{gas}} = 10^{-4} mbar$; parameter is the density $\varrho$.

procedure is equivalent to plotting the electrical heating power $P$ vs. the temperature difference $\Delta T$ [5]. Due to a failure of one temperature sensor in the second guard ring significant radial heat flows occurred during the gas pressure dependent measurements. Therefore the correcture term in the determination of $\lambda$ was rather large and the uncertainty of the data was increased. The depicted error bars in figs. 5 a and b show the size of this correction. The actual errors are far smaller (about 5 %). The measurements in air and in the evacuated state have been confirmed after installing a new sensor.

The uncertainty in the thermal conductivity measured by the hot-wire technique is generally less than 5%, if the new solution is used [6].

## INTERPRETATION

### SOLID AND RADIATIVE THERMAL CONDUCTIVITY

The solid thermal conductivity can be derived from conductivity measurements under vacuum if the radiative thermal conductivity (according to eq.(1)) is known. The latter can be calculated from the Rosseland averaged mass specific extinction from fig. 4 using equation (5). The results are shown (for the different temperatures) in table I.

At room temperature the derived solid thermal conductivities are in the

TABLE I: Solid and radiative thermal conductivity of evacuated monolithic RF-aerogels.

| sample density $\varrho$ | 20°C | | 50°C | | 80°C | |
|---|---|---|---|---|---|---|
| | $\lambda_{\text{rad}}$ | $\lambda_{\text{sol}}$ | $\lambda_{\text{rad}}$ | $\lambda_{\text{sol}}$ | $\lambda_{\text{rad}}$ | $\lambda_{\text{sol}}$ |
| $kg/m^3$ | $10^{-3}W/(m \cdot K)$ | | $10^{-3}W/(m \cdot K)$ | | $10^{-3}W/(m \cdot K)$ | |
| 158 | 1.0 | 4.5 | 1.4 | 5.2 | 2.0 | 6.5 |
| 180 | 0.9 | 5.1 | 1.2 | 5.9 | 1.7 | 6.8 |
| 205 | 0.8 | 6.3 | 1.1 | 6.9 | 1.5 | 7.5 |
| 236 | 0.7 | 7.0 | 0.9 | 7.4 | 1.3 | 8.3 |

range from 4.5 (for a density of $158\,kg/m^3$) to $7 \cdot 10^{-3}\,W/(m \cdot K)$ (for a density of $236\,kg/m^3$). Compared to monolithic silica aerogels of similar density [2] [9] the absolute value of the solid thermal conductivity of RF-aerogels is lower. Compared to the conductivity of the respective bulk material (i.e. fused silica and resorcinol-formaldehyde resin), however, the reduction in thermal conductivity due to the nanoporous network is less (a factor of 10 to $10^2$ for RF compared to a factor of $10^2$ to $10^3$ for silica). This indicates stronger contacts between the primary particles in the RF-aerogel skeleton compared to those in silica aerogels.

The solid thermal conductivity is temperature dependent, it increases by more than 20 % if the temperature is raised from 20 to 80°C. The radiative thermal conductivity increases (at a temperature of 20°C) from $0.7 \cdot 10^{-3}\,W/(m \cdot K)$ for a density of $236\,kg/m^3$ to $1.0 \cdot 10^{-3}\,W/(m \cdot K)$ for $\rho = 158\,kg/m^3$.

## ESTIMATION OF $p_{1/2}$ AND PORE SIZE $\Phi$

With equations (3) and (4) the gas pressure $p_{1/2}$ for which $\lambda = \lambda_{\text{gas},0}/2$ and a typical pore size $\Phi$ can be calculated from the gas pressure dependent thermal conductivities. According to table II the derived values of $p_{1/2}$ increase, as expected, from 3.6 to 5.8 *bar* with increasing density, i.e. at normal air pressure the gaseous thermal conductivity is considerably smaller than $\lambda_{\text{gas},0}/2$. The calculated mean pore sizes $\Phi$ are around $25\,nm$ (table 2). These values are typical for nanostructured, supercritically dried aerogels. As expected, the effective pore diameters decrease with increasing density of the RF-aerogel.

TABLE II: Summary of conductivity results (guarded hot plate measurements) and derived quantities

| sample density $\varrho$ | $\lambda_{\text{tot}}$ at 20°C in air | $\lambda_{\text{sol}} + \lambda_{\text{rad}}$ at 20°C | $\lambda_{\text{sol}}$ at 20°C | $p_{1/2}$ | $\Phi$ |
|---|---|---|---|---|---|
| $kg/m^3$ | $10^{-3}W/(m \cdot K)$ | $10^{-3}W/(m \cdot K)$ | $10^{-3}W/(m \cdot K)$ | $mbar$ | $nm$ |
| 158 | 11.4 | 5.5 | 4.5 | 3600 | 30 |
| 180 | 10.9 | 6.0 | 5.1 | 4400 | 25 |
| 205 | 11.6 | 6.7 | 6.3 | 4400 | 25 |
| 236 | 12.0 | 7.5 | 7.0 | 5800 | 20 |

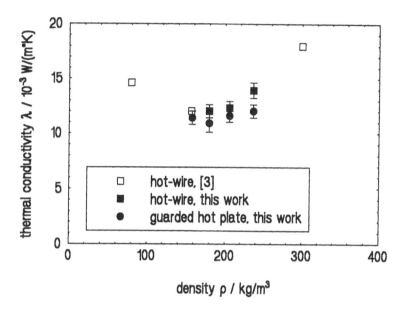

Figure 7: Total thermal conducitivity $\lambda_{\text{tot}}$ of RF-aerogels as a function of density. The molar ratio R/F was 200 for all samples. The temperature was $20°C$, the gas pressure was 1 bar (air). Data from this work and from [3] are shown.

## OPTIMAL DENSITY FOR MINIMIZED THERMAL CONDUCTIVITY IN AIR

The plot $\lambda_{\text{tot}}(\varrho)$ (fig. 7) shows a flat minimum within the density range covered by the provided samples. The optimal density for minimized thermal conductivity in air at ambient temperatures should be in the range of $180\pm10kg/m^3$. At higher densities the total conductivity is dominated by the solid contributions, at lower densities gaseous and radiative conductivities become dominant. These data verify the findings of Lu et al. [3] who suggested a minimum of total thermal conductivity in air-filled RF-aerogel for densities between 150 and 200 $kg/m^3$.

## CONCLUSIONS

We could demonstrate that hot-wire and guarded hot plate measurements yield the same conductivity values within the error range for all investigated aerogel specimens.

The results show that RF-aerogels are highly thermal insulating materials both in air and in the evacuated state. As the thermal conductivity does not increase for gas pressures up to 100 $mbar$, they (as well as silica aerogels) are well-suited as insulants in technical vacuum.

Acknowledgements

This work was supported by Bayerisches Staatsministerium für Wirtschaft, Verkehr und Technologie, Munich, Germany.

## REFERENCES

1. X.Lu, M.C.Arduini-Schuster, J.Kuhn, O.Nilsson, J.Fricke, R.W.Pekala (1992). "Thermal Conductivity of Monolithic Organic Aerogels." *Science* **255**: 971-972.

2. X.Lu, P.Wang, M.C.Arduini-Schuster, J.Kuhn, D.Büttner, O.Nilsson, U.Heinemann, J.Fricke, (1992). "Thermal transport in organic and opacified silica monolithic aerogels." *J. Non-Cryst. Solids* **145**: 207-210.

3. X.Lu, R.Caps, J.Fricke, R.W.Pekala, C.T.Alviso, L.W.Hrubesh, (1995). "Correlation between structure and thermal conductivity of organic aerogels." *J. Non-Cryst. Solids* **188**: 226-234.

4. R.W.Pekala, C.T. Alviso, F.M. Kong, S.S. Hulsey(1992). "Aerogels derived from multifunctional organic monomers." *J. Non-Cryst. Solids* **145**: 90-98.

5. U. Heinemann, J. Hetfleisch, R. Caps, J. Kuhn, J. Fricke, (1995). "Evacuable guarded hot plate for thermal conductivity measurements between $-200°C$ and $800°C$." *Proceedings of Eurotherm Seminar* **44**, Porto, Portugal.

6. H.P.Ebert, V.Bock, O.Nilsson, J.Fricke, (1993). "The hot-wire method applied to porous materials of low thermal conductivity." *High Temp.-High Press.* **25**: 391.

7. J.Blumm, Netzsch Gerätebau, Selb, Germany (1995). *Private communication.*

8. J.Kuhn, S.Korder, M.C.Arduini-Schuster, R.Caps and J.Fricke, (1993). "Infrared-optical transmission and reflection measurements on loose powders." *Rev. Sci. Instrum.* **64** (9): 2523-2530.

9. P.Scheuerpflug, H.J.Morper, G. Neubert, J. Fricke, 1991. "Low Temperature Thermal Transport in Silica Aerogels." *J. Phys. D* **24**: 1395.

# Reference Materials and Transfer Standards for Use in Measurements on Low Thermal Conductivity Materials and Systems

R. P. TYE and U. S. RIKO

## ABSTRACT

Reference materials and transfer standards are essential both to validate thermal conductivity techniques and apparatus based on primary methods and for use in the calibration of equipment based on secondary methods. A variety of materials and specimens are required to satisfy the overall needs which cover wide ranges of thermal conductivity and temperature.

The current available materials and standards for use as specimens of low thermal conductivity (<5W/m.K) are reviewed and examples provided to illustrate their present usefulness and the need for others to be made available. Related on-going and planned work, both nationally and internationally, is described and discussed. Areas where significant gaps exist and where such materials and specimens are needed are identified. Recommendations are made concerning candidate materials for use in specific areas. There is a continuing unfilled need for more international cooperative work to develop several standards for different applications.

## INTRODUCTION

An accurate knowledge of the thermal conductivity of all types of materials is an essential requirement for all applications where transfer of heat is concerned. Overall, the range of thermal conductivity for solids at all temperatures covers approximately six orders of magnitude. This becomes even greater if porous materials in high vacuum are included. This factor necessitates different steady-state and transient state methods and techniques be available for measurements over the complete range (1,2).

The measurement of thermal conductivity and its related properties, although simple in principle, is a difficult one to make accurately. This is due essentially to the problem of ensuring that heat losses are eliminated or can be minimised and/or allowed for by correction. This is particularly true for materials having a low thermal conductivity defined here as <5 W/m.K and especially so for

Ulvac Sinku Riko Inc., 6 Riverside Drive, Andover, MA 08810

thermal insulation defined here as <0.2 W/m.K. In this case, heat fluxes are generally small and any extraneous heat loss or gain becomes significant compared to the total amount.

Clearly, materials or specimens having established or accepted values are necessary to serve the needs of the experimentalist who is trying to establish reliable materials performance for the range of materials used by those working in the field of heat transfer. Specifically, they are necessary for the following requirements:

- to validate the performance of new techniques or apparatus.

- to verify existing absolute apparatus such as guarded hot plate, e.g. ISO 8302, ASTM C-177 and thermal diffusivity systems, e.g. ASTM E-1461.

- to calibrate test apparatus based on secondary methods such as the heat flow meter, e.g. ISO 8301, ASTM C-518 and comparative techniques, e.g. ASTM E-1225.

- to serve as reference points in round robin and interlaboratory studies.

- to be used by manufacturers and users of materials and products as referee materials in order to question and/or verify competitive claims.

In general, reference materials fall into three categories.

- Standard (or Certified) Reference Material, SRM, obtained from a well characterised stock held be a national standards laboratory or similar organisation, and having very reproducible values. The latter are certified by the organisation and are based usually upon the results of a significant testing and evaluation programme.

- Reference material, RM, usually a material available commercially which has been measured by a number of different organisations, often internationally, and having a consensus value.

- Transfer Standard, TS, a specimen of a material and a given geometry having a thermal conductivity or thermal resistance which has been measured using an absolute method by a national standards laboratory or equivalent organisation. This is a unique property value which cannot necessarily be applied to the material itself.

The major criteria for a reference material are that it:

- be uniform in its properties, e.g. composition, density thickness etc.
- be durable when handled.
- be stable over the temperature range of application.
- have properties unaffected be external variables.

For the wide ranges of property and temperature involved in different applications, a number of such materials are necessary to fulfil the needs.

Historically, most attention has been focused on the need for reference materials having high thermal conductivity values (>10 W/m.K). As a result of these international efforts through various national standards laboratories, and international data generation organisation etc. a limited number of metals both in the SRM and RM categories are available. These include for example, electrolytic iron, a stainless steel, tungsten, Inconel 718 and an isotropic graphite. However, the stocks of some of these are now dangerously low or depleted and further work will be necessary to renew them.

In the range between 1 and 5W/m.K, fused silica, Pyrex 7740 glass and Pyroceram 9606 have been studied extensively since the mid sixties and are used as RM materials (3). However, the useful range of the first two is limited to below 700K, where the effects of radiation transmission are relatively small, while the third material is still of relatively high thermal conductivity particularly below 300K.

An interlaboratory study involving eight laboratories and three different steady-state and transient methods was carried out on specimens of Pyrex 7740 and Pyroceram 9606 cut from the same batches (4). The results confirmed the previously accepted values to better than 1%, but the overall uncertainty of some +/- 6 to 7% remained. The National Institute of Standards and Technology, NIST, currently retains the specimens which consist of appropriate pairs of 200mm diameter guarded hot plate pieces, several approximately 50mm diameter comparative pieces and 10 and 12mm diameter thermal diffusivity pieces. They are available for loan upon request.

In the range of below 0.1W/m.K, the only materials of wide use are a high density 150kg/m$^3$ glass fibreboard such as SRM 1450b and a lower density, 50kg/m$^3$, glass fibre blanket such as the SRM 1451, both of which are available from NIST in the USA. Similar materials are available from other national standard organisations in a number of European countries. However, the values of thermal conductivity are very similar and the useful temperature range while recently extended to 100 to 400K, is still limited due essentially to the fact that the contained phenolic binder burns off at elevated temperatures. For the blanket material, there is a large radiation contribution at the higher temperature. Thus, the choice of materials having values of below 2 W/m.K is still very restricted. Furthermore, the stock of 1450b in the USA is now depleted, as is that of a similar French material that was available from the European Reference Bureau in Brussels. New stocks of both materials will be characterised in the near future by NIST in the USA and Laboratoire National D'Essai in France.

The following discussion elaborates on the ever increasing urgency for further matrials to be made available in the range of thermal conductivity below 2 W/m.K. It also includes a summary of some of the current work which is known to be on-going in order to help satisfy these needs.

## ESTABLISHMENT OF NEEDS

Some 20 years ago, the present author drew attention to the need in this area and indicated that in the USA, the ASTM Committee C16 on Thermal Insulation was addressing the problem through an active task force (5). By 1977, a programme plan, including some suggested materials, had been developed by this group together with a suggested timetable. Despite the urgency, only a relatively small amount of work has been carried out due, in general, to limited funding. While the need is obvious to those working in the field, it is often difficult to persuade management and funding agencies to support work which may be viewed as somewhat esoteric, especially as the sums of money to cover research costs of initial characterisation and subsequent measurements programmes are considerable.

Over the past 20 years or so, several factors have exacerbated the needs for reference materials, especially for the thermal insulation regime. These include:

- increased use of thermal insulations because of energy conservation and environmental quality issues.

- development of much larger dimension apparatus (6) especially those using heat flux transducers and requiring calibration with thicker SRM or TS specimens having both the range of property and heat transmission mechanisms similar to the test materials.

- materials being used at much higher temperatures.

- significant differences in property of 30% or more, obtained for the same material by different apparatus or different organisations using the same methods, even at moderate temperatures (6).

- development of new steady-state and transient methods of measurement.

- national and international accreditation of laboratories and certification of products.

- national or local regulations and energy codes implying the needs for accurate energy savings figures based on measured values.

- international standardisation and designs of "standard" apparatus.

- recognition of the thickness effect and other issues relating to variability and inhomogeneity in low density materials and products. The former issue has been addressed partially for building insulation materials but not at temperatures much above 300K.

- increasing use of very high thermal resistivity inhomogeneous panels based on containment of a low conductivity gas or using a high or partial vacuum.

- increased modelling and analysis of materials and systems performance that requires validation.

## CURRENT AND FUTURE WORK

### A.    New Materials

The original plan listed 14 recommended potential reference materials. However, it was not exhaustive in scope.  Based upon work with which the present author has been associated, it is suggested that the following materials could now be included, especially as replacements for some of the earlier materials.

1.    A low density, blown, stabilised aluminosilicate fibre (Nextel[R]) blanket of about 12 kg/m$^3$ and having a fibre diameter of the order of 3 microns.

   This appears to be very uniform and is stable up to 1000K and possibly higher.  It can be readily compressed as blankets to uniform densities up to the order of 120 kg/m$^3$.  These blankets have much superior thermal performance to equivalent density 7 micron alumiosilicate products and are much more homogeneous.  Thus, the material also appears to be a valuable subject for modelling of heat transfer mechanisms at elevated temperatures.

2.    A closed cell, low density phenolic foam containing low conductivity blowing agent.

   This material, unlike polyurethanes and polyisocyanurates does not appear to age significantly.  If thin, very low permeance skins could be applied it would be ideally suited as a material having a very low value of thermal conducitivity ~0.016 W/m.K.  It has a further advantage in that it can be used to higher temperatures (~400K) than some other cellular plastics.

3.    A high density (>150kg/m$^3$) pure rock fibre product manufactured in Europe.

   This is uniform and stable to 1200K and does not appear to be affected by cyclic heating and cooling.  It has much lower unfiberised solid ("shot") content than other similar slag or rock mineral fibre products. A significant advantage is that it covers a range of thermal conductivity similar to that of many of the current fibrous materials used in common commercial and industrial applications without having a very large radiation component at the temperatures of use.

## B. On-Going Work

1. North America

As discussed earlier, the range of application of SRM 1450b and 1451 has been extended to approximately 110 to 400K. During these studies, it was found that the phenolic binder is sensitive to humidity, and rapidly absorbs up to 0.4wt% of water after drying. As a consequence, abnormal heat capacity values can be measured due to this water intake. It is essential therefore, that those involved in developing transient methods and using such SRM materials for verification be aware of such problems, which may be introduced by external issues.

To serve the needs of those workers using large guarded hot plate and heat flow meter apparatus to evaluate building insulation performance, both NIST in USA and the National Research Council, NRC, in Canada now provide individual transfer standards. For example, NIST provides a closely matched pair of low density, ~12 kg/m$^3$, fibrous glass blankets approximately 150mm in thickness.

The values of apparent thermal conductivity are determined at 24°C with the NIST 1m diameter line source hot plate. In this context, the plate can also be used to characterise similar types of insulation materials supplied by clients requiring their own set of TS specimens.

Following the C-16 suggestions, NIST established a programme of somewhat limited scope to develop SRM materials suitable for the low end of the range and for temperatures up to 800K as a starting point. As a result, a microporous material, one of those suggested by the Task Group, has now become SRM 1449 for a limited temperature range around 300K. This is a fumed amorphous silica (10$^{-8}$ diameter particles) material manufactured in Germany by Wacker Chemie. It is hoped to extend the range following completion of an interlaboratory study of high temperature guarded hot plates which is now underway. As part of this study, a high density calcium silicate block material and a high density aluminosilicate blanket material may also become reference materials.

This SRM has a density in the range 200kg/m$^3$ and apparent thermal conductivity at 300K of approximately 0.022 W/m.K. It was chosen first because of this low value and also because it appeared to be stable up to at least 800K. A considerable number of measurements have been made on this material to elevated temperatures and at different pressures. However, NIST has not yet extended the temperature range of the SRM Certificate, but should do so once the guarded hot plate comparison measurements programme has been completed.

Results indicated that it is stable although somewhat friable and has uniform reproducible thermal conductivity values at 300K and up to 700K. These are affected by barometric pressure, as would be expected for microporous materials, and this point must be borne in mind when being used. Thus reported values must also include the ambient pressure at the time of measurement.

It is still realised that this material, while having a low value of thermal conductivity, has a very much higher density and different thermal diffusivity to other insulation materials such as cellular plastics. Furthermore, the heat transmission mechanisms are much different, since radiation has been minimised and the gas conduction is much reduced. In order to satisfy the need for a similar type of material to existing cellular plastics, NIST is now examining the possibility of obtaining a special lot of the fluorocarbon blown phenolic foam mentioned earlier.

During the past several years, an unfunded international programme developed jointly by ASTM and NIST has been carried out. The purpose was to produce an SRM for use by the glass and ceramics industries. The material is the original stock of Pyroceram 9606 purchased in 1963. Measurements of thermal diffusivity and thermal conductivity were to be made by several methods to 1300K where possible. Not all participants could undertake measurements to this temperature. At this time, the various sets of thermal diffusivity results are under analysis, but several laboratories have yet to complete thermal conductivity measurements. Because of the fact that not all results cover the total range of temperature, the decision on how best to analyse results becomes an important factor. However, it is hoped that this material (to be designated SRM 1415) can be given certification within the next year.

## 2.    Europe

Work on standard reference materials, particularly for use over a limited temperature range around 300K, is being undertaken in several different countries and under the auspices of central organisations in CEN and former East Europe. In all cases, there is one common material type, namely a high stable, high density fibrous glass board similar to the SRM 1450b. The densities vary between 90 and 144 $kg/m^3$ and the materials are available with certified values for the approximate range 150 to 370K.

A great deal of work has also been carried out under a Bureau Communitaire de Reference, BCR, programme on a Pyrex glass, very similar to Pyrex 7740. This was chosen because it has a value of thermal conductivity similar to many construction materials such as light-weight concretes and is not affected by moisture. A very careful series of measurements has been undertaken by four laboratories in France, Germany, Italy and the UK with special attention being paid to temperature measurement of the surfaces. Agreement between the laboratories was better than 2% overall for the temperature range 170 to 400K.

This group has also evaluated a 10 $kg/m^3$ fibrous material specifically to assist in analysing and understanding heat transmission mechanisms and especially the thickness effect. The material is a very uniform, low density polyester, hollow, 4 micron fibre used in the clothing and furnishing industries. Results indicate that at 300K, it has a thickness effect of some 15% between 30 and 100mm. It has an extremely high radiation contribution to the total apparent thermal conductivity with a reduction of 0.5% for each 0.1% increase in density. It has not yet been certified as a reference material. Although this material has a limited range of applicability, it should prove a very useful tool in current modelling, especially if complemented with a higher temperature material such as the Nextel[R].

In the UK, individual TS specimens are made available by the National Physical Laboratory, NPL. These have measured values with an accuracy of +/- 2% over the range 270 to 350K. They consist of various sizes of 50mm thick pieces of Perspex, a transparent plastic, a Nylon, and a high density polyethylene. Each specimen is cut from a lot of a specific material purchased especially for use as transfer standards. They were chosen particularly to assist all of those involved in undertaking measurements on hard, higher thermal conductivity materials such as concretes since in the seventies there had been considerable discrepancies (>30%) in results between laboratories on such materials. Due to the availability of these transfer standards, significant improvement in measurement technology has been achieved with some consequent reduction in the discrepancies.

Recently NPL has completed measurements up to 900K on a high density calcium silicate block material and this is now available as an RM for those interested in high temperature measurements. This material should assist in reducing current uncertainties in values obtained by guarded hot plate methods which are in the +/- 10% range.

Other materials available in Europe include three developed in Germany, supplemented by a medium and a high density fibreglass board from France and Hungary, respectively. These are listed in Table I. They cover a useful range but are orientated towards building applications only.

**TABLE I**

**REFERENCE MATERIALS IN EUROPEAN COUNTRIES**

| Material | Thermal Conductivity at 300K (W/m.K) | Accuracy (%) | Temperature Range (K) |
|---|---|---|---|
| Polymethacrylsauremethylester | 0.2 | +/-1 | 240-350 |
| Optical glass BK7 | 1.0 | +/-1 | 240-350 |
| Schaumpolystyrene | 0.04 | +/-3 | 270-330 |
| Medium density fibreglass | 0.030 | +/-2 | 270-330 |
| High density fibreglass board | 0.029 | +/-2 | 270-330 |

At high temperatures, one study was carried out during the approximate period 1987 - 1992 on a well characterised high density ($300kg/m^3$) calcium silicated material under the European Community Reference Materials Programme. In total, 16 organisations were involved, 8 used the parallel wire version of the line source method, 5 used the guarded hot plate, 4 used the ASTM C-182 brick calorimeter method and 3 also used different thermal diffusivity techniques during the initial characterisation.

Measurements were made or attempted to a mean temperature of 1200K. Overall, the results were most disappointing. Considerable differences of +/-10% and greater were obtained using the hot wire method, 3 sets of guarded hot plate results were good agreement but two others were found to be bad outliers and 3 of the 4 calorimeter test results were also very significant outliers, as were all of the thermal diffusivity results.

As a result of the overall analysis of the results, the BCR decided that the material could not become a Certified Reference Material. However, the stock of material has been retained by BCR as a Reference material only for the hot wire method and having the values of uncertainties shown in Table 2.

## TABLE II

Reference values with 95% confidence intervals for thermal conductivity of calcium silicate bricks using the parallel hot wire method.

| TEMPERATURE (K) | THERMAL CONDUCTIVITY $10^3$(W/m.K) |
|---|---|
| 300 | (97 +/- 8) |
| 523 | (107 +/- 7) |
| 773 | (132 +/- 9) |
| 973 | (155 +/- 10) |
| 1173 | (184 +/- 12) |

In the preliminary characterisation of this material, it exhibited approximate 15% anisotropy in the measured thermal conductivity. However, in the measurements programme, there appeared to be good agreement between the above averaged results and those of the three guarded hot plates over the range 523 to 973K. Because of the anisotropy, it would be expected that the hot wire results (some intermediate value between parallel and perpendicular direction of heat flow) would be some 5 to 10% above the guarded hot plate (heat flow perpendicular). This discrepancy has not been resolved.

Currently there is a proposal to the European Union to develop a high temperature reference material(s) for use by the thermal insulation industry. It is possible that this will include a new series of measurements on this well characterised calcium silicate material using only the guarded hot plate.

In the UK, the National Physical Laboratory has a small stock of a very high density (>800 kg/m$^3$) calcium silicate material. Measurements have been made on several specimen pairs from the stock. As a result of these measurements it is now a Certified Reference Material for the approximate temperature range $200 < T < 800°C$. The thermal conductivity can be expressed by the cubic equation:

$$\lambda = 0.2096 - 1.479 \times 10^{-4}T + 2.5108 \times 10^{-7}T^2 - 1.1914 \times 10^{-10}T^3 \quad (1)$$

with a total determinate error uncertainty estimated to be +/- 4% at the 95% confidence level. From a practical point of view, the thermal conductivity is almost independent of temperature over the above range and this is an important attribute for a reference material.

A separate European programme involved measurements being undertaken in several different countries on a French manufactured cordierite ceramic. This is believed to be quite stable to at least 1000K and has a value of thermal conductivity at 300K of approximately 1.5 W/m.K with a relatively small temperature coefficient up to 450K. The final results have not yet been published. Finally, NPL in the UK also has a limited supply of a microporous material (Microtherm) and a cellular glass from which it will supply individually measured transfer standards to appropriate temperature requirements.

## COMPOSITE BUILDING PANELS

The previous discussion has concentrated upon "homogeneous" materials or products where the term thermal conductivity can be applied. However, there has been significant and growing interest in evaluating the thermal performance of building envelope components, including fenestrations, using guarded or calibrated hot boxes. Much work has been ongoing in North America and Europe to provide transfer standards to assist in the calibration and validation of the hot boxes used for measurement of thermal resistance and overall heat transmission coefficient. Studies in each geographical region during the eighties and nineties indicated that the precision of hot box methods was the order of +/- 8% for opaque systems and +/-10% to 20% for windows.

In each case, the problem has been addressed by developing transfer standards and undertaking interlaboratory studies or round robin measurement programmes. It is impractical to consider fabrication of one panel for interlaboratory comparison. The philosophy for opaque systems has been to design a specific insulated system using selected, easily obtained, and subsequently well characterised component materials. These are fabricated in accordance with a set of selected criteria and careful instructions. Two round robins on different panels carried out within the US Department of Commerce, National Voluntary Laboratory Programme, NVLAP, have now established thermal transmittance values to the order of +/- 5% for two panel types with different thermal resistance values. Designs and instructions for these composite panels are available for organisations wishing to obtain some idea of the precision of its hot box in measuring wall systems.

For windows, there have been many controversies regarding various measurement techniques used and the performance values quoted by manufacturers. To address these problems, certified transfer standards are available from NPL in the UK and the National Research Council, NRC, in Canada. These panels consist essentially of a uniform thickness piece of a stabilised polystyrene material sandwiched between toughened glass or polymeric layers. They are instrumented with arrays of thermocouples on each internal surface should these be required by the requisite national standard test method(s).

The availability of the above transfer standards has improved measurement technology in both areas and thus reduced the controversies significantly.

## SUMMARY

A few small steps have been taken internationally towards improving thermal measurement technology and low thermal conductivity conductance precisions. However, more materials, more efforts, and more funding are required before one can be assured that current and new methods, apparatus and techniques are providing highly reliable results. In particular, outstanding needs in the low conductivity area which must be addressed are:

- a loose fill material, particularly one which can be used to evaluate radial flow methods, especially the pipe test method. Most of the insulations described above are anisotropic to some extent or cannot be used readily in circular configurations. Thus, a material having all the attributes of a SRM plus the ability to pack to a reproducible density is essential. Glass or ceramic beads or similar types of microspheres is one suggestion.

- a high thermal resistivity material or inhomogeneous evacuated system. This is required in order to show that current methodology is providing reliable results on such types of specimens where the measured thermal conductivity properties can be several orders of magnitude below those currently evaluated with most insulation materials.

## REFERENCES

1.    Thermal Conductivity, Volumes I and II, Ed., R.P. Tye, Academic Press London and New York, 1969.

2.    Compendium of Thermophysical Properties Measurements Methods Volumes 1 and II, Eds., K.G. Maglic, Y. Cezairliyan and V.E. Peletsky, Plenum Press, New York, 1988 and 1992.

3.    Powell, R.W., Ho, C.Y., and Liley, P.E., "Thermal Conductivity of Selected Materials", Special Publication NSRDS-NBS8, National Bureau of Standards, Washington, DC., 1966.

4.    Hulstrum, L.C., Tye, R.P. and Smith, S.E., Thermal Conductivity 19, Ed. D.W. Yarbrough, Plenum Press, New York, 1986, pp. 199-211.

5.    Thermal Transmission Measurements on Insulation, Ed., R.P. Tye, ASTM STP 660, ASTM, Philidelphia, 1978.

6.    Tye, R.P., Coumou, K.G., Desjarlais, A.O. and Haines, D.M., Thermal Insulation Materials and Systems, STP 922, Eds., F.J. Powell and S.L. Matthews, ASTM, Philadelphia, 1987, pp 651-670.

7.    Tye, R.P., and Desjarlais, A.O., Thermal Insulation Materials and Systems for Energy Conservation in the '80's, Eds., F.A. Govan, D.M. Greason and J.D. McCallister, ASTM STP 789, ASTM, Philadelphia, 1983, pp. 733-748.

8.    Smith, D.R., NISTIR 89-3919, National Institute of Standards Technology, August, 1989.

9.    M. C. Franken, March, 1995, Private Communication.

# The NPL High Temperature Guarded Hot-Plate

D. R. SALMON

## ABSTRACT

This paper describes the design and construction of a guarded hot-plate apparatus for the measurement of insulations and refractories in the temperature range 100°C to 850°C having thermal conductivities of up to 0.5 W/m.K and thermal resistances in the range 0.05 to 3.2 m$^2$ K/W. Information is given on measurements carried out to assess the performance and uncertainty of the apparatus using calcium silicate and microporous insulation. Details are also given of an international intercomparison on a low density calcium silicate carried out as part of a BCR programme for the European Community.

## INTRODUCTION

The steady-state guarded hot-plate method is capable of providing very accurate thermal conductivity values. Well designed guarded hot-plates are essential not only for measurements of the highest accuracy but also as a means of checking and developing secondary methods such as the line source and panel test methods. It is important, therefore, that users of these secondary methods should check their experimental techniques and procedures very carefully using appropriate certified reference materials similar in type to the products they measure routinely.

The NPL high temperature guarded hot-plate has been designed to provide a suitable source of accurate thermal conductivity measurements in the UK and of individually calibrated NPL transfer standards for performance checking of the type referred to above. An important feature of the NPL apparatus, as with all guarded hot-plates, is its ability to determine the thermal conductivity of specimens using small temperature differences across the specimen throughout its measurement range of up to 850°C, thereby reducing to a minimum any uncertainties associated with non-linear thermal conductivity versus temperature behaviour.

Centre for Quantum Metrology, National Physical Laboratory, Teddington, Middx. TW11 0LW, UK.

## DESIGN CRITERIA

The basic requirements for the proposed apparatus were:

1. To measure solid insulations and refractories having thicknesses in the range 25 mm to 50 mm
2. To measure materials with thermal conductivities of up to 1 W/m.K and thermal resistances from 0.05 m $^2$ K/W to 3.2 m $^2$ K/W
3. To cover the temperature range from 100°C to 850°C
4. To have a measurement uncertainty of about ±5% over the whole range of temperature and thermal resistance
5. To fully automate the apparatus.

With these basic design criteria in mind it was decided to build a guarded hot-plate apparatus. The guarded hot-plate method is an absolute method which is recognised as the most reliable for measurements on insulating materials with thermal conductivities of up to 2 W/m.K. It is a long established method which over the years has been developed for use at temperatures between -196°C and 500°C. The method has been standardised by most national and international standardising bodies, ISO[1], BSI[2], DIN, ASTM etc, and is well established in many countries, particularly for measurements near ambient temperature. Its value has grown as the requirements for energy efficiency have become more important.

## APPARATUS DESIGN

A cylindrical double-sided design of 305 mm diameter was chosen in preference to the conventional square design prevalent near ambient temperatures. The radial symmetry of the apparatus greatly simplifies the design of the lateral guard and the edge guard system required to minimise edge heat losses at high temperatures. In addition, the number of power supply units and temperature controllers is significantly reduced, lowering construction costs. Figure 1 shows a line drawing of the basic design for the high temperature guarded hot-plate. Details of the specific elements of the apparatus are discussed below.

### GUARDED HEATER PLATE

The guarded heater plate consists of two 305 mm diameter plates, about 10 mm thick, bolted together. The plates are machined from Inconel 600, a high temperature nickel alloy with very good dimensional stability at temperatures up to at least 900°C. As with the heated cold plates, one plate is machined with grooves to provide a location for the guard and metering area heating elements. The second plate has shallow grooves cut into one face to allow a thin wire, differential thermocouple to be wound across the guard-metering gap.

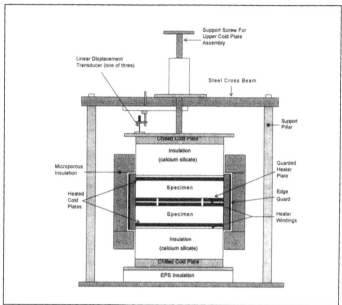

Figure 1.    Schematic drawing of the NPL high temperature guarded hot plate

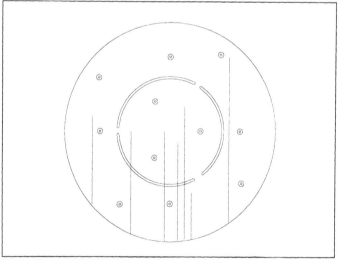

Figure 2.    Detail of guarded heater plate showing guard-metering area gap and location of thermocouple grooves

Both plates have a narrow slot centred on a 150 mm diameter cut through them to separate the central metering area of the plate from the guard area, three small bridging pieces being left in position to provide mechanical support for the centre area, (Figure 2). The outer surface of each plate is machined flat to better than 0.05 mm across the whole face. Grooves are machined into the face to take sheathed mineral-insulated thermocouples.

## COLD PLATE SYSTEM

The two chilled plates act as heat sinks for heat flowing through the specimen, and each consists of a 30 mm thick aluminium alloy plate with cooling channels cut concentrically into the top surface. A second aluminium alloy plate is bolted to the top of the plate with a gasket between them to form enclosed cooling channels through which cooling fluid from a temperature controlled bath is pumped to maintain the plate temperature at 20°C. The working face of the chilled plate, adjacent to the insulation, is machined flat to better than 0.1 mm across the whole surface.

Heated cold plates are separated from the chilled cold plates by a thick block of calcium silicate insulation. Since the chilled cold plate runs at 20°C the thermal conductivity of insulation and specimen and the temperature drop through the specimen determine the lower temperature limit of the heated cold plate.

The two heated cold plates are made to a guarded design with the central area surrounded by a concentric lateral guard area. The plates are made from Inconel alloy and measure 305 mm diameter by 10 mm thick. The back surface of each plate has a concentric groove machined into the central area and similar concentric grooves machined into its guard section to locate the sheathed heating elements. The surface adjacent to the specimen is machined flat to better than 0.05 mm across the face and grooves are machined into this face to take four mineral-insulated nicrosil-nisil (Type N) thermocouples and a differential thermocouple for balancing the temperatures of the guard and central areas.

## EDGE GUARD ASSEMBLY

The edge guard assembly consists of two insulated semi-cylindrical Inconel 600 plates which surround the specimen stack to leave a gap of about 25 mm between the specimens and edge guard. The edge guard plates extend at least 30 mm beyond the heated cold plates when 50 mm thick specimens are being measured, the overlap obviously being greater for thinner specimens. Separate heating elements are wound on the outside of each guard plate and connected in series to a single power supply. The outer surfaces of the guard plates are insulated with microporous insulation encased in a stainless steel lining, the thickness of insulation being sufficient to keep the temperature of the stainless steel below about 100°C when the edge guard is at 800°C.

## TEMPERATURE MEASUREMENT

Type N thermocouples are used throughout the apparatus in the form of mineral- insulated thermocouple wire in a Nicrosil sheath of 1.5 mm diameter, and of sufficient length to allow the sheath terminations to be made at room temperature. The thermocouple wires terminate in an isothermal block where connections are made to copper wires from the scanning unit. Samples from the ends of the sheathed thermocouples were calibrated by the NPL Temperature Standards Section and a least-squares polynomial was fitted to the calibration data. The output voltage from a cold-junction compensation unit, referenced to the temperature of the isothermal block, is connected to the scanner. The computer adds this reference voltage to the scanned thermocouple voltages and then derives temperatures from these using polynomial coefficients obtained for the calibrated wire.

## TEMPERATURE CONTROL

Temperature control of the various heater plates is achieved using Eurotherm controllers with an analogue output of between 0 and 5 Volts d.c. which is used to operate voltage programmable Kingshill power supply units. The controllers for the metering area and the heated cold plate heaters are linked to the computer through an RS422 interface. The set point for the edge guard controller is obtained by remote transmission of the set point from the metering area temperature controller. The transmitted set point can be offset locally to change the temperature of the edge guard relative to the main heater plate. For most of the heaters individual thermocouples are used as the sensing elements for the temperature controllers on the respective heater plates. However, for the lateral guard of the guarded heater plate the sensing element is a differential thermocouple wound across the guard-metering gap. The differential thermocouple has a sensitivity of about 230 µV per degree C over most of the temperature range of the apparatus.

## AUTOMATION

Acquisition and recording of data during a measurement are achieved using a data logging system consisting of a Keithley 706 scanner and a Keithley 195A digital voltmeter controlled by an Olivetti PC compatible computer operating on an IEEE bus. The computer is also interfaced via RS422 with the temperature controllers that maintain the temperature of the metering area and the two heated cold plates,  providing the means for complete automation of the apparatus. A computer program has been written (in HTBasic) to set the various plate temperatures, check for equilibrium conditions, record and analyse the transducer outputs and provide a detailed print-out of thermal conductivity results. This automation has reduced the staff time required to monitor instruments and check for equilibrium conditions, and the time the apparatus remains at high temperatures, thereby extending the lifetime of heating elements and thermocouples.

## SPECIMEN MOUNTING AND THICKNESS CORRECTION

The edge guard assembly incorporates a hinge system which allows the two semi-cylindrical parts of the guard to swing open for easy access to the specimen stack for changing specimens. When assembled, the cold and chilled plate assembly rest on top of the specimen stack and moves up and down with the expansion and contraction of the apparatus during a measurement. Movement of the upper chilled plate is monitored with reference to the base plate of the apparatus using three linear displacement transducers. The transducer system is calibrated to remove the effects of expansion of the apparatus from the expansion of the specimens using foam glass specimens which have a comparatively low expansion coefficient. Checks are made using specimens whose expansion coefficient is known and it is found that the method provides a good first order correction to take into account the expansion of the specimens.

## ASSESSMENT OF PERFORMANCE

After construction, the performance of the apparatus was assessed mainly using specimens of high density calcium silicate having a thermal conductivity of about 0.2 W/m.K. Additional tests were carried out using specimens of Microtherm with a thermal conductivity of about 0.029 W/m.K. Tests were carried out to check that the uncertainties associated with (i) lateral guard to metering area balance and (ii) edge heat losses are small, by measuring the specimen thermal conductivity at constant temperature with increasing temperature difference, $\Delta\theta$, between the hot and cold plates. An examination was also made of the sensitivity of the thermal conductivity values to lateral guard/metering area temperature imbalance and edge guard temperature imbalance.

## ERRORS ASSOCIATED WITH NON UNIFORM HEAT FLOW

Variation of Thermal Conductivity with $\Delta\theta$

To check that guard-metering area imbalance errors and edge heat losses are suitably small, measurements were carried out with high density calcium silicate at constant mean temperatures of 400°C and 700°C with increasing Δθ. The metering area power applied to the calcium silicate specimens to achieve various temperature gradients through the specimens is listed in Table 1 and presented graphically in Figure 3. At both temperatures a linear fit through the experimental points gives a very good fit with a small intercept on the heat flow axis for zero temperature gradient through the specimen. In these measurements the thermal conductivity remained constant to within 0.5% over the entire range of temperature gradients.

TABLE I.     VARIATION OF METERING AREA POWER WITH TEMPERATURE DROP ACROSS A HIGH DENSITY CALCIUM SILICATE SPECIMEN.

| Mean specimen temperature /°C | Temperature drop /°C | Measured heater power /W |
|---|---|---|
| 400 | 23.8 | 3.1 |
|  | 48.4 | 6.39 |
|  | 72.6 | 9.71 |
| 700 | 24.1 | 3.32 |
|  | 48 | 6.57 |
|  | 71.9 | 9.93 |

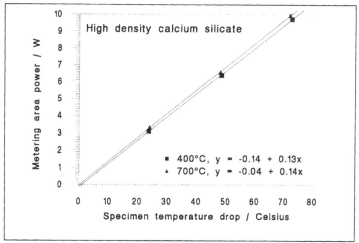

Figure 3.     Metering area power as a function of the temperature drop through specimens of high density calcium silicate.

Variation of Thermal Conductivity with Lateral Guard Temperature Imbalance

Measurements were carried out on microporous insulation and calcium silicate to examine the effect of guard to metering area temperature imbalance on the measured thermal conductivity. The measurements showed that for microporous insulation a thermopile imbalance of 1 μV produces less than 0.5% variation in the measured thermal conductivity. Since these measurements were made the guarded heater plate has been dismantled to repair a broken differential thermocouple. It was decided to double the number of junctions in the differential thermocouple to reduce the sensitivity of thermal conductivity to the temperature imbalance.

Variation of Thermal Conductivity with Edge Guard Temperature Balance

The thermal conductivity of 50 mm thick calcium silicate specimens was measured at a mean specimen temperature of 300°C with the edge guard at temperatures of 269.5, 279, 293, 307.5, 317, and 326.5°C. It was found that the thermal conductivity only changed by 0.04% for a one degree change in the edge guard temperature, ie at this specimen thickness the lateral guard of the main heater plate is very effective.

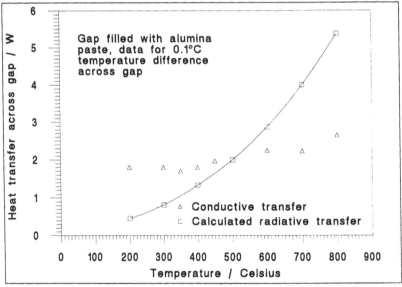

Figure 4.    Heat transfer by radiation across an unfilled gap and conduction across a filled gap for a 0.1°C temperature difference across the guard and metering area gap.

Heat Flow Across Guard Metering Area Gap

For guarded hot-plates designed to operate close to ambient temperatures it is normal practice for the guard-metering area gap to be left empty, the air gap acting as a good insulator. At higher temperatures, however, radiative heat exchange across the gap becomes significant and in this apparatus the gap has been filled with alumina cement to reduce heat flow across the gap for temperatures above about 400°C. Figure 4 shows the calculated radiative heat transfer for a 0.1°C temperature difference across the unfilled gap. It also shows the conduction across the filled gap determined from measurements of the sensitivity of thermal conductivity to guard metering area temperature balance. Ideally, a fibrous material of lower conductivity than alumina cement should be used to reduce the heat flow across the gap.

## INTERNATIONAL INTERCOMPARISONS

The guarded hot-plate was used to participate in a BCR programme designed to certify a low density calcium silicate (density about $300\,\mathrm{kg/m^3}$) as a thermal conductivity reference material for temperatures up to 900°C. Measurements were carried out on the material by several European laboratories using both hot-wire and guarded hot-plate methods. Guarded hot-plate measurements were carried out at five well established laboratories including NPL. The results from two of the laboratories were rejected as outliers, the remaining three sets of results, Figure 5, show that very good agreement was achieved between these laboratories. Although the experimental work on the programme has been completed the full results and detailed assessment of the programme have not yet been published.

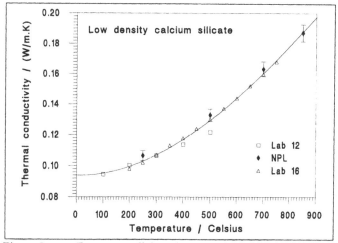

Figure 5.     European collaborative guarded hot-plate measurements on low density calcium silicate funded by BCR

OVERALL UNCERTAINTY

One of the major sources of uncertainty in measuring relatively dense high conductivity materials in a hot-plate apparatus with thermocouples embedded in the plates is associated with the presence of air at the interfaces between the specimen and plates. Air has a relatively low thermal conductivity, so the presence of air films at the interface can greatly affect both the uniformity of heat flux distribution through the material and its apparent thermal resistance. However, the effect becomes progressively less severe at higher temperatures due to increasing radiative heat transfer across the air gap. In this apparatus good thermal contact between specimens and plates is achieved by ensuring that all the plate surfaces are flat to better than 0.05 mm and that specimen surfaces are as flat as possible.

The main sources of uncertainty arise from measurement of the temperature drop through the specimens, specimen thickness and edge heat losses. Minor sources of uncertainty arise from measurement of the heater power delivered to the metering area, and the dimensions of the latter.

The uncertainties in determining the basic parameters measured directly, such as heater power, temperature and specimen dimensions, have been calculated. Factors taken into account in the calculation include associated uncertainties in the calibration of the appropriate measuring instruments and transducers. Estimates of the uncertainties associated with lateral and edge guard imbalances have been determined experimentally. By adding the individual uncertainties in quadrature an overall uncertainty of ±5% within 95% confidence limits has been determined for measurements covering the temperature range from 100°C to 850°C. Repeatability tests carried out on samples re-assembled and measured in the apparatus show agreement with earlier values to better than 1%.

**CONCLUSIONS**

An apparatus has been designed and constructed for the measurement of insulation and refractory material in the temperature range 100°C to 850°C. The performance of the apparatus has been extensively checked through measurements on microporous insulation and calcium silicate. It was found that the heat transfer across the guard metering area gap and the edge heat losses were acceptably small, being less than about 0.5%. The measured thermal conductivity was also found to be insensitive to the temperature of the edge guard, an indication of the low edge heat losses.

An intercomparison of the apparatus with other guarded hot-plate apparatus in Europe, as part of a BCR programme to certify a low density calcium silicate as a reference material, showed agreement of better than 3% between the results from participating laboratories.

An uncertainty of ±5% within a confidence limit of 95% has been assigned to the measurements obtained using the apparatus over the entire temperature range.

The apparatus is now being used for the calibration of transfer standards of high density calcium silicate, microtherm and foamed glass.

## REFERENCES

1.    BS874 : British Standard Methods for:- Determining thermal insulating properties. Section 2.1 : Guarded hot-plate method. Published 1986
2.    ISO 8302:1991(E). International standard:- Determination of steady-state thermal resistance and related properties - Guarded hot-plate apparatus.

# Combined Radiation and Conduction in Fibrous Insulation (from 24°C to 400°C)

P. BOULET, G. JEANDEL, P. DE DIANOUS, F. PINCEMIN

## ABSTRACT

Heat transfer by combined radiation and conduction through thermal insulation is treated. A numerical solution based on a comprehensive theoretical study is presented. We paid particular attention to high temperature applications. We focused on the effect of variations of the material refractive index with temperature in order to obtain accurate results. We then studied the degradation of thermal properties with temperature. The strong non-gray behavior of glassfiber-like materials, along with the shift of the blackbody emission spectrum towards shorter wavelengths, leads to a modification of the spectral radiative flux profile. Moreover, the temperature profile, linear at room temperature, becomes strongly curved at higher temperature levels. Theoretical predictions of the total heat transfer through an industrial mineral wool product made of soda-lime type glass are presented. They are validated by experimental measurements at both room and elevated temperatures.

Pascal Boulet, Gérard Jeandel : LEMTA, Université de Nancy I, Faculté des Sciences, BP 239, 54506 Vandoeuvre les Nancy Cedex, France

Philippe de Dianous : Isover Saint-Gobain, CRIR, BP 19, 60290 Rantigny, France

François Pincemin : Saint-Gobain Recherche, 39 quai Lucien Lefranc, BP 135, 93303 Aubervilliers Cedex, France

# NOMENCLATURE

E : medium thickness
$f(r)$ : length weighted radius distribution
$F_v$ : fibre volume fraction
$L_\lambda$ : spectral intensity
$L^o{}_\lambda$ : blackbody spectral intensity
$n_\lambda$ : refraction index
$k_\lambda$ : extinction index
$Q_c$ : heat flux by conduction
$Q_r$ : heat flux by radiation
$Q_T$ : calculated total heat flux
$Q_\lambda$ : spectral efficiency (of absorption, scattering or extinction)
r : fibre radius
$R_\lambda$ : intensity reflection factor
T : temperature
$T_\lambda$ : intensity transmission factor
$\lambda$ : wavelength
$\lambda_c$ : conductive part of the effective thermal conductivity
$\rho$ : medium density
$\sigma_{a\lambda}$ : spectral absorption coefficent
$\sigma_{s\lambda}$ : spectral scattering coefficent
$\sigma_{e\lambda}$ : spectral extinction coefficent
$\mu$ : direction of discretization in space
$\theta_i$ : polar angle
$*$ : complex variable

# INTRODUCTION

Over the last decade, many works have been devoted to the theoretical study of heat transfer by combined radiation and conduction in participating media. The progress achieved, especially thanks to numerical computing, led to detailed and accurate descriptions of the thermal phenomena encountered in many applications.

In this paper, we focus on heat transfer through fibrous media used for thermal insulation. The main goal is to predict the thermal characteristics of a fibrous material, given the chemical composition and the exact morphology (size and orientation) of the fibers.

Several studies have been undertaken by Saint-Gobain and the LEMTA to determine the influence of different parameters that affect heat transfer in mineral wool products. In previous papers[1,2,3,4], we focused on the effect of strong non-gray radiative behavior of glass, anisotropic scattering, fiberglass composition and more recently non-fibrous particles in the material. A numerical model was developed to account for these features. It enables us to calculate the heat flux across the material as a function of fiber characteristics, temperatures at the boundaries and thickness of the medium[2].

The purpose of the present work is to investigate the thermal behavior of industrial mineral wool products at high temperatures. The dispersed diameter spectrum of such products has been taken into account. We especially studied the variation of the optical properties of the bulk glass with temperature and its influence on the spectral radiative properties of the material. Coupling between radiation and conduction is considered, and results are compared to others obtained using single additivity of fluxes.

In the following, we consider an industrial glassfiber material, whose characteristics are given below :
    i) density : 50 kg/m³
    ii) thickness : 100 mm
    iii) black boundaries
    iv) fibers made of industrial soda-lime type glass, used by Isover Saint-Gobain

Fibers are assumed to be lying randomly in planes parallel to the boundaries.

In the next section, we present the theoretical background used. From the optical indices of the bulk material and the length weighted fiber diameter spectrum, a comprehensive simulation of the radiative transfer is possible through the solution of the radiative transfer equation (R.T.E.). The solving scheme is briefly presented.

The last two sections present the numerical results of our study. In particular, a comparison between characteristics at 24°C and 400°C is presented. Numerical results are then compared to experimental ones. To our knowledge, it is the first time a multiflux model is validated within a large temperature range using industrial, medium density, mineral wool insulations, characterized by a dispersed diameter spectrum.


## THEORETICAL BACKGROUND


The main steps needed for the theoretical characterization of a fibrous medium are reviewed in this section. We have to solve a problem of combined conduction and radiation in an absorbing, emitting and anisotropically scattering medium, with a strong non-gray behavior. Considering the small size of the medium in the heat transfer direction compared with the other dimensions and the random distribution of the fibers in the plane normal to the transfer direction, we can assume one-dimensional transfer with azimuthal symmetry.

In order to solve the problem, one first has to determine the radiative properties of the medium, which requires good knowledge of both the optical properties of the bulk glass and the fiber morphology. Then the heat flux can be calculated by solving the R.T.E. and the energy equation.

The length weighted diameter spectrum of standard fibers (mean diameter around 4 μm), was measured with a scanning electronic microscope.

The complex refractive index $n_\lambda^* = n_\lambda - i.k_\lambda$ ($i^2 = -1$) was determined from the reflection spectrum $R_\lambda$ of a flat glass sample in air[2,5] and from the transmission spectrum $T_\lambda$ of a slab[2].

Figure 1 displays the optical indices of the glass at room temperature and at 400°C. We see almost no influence of temperature on the optical indices. However we notice a decrease of the absorption peaks at 10 μm and 22 μm, and a broadening of those peaks when the temperature rises.

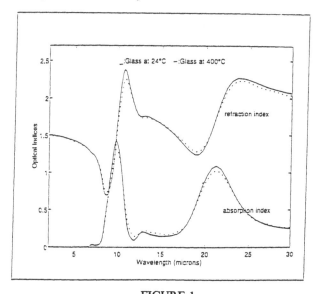

FIGURE 1

Optical indices the glass at 24°C and 400°C

RADIATIVE PROPERTIES

Using these fiber characteristics, one can calculate the spectral attenuation coefficients on the basis of the work by Lee[5], for monodispersed fibers :

$$\sigma_\lambda^r(\mu) = \frac{2F_v}{\pi\, r} Q_\lambda \tag{1}$$

where $\sigma_\lambda^r$ stands for either spectral absorption, scattering or extinction coefficients, $F_v$ is the fiber volume fraction of the medium, r is the fiber radius, $Q_\lambda$ is the corresponding efficiency as defined by Kerker [6]. The superscript r stands for radius r.

From the fiber diameter distribution, we derive the mean spectral coefficients of the whole medium using the following integration[1] :

$$\sigma_\lambda(\mu) = \frac{\int_0^\infty \sigma_\lambda^r(\mu)\, f(r)\, r^2\, dr}{\int_0^\infty f(r)\, r^2\, dr} \tag{2}$$

Following Boulet *et al.*[3], we calculate in the same way the mean bi-directional scattering coefficients, to accurately characterize anisotropic scattering.

## TRANSFER EQUATION

The R.T.E. for a one-dimensional problem with azimuthal symmetry using the multi-flux approximation[3] is :

$$\mu_i \frac{\partial L_\lambda(y,\mu_i)}{\partial y} = -\sigma_{e\lambda}(\mu_i)\, L_\lambda(y,\mu_i) + \sigma_{a\lambda}(\mu_i)\, L_\lambda^\circ(T(y)) + \frac{1}{2}\sum_{j=1}^m P_\lambda(\mu_j,\mu_i)\, L_\lambda(y,\mu_j) \tag{3}$$

$L_\lambda(y,\mu_i)$ is the spectral intensity of radiation at abscissa y, assumed constant over the interval $\Delta\mu$ around the $\mu_i$ direction (Figure 2) for the multi-flux model. $L_\lambda^\circ(T(y))$ is the blackbody emission at temperature T(y) and wavelength $\lambda$, $\sigma_{e\lambda}$ and $\sigma_{a\lambda}$ are the spectral extinction and absorption coefficients respectively. $\mu_i$ is the cosine of the polar angle $\theta_i$. $P_\lambda(\mu_j,\mu_i)$ is the bi-directional spectral scattering coefficient characterizing scattering of radiation around $\mu_j$ towards the $\mu_i$ direction. m is the number of discrete directions considered (currently 12 in our applications).

This equation is coupled with the energy equation, which is given below for steady-state conditions with no energy source[7] :

$$\text{div}(Q_c + Q_r) = 0 \tag{4}$$

$Q_c = -\lambda_c\, \partial T/\partial y$ is the conductive flux, with $\lambda_c$ designating the thermal conductivity.):

$$\lambda_c = 0.2572\, T^{0.81} + 0.0527\, \rho^{0.91}\, (1+0.0013T) \quad (mW/m.K,\ T\ in\ Kelvin) \tag{5}$$

This semi-empirical formula has been determined by Isover Saint-Gobain[2] , and has been validated in a wide range of density and temperature (5-70kg/m³, 10-400°C).

The radiative flux is derived from a double numerical integration of the flux with respect to $\mu$ and to the wavelength domain.

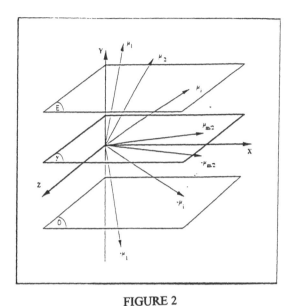

FIGURE 2

Geometry of the problem

Equations (3) and (4) are solved numerically by an iterative scheme, using a linear initial temperature profile. Convergence is reached when the temperature profile does not change anymore between two iterations. Stability is achieved using an under-relaxation scheme, especially required for high temperature applications.

## NUMERICAL RESULTS

Spectral absorption and scattering coefficients are plotted in Figure 3, at 24 and 400°C, for the same medium (as described in the introduction, spectral radiative properties at 24 and 400°C are calculated from optical properties of the bulk glass determined at 24°C and 400°C respectively). We see in Figure 3 that a temperature variation does not affect the radiative coefficients. Actually, this was expected from the complex indices, since they do not change with temperature. However, absorption seems slightly higher at 400°C, whereas the opposite is observed for scattering.

On the basis of these calculated properties, we used our numerical model to determine heat fluxes for several temperature conditions.

We first considered a medium at a mean temperature of 24°C, with a temperature difference of 10°C between its surfaces. Results are given in Table I (left column). Total heat flux conservation is achieved over the whole medium. Results are given in Table I (left column). The conductive part of the flux is predominant, because of the high density of the medium. The ratio of conductive to radiative contributions to the total heat transfer is about 10 in the middle of the medium. The temperature profile is nearly linear, with the deviation from linearity not exceeding 0.04°C.

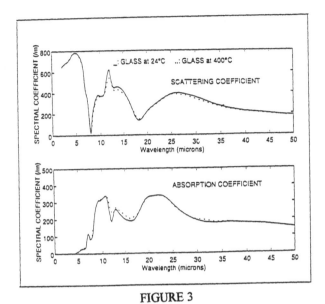

FIGURE 3

Radiative coefficients of the medium at 24°C and 400°C

The second application deals with the same medium, but at a mean temperature of 400°C and with a temperature difference of 10°C. The radiative properties used for the calculation were determined at 400°C. Results are given in table I (right column). The total heat flux is three times higher than in the previous case. The radiative part is nearly equal to the conductive one, with the ratio of conduction to radiation being equal to 1.2 in the middle of the medium. Despite this radiative contribution, the temperature profile remains nearly linear. As shown in Figure 4, deviations from linearity never exceed 0.05°C.

Table I

Radiative and conductive flux densities at 24°C and 400°C, with $\Delta T = 10°C$

| flux density (W/m²) | 24°C | 400°C |
|---|---|---|
| radiative part | 0.29 | 4.54 |
| conductive part | 2.85 | 5.36 |

The increase of radiative flux is mainly due to the increase of mean temperature, which leads to a higher blackbody emission. The spectral radiative flux profile is plotted in Figure 5 at both mean temperatures. In order to compare respective variations, we normalized the results from 0 to 1, dividing by the maximum heat flux value for each case (at 8 μm and 3.5 μm for mean temperatures of 24 and 400°C respectively). As expected from the blackbody emission curves, the heat flux profile shifts       toward       the       shorter       wavelengths       as       the       temperature       rises.

As a consequence, the so-called Christiansen peak,[8] observed in the heat flux profile near 8 μm at ambient temperature, almost vanishes at 400°C. Actually, near 8 μm, where the scattering coefficient is small, extinction of thermal radiation is mainly due to absorption, which is also small : this is the Christiansen effect, important at 24°C. When the temperature rises, the blackbody emission shifts towards shorter wavelengths, where absorption becomes negligible compared to scattering. However, the large scattering coefficient in the wavelength domain around 4 μm (see Figure 5), is not sufficient to prevent radiative heat transfer. Indeed, since the fibers are larger than, the wavelength, the scattering process leads mainly to forward scattering.

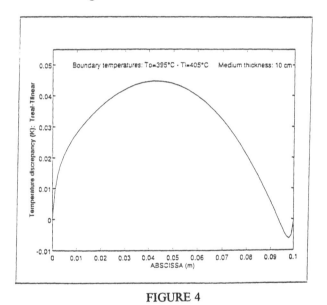

FIGURE 4

Temperature profile in the medium with $\Delta T = 10°C$

Using the complex refractive index of the glass measured at 24°C as input data for the heat transfer calculation at 400°C yields a difference of 0.5 % for the total flux. This means that the complex index of glass only needs to be determined at ambient temperature. This index can then be used at other temperatures up to 400°C.

In order to assess the effect of a larger temperature gradient across the medium, the same numerical simulation has been carried out with a boundary temperature of 20°C on one side and 400°C on the other side of the medium. As expected, a higher heat flux is obtained (228 W/m²).

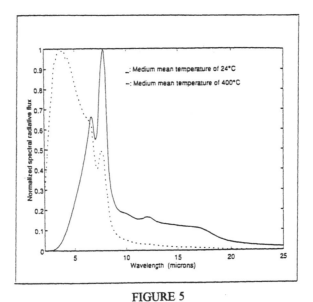

FIGURE 5

Spectral radiative flux profile at 24°C and 400°C, with ΔT = 10°C

The ratio of conduction to radiation is equal to 2 in the middle of the medium, which indicates a lower radiative contribution to the total heat transfer compared to the previous calculation. The temperature profile, however, is far from linear (Figure 6). The maximum deviation from linearity is 52°C at abscissa of 0.4 E, with E being the thickness. This indicates a strong coupling between conduction and radiation.

The large effect of radiation on temperature profile across the medium is probably due to the strongly non-gray behavior of glass. Indeed, one can show[12] that the temperature profile is not only dependent on the ratio of conductive to radiative heat flux, but also on the albedo (ratio of scattering to extinction coefficients). This coefficient has large variations with wavelength, caused by spectral variations of optical indices of glass.

If we do not take coupling into account in the calculation*, we get the flux densities as given in Table II, in the middle of the medium, with a temperature difference of 10°C. The results are very close to the ones given in Table I, with coupling accounted for.

---

* The radiative part is then calculated using a radiative transfer model without conduction, with the multi-flux approximation. Linear temperature profile is assumed. The unknown is then the intensity of radiation along the thickness, which we integrate to determine the flux. The conductive part is obtained from relations (4) and (5).

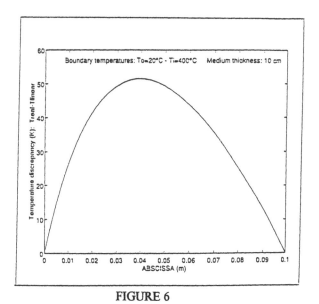

**FIGURE 6**

Temperature profile in the medium with high temperature difference

TABLE II

Radiative and conductive fluxes at 24°C and 400°C,
with $\Delta T = 10°C$, with no coupling between conduction and radiation

| flux density (W/m²) | $\Delta T = 10°C$ | |
| --- | --- | --- |
| | T = 24°C | T = 400°C |
| radiative part | 0.29 | 4.55 |
| conductive part | 2.81 | 5.39 |

However, considering high temperature difference (cold temperature at 20°C, hot temperature at 400°C) overestimates radiative flux by 16 %, and underestimates conductive flux by 3 %, in the middle of the medium. A model assuming single flux additivity is therefore no longer valid for high temperature gradients, since coupling between conduction and radiation becomes not negligible. Moreover, it does not allow the evaluation of the temperature profile inside the medium. This last example illustrates the need for a comprehensive model that takes into account coupling phenomena.

## COMPARISON WITH EXPERIMENTAL DATA

The apparent thermal conductivity of industrial products 50 kg/m³ in density has been experimentally determined at 24°C and 400°C.

Measurements at 24°C were performed with a bi-guarded hot plate apparatus[9], according to ISO 8302[10] standard, with a 10°C temperature difference across the blanket. Measurements were performed on industrial specimens 50 mm in thickness. Another bi-guarded hot plate apparatus, specifically designed for high temperature measurements according to ISO 8302, was used for measurements of the reviewer requested use of the apparent thermal conductivity at 400°C. The temperature difference applied across the specimen was 50°C, in order to produce a large enough output signal. As indicated in the previous section, a small temperature difference yields a linear temperature profile inside the medium. We can therefore expect the measured apparent thermal conductivity to be unaffected by a 50°C temperature difference. This has been numerically confirmed at a mean temperature of 400°C. The apparent thermal conductivities calculated with temperature differences of 10 and 50°C were almost identical, respectively equal to 98.7 and 98.8 mW/m.K. A temperature difference of 10°C was then applied for the calculations at both 24 and 400°C.

The experimental and numerical results are given in table III. The calculations of the apparent thermal conductivity have been carried out using the characteristics of the measured specimens, described above (in particular a thickness of 50 mm), at mean temperatures of 24°C and 400°C. The length weighted diameter spectrum of the measured specimens was accurately determined and used for the calculations.

TABLE III

Comparison between numerical and experimental results

|  | $\lambda$ (mW/mK) from Numerical Simulation NS | $\lambda$ (mW/mK) from Experimental Measurement EM | discrepancy (NS-EM)/EM |
|---|---|---|---|
| T = 24°C | 31.4 | 32.3 | -2.8 % |
| T = 400°C | 98.7 | 106.1 | -7.0 % |

These results show good agreement between theory and experiment. Discrepancies are within the experimental uncertainties*. It therefore validates our multi-flux model for both low and high temperature applications.

As noticed in a previous study[1] for low density mineral wool products at room temperature, the multi-flux model systematically underestimates the thermal conductivity. However, while the discrepancy is about 8.5 % for low density products (10 kg/m$^3$), it is limited to 3 % for heavier products, 50 kg/m$^3$ in density.

Indeed, the fiber orientation has an effect on radiative properties of the fibrous medium[9]. The conductivity is minimized when the fibers are all oriented normal to the heat transfer direction. Therefore, since the fibers are more compressed in heavier fibrous products, the assumption of fibers lying in planes parallel to the boundaries is probably more relevant for the present products.

---

* Accuracy and reproducibility of heat flow meter apparatus is generally estimated to ± 4 %. There have been very limited intercomparison measurements in Europe using the guarded hot plate method at high temperature. The results have indicated differences in the order of ± 15 %. An American study confirmed uncertainties of ± 10 % (see Hust, J.G. and Smith, D.R. "Round Robin measurements of the apparent thermal conductivity of two refractory insulation materials using high temperature guarded hot plate apparatus", NBSIR 88-3087, NIST, Boulder, Colorado, April 1988).

The difference between measured and calculated apparent thermal conductivities increases with temperature, because of a greater contribution of radiation to the total flux when the temperature rises. Differences between theoretical and measured results are therefore enhanced.


## CONCLUSION

A radiative model using the multi-flux approximation has been applied to the characterization of heat transfer in fibrous mineral wool products at both room and high temperatures (up to 400°C). This model enables us to describe radiation in an absorbing, emitting and anisotropically scattering medium, with a strong non-gray behavior. Coupling between conduction and radiation is accounted for.

Optical properties of an industrial glass were measured at 24 and 400°C, and showed almost no influence of temperature on the complex index.

This model was applied to an industrial mineral wool product made of this glass, whose diameter spectrum had been accurately determined. It was validated by comparisons with experimental data at 24 and 400°C for medium density blankets (50 kg/m$^3$). Numerical results are in good agreement with experimental measurements, within the experimental uncertainty.

We highlighted the need for a comprehensive model, that takes into account coupling between conduction and radiation, when considering a large temperature gradient across the medium.

More generally, this model can be considered as an efficient tool to predict the thermal performance of fibrous materials at room and elevated temperatures (up to 400°C), provided the complex index of the bulk material and the length weighted diameter spectrum of the fibers are known.

**REFERENCES**

[1] Boulet P. Jeandel. G.,Morlot G., Silberstein A. and de Dianous P., Study of the radiative behaviour of several fibreglass materials, 22nd ITCC, 1993

[2] Banner P., Propriétés radiatives des verres et des fontes de silicates. Modélisation des transferts de chaleur, Thèse, Ecole Centrale Paris, 1990

[3] Boulet P.,Jeandel G. and Morlot G. Model of radiative transfer in fibrous media - Matrix method, Int. J. Heat Mass Transfer, vol. 36 (18), pp. 4287-4297, 1993

[4] Boulet P., Jeandel G., Morlot G. and de Dianous P., Etude théorique de l'influence des infibrés sur le comportement radiatif des isolants fibreux, accepted for publication in Int. J. Heat Mass Transfer, 1995

[5] Bohren C.F. and Huffman D.R., Absorption and scattering of light by small particules, J. Wiley and Sons, New-York, 1983

[6] Lee S.C., Radiation heat transfer model for fibers oriented parallel to diffuse boundaries, J. Thermophys. Heat Transfer, vol. 2 (4), pp. 303-308, 1988

[7] Kerke M., The scattering of light and other electromagnetic radiations, Academic Press, New York, 1969

[8] Ozisik M.N., Radiative Transfer and interactions with conduction and convection, J. Wiley, New York, 1973

[9] Langlais C., Guilbert G. Banner D. and Klarsfeld S., Influence of the chemical composition of glass on heat transfer through glass fibre insulations in relation to their morphology and temperature of use,  1st Conference of the European Society of Glass Science and Technology, Sheffield, England, 9-12 September 1991

[10] de Ponte et al., Reference guarded hot plate apparatus for the determination of steady-statethermal transmission properties in agreement with the new international standard ISO/DIS 8302, Thermal Conductivity, vol. 19, New York : Plenum publishing Co., pp. 232-240, 1988

[11] ISO 8302, Thermal insulation - Determination of steady-state thermal resistance and related properties - heat flow meter apparatus

[12] Viskanta R., Heat transfer by conduction and radiation in absorbing and scattering materials, Journal of Heat Transfer, pp 143-150, 1965.

# Thermoinsulation of Oil- and Gas-Pipelines with Foamed Plastics in Russia

F. SHUTOV

## ABSTRACT

The most common long distance oil- and gas- pipelines in Russia are so-called channelless lines. In view of the fast development and improvement of construction technology, there is a need for more effective materials for thermoinsulation, sufficiently lightweight and strong, and at the same time possessing a combination required for efficient and reliable performance of underground lines. Foamed plastics based on thermosetting phenolic resins have been found to be such a material. Many years of field testing proved it to be effective in terms of thermoinsulation, energy saving, and low cost. New generation of phenolic foams for pipelines is based on in situ (foaming-in-place) technology, having very uniform cellular structure, low brittleness and corrosive activity and much higher mechanical strength. The main technical requirements for plastic foam insulation are discussed: apparent density, compressive strength, moisture content, thermoconductivity, and etc. Results of model laboratory tests under conditions imitating the service of heat insulated pipes in real channelless lines are presented, and these data are compared with foam concrete insulation. Data on heat conductivity of phenolic foams at broad range of moisture content and apparent densities are discussed, as well as results of water and moisture sorption. Briefly discussed is a pouring technique for carrying out of phenolic foam insulation on metal pipes in situ at different ambient temperatures (from -30° C to +30°C). Some recommendations and conclusions based on long-term application of pipe insulation in Russia are given.

## INTRODUCTION

The rapid growth of urban and industrial premises calls for an extensive construction of heat supply networks. The factors important to the economy of heat supply include cost reduction, increase of efficiency and reliability of the supply pipelines. The type of insulation is an important contributor to all three factors.

F. Shutov, Center for Manufacturing Research, Tennessee Technological University, Cookeville, TN 38505, USA

In view of the fast development and improvement of construction technology, there is a need for more effective materials, sufficiently light and strong, and at the same time possessing a combination of specific underground heat lines. Foamed plastics fabricated from thermostable resol-type phenol-formaldehyde (PF) or phenolic resins have been found to be such a material [1].

Over 15 years ago, a principally novel type of cast-in-place heat insulation for underground heat lines are developed in Russia based upon such foamed plastics. Many years of field testing proved it to be effective; the monolithic PF foam insulation based heat insulation (i.e. a "monolithic layer," without any voids and chinks, between the foam and the steel surface of the pipe) has been found to be advantageous with regard to energy saving and its own low cost [2].

The simple PF foam production process, workability of full mechanization of the foam fabrication process and, finally, a range of advantageous properties, such as sufficiently high mechanical strength with the low apparent stability, the ability to protect metals from corrosion, etc., have pushed phenolic foams to the forefront amongst other types of heat insulation.

## REQUIREMENTS OF HEAT INSULATION MATERIALS
## INTENDED FOR CHANNELLESS PIPE LINES

The heat insulation of pipelines in channelless systems has to stand high heat carrier temperatures and support the weight of the overlying ground. The covering has, furthermore, to perform the two-fold function of protecting the steel pipes from environmental corrosive attacks on the one hand, and minimizing heat losses during operation of the heat supply network on the other.

It is a well-known fact that the lifetime and efficiency of heat insulation of channelless underground pipelines depend primarily on the continuity of the casing and the heat insulating layer, i.e. absence of gaps between the exterior surface of the pipe and the plastic foam. Thus the protection of steel pipes from corrosion is best achieved if the insulating layer is impermeable to water, air and stray electrical currents. The maximum admissible values of the corresponding values are:

| | |
|---|---|
| air permeability | $1.9 \times 10^{-6}$ kg/m mm $H_2O$ |
| resistivity | $10^5 \Omega$ cm |
| moisture content in monolithic heat insulation | Not more than 6 percent by volume |

From many years of experience in running heat supply lines it has been established that the above values of air permeability, water permeability and resistivity may be achieved by using the following coating structures: a vapor barrier of a bituminous rubber cement reinforced with glass fibers, a polyethylene film, and two layers of cement-impregnated Kraft paper with a layer of water glass interposed between them [3]. As well as the protective function, the total monolithic cased heat insulated pipe must be mechanically strong. Table 1 gives the standard specifications for the PF foam heat insulation of channelless heat supply pipelines.

Table 1. Requirements for Foam Insulation of Pipelines

| Characteristic | Unit of measurement | Grade FL phenolic foam | Remarks |
|---|---|---|---|
| Apparent density, not less than | kg/m³ | 100 | |
| Ultimate compression strength: | kg/m² (MPa) | | |
| short-term | | 1.9(0.19) | |
| long-term | | 4.0(0.4) | |
| Moisture content | vol % | 6 | At 20°C |
| Thermal conductivity, not more than | kcal/m h °C (W m cal) | 0.06(0.069) | At 100° c and moisture content of 6 vol. |
| Air permeability, less than | kg/m h mm H₂O | 1.9×10⁻⁶ | The same for heat insulation an vapor barrier |
| Resistivity, less than | Ω cm (Ω m) | 10⁵(10⁷) | |

## LABORATORY TESTS

Model tests have been carried out in the laboratory under conditions imitating the service of heat insulated pipes in real channelless lines to establish the performance characteristics of metallic pipes insulated continued until a steady heat flow was established; during the test, soil samples are taken periodically to check moisture content. The results of the long-term laboratory tests (which lasted up to 80 h) are presented in Table 2.

Table 2. Comparison of Concrete Foam and Phenolic Foam for Pipe Insulation

| Insulation property | Unit of measurement | Insulation | |
|---|---|---|---|
| | | Reinforced concrete foam | FL-1 Grade phenolic foam |
| Apparent density | kg/m³ | 132 | 136 |
| Foam moisture content prior to tests | vol % | 9.0 | 9.8 |
| Same in steady operation | vol % | 7.9 | 9.1 |
| Pipe surface temperature | °C | 150 | 74 |
| Heat losses | kcal/m h | 217 | 217 |

## PROCESS OF APPLICATION OF FOAM HEAT INSULATION ONTO METALLIC PIPES

In the three-component pour machine all three components, the phenolic resin, urea formaldehyde (UF) resin and the acid catalyst, are mixed in the mixing head during the few seconds immediately preceding the foaming. During so short a time period the viscosity of composition is not increased; on the contrary, owing to a minor increase of temperature when the gas-generating reaction has begun, the system viscosity is decreased which facilitates mixing of the composition (Table 3).

After coating the insulation onto the pipes, they are delivered to the assembly site where the separate lengths are joined and welded together. One of the decisive technological factors in providing a monolithic channelless underground pipeline as a whole is the development of reliable joint-sealing methods. The joints may be insulated in two different ways. The first is by applying the heat insulating coat with the help of prefabricated shells; however, because the insulation of the joint is not continuous in this case, the continuity of the whole insulation of the pipeline cannot be achieved, and air and water may penetrate through to the heat insulation layer which will inevitably result in corrosion of steel pipes.

A more advanced method of sealing joints between steel pipes is the in situ pouring of the composition into a prefabricated mold. In the simplest case the mold may be a ply of roofing paper or asphalt sheathing paper wrapped around the joint; to prevent tearing of the paper (due to excess pressure during foaming) it is externally reinforced with a band of wooden strips, for example.

The FL-2 (based on PF resin) or FL-3 (based on mixture of PF and UF resins) phenolic compositions are poured through an opening in the top portion of the mold. Thirty to forty minutes after foaming, the banding and the paper are removed and the water barrier is applied. The amount of composition required per joint is calculated beforehand on the basis of the length and diameter of the joint.

Table 3. Foaming Parameters and Properties of the FL-3 Grade Foam

| UF resin content (parts by weight) | Parameter/Properties | | | | |
|---|---|---|---|---|---|
| | Foaming time (min) | Maximum foaming temperature ($T°C$) | Polymer matrix density ($g/cm^3$) | Color | Cellular Structure |
| 0 | 3.5 | 105 | 1.20 | Red | Large-cell, non-uniform size |
| 10 | 4.0 | 85 | 1.34 | Orange | Large-cell, non-uniform size |
| 20 | 4.5 | 82 | 1.39 | Light Yellow | Small-cell, non-uniform size |
| 40 | 4.5 | 75 | 1.41 | Yellow | Small-cell, non-uniform size |

## MECHANICAL PROPERTIES

As already mentioned, the heat insulating layer of a channelless underground pipeline has to resist the pressures of the ground, traffic and the pipe with water. These loads produce compressive and tensile stresses in the thermo-insulating layer, it is necessary that the insulation material be integral (continuous, monolithic) and the phenolic foam must have a strength generally higher than the expected loads.

## WATER ABSORPTION

Phenolic foams have up to 90 percent open cells, hence moisture can easily penetrate the cellular material filling the cells to complete saturation. Cell size is the most important single factor determining water absorption. In a large-cell phenolic foam the water uptake is higher than in a small-cell one. The presence of a denser skin on the heat insulation surface reduces the absorptivity of plastic foams (Figure 1).

## HEAT CONDUCTIVITY

Owing to the cellular structure, the heat transfer in the material occurs because of the heat conductivity of the polymer phase and of the gas; the other, less important, factors are convection of the gas phase and radiation of heat by cell walls. Furthermore, the heat conductivity of cellular plastics depends on the composition formulation, the presence of various adjutants, the type of blowing agent, degree of dispersion and structure; however, the major factor is the apparent density, i.e. the volume fraction occupied by the gas phase. As the experiments with FL-2 and FL-3 have shown, the cellular plastics in question have practically the same coefficient of heat conductivity depending only on the density and moisture content of the material (Figure 2).

A study of the sorption curves of phenolic foams of different densities shows that capillary condensation of moisture in the pores of the material begins as soon as the moisture content in the material attains a value of 10-20 percent by mass. Below this value the absorbed moisture in the material is in the form of water vapor and adsorbed layers on the cell walls and actually has a very small effect on the coefficient of heat conductivity. When the absorbed water weight exceeds 20 percent, the capillary condensation of moisture in the cells begins and causes a sharp increase in the coefficient of heat conductivity of the material, because the heat conductivity of water which has replaced air in the cells is about 20 times as great as that of air.

Although the water absorption of laboratory specimens is rather small, the coefficient of heat conductivity of phenolic foams FL-2 and FL-3 within the range of apparent densities under consideration ($\gamma = 20\text{-}200$ kg/m³) is between 0.04 and 0.06 kcal/m h° C. Nevertheless, considering the sharp increase of heat conductivity that may be caused when the water content or temperature increases, the insulation material must be provided with a water barrier.

## EFFECT OF LONG-TERM EXPOSURE

Based on the reduction of strength of foams as functions of temperature and time of exposure in semilogarithmic coordinates and extrapolating the strength reduction to a given limit of 10,050 days (steel pipe depreciation period), the service lifetime of the material in question can be calculated (Figure 3). As can be seen, the strength reduction is approximately 7-10 percent after exposure for 4000 h to 150° C and 28-29 percent after exposure for the same time to 200° C. Most of the strength loss occurs

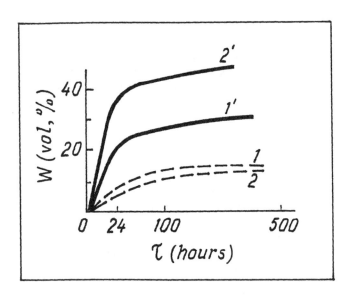

Figure 1. Water Absorption of Phenolic Foam Grades FL-2 (1) and FL-3 with Density 100 kg/m³: 1,2 - After Surface Contact with Water, 1', 2' - After Submerging

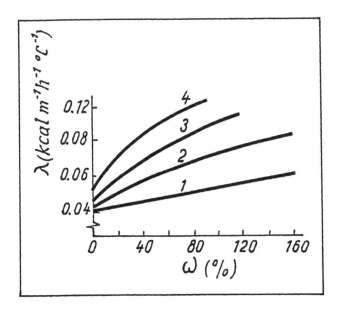

Figure 2. Thermoconductivity of various of phenolic foam FL-3 moisture of different foam density: 1-60, 2-100, 3-150, 4-200 kg/m³

during the first 500 h of thermal aging, which is also indicated by the slope of the strength reduction curve representative of the rate of thermal degradation. After that, the graphs representing the rate of degradation and loss of mechanical strength become flatter.

## WATER ABSORPTION AFTER THERMAL AGING

The amount of absorbed water is known to depend on the apparent density of foam and decreases when the latter increases. As the temperature is raised, water absorptivity always increases, which may result in undesirable corrosion of steel pipes insulated with this material.

Thermal aging also affects water absorptivity (Figure 4). Two factors are, apparently, responsible for the increase of water absorption by the phenolic foams FL-2 and FL-3 as thermal oxidation develops: the first is physical, the generation of defects in the original macrostructure (microcracking and burn-up of material) which facilitate and accelerate saturation of samples with water; the second is chemical, formation of strongly polar hydrophilic chemical groups during the thermo-oxidative degradation which results in the build-up of water molecules in the material.

## FULL-SCALE TESTS OF CAST-IN-PLACE FOAM INSULATION

The coefficient of thermal conductivity was measured under real operating pipeline conductions with the aid of Schmidt belts. The method of determining the thermal conductivity consisted of measuring the heat flux through the measuring belts (Schmidt belts) which were comprised of a plurality of series- connected thermocouples enclosed within a sealed flexible rubber sheath in the form of a long plate with two terminals for the connection of copper conductors. The heat flux measuring belts were mounted on the smoothed heat insulation surface (smoothed with the help of gypsum).

The coefficient of thermal conductivity of cast-in-place phenolic insulation was found from the formula:

$$\lambda = q/(T_{in} - T_{out})$$

where $q$ is specific heat flux. Here $T_{in}$ is the temperature on the pipe surface (°C) and $T_{out}$ is the temperature on the heat insulation surface (°C).

The coefficient of thermal conductivity of foam FL-3 was calculated to be 0.046 W/m °C for averaged apparent density of foam of 100 kg/m$^3$ and moisture content of 6 percent. The results not only showed the heat insulating coat to be highly effective but also confirmed that the water barrier was a reliable protection (Table 4).

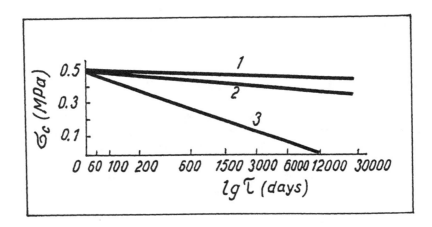

Figure 3. Relationship Between Compression Strength of Phenolic Foam FL-3 and Exposure Time after Thermal Aging at 100° (1), 150° (2) and 200° C(3)

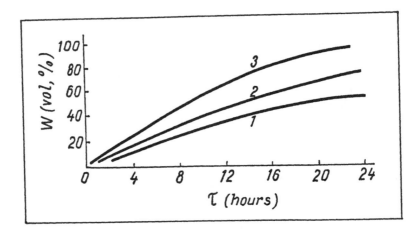

Figure 4. Kinetics of Water Absorption of Phenolic Foam FL-3 and Exposure Time of Thermal Aging at 150° C for 100 (1), 300 (2) and 500 Hours (3) of Aging

462

Table 4.  Long-term Aging of Phenolic Foam Insulated Pipelines

| Characteristics of pipeline with insulation | Service time of channelless pipeline | | |
|---|---|---|---|
| | 1 year | 4 years 9 months | 5 years 10 months |
| Pipe diameter (mm) | 219 | 219 | 219 |
| State of routing | Periodically flooded | Periodically flooded | Periodically flooded |
| Protective coating of insulated pipes | Water barrier of 2-ply glass cloth on bituminous rubber cement | Bituminous rubber cement reinforced with two layers of glass cloth | Bitumen plus two layers of glass cloth |
| Heat carrier parameters: | | | |
| hot pipe (°C) | 84 | 180 | 84 |
| return pipe (°C) | 60 | 70 | 60 |
| Coefficient of thermal conductivity (kcal/m h °C) | — | 0.056 | 0.048 |
| Moisture content of foam (vol percent) | 3 | 23.6 | 3 |
| Corrosion on pipes | None | Weak traces | 0.00253 mm/year |

## TECHNICO-ECONOMIC EFFECT OF FOAM INSULATION

Analysis of the effectiveness of the heat insulation is difficult for several reasons [5]. The principal ones are given below.

On the one hand, the cost of the heat insulation and protective coatings for channelless heat lines does not exceed 8-10 percent of the total cost and cannot have a large effect on the capital investment. However, the quality of the coatings is decisive as regards the technicoeconomic indices of a heat pipeline after a long service, because of both reduction of heat losses and increase of the service lifetime of the structures. Thus when seeking means to improve the heat insulation coatings one cannot ignore their actual cost, including the cost of materials and labor.

Despite the higher cost of phenolic heat insulation, for which the higher cost of the raw material is responsible, the overall economic effect of the use of phenolic heat insulation is, nevertheless, better than of reinforced foamed concrete insulation because phenolic foam insulation in a channelless heat line reduces the cost of construction of heat lines by as much as 16.9-27.8 percent and provides an especially high economic effect for 200-800 mm pipes. In the latter case the difference between the capital investments per cubic meter for reinforced foamed concrete and phenolic foam may be as high as 30 percent.

The higher capital cost of the production of pipelines insulated with reinforced foamed concrete is due to the complex technology of applying the insulation layer, the cumbersome and costly equipment required for the purpose, and the large number of manufacturing process steps, of which the principal ones are: preparation of the foamed concrete mixture, filling the molds, fabrication of the reinforcement and placement in the molds, heat treatment of the foamed concrete mixture in autoclaves for periods as along as 14-16 h; delivery of the heat insulated pipes to the water barrier application station.

The process of preparation and application of phenolic heat insulation is very simple and, in contrast to reinforced foamed concrete and bitumen-perlite, does not require cumbersome equipment. Power requirements are minimal. No extra power is consumed at any stage of the preparation of the heat insulation layer of PF resin foam on a pipe. Besides, the process of application of the phenolic insulation and water barrier may be fully automated.

Owing to the valuable mechanical and thermophysical properties and the markedly lower apparent density of phenolic foams compared with other insulation materials (the latter being a valuable property in itself, e.g. in transportation and assembly of pipes), the heat insulation made from this material is most effective and specifically advantageous in channelless heat lines.

## CONCLUSIONS

Investigations of the properties and development of a process for manufacturing novel phenolic foams, modified with urea-formaldehyde resin, have shown it to be an effective heat insulation material for insulating channelless pipelines.

Physico-mechanical and thermo-physical tests of these phenolic foams have demonstrated unambiguously that these materials present an all-round solution to the problem of reliable protection of metallic pipes of heat lines under the long-term action of elevated temperatures.

Considering the necessity of using phenolic insulation for high temperature pipelines (200° C and above), an investigation was undertaken towards upgrading the heat resistance of the existing phenolic foam grades. The result was a positive improvement and the method is recommended for commercial-scale production.

The principal advantages of using phenolic foams as insulation materials for channelless pipelines are:
- Highest possible automation of the pipe insulation process with minimum material and labor consumption.
- Simple process of insulation application, easiness of its control and adjustment.
- Compared to channeled pipelines, the amount of reinforced concrete components required is cut by as much as 70-85 percent.
- Reduction of the pipeline construction work in assembling prefabricated pipe lengths and filling the joints between them.
- Cost reduction and complete industrialization of the heat pipeline construction.
- Improvement of durability and performance reliability of heat lines.
- Reduction of heat losses in channelless pipeline and reduction of maintenance costs.

The long-term operation of pipelines insulated with phenolic foams has allowed the reassessment of the possibility of reliable performance of pipelines over periods exceeding the standard steel pipe depreciation time.

## ACKNOWLEDGMENT

I would like to express my deep gratitude to my colleagues from St. Petersburg, Russia, Dr. G. Ivanov (Polymer Laboratory) Civil Research Institute, and Dr. I. Chaikin (Department of Physics, University of Civil Engineering) for their many years of cooperation and intellectual contribution to develop this project.

## REFERENCES

1. Berlin, A., Shutov, F., Zhitinkina, A., 1982, "Foam Based on Reactive Oligomers," Technomic, pp. 145-214.

2. Shutov, F., Ivanov, V., 1984, 1985, "Foams Based on Phenolic and Urea-Formaldehyde Mixtures," *Cellular Polymers*, 359-382, 411-432, 27-53.

3. Shutov, F., 1984, "Phenolic Foams in the USSR," *Cellular Polymers*, 95-104.

4. Shutov, F., 1991, "Cellular Structure and Properties of Foamed Polymers," in *Handbook of Polymeric Foams and Foam Technology*, D. Klempner and K. Frisch eds, Hanser, pp. 17-45.

5. Chaikin, I., Shutov, F., 1993, "Quality Control of Foam Insulation of Pipelines using Dielectrical Method," Annual Technical Conference, University of Civil Engineering, St. Petersburg, Russia.

6. *Fiberglass Pipe Handbook*, 1989, *SPI* Composites Institute, Society of Plastics Industry.

# Effect of Sub-Minute High Temperature Heat Treatments on the Thermal Conductivity of Carbon-Bonded Carbon Fiber (CBCF) Insulation

R. B. DINWIDDIE, G. E. NELSON and C. E. WEAVER

## ABSTRACT

The thermal conductivity of carbon-bonded carbon fiber insulation in vacuum was determined before and after short duration heat treatments to model the effect of varying degrees of graphitization. Specimens of the insulation were heat-treated for 10, 15 and 20 seconds at 2673, 2873, 3073, and 3273 K. The dimensions and mass of each specimen was recorded before and after heat treatments. The thermal conductivity of the heat treated specimens was measured as a function of temperature up to 2273 K. These data are compared with previously measured specimens heat treated at the same temperatures for 1 minute and one sample heat-treated at 3273 K for 1 h. The thermal conductivity increases with both the heat treatment temperature and time at temperature. The thermal conductivity data has been modeled to obtain equations that predict the thermal conductivity of the insulation as a function of temperature (673 K $\leq$ T $\leq$ 3773 K) and heat treatment conditions of time (0 $\leq$ t $\leq$ 20 s) and temperature (2673 K to 3773 K).

## INTRODUCTION

Carbon-bonded carbon fiber insulation possesses a unique set of thermal and mechanical properties that permits use as a high-temperature insulating material in vacuum environments. Fabrication of the material has been described previously in detail [1]. During certain application conditions, the insulation may be exposed to thermal ramps (heating and cooling) with time spans ranging up to 20 seconds and peak temperatures in the range of 2200 K to about 4100 K. Mizushima [2] has shown that heat treatments as short as four minutes can result in partial graphitization. Thus, short duration high-temperature exposure may be expected to cause varying degrees of graphitization of the carbon fibers in the insulation. Such changes in the fiber microstructure would have a profound impact on the thermal conductivity of the insulation. The purpose of this study is to experimentally measure, explain and model the effect of shorter duration (up to 20 seconds) high-temperature exposures on the thermal conductivity of the insulation.

A total of 36 insulation specimens were heat treated in this study. Each specimen was heat treated once. There were twelve heat treatment runs, each beginning with three as-fabricated CBCF insulation specimens. Heat treatments

R. B. Dinwiddie, G. E. Nelson, and C. E. Weaver
Oak Ridge National Laboratory, P.O. Box 2008, Oak Ridge, TN 37831

lasted for either 10, 15 or 20 seconds at temperatures of 2673, 2873, 3073 or 3273 K. The thickness, diameter, mass, and room-temperature thermal diffusivity were measured for each test specimen before and after heat treatment. The thermal diffusivity of two specimens from each heat treatment run was measured up to 2273 K after heat treatment. The third specimen was archived for future studies.

The thermal conductivity was calculated from the specific heat of carbon, bulk density of the test specimen, and thermal diffusivity data. The thermal conductivity data were then modeled to obtain equations that predict the thermal conductivity of the insulation as a function of temperature (673 K ≤ T ≤ 3773 K) and heat treatment conditions of time (0 ≤ t ≤ 20 s) and temperature (2273 K to 3773 K). It is important to note that these models were developed from measurements in the temperature range 673 to 2273 K on insulation exposed for 10, 15 or 20 seconds at temperatures in the range 2673 to 3273 K. The accuracy of these equations is typically within ±10% of the experimental data under these conditions. Extrapolation of these equations to higher heat treatment and measurement temperatures could result in a higher, and yet undetermined, uncertainty. Accuracy for the as-fabricated material (typically ±20%) is lower due to a very low thermal diffusivity. The model developed in this report is not suitable for exposure times greater than 20 seconds.

## HEAT TREATMENTS

Heat treatments were carried out in a graphite tube furnace with flowing ultrahigh-purity argon gas (Oxygen: 0.5 ppm, Moisture: 0.0 ppm, Hydrocarbon: 0.0 ppm). Temperatures were determined by correlating furnace power levels with pyrometer readings. The pyrometer was carefully calibrated prior to the heat treatments in an arrangement designed to simulate the actual furnace setup. A diagram of the graphite furnace is shown in Fig. 1. In order to make short duration heat treatments, the furnace was modified to allow the addition of a push-rod . This push-rod was used to quickly slide a specimen capsule from the cool zone at one end of the furnace into the hot zone and then into the cool zone at the opposite end of the furnace. The exposure time was defined to be the time the specimen was in the hot zone. The time required for the specimen to heat up and cool down were initially assumed to be negligible due to the effectiveness of radiation heat transfer at these temperatures. The time required to reach the heat treatment temperature was later determined to be approximately 4.3 seconds, as discussed below.

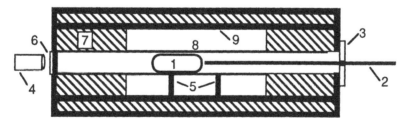

Figure 1. Diagram of furnace used for heat treating the insulation. 1) graphite specimen capsule, 2) graphite push-rod , 3) carbon-felt push-rod bushing, 4) pyrometer, 5) graphite tube support, 6) window, 7) carbon furnace insulation, 8) hot zone of furnace 9) graphite heater element. (Drawing not to scale)

A diagram of the specimen capsule is shown in Fig. 2. The capsule was machined from graphite to hold seven specimens of the insulation (nominally 12.5-mm-diameter. by 1-mm-thick disks). The capsule is open at the leading end, and the specimens are held in place by an open-end cap at the other end. The three test specimens are surrounded by four dummy specimens (two on the leading end and two on the cap end). The specimens were separated by thin graphite disks to uniformly distribute the furnace heat throughout the capsule. Following heat treatment, specimens were removed and characterized for thickness, diameter, mass, and density. Two specimens were inserted into a vacuum furnace for determination of thermal diffusivity as a function of temperature. The third test specimen has been reserved for x-ray diffraction analysis and microscopy studies.

## DIMENSIONAL AND WEIGHT CHANGES

The thickness and diameter of each test specimen were measured before and after heat treatment using a digital caliper. The mass of each specimen was measured on an analytical balance before and after heat treatment. All measurement instruments have been calibrated within the past six months using standards from the National Institute of Standards and Technology. These data show a general trend, the fraction of weight loss increases with both heat treatment temperature and time. However, the total weight change is very small, typically less than 2%, and the scatter in this data is large.

## THERMAL DIFFUSIVITY/CONDUCTIVITY

Thermal diffusivity measurements were made under high vacuum conditions ($10^{-6}$ Torr) using the flash diffusivity system operated by the Carbon and Insulation Materials Technology Group in the Metals and Ceramics Division at Oak Ridge National Laboratory. Thermal conductivity values were calculated from the relationship $\mathbf{k} = \alpha \rho C_p$, where $\mathbf{k}$ is the thermal conductivity, $\alpha$ is the thermal diffusivity, $\rho$ is the bulk density, and $C_p$ is the specific heat at constant pressure. Thermal diffusivity values were calculated using a parameter estimation technique developed by Koski [3]. The specific heat values used in these calculations are given in Table I.

The thermal conductivity of four specimens of the as-fabricated insulation is shown in Fig. 3. This plot demonstrates the typical specimen to specimen variability in the insulation and the typical experimental scatter observed in the thermal conductivity data. The thermal conductivity increases with increasing measurement temperature due to the contribution of radiation within the insulation.

The thermal conductivity increases as the furnace hot zone exposure time is increased. This is shown clearly in Fig. 4, where the thermal conductivity measured at 1073 K is plotted as a function of the time the specimen was exposed to 3273 K. The solid horizontal line represents the value of thermal conductivity after a 1 hour thermal exposure at 3273 K.

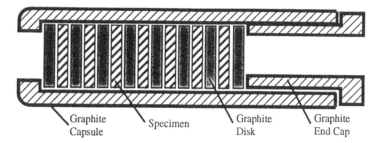

Figure 2. Diagram of graphite specimen capsule used for heat treating specimens of the insulation. (Drawing not to scale)

TABLE I.  The temperature dependence of specific heat of the insulation.

| Temperature (K) | Specific Heat J/(kg · K ) |
|---|---|
| 298 | 733.0 |
| 373 | 930.3 |
| 473 | 1155 |
| 573 | 1338 |
| 673 | 1484 |
| 773 | 1601 |
| 873 | 1696 |
| 973 | 1772 |
| 1073 | 1865 |
| 1173 | 1886 |
| 1273 | 1929 |
| 1373 | 1965 |
| 1473 | 1995 |
| 1573 | 2021 |
| 1673 | 2043 |
| 1773 | 2062 |
| 1873 | 2078 |
| 1973 | 2093 |
| 2073 | 2105 |
| 2173 | 2116 |
| 2273 | 2125 |

## IMPROVED MODEL

The temperature dependence of thermal conductivity ($\lambda_T$) for fibrous insulation may be modeled using the relationship

$$\lambda_T = A\, f_s\, \lambda_S + \lambda_R + \lambda_G, \qquad (1)$$

where A is the fraction of heat transfer in the parallel mode, $f_s$ is the volume fraction of solid (approximately 0.1 for the insulation), $\lambda_S$ is the thermal conductivity of the solid phase, $\lambda_G$ is the thermal conductivity contribution due to the presence of a gas phase within the specimen(assumed here to be zero), and $\lambda_R$ is the effective thermal conductivity due to radiation given by

$$\lambda_R = (16\, n^2\, \sigma/3\, \sigma_e)\, T^3, \qquad (2)$$

where n is the index of refraction (n=1), $\sigma$ is the Stefan-Boltzmann constant [5.6697 x $10^{-8}$ W/(m$^2\cdot$K$^4$)], $\sigma_e$ is the extinction coefficient for the insulation, and T is the absolute temperature. The solid-phase thermal conductivity must be scaled to account for density and the fact that not all the solid phase contributes to the solid thermal conductivity. The temperature dependence of the thermal conductivity of graphite is typically modeled by

$$\lambda = A_1 \cdot T^{-\varepsilon}, \qquad (3)$$

where $A_1$ and $\varepsilon$ are fitting parameters with $\varepsilon$=1 for ideal graphite. For polycrystalline graphite with point defects, $\varepsilon$ is between 0.5 and 1. Poor quality graphite with a significant population of extended defects has a value of $\varepsilon$ of less than 0.5.

To model the sub-minute thermal conductivity data, it was decided to keep the radiation contribution consistent with previous results ($\sigma_e$ = 14000 m$^{-1}$) [4], and to modify the solid contribution with the relationship

$$\lambda_S = X\, A_1\, T^{(\varepsilon + Z)} \qquad (4)$$

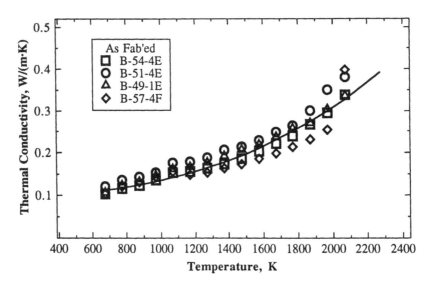

Figure 3. The thermal conductivity increases with increasing measurement temperature due to the contribution of radiation within the insulation.

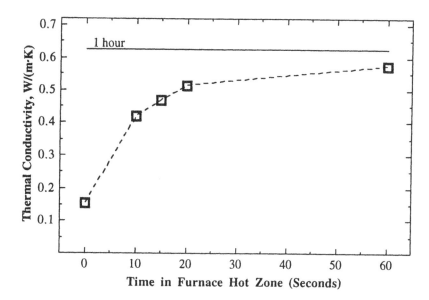

Figure 4. The thermal conductivity of the insulation measured at 1073 K is shown as a function of heat treatment time at 3273 K.

where $A_1$ and $\varepsilon$ are determined by modeling the thermal conductivity of as-fabricated specimens ($A_1 = 0.0232747$ and $\varepsilon = 0.237966$), X and Z are (to be determined) functions of heat treatment time and/or heat treatment temperature. Thus, the functional form of the equation used to model the thermal conductivity of the insulation is given by

$$\lambda = X \, A_1 \, T^{(\varepsilon + Z)} + 2.1599 \times 10^{-11} \, T^3 \qquad (5)$$

Equation 5 was fit to the thermal conductivity data for specimens heat treated at various temperatures for 10, 15 and 20 seconds. Figure 5 shows a plot of the resulting fitting parameter, Z, as a function of the heat treatment temperatures. Equation 6 gives the linear relationship between Z and the heat treatment temperature, $T_m$.

$$Z = 1.276492 - 0.0005659 * T_m \qquad (6)$$

Equation 6 was substituted back into Eq 5 which was used to refit the data to find the relationship between X and the heat treatment conditions (Eqs 7 and 8). Figure 6 shows a plot of the fitting parameter X as a function of the heat treatment temperature for the 10, 15 and 20 second thermal conductivity data. The fitting parameter X was found to contain a time dependent term R. By definition, we know that X=1 when t=0, therefore, R must equal $1.26 \times 10^{-5}$ when t=0. Initially, it was found that R equals this value for a time of 4.3 seconds. We were able to bring the R parameter data in line with the as-fabricated value by subtracting 4.3 seconds from the time in the furnace hot zone. This value of 4.3 seconds can be thought of as the time required by the specimen to reach the heat treatment temperature. Figure 7 shows the relationship between the parameter R, and the corrected time at temperature. Equations 7 and 8 give the relationship between X and the heat treatment temperature and time, t, of exposure.

$$X = R \ exp(T_m * 0.004963) \tag{7}$$

$$R = 1.26115 \times 10^{-5} + 3.6166 \times 10^{-7} * t \tag{8}$$

Figures 8, 9 and 10 show the results of combining Eq 5, 6, 7 and 8 and refitting the thermal conductivity data for specimens heat treated for 5.7, 10.7, and 15.7 seconds, respectively. The prediction for as-fabricated thermal conductivity is included for comparison. Figure 3 shows the fit of this model to the as-fabricated data.

The quality of the fit of this model can be determined by plotting the residuals. Figures 11 and 12 show plots of the residuals as a function of temperature for heat treated specimens and as-fabricated specimens, respectively. Typical scatter in the residuals for the heat treated specimens lies between ±10%. The standard deviation of the percent residuals for as-fabricated and heat treated specimens are shown in Tables II and III, respectively. The scatter in the residuals for as-fabricated specimens lies between ±20%. This larger scatter is due to the allowable variations in the CBCF insulation, which results in a specimen-to-specimen variation in thermal conductivity.

## CONCLUSIONS

The overall thermal conductivity of the insulation increases with heat treatment temperature and time at temperature. The functional form for the solid-phase thermal conductivity, used to model the temperature dependence of thermal conductivity of the insulation is $X A_1 T^{(\varepsilon+Z)}$. The constants $A_1$ and $\varepsilon$ are determined from modeling the thermal conductivity of as-fabricated specimens, where $X = 1$ and $Z = 0$. The fitting parameter, X, is a function of both time and heat treatment temperature. The parameter X may be calculated using Eq. 7 and 8. The fitting parameter Z is a function only of the heat treatment temperature and may be calculated from the linear relationship Eq. 6. The predicted values are typically within ±10% of the measured values except for as-fabricated specimens where the agreement is within ±20%. The larger uncertainty in the as-fabricated data is due to specimen-to-specimen variability in the material and the low thermal conductivity.

## ACKNOWLEDGMENTS

Research sponsored by the Space and National Security Programs, Office of Nuclear Energy, Science and Technology, and the Assistant Secretary for Energy Efficiency and Renewable Energy, Office of Transportation Technologies, as part of the High Temperature Materials Laboratory User Program, Oak Ridge National Laboratory, managed by Lockheed Martin Energy Research Corp. for the U.S. Department of Energy under contract DE-AC05-96OR22464

## REFERENCES

1. G. C. Wei and J M Robbins, "Carbon-Bonded Carbon Fiber Insulation for Radioisotope Space Power Systems," Am. Ceram. Soc. Bull., 64[5] pp691-99 (1985)
2. Sanchi Mizushima, "Rate of Graphitization of Carbon," Proc. 5th Conf. on Carbon, Vol. II, The Macmillan Co. (1962)
3. J.A. Koski, "Improved Data Reduction Methods for Laser Pulse Diffusivity Determination with the Use of Minicomputers," pp. 94-103 in *Proceedings of the Eighth Symposium on Thermophysical Properties*, Vol. 2, ed. Jan V. Sengers, The American Institute of Physics, 1981.
4. W. P. Eatherly, Some Considerations on the Thermal Conductivity of CBCF, Internal Report for Technology for Special Nuclear Projects, October 31, 1983.

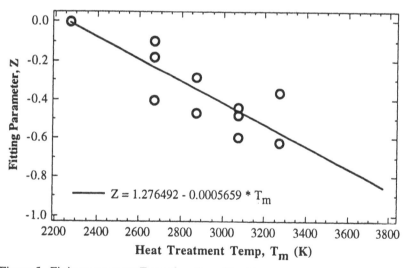

Figure 5. Fitting parameter Z as a function of heat treatment temperature.

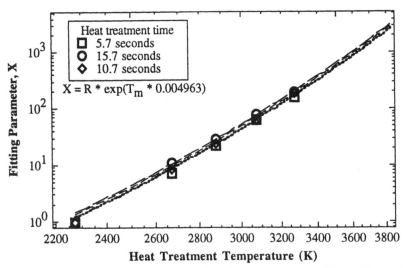

Figure 6. The relationship between X and the heat treatment conditions ($T_m$ and time) determined through a non-linear least squares fitting of Eqs. 7 and 8.

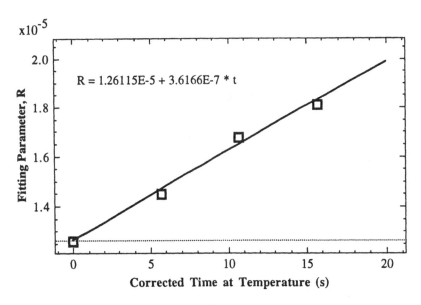

Figure 7. Fitting parameter R as a function of time at heat treatment temperature. A constant value of 4.3 seconds has been subtracted from the time in the hot zone of the furnace in order to bring the data in line with the as-fabricated value of R.

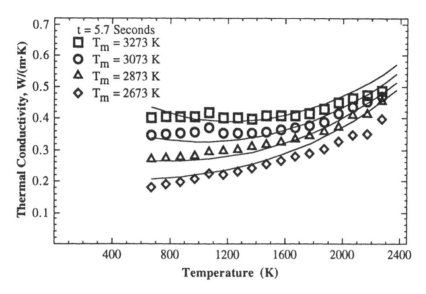

Figure 8. Fit of model to thermal conductivity data for specimens heat treated for 5.7 s (furnace hot zone exposure = 10 s).

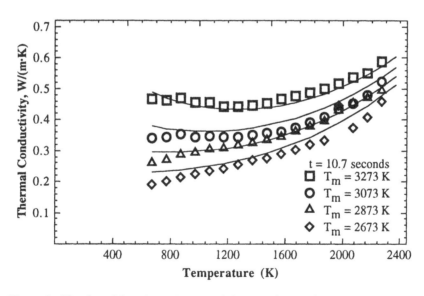

Figure 9. Fit of model to thermal conductivity data for specimens heat treated for 10.7 s (furnace hot zone exposure = 15 s).

475

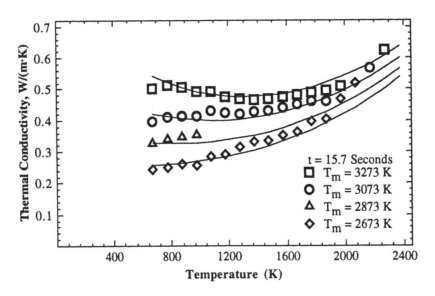

Figure 10. Fit of model to thermal conductivity data for specimens heat treated for 15.7 s (furnace hot zone exposure = 20 s).

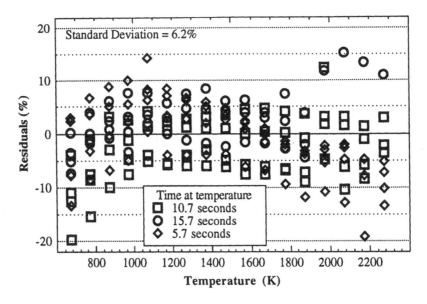

Figure 11. Percent residuals for all heat treated data

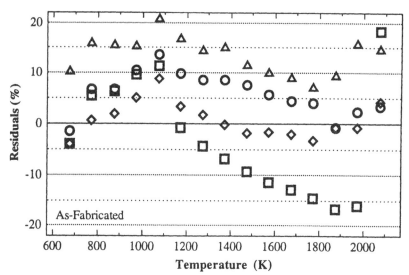

Figure 12. Percent residuals for as-fabricated data

TABLE II.  The standard deviation of the percent residuals from the fit of Eq. 5 to the thermal conductivity data for as-fabricated specimens.

| Specimen ID | Standard Deviation (%) |
|---|---|
| B-54-4E | 3.47 |
| B-51-4E | 3.68 |
| B-49-1E | 4.13 |
| B-57-4F | 11.0 |

TABLE III.  The standard deviation of the percent residuals from the fit of Eq. 5 to the thermal conductivity data for heat treated specimens.

| Heat Treatment Temperature (°C) | Heat Treatment Time(s) | Standard Deviation (%) |
|---|---|---|
| 2673 | 5.7 | 4.86 |
| 2873 | 5.7 | 4.96 |
| 3073 | 5.7 | 5.99 |
| 3273 | 5.7 | 4.53 |
| 2673 | 10.7 | 6.55 |
| 2873 | 10.7 | 3.84 |
| 3073 | 10.7 | 2.03 |
| 3273 | 10.7 | 2.19 |
| 2673 | 15.7 | 5.45 |
| 2873 | 15.7 | 3.29 |
| 3073 | 15.7 | 4.33 |
| 3273 | 15.7 | 4.11 |

**SESSION 6**

---

# FLUIDS

SESSION CHAIR
D. W. Yarbrough

# Initial Density Dependence of the Thermal Conductivity of Polyatomic Gases

A. BERNNAT, M. PAPADAKI and W. A. WAKEHAM

## ABSTRACT

New measurements of the thermal conductivity of argon, nitrogen and carbon monoxide within the temperature range 83 K to 387 K and for pressures up to 10MPa have been performed. The experimental data have an estimated accuracy of ±0.5% and are used in this paper to derive values of the first coefficient of the density expansion of the thermal conductivity of unprecedented accuracy over such a wide range of temperature. The results are compared with a theoretical evaluation of the coefficient for argon and with several empirical estimations for nitrogen and carbon monoxide. The agreement with the theoretical calculation of Rainwater and Friend for argon is good while for nitrogen and carbon monoxide the same theory with a semi-empirical addition to account for the internal energy proves more successful than the Modified Enskog Theory.

## INTRODUCTION

In the limit of zero-density the theory of the transport properties of monatomic and polyatomic gases is well established [1,2]. However, at moderate densities the formal theory has scarcely been developed for a series of fundamental reasons. This is unfortunate, not only because of the scientific interest in such a theory, but also because the estimation of the properties of dense gaseous mixtures crucially depends upon a knowledge of the behaviour of the pure components. For industrial purposes the prediction of the properties of such mixtures is vital and the experimental information available always inadequate, so that the prediction of the properties of both pure components and mixtures at elevated densities is important.

Anke Bernnat, Institut fur Technische Thermodynamik und Thermische Verfahrenstechnik, University of Stuttgart, Stuttgart, Germany
Maria Papadaki and William A. Wakeham, Department of Chemical Engineering and Chemical Technology, Imperial College, Prince Consort Road, London SW7 2BY, UK

In this paper we use recent measurements of the thermal conductivity of one monatomic and two diatomic gases over a wide range of temperatures to evaluate the first density coefficient of thermal conductivity, $\lambda_1$ in the expansion [3]

$$\lambda = \lambda_0 + \lambda_1 \rho + \lambda_2 \rho^2 + \lambda_2' \rho^2 \ln \rho + ... \qquad (1)$$

Among these coefficients the available evidence [4] suggests that $\lambda_2'$ is very small so that for all practical purposes it may be ignored. The remaining coefficients are always significant although it is so far only practicable to measure $\lambda_0$ and $\lambda_1$.

These data provide, for the first time, values of $\lambda_1$ over a wide range of temperature and density. We use the experimental results for an examination of recent means of evaluating the same coefficient, for argon on the basis of the theory of Rainwater and Friend [5], and for nitrogen and carbon monoxide on the basis of one form of the Modified Enskog theory [6] and an empirical extension of the Rainwater and Friend approach to polyatomic systems.


**EXPERIMENTAL**

The measurements of the thermal conductivity have been carried out by means of the transient hot-wire apparatus described in detail elsewhere [7]. In the present measurements we have employed a different set of hot-wires and the modified set of working equations described by Li *et al.* [8].

The measurements extend over the temperature range 83 to 387 K and were performed along isotherms for pressures from 0.01 to 10 MPa. Great care was taken to extend the range of pressures, and therefore densities, to sufficiently low values so as to always be able to determine the first density coefficient by means of an appropriate statistical analysis [4]. The gas samples were provided by BOC plc with a purity in excess of 99.998%. For argon the measurements were carried out along seven isotherms in the temperature range 135 to 320 K, for nitrogen along twelve isotherms in the range 85 to 387 K and for carbon monoxide along six isotherms in the range 83 to 297 K. The results of the measurements are contained in Figures 1 to 3 for each of the gases.

It is estimated that along all but the lowest isotherms the uncertainty of the thermal conductivity data is one of ±0.5 %. For the lowest isotherms at around 85 K it is estimated that the uncertainty in the thermal conductivity may rise to about 1% as measurements had to be extended to pressures as low as 0.01MPa [8].


STATISTICAL ANALYSIS

The regression technique whereby statistically and physically significant values of the first two coefficients of the density expansion of the thermal conductivity of equation (1) can be derived from experimental data have been described elsewhere [3]. Here it is sufficient to record that the statistical analysis of data along each isotherm supports the fact that equation (1) does represent the experimental data adequately although the logarithmic term is not statistically significant.

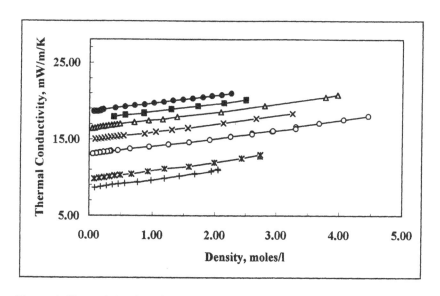

**Figure 1.** Thermal Conductivity of Argon. (●) 320 K, (■) 298 K, (Δ)275 K, (X) 247 K, (o) 211.5 K, (✳) 155 K, (+) 136.5 K

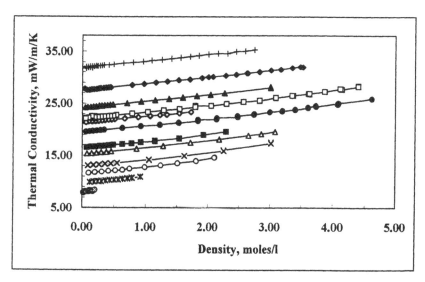

**Figure 2.** Thermal Conductivity of Nitrogen. (+) 387 K, (◆) 322 K, (▲) 275 K, (□) 247 K, (◊) 237 K, (●) 213 K, (■) 177 K, (Δ) 164.5 K, (X) 139 K, (o) 123 K, (✳) 104.5 K, (☉) 85 K

**RESULTS**

ZERO-DENSITY BEHAVIOUR

The first item of interest, which is important to the subsequent discussion of the first density coefficient, $\lambda_1$, concerns the thermal conductivity $\lambda_0$ in the limit of zero-density. Figure 4 compares the zero-density thermal conductivity of argon, nitrogen and carbon monoxide with the values calculated on the basis of the best available intermolecular pair potential surfaces for these gases. In the case of argon, the pair potential employed and the calculation procedure employed is summarised by Maitland *et al.* [1]. For nitrogen we employ the results of the classical trajectory calculations of Heck and Dickinson [9] for the intermolecular potential of van der Avoird *et al.* [10]. For carbon monoxide classical trajectory calculations have again been performed by Dickinson and his co-workers [11] using a potential energy surface of van der Avoird *et al.* [12].

Figure 4 demonstrates that for argon the agreement between theory and experiment is extremely good providing mutual confirmation of the accuracy of the pair potential and of the experimental data. For nitrogen, the agreement is slightly worse and evidently the theoretical and experimental thermal conductivity have slightly different temperature dependencies. Nevertheless the discrepancies do not amount to more than 2.5% and these are at the lowest temperature where the experimental uncertainty is greatest. It is therefore reasonable to assert that the intermolecular pair potential surface employed for nitrogen must be quite realistic.

In the case of carbon monoxide the disagreement between the experimental and theoretical values of the thermal conductivity in the limit of zero-density is quite marked and this suggests that the intermolecular pair potential in this case needs further refinement.

EXCESS THERMAL CONDUCTIVITY

When examining the density dependence of the thermal conductivity of gases it is both instructive and useful to consider the excess thermal conductivity [6] defined by the equation,

$$\Delta\lambda(T,\rho) = \lambda(T,\rho) - \lambda_0(T) \qquad (2)$$

It has often been reported [6] that the excess thermal conductivity is independent of temperature in the gas phase over a wide range of temperature except near the critical temperature. Figures 5, 6 and 7 indicate that the present results support this conclusion very strongly. It can be seen that except for nitrogen in the neighbourhood of the critical temperature (126.2 K) there is no systematic trend of the excess thermal conductivity with temperature. This provides an important empirical means to evaluate the thermal conductivity for pure gases over ranges of temperature where it has not been measured by simply adding a representation of the excess property determined under one set of conditions to the zero-density property determined under the conditions of interest.

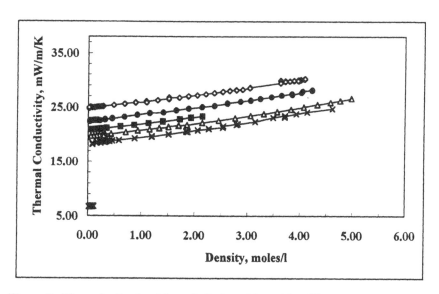

**Figure 3.** Thermal Conductivity of Carbon Monoxide. (◊) 297 K, (●) 263 K, (■) 240 K, (Δ) 225 K, (X) 210 K, (✳) 83 K

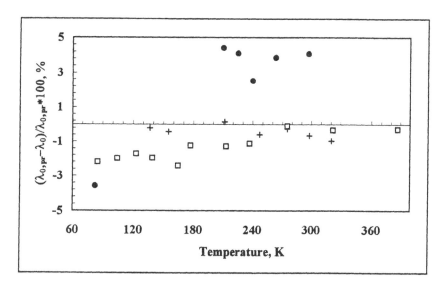

**Figure 4.** Comparison of the experimentally deduced thermal conductivity in the limit of zero density ,$\lambda_0$, with theoretical models, $\lambda_{0,pr}$, (+) Argon with Ref. [1], (□) Nitrogen with Ref. [9], (●) Carbon Monoxide with Ref. [11]

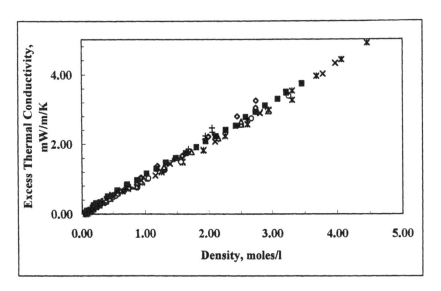

**Figure 5.** Excess Thermal Conductivity of Argon. (■) 320 K, (Δ) 298 K, (X)275 K, (o) 247 K, (✳) 211.5 K, (◊) 155 K, (+) 136.5 K

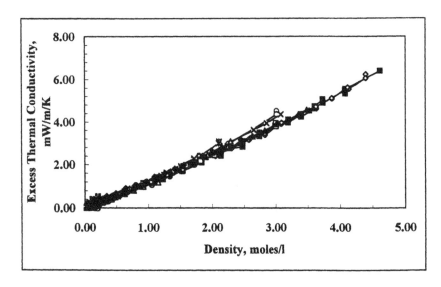

**Figure 6.** Excess Thermal Conductivity of Nitrogen. (◆) 387 K, (▲) 322 K, (□) 275 K, (◊) 247 K, (●) 237 K, (■) 213 K, (Δ) 177 K, (X) 164.5 K, (o) 139 K, (✳) 123 K, (+) 104.5 K, (⊕) 85 K

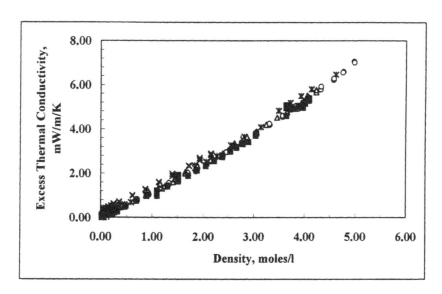

**Figure 7.** Excess Thermal Conductivity of Carbon Monoxide. (■) 297 K, (Δ) 263 K, (●) 240 K, (o) 225 K, (✳) 210 K, (+) 83 K

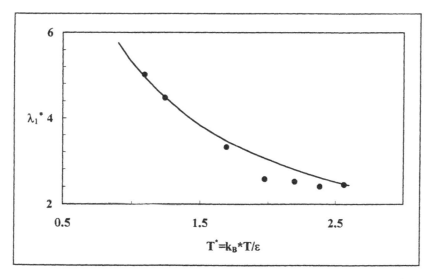

**Figure 8.** Comparison of the experimental, reduced first density coefficient $\lambda_1^*$ for argon,(●), with the optimised Rainwater & Friend theory, (—), Ref. [6]

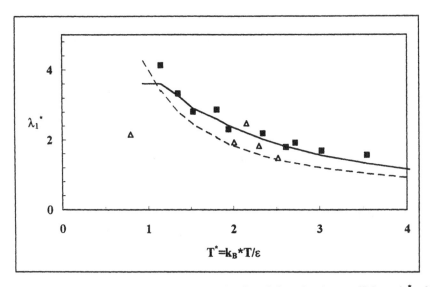

**Figure 9.** Comparison of the experimental reduced first density coefficient $\lambda_1^*$ of Nitrogen, (■), and Carbon Monoxide ,(Δ), with MET-II, (---), Ref. [13] and an empirical extension of the Rainwater and Friend approach, (——), Ref. [6]

## FIRST DENSITY COEFFICIENT OF THERMAL CONDUCTIVITY

In order to compare the predicted and experimental values of the thermal conductivity of the three gases of interest it is convenient to make the quantity dimensionless. For this purpose we make use of the definition of a dimensionless coefficient given by Bich and Vogel [6],

$$\lambda_1^* = \frac{\lambda_1}{\lambda_0 N_A \sigma^3} \qquad (3)$$

in which $N_A$ is Avogadro's constant and $\sigma$ is a length scaling parameter. It is also convenient to use a dimensionless temperature defined by $T^* = k_B T / \varepsilon$ where $\varepsilon$ is an energy scaling parameter and $k_B$ is Boltzmann's constant. Since we wish to make comparisons with the semi-theoretical models of Bich and Vogel we use uniformly the scaling parameters that they have listed [6]. No special significance is ascribed to them in terms of real characteristics of any intermolecular potential.

Figure 8 contains a plot of the reduced first density coefficient of thermal conductivity for argon with the optimised implementation of the Rainwater and Friend theory reported by Bich and Vogel [6]. It can be seen that this theory, which is based upon a Lennard-Jones 12-6 potential model, gives a good account of the experimental data over quite a wide temperature range with no adjustable parameters. This suggests that the theory can be used to extend the temperature range of the evaluation of the first density coefficient of thermal conductivity quite reliably.

Figure 9 contains a plot of the same reduced density coefficient of thermal conductivity as deduced from experiment for nitrogen and carbon monoxide and compares it with the results of two predictive schemes. Both nitrogen and carbon monoxide are diatomic molecules so that in order to treat the transport of energy in such gases it is necessary to consider not just the translational component of the energy transport but also that of the internal energy. It is logical to treat the transport of translational energy by the means employed for monatomic systems where this is the only available form of energy. For the internal energy component it is conventional to treat it as a process of diffusion of internal energy. Both of the predictive schemes to be tested have this as a common basis but they differ in the way that the translational component of the energy is handled at elevated density.

In the scheme described by Bich and Vogel, the evaluation of $\lambda_1^*$ is performed using the Rainwater-Friend result for the translational contribution and for the internal energy term making use of the diffusive contribution proposed by Mason and Monchick together with two additional terms proposed by Stogryn and Hirschfelder [6].

In the scheme proposed by Ross *et al.* [13] and denoted by MET-II [6] the evaluation of the translational contribution to the first density coefficient is again treated in the same way as for monatomic systems but both are this time treated within the Modified Enskog theory rather than that of Rainwater and Friend. According to this theory the Enskog results for dense fluids of hard spheres can be applied to real fluids if a suitable interpretation is made of the effective hard-sphere diameter and of the pair distribution function for the gas. In particular, within the MET-II version of this method the effective hard-sphere diameter of the gas is to be deduced from the real second virial coefficient of the gas while the hard-sphere equation of state provides the pair distribution function. For the internal energy contribution the MET-II theory, as proposed by Ross *et al.* [13], simply accounts for the hard sphere diffusion term and ignores the other effects introduced by Bich and Vogel [6].

Figure 9 compares the predictions of these two methods for evaluating the reduced density coefficient of thermal conductivity with the experimental results for nitrogen and carbon monoxide. It is evident that the method proposed by Bich and Vogel provides a significantly better representation of the data than does that of the MET-II model. At this stage it is not possible to assert whether this improvement results from the use of a better model for the translational part of the coefficient or from the additional terms accounted for in the internal contribution. Further investigations making use of both the first density coefficient of viscosity and of thermal conductivity will be necessary to resolve this.

## CONCLUSIONS

The new experimental data for the first density coefficient of thermal conductivity over a wide range of temperature are able to provide a discriminant between two means of prediction for polyatomic gases. The method proposed by Bich and Vogel [6] that uses the theoretical results of Rainwater and Friend for the translational contribution and a terms in the internal contribution that accounts for the heat transport of dimers in addition to a simple diffusive contribution proves superior to earlier proposals.

## REFERENCES

1. Maitland, G.C., Smith, E.B., Rigby, M.R. and Wakeham W.A., 1981. *Intermolecular Forces: Their Origin and Determination*, Oxford, Clarendon Press.

2. McCourt, F.R.W., Beenakker, J.J.M., Kohler, W.E. and Kuscer, I. ,1990. *Nonequilibrium Phenomena in Polyatomic Gases*, Vol I, Oxford, Clarendon Press.

3. Dorfman J.R. and Cohen E.G.D, 1967. " Difficulties in the kinetic theory of dense gases." *Journal of Mathematical Physics*, 8:282-295.

4. Groot J.J, Kestin J., Sookiazian H., 1978. "The thermal conductivity of four monatomic gases as a function of density near room temperature." *Physica*, 92a:177-244.

5. Rainwater, J.C.and Friend, D.G., 1987. "Second viscosity and thermal conductivity coefficients of gases: Extension to low reduced temperature." *Physical Review A*, 36:4062-4066.

6. Bich, E. and Vogel, E.,1996. *Chapter 5, Transport Properties of Fluids, Theory, Prediction and Estimation*, Millat, J., Dymond, J.H. and Nieto de Castro, C.A., eds. Cambridge: Cambridge University Press.

7. Maitland, G.C., Mustafa, M, Ross, M., Trengove, R.D., Wakeham, W.A. and Zalaf, M., 1986. " Transient hot-wire measurements of the thermal conductivity of gases at elevated temperatures." *International Journal of Thermophysics*, 7:245-258.

8. Li, S.F.Y., Papadaki, M. and Wakeham ,W.A., 1993. "The measurement of the thermal conductivity of gases at low density by the transient hot-wire technique." *High Temperatures-High Pressures*, 25: 451-459.

9. Heck, L and Dickinson, A.S.,1994. " Transport and relaxation properties of $N_2$." *Molecular Physics*, 81:1325-1352.

10. Van der Avoird, A., Wormer, P.E.S. and Jansen, A.P.J., 1986. "An improved intermolecular potential for nitrogen." *Journal of Chemical Physics*, 84:1629-1638.

11. Heck, L., Dickinson, A.S. and Vesovic, V., 1995, " Testing the Mason - Monchick approximation for transport proeprties of Carbon Monoxide." *Chemical Physics Letters,* 240:151-156.

12. Van der Pol, A., Van der Avoird, A. and Wormer, P.E.S., 1990. "An ab initio intermolecular potential for the carbon monoxide dimer $(CO)_2$." *Journal of Chemical Physics,* 92:7498-7503.

13. Ross, M.R., Szczepanski, R., Trengove, R.D. and Wakeham, W.A., 1986. "The initial density dependence of the transport properties of gases." *AIChE Annual Winter Meeting, Paper WA86/23.*

# Measurement of Thermal Diffusivity and Conductivity of a Multi-Layer Specimen with Temperature Oscillation Techniques

W. CZARNETZKI and W. ROETZEL

## ABSTRACT

Temperature oscillation techniques are proposed for the simultaneous measurement of thermal diffusivity and conductivity of liquids. Temperature oscillations are generated at the surface of a liquid specimen which is separated into several layers by reference layers. A temperature wave propagates through the layers and gets attenuated. Measuring and evaluating the amplitude attenuation and/or the phase shift between the temperature oscillations at several appropriate locations make it possible to simultaneously determine the thermal diffusivity and thermal conductivity.

The direct heat conduction problem is solved to specify the optimum design of the measurement apparatus by means of parameter studies. Measurement cells are designed and experiments are carried out with water, ethanol and heptane. The measured values of thermal diffusivity and thermal conductivity agree well with data obtained from the literature.

## INTRODUCTION

Thermal diffusivity is the important thermophysical property to describe transient heat conduction in a solid or a stationary liquid. Therefore, a non-steady state measurement technique is applicable. The presented temperature oscillation techniques, based on a previously proposed method, combines the advantages of a quasi-steady state measurement with the possibility to measure a property describing a non-steady state [1-5]. Earlier applications were made only for solid materials [6]. To measure thermal diffusivity and conductivity of fluids, convection must be avoided by using thin specimens. If the specimen becomes too thin, no amplitude attenuation will be measured. To achieve a reasonable amplitude attenuation, specimen layers are placed behind each other separated by reference layers.

Walter Czarnetzki and Wilfried Roetzel
Universität der Bundeswehr Hamburg, D-22039 Hamburg, Germany

To confirm the practical applicability, measurement cells are designed and experiments are carried out with different liquids. By computerised operation the measurement can be performed without attendance.

## MEASUREMENT PRINCIPLE

The energy equation

$$\frac{\partial T_j}{\partial t} = a_j \nabla^2 T_j \quad j = 1...n \tag{1}$$

describes heat conduction in an isotropic solid layer or stationary liquid layer with constant thermal conductivity. At the non-adiabatic surfaces of the multi-layer specimen periodic temperature oscillations are generated with the period $t_p$ and the constant angular frequency

$$\omega = \frac{2\pi}{t_p}. \tag{2}$$

## TEMPERATURE OSCILLATIONS IN A MULTI-LAYER-SLAB

In this case, the initial and boundary conditions are independent of the coordinates y and z. Thus, the temperature will be a function of x and t only. Assuming no thermal contact resistance at each contacting surface the temperature and the heat flux must be equal

$$T_j(\delta_1) = T_{j+1}(\delta_j), \quad j = 1...n-1 \tag{3}$$

$$\lambda_j \left( \frac{\partial T_j}{\partial x} \right)_{x=\delta_j} = \lambda_{j+1} \left( \frac{\partial T_{j+1}}{\partial x} \right)_{x=\delta_j}, \quad j = 1...n-1. \tag{4}$$

In the centre of the multi-layer specimen the heat flux must be zero due to the symmetry of the problem

$$\left( \frac{\partial T_1}{\partial x} \right)_{x=0} = 0. \tag{5}$$

A temperature oscillation must be generated in the centre at steady periodic condition

$$T|_{x=0} = b_o \cos(\omega t - \varphi_o). \tag{6}$$

Figure 1 shows the geometry and the coordinates for the one-dimensional physical problem considered here. Laplace transform techniques are applied to obtain the steady periodic solution of the direct problem given by Eq. (1) through Eq. (6). For convenience the complex solution is presented:

$$T_j = b_o e^{i\cdot(\omega t - \varphi_o)}\left\{C_{j-1}\cosh\left(\sqrt{\frac{i\omega}{a_j}}\left(x-\delta_{j-1}\right)\right)+D_{j-1}\sinh\left(\sqrt{\frac{i\omega}{a_j}}\left(x-\delta_{j-1}\right)\right)\right\} \quad j=2,\dots n \quad (7)$$

with

$$C_j = C_{j-1}\cosh\left(\sqrt{\frac{i\omega}{a_j}}\underbrace{\left(\delta_j-\delta_{j-1}\right)}_{d_j}\right)+D_{j-1}\sinh\left(\sqrt{\frac{i\omega}{a_j}}\left(\delta_j-\delta_{j-1}\right)\right), j=2,\dots n \quad (8)$$

$$D_j = \frac{\lambda_2}{\lambda_1}\sqrt{\frac{a_1}{a_2}}\left\{C_{j-1}\sinh\left(\sqrt{\frac{i\omega}{a_j}}d_j\right)+D_{j-1}\cosh\left(\sqrt{\frac{i\omega}{a_j}}d_j\right)\right\}, j=2,\dots n \quad (9)$$

$$C_1 = \cosh\left(\sqrt{\frac{i\omega}{a_1}}d_1\right) \quad (10)$$

$$D_1 = \frac{\lambda_1}{\lambda_2}\sqrt{\frac{a_2}{a_1}}\sinh\left(\sqrt{\frac{i\omega}{a_1}}d_1\right) \quad (11)$$

The complex temperature distribution in the first layer becomes

$$T_1 = b_0 e^{i\cdot(\omega t - \varphi_0)}\cosh\left(\sqrt{\frac{i\omega}{a_1}}x\right). \quad (12)$$

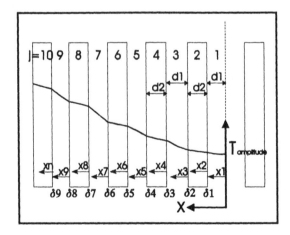

Figure 1. Physical problem

## TEMPERATURE OSCILLATIONS IN A MULTI-LAYER-CIRCULAR-CYLINDER

Describing Eq. (1) with cylindrical coordinates, assuming that the flow of heat takes place in planes perpendicular to the axis, and that the lines of flow are radial, one obtains

$$\frac{\partial T_j}{\partial t} = a_j \left( \frac{\partial^2 T_j}{\partial r^2} + \frac{1}{r} \frac{\partial T_j}{\partial r} \right), j = 1...n. \tag{13}$$

The boundary conditions are given by Eq. (3) through Eq. (6) replacing the coordinate x by r. The steady periodic solution of the cylindrical problem becomes

$$T_j(t) = b_0 \, e^{i(\omega t - \varphi_o)} \left[ A_j \cdot J_0 \left( i \sqrt{\frac{i\omega}{a_j}} r \right) + B_j \cdot N_0 \left( i \sqrt{\frac{i\omega}{a_j}} r \right) \right] \tag{14}$$

with

$$A_j = -\frac{A_{j-1}((\lambda_1 \sqrt{a_2} / \lambda_2 / \sqrt{a_1})^{(-1)^j} J_1(iq_{j-1}R_{j-1}) N_0(iq_j R_{j-1}) - J_0(iq_{j-1}R_{j-1}) N_1(iq_j R_{j-1}))}{J_0(iq_j R_{j-1}) N_1(iq_j R_{j-1}) - J_1(iq_j R_{j-1}) N_0(iq_{j-1}R_{j-1})}$$

$$-\frac{B_{j-1}((\lambda_1 \sqrt{a_2} / \lambda_2 / \sqrt{a_1})^{(-1)^j} N_1(iq_{j-1}R_{j-1}) N_0(iq_j R_{j-1}) N_0(iq_{j-1}R_{j-1}) N_1(iq_j R_j))}{J_0(iq_j R_{j-1}) N_1(iq_j R_{j-1}) - J_1(iq_j R_{j-1}) N_0(iq_{j-1}R_{j-1})}$$

j=3,...n  $\tag{15}$

$$B_j = \frac{A_{j-1}((\lambda_1 \sqrt{a_2} / \lambda_2 / \sqrt{a_1})^{(-1)^j} J_1(iq_{j-1}R_{j-1}) J_0(iq_j R_{j-1}) - J_0(iq_{j-1}R_{j-1}) J_1(iq_j R_{j-1}))}{J_0(iq_j R_{j-1}) N_1(iq_j R_{j-1}) - J_1(iq_j R_{j-1}) N_0(iq_{j-1}R_{j-1})}$$

$$+\frac{B_{j-1}((\lambda_1 \sqrt{a_2} / \lambda_2 / \sqrt{a_1})^{(-1)^j} N_1(iq_{j-1}R_{j-1}) N_0(iq_j R_{j-1}) J_0(iq_{j-1}R_{j-1}) J_1(iq_j R_j))}{J_0(iq_j R_{j-1}) N_1(iq_j R_{j-1}) - J_1(iq_j R_{j-1}) N_0(iq_{j-1}R_{j-1})}$$

j=3,...n  $\tag{16}$

$$A_2 = \frac{(J_0(iq_1 R_1) N_1(iq_2 R_1) - (\lambda_1 \sqrt{a_2} / \lambda_2 / \sqrt{a_1}) J_1(iq_1 R_1) N_0(iq_2 R_1))}{J_0(iq_2 R_1) N_1(iq_2 R_1) - J_1(iq_2 R_1) \cdot N_0(iq_2 R_1)} \tag{17}$$

$$B_2 = \frac{(J_0(iq_1 R_1) J_1(iq_2 R_1) - (\lambda_1 \sqrt{a_2} / \lambda_2 / \sqrt{a_1}) J_1(iq_1 R_1) J_0(iq_2 R_1))}{J_0(iq_2 R_1) N_1(iq_2 R_1) - J_1(iq_2 R_1) \cdot N_0(iq_2 R_1)} \tag{18}$$

$$q_j = \sqrt{\frac{i\omega}{a_j}} \quad j = 1, ...n. \tag{19}$$

The complex temperature distribution in the central layer becomes

$$T_1(t) = b_0 \, e^{i(\omega t - \varphi_o)} J_0\!\left( i \sqrt{\frac{i\omega}{a_1}}\, r \right) \tag{20}$$

The complex amplitude ratio $B^*$ between two points i.e.: $x = \delta_3$ and $x = \delta_5$ becomes

$$B^*_{35} = \frac{T_5(t, x = \delta_5)}{T_3(t, x = \delta_3)} \tag{21}$$

The real phase difference $\Delta G_{35}$ and the real amplitude ratio $B_{35}$ is expressed in

$$\Delta G_{35} = \arctan\!\left( \frac{\mathrm{Im}\!\left[ B^*_{35} \right]}{\mathrm{Re}\!\left[ B^*_{35} \right]} \right) \tag{22}$$

and

$$B_{35} = \sqrt{\left( \mathrm{Re}\!\left[ B^*_{35} \right] \right)^2 + \left( \mathrm{Im}\!\left[ B^*_{35} \right] \right)^2} \, . \tag{23}$$

## WORKING ALGORITHM

The measured temperature oscillations are regarded as a superposition of several sinusoids of different frequencies, amplitudes and phases. Each one represents a solution according to Eqs. (7, 13, 15, 20). In the experiments the fundamental oscillation (k=1) is considered. Amplitude and phase can be evaluated by on-line Fourier analysis or by analog multiplier method [9]. Applied to the measured temperature at three well defined positions (i.e. $x = \delta_1$, $\delta_3$, $\delta_5$) of the multi-layer specimen yields the "measured" values of two phase differences and two amplitude ratios, i.e. $\Delta G_{35}$, $\Delta G_{15}$, $B_{15}$ and $B_{35}$.

Introducing the dimensionless thickness of the reference layer with known thermal properties

$$\zeta = r_R \sqrt{\frac{\omega}{\alpha_R}}, \tag{24}$$

the dimensionless thickness of the layer with unknown thermal properties

$$\xi = r \sqrt{\frac{\omega}{\alpha}}. \tag{25}$$

and the dimensionless property ratio

$$C = \frac{\lambda \sqrt{a_R}}{\lambda_R \sqrt{a}} \tag{26}$$

leaves the problem to determine $\xi$ and C. A start-approximation C is chosen. While two amplitude ratios or phase differences at appropriate locations are known, two dimensionless thicknesses $\xi$ are calculated from Eqs. (21) using Newton's method. An iterative varying step method is used to determine C, which gives the same dimensionless thickness $\xi$ for both amplitude ratios or both phase differences, respectively. The thermal diffusivity and conductivity can be determined from Eq. (25) and Eq. (26).

To reduce the number of parameters and for a better visual presentation the dimensionless thickness of the layers and the dimensionless property ratio according to Eqs.(24, 25, 26) is introduced. Plotting the amplitude ratio of the temperature oscillations at the position $r = \delta_4$ and $r = \delta_1$ as a function of the dimensionless property ratio C and the dimensionless thickness of the specimen layer gives Fig. 2. To show the amplitude ratio of the temperature oscillations at the position $r = \delta_4$ and $r = \delta_3$ in one diagram the dimensionless thickness is divided by the respective radius. Illustrating the working algorithm one has to find the intersection point of lines representing the measured amplitude ratio. From the coordinates of the intersection point the thermal diffusivity and the thermal conductivity can be obtained using Eq. (25) and Eq. (26), respectively. If the diagram (Fig. 2) is discretely stored in a table and resorted, the working algorithm can also be implemented by a two-dimensional interpolation.

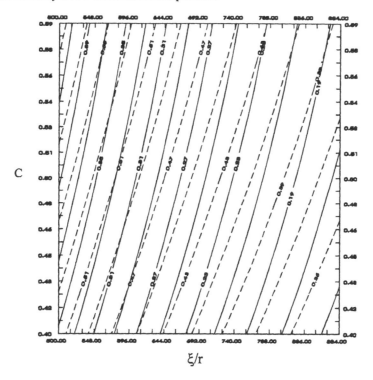

Figure 2. Amplitude ratio at two different positions ($r_1$ solid lines, $r_3$ dashed lines)
as a function of C and $\xi/r$

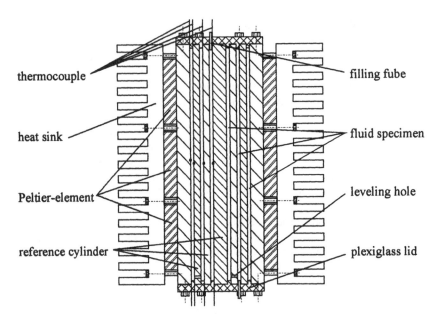

thermocouple

filling fube

heat sink

fluid specimen

Peltier-element

leveling hole

reference cylinder

plexiglass lid

Figure 3. Schematic of the measurement cell

## EXPERIMENTS

### MEASUREMENT CELL

Figure 3 shows a simplified scheme of the cylindrical measurement cell with three specimen layers. The liquid is filled into the annular hollow spaces (gap width 1.5 mm), formed by one reference cylinder (diameter 12 mm) and two hollow reference cylinders (inner diameter 7.5 mm and 15 mm, substance 1.5) made of stainless steel ( X 10 CrNiTi 18 9, No.:1.4541 , $\alpha$=4.43 $\cdot 10^{-6}$ m²/s, $\lambda$=15.63 W/m/K). The reference cylinders are centred by two lids made of Plexiglas in which annular separators are turned. Filling and deaireating is done via the small tubes in the lids. The surface temperatures of the reference cylinders are measured by means of Ni-CrNi thermocouples (diameter 0.1 mm), which are butt welded on the outer surface. Only for the outermost hollow cylinder it is possible to weld the thermocouple junctions on the inside surface. The temperatures of the outermost cylinder is changed periodically by means of Peltier-elements (MELCOR, Trenton, USA, CP 1.4-35-045L, 30 mm x 15 mm) ). The Peltier-element uses the inverse thermocouple effect to generate a temperature difference between its two sides. One side is kept at a constant temperature using a heat sink. The other side periodically changes its temperature, while the Peltier-element is fed with a periodic voltage. The thermal wave propagates through the layers into the centre. To improve heat transfer finned plates or water cooled heat sinks are mounted on top of the Peltier-elements. The Peltier-elements can generate temperature oscillation with a maximum frequency of 0.2 Hz and an corresponding amplitude of 0.5 K. At lower frequencies the Peltier-elements can provide amplitudes up to 30 K. Although the Peltier-elements are an adequate device to generate temperature oscillations in a temperature range from -30°C up to 100°C, this described technique does not require on the use of Peltier-elements. Any periodically-modulated heat source connected with a constant heat sink can be used to drive thermal waves into a test specimen.

## MEASUREMENTS

To avoid convection, the measurement cell is in a vertical position in which no unstable thermal stratification can occur. To agree with the assumption of constant thermal properties, the generated amplitude of the temperature oscillation inside the specimen is less than 1 K. The frequency is chosen such that the greatest amplitude attenuation is about 0.3. The thermocouple signals are amplified and then digitised by an A/D-converter on a PC-board. Relative temperature changes of 0.005 K can be detected by this device. The accuracy of the absolute temperature measurement depends on the accuracy of the calibration. In this case the absolute accuracy is specified to within 0.1 K. To evaluate the thermal diffusivity and conductivity relative temperature changes are necessary. To remove the noise of the temperature signal a capacitor and a resistance are used as a low-pass filter (edge frequency 40 Hz, see Figure 4.). The evaluation is done by a program developed by the authors and the results are visualised on the PC-screen.

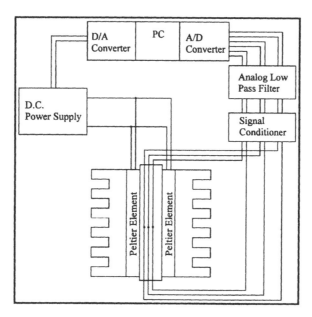

Figure 4. Electrical circuit diagram

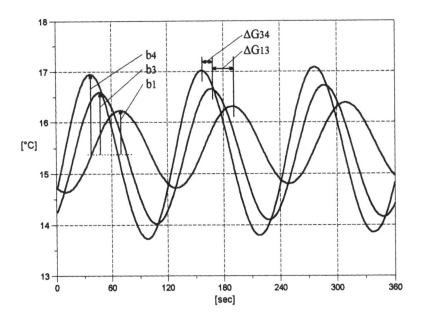

Figure 5. Temperature plot, Water, $t_p$=120 sec.

The mathematical model implies that all initial conditions have faded away. The on-line measurements make it possible to properly detect the steady periodic state. As long as there is a non-steady state, the measurement data will vary with time. After several time periods the 'measured' thermal diffusivity and conductivity do not change. A typical plot of the three measured temperatures is shown in Fig. 5, in which water was measured with a period of 120 seconds. The amplitude attenuation and the phase shift between the oscillations at three different positions in the cell can be clearly seen.

## RESULTS

To show the practical applicability experiments were carried out with three liquids and compared with values from the literature. To reduce the random error, the temperature was measured over two periods and the integration range of the Fourier analysis was shifted. Table 1 shows the measured thermal diffusivity and conductivity and the values available from the literature. It has to be noticed, that the data for thermal diffusivity in the literature were calculated from thermal conductivity, specific heat and density [11]. It has to be emphazised, that no calibration of any kind of the measurement apparatus so far has been done. Even the thermocouples have not been calibrated, only their voltage signals were used in the evaluation to get the results. As shown in Eqs. (22, 23), it is possible to determine the thermal diffusivity and conductivity from the amplitude or phase, respectively. In principle, both evaluations must give the same result. In the real experiment the measurement errors have to be taken into account and the corresponding sensitivity coefficients [9]. The determination of these coefficients shows, that both the thermal diffusivity and conductivity are more sensitive to the amplitude attenuation. Thus, the amplitude information was used to get the experimental results shown in Table 1. The measurement errors depend on the various experimental parameters and on the measured medium itself. The most significant effects on the measurement errors is the uncertainty of the thermophysical properties of the reference layer. Compared to the previous proposed method using temperature oscillations the effects on thermocouple dislocation, onset of free convection and contact resistance of the thermocouples are negligible, the magnitude of measurement errors with cylindrical geometry of thermal diffusivity are less than 6% and of thermal conductivity less than 8% [7,8].

TABLE I. EXPERIMENTAL RESULTS OF DIFFERENT SPECIMENS AT 20°C

| | Diffusivity [$10^{-8}$m²/s] | | Conductivity [W/(mK)] | | Period [s] |
| | this Work | Literature | this Work | Literature | |
|---|---|---|---|---|---|
| Water | 13.89 | 14.34 [11] | 0.629 | 0.601 [10] | 150 |
| Ethanol | 9.43 | 8.97 [11] | 0.178 | 0.173 [11] | 150 |
| n-Heptane | 8.54 | 8.06 [12] | 0.133 | 0.124 [11] | 180 |

## CONCLUSIONS

The described apparatuses perform well during these measurements. They can be used for absolute measurements slightly above or below ambient temperatures. The experiments have demonstrated the applicability of the proposed method for simultaneous determination of thermal diffusivity and conductivity. The apparatus can easily be adapted to measurements at pressures different from ambient pressure. If the finned plates are thermostated, greater temperature differences from ambient temperature are possible. Compared to the established temperature oscillation method the calibration procedure can easily be carried out.

## REFERENCES

1. Ångström, A. J, 1861. "Neue Methode, das Wärmeleitungsvermögen der Körper zu bestimmen." *Annalen der Physik und Chemie*, Band CXIV:513-530.
2. Maglic, K.D., A. Cezairliyan, and V.E. Peletsky 1982. *Compendium of Thermophysical Property Measurement Methods* Plenum Press, New York, pp. 337-365.
3. El Osery, I.A. and M.A. El Osery, 1987."A theoretical aspect of proposed experiment for measurement of material thermophysical properties using periodic heating techniques." *Modelling, Simulation & Control, B*, 12(1):1-6.

4.  Rouault, H., J. Khedari, C. Arzoumanian and J. Rogez, 1987. "High-temperature diffusivity measurements by periodic stationary method." *High Temperatures - High Pressures*, 19:357.

5.  Lopez-Baeza, E.L., J. de la Rubia and H.J. Goldsmid, 1987." Ångström's thermal diffusivity method for short samples." *J. Phys. D: Appl. Phys.* 20:1156-1158.

6.  Bierer, M., A. Erhard and E. Hahne, 1990. *Proceedings of the Ninth International Heat Transfer Conference* Hemishere Publishing Corporation, 3:175-180.

7.  Czarnetzki, W. and W. Roetzel 1994. *Thermal Conductivity 22*, Technomic Publishing, Lancaster, pp 275-286.

8.  Czarnetzki, W. and W. Roetzel, 1995. "Temperature Oscillation Techniques for Simultaneous Measurement of Thermal Diffusivity and Conductivity." *Int. J. Thermophysics*, 16(2):413-422.

9.  Czarnetzki, W. and W. Roetzel, 1995. HTD-Vol. 312, 1995 National Heat Transfer Conference- Vol. 10 ASME.

10. Ramires, M.L.V., C. A. Nieto de Castro, R. A. Perkins, 1993. "New improved recommendations for the thermal conductivity of toluene and water.", *High Temperatures- High Pressures*, 25:269-277.

11. VDI-Wärmeatlas, 1991. 6. Auflage, VDI Verlag GmbH, Düsseldorf.

12. Knibbe P.G. and J. D. Raal, 1987. "Simultaneous measurement of the thermal conductivity and thermal diffusivity of liquids." *Int. J. Thermophysics*, 8(2):181-191.

## NOMENCLATURE

a   thermal diffusivity
b   amplitude
B   amplitude ratio
C   dimensionless property ratio, constant
D   constant
d   thickness of layer/specimen
G   phase difference
J   Besselfunction of the first kind
N   Besselfunction of the second kind
r   cylindrical coordinate
T   temperature
t   time
x   space coordinate
$\varphi$   phase
$\lambda$   thermal conductivity
$\omega$   angular frequency
$\xi$   dimensionless space coordinate
$\zeta$   dimensionless space coordinate
$\nabla$   Nabla operator

Indices

| | |
|---|---|
| * | complex |
| 0 | x=0, |
| p | period |
| R | reference |

# Photoacoustic Measurements to Determine
# Thermal Diffusivities of Gases
# at Moderate Pressures

J. SOLDNER and K. STEPHAN

## ABSTRACT

The photoacoustic effect is presented as a suitable experimental technique to determine thermal diffusivities and hence thermal conductivities of gases at moderate pressure. The theory is developed and an experimental setup is described. Measurements were done in the temperature range from $253K$ to $373\ K$ and in the pressure range from $0.01\ MPa$ to $0.1\ MPa$. Errors caused by damped harmonic oscillations in the photoacoustic cell – so called Helmholtz oscillations – are corrected. A comparison of measurements for gaseous argon with recommended data leads to an estimated accuracy of $\pm\,0.5\,\%$.

## INTRODUCTION

Thermal conductivities are usually determined by means of the transient hot wire method [9] or by means of steady state measurements with a concentric-cylinder apparatus [8], whereas thermal diffusivities can be measured by the dynamic light scattering method [4] or by the transient hot wire technique too. The accuracy of the thermal conductivity data of transient hot wire measurements are in the order of $\pm$ 1% and that of thermal diffusivity data in the order of $\pm$ 5%. The overall error of thermal diffusivities determined by means of the dynamic light scattering method is about $\pm$ 2%. The same accuracy can be attained for thermal conductivities by steady state measurements with a concentric-cylinder apparatus.

The transient hot wire method and the concentric-cylinder apparatus are, however, not applicable in the range of pressures below $0.2\ MPa$, whereas the dynamic light scattering method is appropriate for measurements at high pressures where densities are above $100\ kg\ m^{-3}$. Thermal conductivities at moderate pressures cannot be determined by these methods. They are usually calculated by extrapolation from measurements at higher pressures. An efficient method to determine thermal diffusivities at moderate pressures is the the photoacoustic

Institut für Technische Thermodynamik und Thermische Verfahrenstechnik
University of Stuttgart, 70550 Stuttgart, Germany

technique presented in this paper. The measured thermal diffusivities can be converted subsequently into thermal conductivities with the aid of an equation of state for the density $\rho$ and the isochoric heat capacity $c_v$.

Thermal conductivity data of gases at moderate pressures are interesting e.g. for refrigerants, because below ambient temperature their saturation pressure is low. Furthermore, the temperature dependent thermal conductivity $\lambda_0$ of the dilute gas is of prime importance for thermal conductivity surface fits based upon the residual concept [3, 5].

The photoacoustic effect is based on the conversion of radiation energy into an acoustic signal. In these experiments a periodically modulated laser penetrates the sample chamber, a cylindrical cell charged with the fluid to be tested. Small amounts of the radiation energy are absorbed by ethane added to the test gas. The amount of ethane added is of the order of ppm, so that the thermophysical properties of the test gas are not affected. On the other hand ethane is a strong absorbent at 3.4 $\mu m$, the emission wavelength of the laser. This energy transfer between ethane and the test gas inside the cell provokes a small rise in temperature and pressure in the gas. During the following dark phase the temperature and the pressure decrease to their previous level. The pressure fluctuations thus produced in the gas can be detected by a sensitive microphone. The thermal diffusivity of the sample gas can be determined from the time dependent pressure rise.

## EXPERIMENTAL SETUP

The experimental setup is shown schematically in Figs. 1 and 2. The sample chamber is a stainless steel tube with an inner diameter of 1.5 $mm$ and a length of 100 $mm$ shrunk into a copper cylinder so that a good thermal contact is attained.

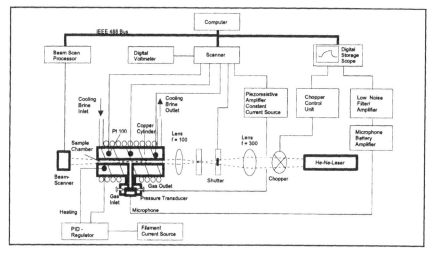

Fig. 1: The experimental setup

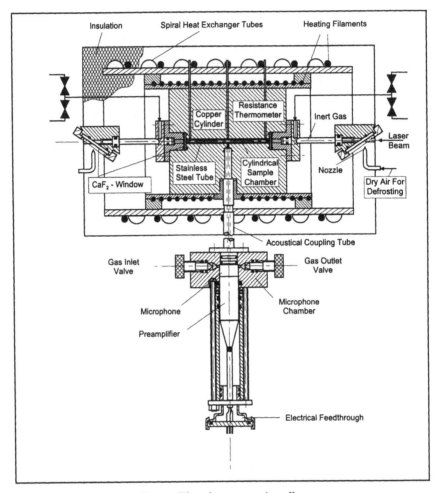

Fig. 2: The photoacoustic cell

The ends are sealed with $CaF_2$- windows which allow nearly 100 % transmission of the infrared laser light. The copper cylinder is thermostated by two controlled filament resistance heaters mounted on concentric cylinders. The outer heater balances the influence of the ambient temperature and the inner control acts as a second thermostat. The whole apparatus is insulated. Thus, the copper cylinder is thermostated with a constancy of $\pm 1\ mK$ and the temperature gradients in the copper cylinder are minimized.

For experiments below ambient temperature the outer heater is replaced by a spiral tube heat exchanger cooled with brine from a cryostat. Then a second pair of $CaF_2$-windows is needed to avoid fogging of the windows. The space between inner and outer windows is filled with inert gas. The outer windows can be defrosted by dry air.

The microphone cannot withstand high temperatures and is therefore located outside the insulation. Microphone and sample chamber are connected by a coupling tube.

A helium-neon laser with an emission wavelength of 3.4 $\mu m$ is used as a radiation source. The laser beam is modulated periodically with a mechanical chopper. The beam is focused by means of two lenses on the center of the sample chamber. The sample chamber had to be adjusted so that its cylindrical axis and the beam axis coincide. The adjustment is accomplished by means of a beam profiler.

The pressure rise detected by a condenser microphone is amplified, digitized and stored in the oscilloscope. Subsequently the data from the photoacoustic signal are transfered to the computer via an IEEE 488 bus.

The system temperature is determined as an average of three temperatures at different locations along the sample chamber. The system pressures are measured with a piezoresistive pressure transducer. All these devices are computer-controlled.

## ANALYSIS

The basic equations for the evaluation of thermal diffusivities from the photoacoustic signal were given in previous papers [6, 7]. Nevertheless, the basic steps are summarized here. The working equation is the equation for heat conduction in an infinite cylinder heated along its axis

$$\rho c_v \frac{\partial T}{\partial t} = \lambda \frac{\partial^2 T}{\partial r^2} + \frac{\lambda}{r} \frac{\partial T}{\partial r} + A(r, z). \tag{1}$$

The assumption of an infinite cylinder is justified because of a length to diameter ratio of 64. Axial temperature gradients are negligible because only about 3% of the radiation energy is absorbed in the fluid.

The heat generation term $A(r, z)$ is given by the radiation energy absorbed by the fluid

$$A(r, z) = \alpha I(r, z) \tag{2}$$

where $\alpha$ denotes the absorptivity coefficient and $I(r, z)$ the intensity.

The ideal laser beam, (Fig. 3), obeys a Gaussian intensity distribution

$$I(r, z) = \frac{W}{\pi w(z)^2} exp \left( -\frac{r^2}{w(z)^2} \right), \tag{3}$$

wherein $W$ is the radiation power.

The beam radius $w(z)$ is defined as radius where the beam intensity reaches $1/e$ of its peak value. The radius $w(z)$ of the beam follows a hyperbolic relationship along its propagation axis

$$w(z) = w_0 (1 + C z^2)^{\frac{1}{2}}, \tag{4}$$

where $w_0$ is the Gaussian radius of the focus of the beam located in the center of the sample chamber. The center of the sample chamber is the zero point of z-axis. Fig. 4 shows the contour of an ideal Gaussian beam.

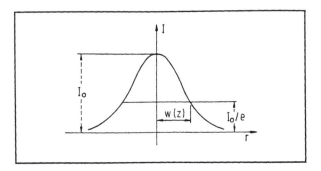

Fig. 3: Intensity distribution of a Gaussian laser beam.

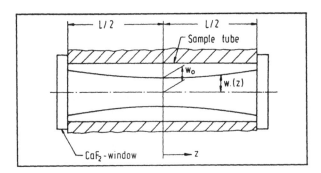

Fig. 4: Contour of a Gaussian laser beam.

The initial condition for the solution of Eqn. (1) is

$$T(t = 0, r, z) = T_0. \tag{5}$$

Because of the thermostated copper cylinder the boundary condition at the wall of the sample chamber is

$$T(t, R, z) = T_0. \tag{6}$$

The second boundary condition results from rotation symmetry of the coordinate system

$$\left(\frac{\partial T}{\partial r}\right)_{t, r=0, z} = 0. \tag{7}$$

An analytical solution $\Delta T(t, r, z) = T(t, r, z) - T_0$ of Eqn. (1) is given in [6]. An average temperature rise $\Delta \overline{T}(t)$ can be determined from the energy balance at the sample chamber

$$Q_W + Q_A = \rho \int_0^V (u(t) - u_0) dV = \rho \int_0^V c_v \Delta T(t, r, z) dV = \rho \bar{c}_v \Delta \overline{T}(t) V, \tag{8}$$

wherein $Q_A$ is the radiation energy absorbed by ethane and $Q_W$ the energy transfered from the gas to the wall of the sample chamber. $u(t)$ is the specific internal energy and $u_0$ denotes the specific internal energy at time $t = 0$ when the pressure rise starts. Assuming constant specific heat capacity $c_v$, which is allowed because of the small temperature rise, leads to

$$\Delta \overline{T}(t) = \frac{1}{V} \int_0^V \Delta T(t, r, z) dV = \Delta \overline{T}_\infty \left[ 1 + \sum_{m=1}^\infty K_m \cdot exp \left( -\frac{\zeta_m^2 a t}{R^2} \right) \right]. \quad (9)$$

$\zeta_m$ are the zeros of the Bessel function of zero order. The coefficients $K_m$ are related to the geometry of the laser beam as expressed in Eqns. (3) and (4) and can be calculated as follows

$$K_m = \frac{D_m}{B}, \quad (10)$$

$$D_m = \frac{8}{\zeta_m^2 L} \sum_{n=1}^\infty \left\{ \left[ 1 - \frac{\zeta_m}{J_1(\zeta_m)} \int_0^1 \left( \frac{r}{R} \right)^{2n+1} \frac{J_0(\zeta_m \frac{r}{R})}{R} dr \right] \int_0^{L/2} \frac{(-1)^n}{nn! \left( \frac{w}{R} \right)^{2n}} dx \right\} \quad (11)$$

$$B = \frac{4}{R^2 L} \int_0^{L/2} \int_0^R \sum_{n=1}^\infty \frac{(-1)^n}{nn! \left( \frac{w}{R} \right)^{2n}} \left[ \left( \frac{r}{R} \right)^{2n} - 1 \right] r dr dx. \quad (12)$$

$J_0$, $J_1$ are the Bessel functions of zero and first order, respectively. The coefficients $\omega_0$, $C$, $B$ and $K_m$ are listed in Table 1.

| coefficient | value |
|---|---|
| $\omega_0$ | 0.2356 |
| $C$ | $0.157094 \times 10^{-3}$ |
| $B$ | 0.8934250006 |
| $K_1$ | −1.0632171051 |
| $K_2$ | 0.0697009353 |
| $K_3$ | −0.0070622140 |
| $K_4$ | 0.0006177456 |
| $K_5$ | −0.0000403785 |
| $K_6$ | 0.0000017035 |

Table 1: Coefficients that depend on the geometry of the laser beam.

Temperature and pressure are related by the equation of state $T(p, v)$, with $v = const$ in this case. A Taylor series yields

$$\Delta T(t) = T - T_0 = \left(\frac{\partial T}{\partial p}\right)_{p_0, T_0} \cdot \Delta p(t) + \dots \tag{13}$$

$$\Delta T(t \to \infty) = \Delta T_\infty = \left(\frac{\partial T}{\partial p}\right)_{p_0, T_0} \cdot \Delta p_\infty + \dots \quad . \tag{14}$$

Neglecting terms of higher than first order due to the small pressure and temperature rise and using Eqn. (8) with $c_v = const$ it follows

$$\frac{\Delta \overline{T}(t)}{\Delta \overline{T}_\infty} = \frac{\Delta p(t)}{\Delta p_\infty}. \tag{15}$$

Finally Eqn. (9) and (15) lead to

$$\Delta p(t) = \Delta p_\infty \left[1 + \sum_{m=1}^{\infty} K_m \cdot exp\left(-\frac{\zeta_m^2 a t}{R^2}\right)\right]. \tag{16}$$

## WORKING EQUATION

Eqn. (16) describes the unit-step response in the cylindrical sample tube for a heat source, switched on at $t = 0$. For $t \to \infty$ the pressure rise $\Delta p_\infty$ reaches its steady state. At steady state all energy absorbed by the fluid is transfered to the wall of the sample tube, the surface temperature of which is held constant.

In the experiment, however, the laser beam is modulated periodically with a finite period. The measurements were carried out when a cyclic steady state was reached. The amplification factor of the microphone is only constant for frequencies above its lower cutoff frequency. The chopper frequency therefore has to be chosen above that value.

As a consequence the pressure fluctuations, dashed dotted curve in Fig. 5, do neither reach the value $p'_\infty$ at the end of a dark phase nor the value $p'_\infty + \Delta p_\infty$ at the end of the light phase. The temperature field at the end of one phase therefore had to be taken as an initial condition for calculating the temperature distribution in the following phase. This leads to

$$
\begin{aligned}
p(t) &= \Delta p_\infty \left[1 + \sum_{m=1}^{\infty} K_m exp\left(-\frac{\zeta_m^2 a\, (t - t_0)}{R^2}\right)\right. \\
&\times \left.\left[1 - \sum_{n=1}^{\infty}\left\{exp\left(-\frac{\zeta_m^2 a\, (2n-1)}{2R^2 f_{ch}}\right) - exp\left(-\frac{\zeta_m^2 a\, n}{R^2 f_{ch}}\right)\right\}\right]\right] + p'_\infty.
\end{aligned}
\tag{17}
$$

$t_0$ is the time when the light phase and hence the pressure rise begins. One obtains $t_0$ from the trigger signal of the chopper.

## TREATMENT OF HARMONIC OSCILLATIONS

The photoacoustic signal is recorded as a voltage signal. The solid curve in Fig. 5 shows the pressure decrease during the dark phase and the subsequent pressure rise during the light phase as registered by the oscilloscope. The dashed curve indicates the pressure rise according to Eqn. (18). The difference between the two curves is caused by the damped harmonic oscillation of the Helmholtz resonator formed by the sample chamber, the acoustic coupling tube and the clearance volume in front of the microphone [2]. This oscillation is overlaid with the pressure rise according to Eqn. (18).

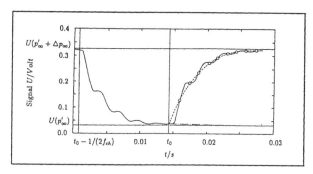

Fig. 5: The photoacoustic signal

In order to determine the thermal diffusivity $a$, the pressure rise $\Delta p_\infty$ and the lower boundary value $p'_\infty$ with the aid of Eqn. (18) from the photoacoustic signal, one has to eliminate the oscillating parts of the signal. Conventional electric low pass filters are not applicable here because the zero points in the denominator of the transfer function provoke a frequency-dependent phase shift.

The pressure rise according to Eqn. (18) intersects the photoacoustic signal at its inflection points, diamond symbols in Fig. 6. These points are first determined from the photoacoustic signal.

Then Eqn. (18) is fitted to the inflection points. Thus, one obtains a first approximation of $a$, $\Delta p_\infty$ and $p'_\infty$. The first fit is then subtracted from the original signal. The difference, dotted line in Fig. 7, contains not only the Helmholtz oscillations, but also noise and low frequency beats resulting from the inaccuracy of the first fit.

The frequency $f_0$ of the harmonic oscillations is then determined by discrete fast-Fourier analysis. Subsequently, the differential signal is treated with a low pass trend filter which eliminates all oscillations with frequencies higher than $f_0$, without any phase shift. As a result the difference between the first fit and the unknown transient pressure rise, indicated by the solid line in Fig. 7, is obtained, which is subsequently added to the first fit. As a result, one attains a second approximation for the pressure rise without any oscillations. In the next step Eqn (16) is fitted to the latter approximation of the pressure rise and one obtains new values for $a$, $\Delta p_\infty$ and $p'_\infty$. This procedure usually converges after four iterations.

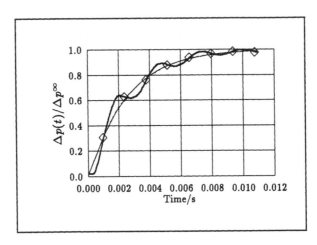

Fig. 6: Normalized pressure rise detected by measurements with argon at $T = 333.13\ K$ and $p = 0.1042\ MPa$. The symbols indicate the numerically computed points of inflections. The thin curve refers to the pressure rise according to Eqn. (17).

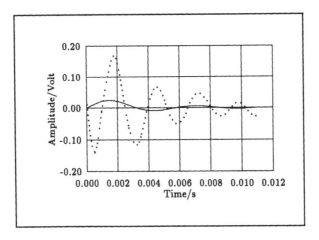

Fig. 7: Amplitudes of Helmholtz oscillations from experiments with argon at $T = 333.13\ K$ and $p = 0.1042\ MPa$. The dotted curve shows the difference between the first fit and the photoacoustic signal. The solid curve shows the dotted signal after using a low pass trend filter.

## RESULTS

For examination of the accuracy of thermal diffusivities determined with the experimental setup, measurements with argon were carried out at a pressure of $0.1\ MPa$ and in the temperature range from $303\ K$ to $373\ K$ in steps of $10\ K$. At each temperature 10 data points were taken. Thermal conductivities of argon are known with an accuracy of $\pm0.5\%$ so that it is a suitable reference fluid. The reference thermal conductivities were defined from a correlation by *Bich et al.* [1] based on the kinetic theory of moderately dense gases.

The experimental thermal diffusivities were converted into thermal conductivities

$$\lambda_{exp} = a_{exp}\rho c_v. \tag{18}$$

The density $\rho$ can be calculated from the equation of state for ideal gases. The isochoric specific heat capacity is obtained from

$$c_v = \frac{1}{\kappa - 1}\frac{\mathcal{R}}{M} \tag{19}$$

with the universal gas constant $\mathcal{R} = 8.3143\ kJ\ kmol^{-1}K^{-1}$, the molar mass of argon $M = 39.948\ kg\ kmol^{-1}$ and the isentropic exponent $\kappa = 1.6667$.

Table 2 contains the experimental results and Fig. 8 presents the experimental data compared to the reference data. The bars indicate the intervals of the 10 measurements taken at given temperature and the diamonds refer to the deviation of the mean values from the reference data. The standard deviation is below 0.8 %. The deviation of the mean values from reference data are within the accuracy of the reference data of $\pm0.5\ \%$

| $T/K$ | $p/MPa$ | $\frac{a_{exp}}{10^{-6}m^2/s}$ | $\frac{\lambda_{exp}}{10^{-3}W/(Km)}$ | $\frac{\lambda_{exp}-\lambda_{lit}}{\lambda_{lit}} \times 100$ | $\frac{\text{standard deviation}}{\%}$ |
|---|---|---|---|---|---|
| 303.13 | 1.026 | 35.31 | 17.93 | -0.06 | 0.28 |
| 313.14 | 1.029 | 37.47 | 18.48 | 0.28 | 0.36 |
| 323.00 | 1.028 | 39.64 | 18.89 | 0.01 | 0.76 |
| 333.13 | 1.036 | 41.46 | 19.34 | -0.26 | 0.32 |
| 343.13 | 1.033 | 43.92 | 19.86 | 0.00 | 0.47 |
| 353.34 | 1.040 | 45.94 | 20.28 | -0.31 | 0.56 |
| 363.11 | 1.040 | 48.55 | 20.86 | 0.35 | 0.52 |
| 373.08 | 1.042 | 50.75 | 21.25 | 0.05 | 0.56 |

Table 2: Mean values of thermal diffusivities and thermal conductivities with deviation from recommended data and standard deviation

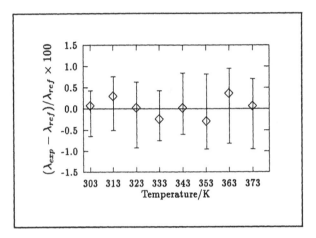

Fig. 8: Average values of thermal conductivities at atmospheric pressure compared to reference data.


## CONCLUSIONS

The photoacoustic technique has proved to be an efficient tool for accurate determination of thermal diffusivities and hence thermal conductivities of gases at moderate pressure. Measurements were carried out with argon as a reference fluid at temperatures ranging between 253 $K$ and 373 $K$. A comparison between measurements and recommended reference data show a deviation which lies within the reported accuracy of the reference data of $\pm 0.5\%$.

The photoacoustic signal is overlapped by Helmholtz oscillations which requires an iteration procedure to evaluate thermal diffusivities from the experiments.

The present results show that the photoacoustic technique may fill the gap in the experimental methods because one obtains thermal diffusivities and hence thermal conductivities in a moderate pressure range where other techniques, e.g. the transient hot wire technique, the concentric-cylinder apparatus or the dynamic light scattering, are not applicable or do not provide accurate results.


## ACKNOWLEDGEMENT

We gratefully acknowledge the financial support by the Deutsche Forschungs-gemeinschaft.

# REFERENCES

[1] E. Bich, J. Millat and E. Vogel. The viscosity and thermal conductivity of pure monoatomic gases from their normal boiling point up to 5000 K in the limit of zero density and at 0.101325 MPa. *J. Phys. Chem. Ref. Data*, 19:1289–1305, 1990.

[2] J. Blitz. *Elements of Acoustics* Butterworths, London, 1964.

[3] J. O. Hirschfelder, C. F. Curtiss and R. B. Bird. *Molecular Theory of Gases and Liquids*. Wiley, New York, 1954.

[4] B. Kruppa and J. Straub. Measurement of thermal diffusivity of the refrigerants R22 and R134 by means of dynamic light scattering. *Paper presented at the 11th Symp. Thermophys. Prop.*, Boulder, 1991.

[5] A. Laesecke, R. Krauss and K. Stephan. *J. Phys. Chem. Ref. Data*, 19:1089, 1990.

[6] K. Stephan and J. Biermann. The photoacoustic technique as a convenient instrument to determine thermal diffusivities of gases. *Int. J. Heat Mass Transfer*, 35(3):605–612, 1992.

[7] K. Stephan, V. Rothacker and W. Hurdelbrink. Thermal diffusivities determined by photoacoustic spectroscopy. *Chem. Eng. Process*, 26:257–261, 1989.

[8] Y. Tanaka, M. Nakata and T. Makita. Thermal conductivity of gaseous HFC-134a, HFC-143a, HCFC-141b, and HCFC-142b. *Int. J. Thermophys.*, 12:949, 1991.

[9] W. A. Wakeham, A. Nagashima and J. V. Sengers. Measurement of the transport properties of fluids. In *Experimental Thermodynamics, Vol. III*, chapter 7. Blackwells Scientific, Oxford, 1991.

## SESSION 7

# METALS

**SESSION CHAIRS**
J. B. Henderson
E. Ashworth

# Thermal Conductivities of Liquid Metals

K. C. MILLS, B. J. MONAGHAN and B. J. KEENE

## ABSTRACT

Data for the thermal conductivity of pure liquid metals which has been published since the review of Touloukian in 1970 has been collated and assessed and recommended values are given. It was found the Wiedemann-Franz-Lorenz (WFL) relation provides accurate values of thermal conductivities of liquids and solids at the melting point.

## 1 INTRODUCTION

Mathematical modelling has become an established tool for improving process control, decreasing energy costs and product quality. Over the last two decades mathematical models have evolved from programs encapsulating empirical knowledge to complex descriptions of the fundamental phenomena influencing the process, such as the fluid flow and heat transfer. As a consequence of this, in the on-going development of these models, one of the primary requirements at the present time is for accurate values for those physical properties of the materials which affect the heat and fluid flow in the process (eg heat capacity, enthalpy of fusion, emissivity, thermal conductivity, density, viscosity and surface tension). The absence of reliable physical property data for the commercial materials involved in these processes is a reflection of the difficulties of obtaining accurate experimental values, especially at high temperatures. This need for reliable physical property data has been recognised footby the Department of Trade and Industry who are currently funding a measurement programme at the National Physical Laboratory to obtain property values for a wide range of commercial alloys. We have adopted a 3-pronged approach viz (i) development of experimental methods to provide accurate physical property data, (ii) correlation and critical evaluations of property data in the literature and (iii) the estimation of physical properties from a knowledge of chemical composition and melting temperature.

All authors, National Physical Laboratory, Teddington, Middlesex, United Kingdom, TW11 OLW

As part of our data evaluation activities we have collated and assessed the thermal conductivity data reported for pure molten metals. However, such a review has importance to (i) the measurement programme since recommended values are needed for calibrant materials to check the reliability of the technique adopted and (ii) to the estimation programme since (a) these models use data for the pure metals and (b) checks can be made on the validity of the approach adopted which makes use of the Wiedemann-Franz-Lorenz (WFL) rule.

The WFL rule relates the thermal ($\lambda$) and electrical ($\sigma$) conductivities and is shown in Equation 1 where T is the thermodynamic temperature (K) and $L_o$ is a constant with a theoretical value of $2.445 \times 10^{-8}$ W$\Omega$ K$^{-2}$.

$$\lambda = L_o \, T \, \sigma \tag{1}$$

It is exceedingly difficult to obtain accurate values of the thermal conductivity of liquids since convection can contribute significantly to the transport of heat. Convection can lead to erroneously high values for the thermal conductivities. Convectional flows present a problem even at ambient temperature but are a much more severe problem at high temperatures because of the difficulties in eliminating thermal gradients (and hence fluid flow) in the liquid sample. In contrast, accurate electrical conductivities are simpler to obtain experimentally and, more importantly, are unaffected by fluid convection. Thus, if the heat transport mechanism in molten metals and alloys occurs predominantly through the transfer of electrons, the WFL rule should be valid. Consequently, thermal conductivity of liquid metals can be calculated from electrical conductivities. Thus the important objective of this investigation was to determine how valid the WFL Rule is for the solid and liquid states of a metal in the region of the melting point.

Mott [1] proposed that the electrical conductivities of the solid ($\sigma_s^m$) and the liquid ($\sigma_\ell^m$) metals at the melting point (denoted by the superscript m) were related to the entropy of fusion ($\Delta S^{fus}$) where C is a constant.

$$\ln \left( \frac{\sigma_s^m}{\sigma_l^m} \right) = C \, \Delta S^{fus} \tag{2}$$

If the WFL rule (Equation 1) is valid then Equation 3 can be obtained from Equation 2, where K is a constant

$$\ln \left( \frac{\lambda_s^m}{\lambda_l^m} \right) = K \, \Delta S^{fus} \tag{3}$$

If this equation is valid the constant K would have a uniform value for all metals and such a rule would be useful for estimating the thermal conductivity of a liquid alloy ($\lambda_\ell^m$) from a knowledge of $\lambda_s^m$ and $\Delta S^{fus}$ which can be measured more easily. Thus a secondary objective of this study was to check whether K had a constant value for all pure metals.

## 2    DATA SOURCES

Touloukian [2] has reported thermal conductivity data for liquid metals and proposed recommended values. Consequently, the survey of literature data [3] has been confined to those data published since 1970.

The electrical resistivity (or conductivity) data for the pure metals used in the calculations based on the WFL rule were taken from Iida and Guthrie [4] and additional data [5] was used for the refractory elements.

Entropies of fusion given in Table I were used to check the validity of Equation 3; these and the melting points were taken from the recent review of Dinsdale [6].

## 3    CRITERIA USED TO EVALUATE THERMAL CONDUCTIVITY DATA

In molten metals, the existence of fluid flows which assist in the transport of heat can result in erroneously high values for the thermal conductivity. It is exceedingly difficult to eliminate all fluid flows which arise from small temperature gradients in the melt. This is true even for temperatures close to the ambient but the gradient becomes progressively more difficult to minimise as the temperature of the melt increases. These difficulties have led to a move from *static* to *transient* methods in recent years [7] since in the latter techniques the measurements are completed in a few seconds before (hopefully) convective flows have had time to become established. Recent line source experiments suggest that convection starts to occur after about 1 second in these experiments [8,9]. In transient techniques these convective flows refer to those arising from the input of energy but do not refer to the convective flows already established in the melt as a result of temperature gradients or electromagnetic stirring. Such flows will also lead to erroneously high values of the thermal conductivity. Thus it is important that all steps to minimise convection should have been taken by

(a)    minimising temperature gradients within the melt
(b)    maintaining the free surface of the melt at a fractionally higher temperature than the bottom
(c)    avoiding electromagnetic stirring
(d)    using electromagnetic forces to suppress convectional forces.

TABLE I  Thermal conductivities of pure metals at the melting point

| | $T^m$ (K) | $\lambda_s^m$ ($W^s m^{-1} K^{-1}$) | $\lambda_l^m$ | $10^2 b$ ($Wm^{-1}K^{-2}$) | $(L^m/L_o^m)$ solid | liquid | $\Delta S^m$ $J.K^{-1} mol^{-1}$ | K |
|---|---|---|---|---|---|---|---|---|
| Ag | 1235 | 362 | 175 | 2.0 | 1.0 | 1.0 | 9.15 | .0794 |
| Al | 933 | 211 | 91 | 0.34 | 1.0 | 1.0 | 10.71 | .0785 |
| Au | 1090 | 247 | 105* | 3.0 | 1.0 | 1.0 | 9.39 | .0911 |
| Bi | 545 | 7.6 | 12 | 1.0 | - | - | 20.75 | .022 |
| Cd | 594 | 37+/-3 | 90 | 1 | 1.03 | - | 10.42 | |
| Ce | 1072 | 21 | 22 | 1.25 | 1.05 | 1.0 | 2.99 | |
| Co | 1768 | 45 | 36 | - | - | - | 9.16 | .0243 |
| Cr | 2180 | 45 | 35 | - | - | - | 9.63 | .0261 |
| Cs | 302 | 35.9 | 20 | 0.27 | 1.05 | 1.0 | 6.95 | |
| Cu | 1358 | 330 | 163 | 2.0 | 1.0 | 1.0 | 9.77 | .0722 |
| Fe | 1811 | 34 | 33 | - | 0.95 | 1.06 | 7.62 | .0039 |
| Ga | 303 | - | 25 | 4.5 | - | 1.0 | 18.45 | |
| Ge | 1211 | 15 | 39 | - | - | - | 30.5 | |
| Hf | 2506 | 39 | - | - | - | - | 10.9 | |
| Hg | 234 | - | - | 0.9 | - | - | 9.80 | |
| In | 430 | 76 | 38 | 2 | 1.1 | 1.1 | 7.63 | .0908 |
| Ir | 2719 | 95 | 76 | - | - | 1.0 | 15.1 | .0148 |
| K | 337 | 98.5 | 56 | -3.8 | 1.0 | 0.9 | 6.90 | .0818 |
| La | 1193 | - | 17 | - | - | 1.0 | 5.19 | |
| Li | 454 | 71 | 43 | 2 | - | - | 6.6 | .076 |
| Mg | 923 | 145 | 79 | - | - | - | 9.18 | .066 |
| Mn | 1519 | 24 | 22 | - | 1.0 | - | 8.50 | .0102 |
| Mo | 2896 | 87 | 72 | - | 1.0 | 1.03 | 12.95 | |
| Na | 371 | 120 | 88 | -5 | 0.85 | 1.05 | 7.00 | .044 |
| Nb | 2750 | 78 | 66 | - | 1.0 | - | 10.90 | |
| Nd | 1289 | - | 18.4 | 0.53 | - | - | 5.54 | |
| Ni | 1728 | 70 | 60 | - | 1.03 | - | 10.11 | .0152 |
| Pb | 601 | 30 | 15 | 0.75 | 1.0 | 1.0 | 7.95 | .0872 |
| Pd | 1828 | 99 | 87 | - | - | 1.05 | 9.15 | |
| Pr | 1204 | - | 22 | 1.4 | 1.0 | - | 5.72 | |
| Pt | 2042 | 80 | 53 | - | 1.01 | - | 10.86 | .0392 |
| Rb | 312 | 58 | 33.2 | -1.9 | - | - | 7.01 | .080 |
| Re | 3459 | 65+/-5 | 55 | - | - | - | 17.5 | |
| Rh | 2237 | 110 | 69 | - | - | 0.96 | 11.9 | .039 |
| Sb | 904 | 17 | 25 | 1.0 | - | - | 22.0 | |
| Sc | 1814 | 24.5 | 22.5 | - | - | 1.0 | 7.77 | |
| Si | 1687 | 25 | 56 | - | - | 1.04 | 29.8 | |
| Sn | 505 | 59.5 | 27 | - | 0.90 | 1.05 | 13.9 | .0567 |
| Ta | 3290 | 70 | 58 | - | - | 1.0 | 11.1 | .0169 |
| Ti | 1941 | 31 | 31 | - | - | 1.1 | 7.28 | .0 |
| V | 2183 | 51 | 43.5 | - | - | 1.0 | 9.85 | .0161 |
| W | 3695 | 95 | 63 | - | - | 0.98 | 14.2 | .029 |
| Zn | 693 | 90 | 50 | 6 | 0.90 | 1.04 | 10.6 | .055 |
| Zr | 2128 | 38 | 36.5 | - | - | - | 9.87 | .0041 |

## 4     MEASUREMENT TECHNIQUES

Much of the thermal conductivity data reported in the last 20 years for the high-melting metals in the solid and liquid states has been obtained by two Russian groups using *temperature wave methods*. Filippov and co-workers [10] used the *radial temperature wave method* (RTW) and Zinovyev and his co-workers [11] used the *plane temperature wave method* (PTW) which involved the use of high heating rates (up to 1000 Ks$^{-1}$). For the assessment of data it is important to establish how accurate these techniques are. The evaluation was carried out by comparing thermal conductivity results obtained by the *temperature wave method* for

(i)     the solid state with those obtained with conventional techniques and

(ii)    the liquid phase for low melting metals obtained with conventional methods, where convection can be more easily be minimised

Inspection of the data reported for a large number of pure metals in the solid state indicated that the data obtained by both the Filippov and Zinovyev groups were within 10% of those obtained with conventional techniques (eg Binkele [12]). Thus the results are within the combined experimental uncertainties of the temperature wave and conventional methods. Furthermore, the values for the solid tend to be dependent upon the mechanical and thermal histories of the specimens and this could also account for any differences in the results.

The comparison is more difficult for the liquid phase measurements. Values reported by Filippov for Bi($\ell$), Cu($\ell$), Pb($\ell$), and Sn($\ell$) are within 10% of the recommended values. Zinovyev et al [13] have reported values for Ag($\ell$) using the RTW method, these were in excellent agreement with the recommended values [2] which suggests the RTW method provides reliable values for the liquid phase. It could be argued that the convectional contributions in Zinovyev's experiments would be small because the rapid heating means that there is little time for the convectional flows to be established. Conversely it could be argued that the large thermal gradients produced might result in the rapid establishment of buoyancy and thermocapillary-driven flows. It was concluded that the measurements due to the Filippov and Zinovyev groups were probably accurate to within ± 10% but it would be reassuring to have data from the Zinovyev group for Cu($\ell$), Pb($\ell$) and Zn($\ell$) where values of reasonable reliability are available.

In recent years the laser pulse method has been used to derive thermal diffusivities and conductivities of liquid metals by holding the melt in sapphire or silica cassettes [14,15]. This is a well established method for the solid phase but for the liquid phase the following problems could be encountered (a) convectional flows could be operative (b) non-wetting conditions tend to cause 'balling-up' for some of the specimens and (c) carbon on the surface is used to eliminate reflection of the laser beam would in some cases (eg Fe) dissolve in the metal and thereby affect the thermal conductivity [16]. Care must be taken to ensure that these problems are overcome.

In summary, the following criteria were used to evaluate the data:

- the nature of the measurement method and the temperature range involved
- the steps taken to minimise buoyancy-drive convection by (a) reducing thermal gradients, (b) maintaining the free surface at a slightly higher temperature and (c) the use of electromagnetic fields to suppress convection
- the steps taken to avoid thermocapillary convection and electromagnetic stirring.

## 5        RESULTS AND DISCUSSION

### 5.1     THERMAL CONDUCTIVITY DATA

The recommended values of the thermal conductivity at the melting point of the solid ($\lambda_s^m$) and liquid phase ($\lambda_\ell^m$) and the temperature coefficient (b) of $\lambda_\ell$ (Equation 4) are given in Table 1.

$$\lambda_l = \lambda_l^m + b\ (T - T^m) \qquad (4)$$

since the most important commercial alloys are based on Al, Cu, Fe, Ni and Ti, values for the thermal conductivities of these metals are shown in Figures 1 to 5, respectively.

<u>Aluminium</u>

Values have been derived from two recent laser pulse investigations [14,15] (Figure 1) are in good agreement with values given by Touloukian [2] and those calculated by the WFL rule.

<u>Copper</u>

Thermal conductivities for molten Cu are given in Figure 2, it can be seen that the results obtained by Tye and Hayden [17] using the axial heat flow method and Filippov using the radial temperature wave method are in excellent agreement with the values given by Touloukian [2]. Values calculated from the WFL are about 3% lower. Copper would appear to be a good system for the calibration of different methods.

<u>Iron</u>

Thermal conductivities data for Fe($\ell$) have been reported by Ostrovskii [18] and by Zinovyev et al [11,19] and are shown in Figure 3. It can be seen that the latter results indicate a slightly higher thermal conductivity value for the liquid than for the

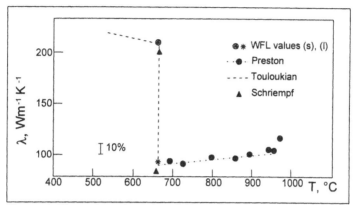

Figure 1. Thermal conductivity of Al as a function of temperature

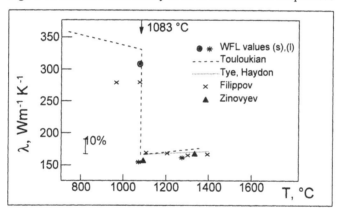

Figure 2. Thermal conductivity of Cu as a function of temperature

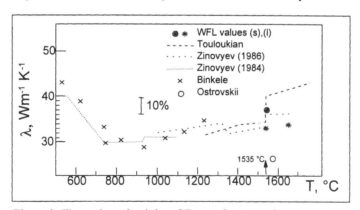

Figure 3. Thermal conductivity of Fe as a function of temperature

solid whereas WFL values based on electrical resistivity values indicate a slight decrease.

### Nickel

Thermal conductivity data for Ni are given in Figure 4 and it can be seen that for the measurements, due to Zinovyev et al [11,19] for (i) the solid phase are in good agreement with those reported by Binkele [12 ] and the $\lambda_s^m$ value calculated by the WFL Rule and (ii) for the liquid phase are about 10% higher than those calculated by the WFL Rule.

### Titanium

Thermal conductivity data for Ti are given in Figure 5. The results obtained for the solid state are within ± 10% [11,12,19] ie the combined uncertainty of the methods. Three investigations have been carried out on liquid Ti, all from the same laboratory, it can be seen that (i) the most recent investigation [11] shows a slight increase in conductivity on melting and (ii) the measured values are about 10% higher than values calculated by the WFL Rule. However the electrical conductivity data suggest that the thermal conductivity should decrease on fusion.

### 5.2    VALIDITY OF WIEDEMANN-FRANZ-LORENZ (WFL) RELATION

Inspection of Table I and the figures shows that the ratio of $(L/L_o)$ is close to unity for all metals for both the solid and liquid metals at the melting point. Since the experimental uncertainties in the thermal and electrical conductivities are $(> \pm 5\%)$ and $(\pm 3\%)$ respectively, small departures from $(L/L_o)$ could be ascribed to experimental uncertainty especially in the thermal conductivities which may contain convectional contributions.

### 5.3    VALIDITY OF EQUATION (3)

Inspection of Table I shows that the constant for low melting metals has a value of about 0.075 eg Al, 0.079; In, 0.091; K, 0.082; Li 0.076; Mg 0.066; Na, 0.044; Pb, 0.087; Rb 0.080; Sn 0.057; Zn, 0.055. Although K values for Cu, Au and Mn (based on electrical conductivity data) are in reasonable agreement with K ≥ 0.075 ± 0.02, it is obvious that K values for Fe, Co and Ni, 0.0039, 0.024, and 0.015 respectively, and the refractory metals are very much smaller than K = 0.075 ± 0.02.

Consequently the Mott relation given in Equation 2 was checked using electrical conductivity data and it was found that there were differences in the values of the constant (C) for low melting alloys and those for Fe, Co and Ni.

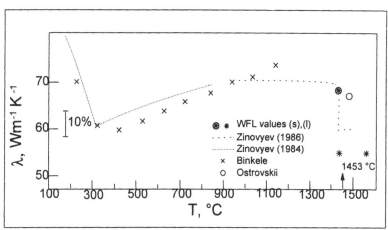

Figure 4. Thermal conductivity of Ni as a function of temperature

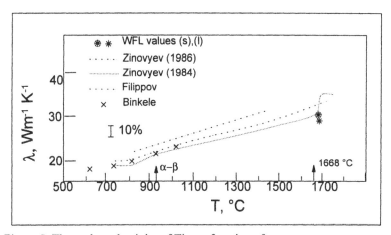

Figure 5. Thermal conductivity of Ti as a function of temperature

## CONCLUSIONS

1)    The Wiedemann-Franz-Lorenz rule is valid for all pure metals at the melting point and in the liquid region.

2)    The proposed relation between the ratio of the thermal conductivities of the solid and liquid at the melting point ($\lambda_s^m/\lambda_\ell^m$) and the entropy of fusion does not yield a uniform proportionality constant.

## ACKNOWLEDGEMENTS

This work was carried out as part of the Measurements for Materials Processing Programme, a programme of underpining research financed by the U.K. Department of Trade and Industry.

## REFERENCES

1.  Mott, N.F.,1934. " The Resistance of Metals," *Proc Royal Soc.,* A146:465-472.

2.  Touloukian, Y. S., R. W. Powell, C. Y. Ho, P. G. Klemens, 1970. Thermophysical Properties of Matter: Vol 1 Thermal Conductivity of Metallic Elements and Alloys Publ. IFI/Plenum Press, New York.

3.  Mills, K. C., B. J. Keene, and B. J. Monaghan. "Thermal Conductivity of Molten Metals," to be published as an NPL report.

4.  Iida, T. and R. I. L. Guthrie, 1993. The Physical Properties of Liquid Metals publ. Oxford Scientific.

5.  Desai, P.O., T. K. Cho, H. M. James, and C. Y. Ho, 1984. "Electrical Resistivity of Selected Elements," *Journal of Phys. Chem. Ref. Data*, 13:1069-1172.

6.  Dinsdale, A.T., 1991. "SGTE Data for Pure Elements," *Calphad*, 15 (4):317-425.

7.  Mardolcar, U.V., and C. A. Nieto de Castro, 1992. "The Measurement of Thermal Conductivity at High Temperature," *High Temperatures-High Pressures*, 2:551-580.

8.  Powell, J. S, 1991." An Instrument for the Measurement of the Thermal Conductivity of Liquids at High Temperature" *Measurement Science Technology*, 2:111-117.

9.  Nakamura, S. and T. Hibiya, 1992."Effect of Convective Heat Transfer on

Thermal Conductivity Measurement Under Microgravity," *Microgravity Science Technology*, 3:156-159.

10. Filippov, L.P., 1973. "Research of Thermophyscial Properties at the Moscow State University," *Journal of Heat Mass Transfer*, 16: 865-885.

11. Zinovyev, V.Y., V. F. Polev, S. G. Taluts, G. P. Zinovyeva, and S. A. Ilinykh, 1986. "Diffusivity and Thermal Conductivity of 3-d Transition Metals in the Solid and Liquid States," *Phys. Met. Metall.*, 61 (6):85-92.

12. Binkele, L., 1985. "Thermal Conductivity, Electrical Resistivity and Lorentz Function for Some Transition Metals Measured by the Direct Electric Heating Technique,"*High Temp-High Pressures*, 17:437-445.

13. Zinovyev, V. E., S. G. Taluts, M. G. Kamashev, B. V. Vlasov, V. P. Polyakova, N. I. Korenovskii, L. I. Chupina, and L. D. Zagrebin, 1994. "Thermal Properties and Lorentz Numbers of Copper and Silver at High and Average Temperatures." *The Physics of Metals and Metallography*, 77(5):492-497.

14. Schriempf, J.T., 1972. "A Laser Flash Technique for Determining Thermal Diffusivity of Liquid Metals at Elevated Temperatures," *High Temperatures-High Pressures*, 4: 411-416.

15. Preston, S.D. and K. C. Mills, to be published *High Temperatures-High Pressures*.

16. Szelagowski, H. and R. Taylor, 1995. "Thermal Diffusivity of Some Casting Alloys." Proceedings of The Fourth Asian Thermophysical Properties Conference, Tokyo, 3:571-574.

17. Tye, P.R. and R. W. Hayden, 1979. "The Thermal Conductivity and Electric Resistivity of Copper and Copper Alloys in the Molten State," *High Temperatures-High Pressures*, 11: 597-600.

18. Ostrovskii, O.I., V. A. Ermachenko, V. M. Popov, V. A. Grigorian, and L. E. Kogtan, 1980. "Thermophysical Properties of Liquid Iron, Cobalt and Nickel" *Russian Journal of Phys Chem.*, 5: 739-741.

19. Zinovyev, V.E., 1984. Kinetic properties of Metals at High Temperatures publ. Metallurgiya, Moscow.

# Thermophysical Property Measurements on Single-Crystal and Directionally Solidified Superalloys into the Fully Molten Region

J. B. HENDERSON and A. STROBEL

## ABSTRACT

The thermal diffusivity was measured for a directionally-solidified IN-718 superalloy and a single-crystal superalloy. In addition, the specific heat and the solidus and liquidus temperatures were measured for IN-718. The measurements were conducted as a function of temperature in the solid, liquid and mushy regions. Multiple transitions were detected in both materials and, in the case of IN-718, confirmed by cross-correlation of the thermal diffusivity and specific heat data. Additionally, comparison of the thermal diffusivity curves revealed considerable differences in the two materials, especially in the mushy region.

## INTRODUCTION

Because of the highly competitive nature of the casting industry, there is ever-increasing pressure to reduce mold design times and production costs and to increase product quality. The numerical casting simulation models are one of the primary tools used to achieve this end. The models currently in use are able to predict quantities such as time-dependent temperature profiles in the solid and liquid domains, solidification rates, etc. for complex geometries. The new generation models currently under development contain many more of the actual physical processes involved during solidification, but until now have been geometrically simpler. These models not only predict solidification rates and temperature profiles, but yield valuable information regarding macro- and micro-segregation, freckling and channeling in complex alloys. These advanced models should eventually allow relationships between microstructure and processing history to be developed and lead to a far better understanding of the casting process.

J. B. Henderson and A. Strobel, NETZSCH-Gerätebau GmbH, Wittelsbacherstraße 42, D-95100 Selb/Bavaria, Germany

Even though the numerical models are quite advanced and have a high potential for improving the quality of cast parts and leading to a better understanding of material behavior, their success depends, in large part, on an accurate thermophysical property data base. Simply stated, the predictions of the models are no better than the input data. The data required by the current models are temperature-dependent thermal conductivity, specific heat, and density in the molten and solid regions as well as in the mushy zone. Additionally, quantities such as transformation energetics and viscosity in the molten region are required. Further, the thermophysical properties of the mold materials are necessary for accurate modeling.

Clearly, the generation of accurate thermophysical property data is an integral part of the modeling effort. There is, however, an additional reward which results from the measurement of these properties. That is, quantities such as the specific heat, thermal conductivity and density when measured through the mushy zone into the fully molten region can be combined to yield a better characterization of the material. The purpose of the work presented in this paper was to measure some of the thermophysical properties required for numerical models and to demonstrate the power of combining bulk data to help characterize complex metal alloys. The measurements were conducted on a directionally-solidified IN-718 superalloy and a proprietary single-crystal superalloy.

## EXPERIMENTAL

The thermal diffusivity of both materials was measured in an inert atmosphere using a Netzsch Model 427 laser flash diffusivity apparatus. The unit used in this work was equipped with a high-temperature, water-cooled furnace capable of operation from 25 to 2000°C. The sample chamber is isolated from the graphite heating element by a protective tube, allowing samples to be tested under a vacuum or in an oxidizing, reducing, or inert atmosphere. The instrument and the laser flash method have been described in detail by Bräuer, et al. [1]. The reader is referred to this publication for further details.

Because the thermal diffusivity was measured in the mushy and fully-molten regions, a special sample holder had to be designed for the laser flash system. The sample holder consists of a sapphire crucible and lid mounted on the standard sample carrier tube. This construction permits the measurement of both liquids and powders without complicated and time-consuming carrier tube realignment. In addition, the sapphire construction eliminates the need to use troublesome 2- and 3-layer models to compute the thermal diffusivity from the experimental data.

The specific heat and solidus and liquidus temperatures were measured for IN-718 using a Netzsch Model 404 differential scanning calorimeter capable of operation from 25 to 1500°C. The measurements were carried out at a heating rate of 20°C/min in an inert atmosphere. The data were reduced by the well-known ratio method using sapphire as the reference material. The instrument and measurement techniques have been described by Henderson, et al. [2].

**RESULTS AND DISCUSSION**

Figure 1 shows the thermal diffusivity of IN-718 as a function of temperature for both the heating and cooling cycles. First it is clear that whatever changes that occurred in the microstructure did not significantly affect the thermal diffusivity, since the heating and cooling data in the solid region are almost identical. Further examination of this figure reveals a slope change in the thermal diffusivity between 700 and 800°C. Also obvious is a change in the diffusivity of ≈5% over the temperature range of 1200-1350°C. The reasons for this behavior can be explained with the aid of additional data.

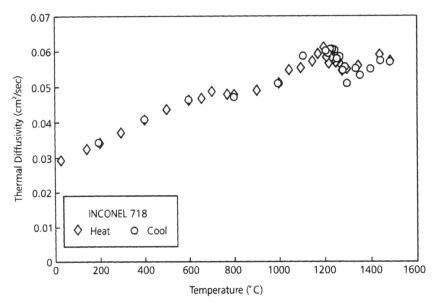

Figure 1. Thermal diffusivity of IN-718

Figure 2 shows the thermal diffusivity and specific heat data for the heating cycle. Clearly, the slope change in the thermal diffusivity data between 700 and 800°C fits with the slope change in the specific heat curve and is due to a solid-solid transition. Also, the peak in the thermal diffusivity curve between 1150 and 1200°C coincides with the small laves phase peak seen in the specific heat data. The 5% drop in the thermal diffusivity occurs across the mushy region and is due largely to the drop in the electronic component of the thermal conductivity. It should be obvious that the large peak in the specific heat curve is a result of melting and integration thereof yields the heat of fusion.

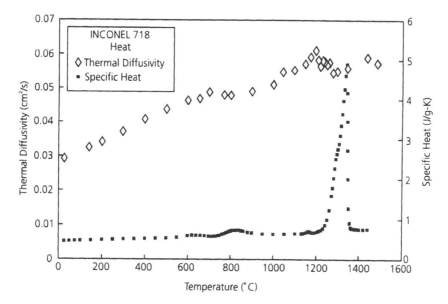

Figure 2. Thermal diffusivity and specific heat of IN-718 for the heating cycle

The thermal diffusivity and specific heat data for the cooling cycle are presented in figure 3. In this case, however, the laves phase peak which is so obvious in the specific heat data is not apparent in the thermal diffusivity curve. The solidus and liquidus temperatures at 1160 and 1320°C, respectively, were determined from the specific heat curve. Finally, the 5% increase in the thermal diffusivity from the fully-molten to the fully-solid region fits well with the change in the electrical resistivity measured by Kaschnitz, et al. [3] for a similar superalloy.

Depicted in figure 4 is the thermal diffusivity of the single-crystal material for both the heating and cooling cycles. As can be seen, the thermal diffusivity increases as a relatively linear function of temperature up to ≈800°C. Above 800°C a slope change occurs which is reminiscent of that for IN-718, and is most probably due to a solid-solid transition. Also of interest is the fact that no significant drop occurs in the thermal diffusivity across the mushy zone upon heating, as was the case with IN-718. It is clear to see, however, that upon cooling, the large increase in the diffusivity across the mushy region is present. Finally, the fact that the thermal diffusivity for the heating and cooling cycles in the solid region below ≈800°C is almost identical suggests that the microstructural changes which occurred had no significant influence on the thermal transport mechanism.

A comparison of the thermal diffusivities of IN-718 and the single-crystal alloy for the heating and cooling cycles is presented in figures 5 and 6, respectively. Examination of figure 5 reveals that the thermal diffusivity of IN-718 is a few percent higher than that of the single-crystal alloy in the solid region, and that they are almost identical in the fully-molten region. Further, the drop in the thermal diffusivity across the mushy zone seen in the IN-718 data is replaced by only a slope

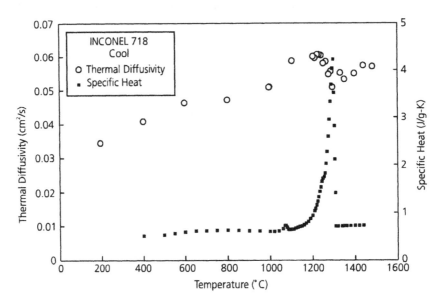

Figure 3. Thermal diffusivity and specific heat of IN-718 for the cooling cycle

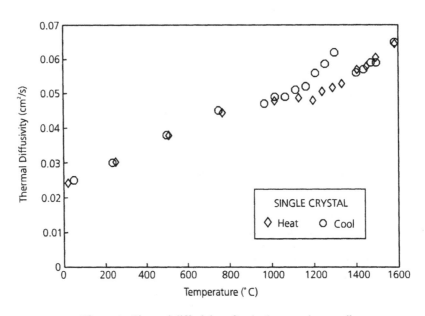

Figure 4. Thermal diffusivity of a single-crystal superalloy

534

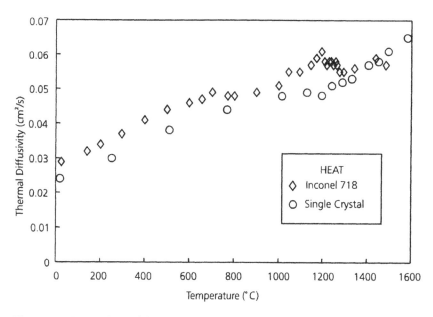

Figure 5.   Comparison of thermal diffusivity curves for IN-718 and a single-crystal superalloy during the heating cycle

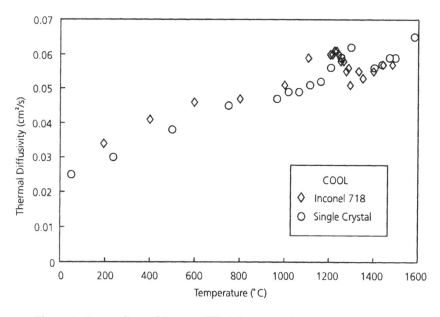

Figure 6.   Comparison of thermal diffusivity curves for IN-718 and a single-crystal superalloy during the cooling cycle

change in the single-crystal data. The fact that this same behavior was observed by Taylor, et al. [4] for a similar single-crystal superalloy during the heating cycle will be discussed later. The data in figure 6, however, show that once the initial structure has been changed by melting, the expected "step" increase in the thermal diffusivity from the fully-liquid to the fully-solid region reappears. As yet, the authors have not been able to account for this behavior, but it could be linked to differences in the density change of the two materials across the mushy zone. It must be pointed out that other directionally-solidified superalloys similar to IN-718 have been measured by the authors and have been found to display the same general "step" change in the thermal diffusivity across the mushy zone for both the heating and cooling cycles. This would tend to rule out the possibility of anomalies in the IN-718 thermal diffusivity data.

Finally, the thermal diffusivity values published by Taylor, et al. [4] are compared with the data presented in this paper in figure 7. Examination of this figure clearly reveals that the behavior of the two single-crystal materials across the mushy zone is quite similar. In addition, the absolute values of the thermal diffusivity are in good agreement even though the measurements were conducted on two different superalloys.

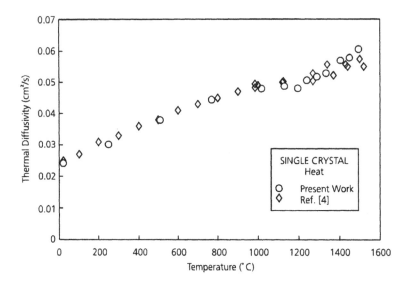

Figure 7.  Comparison of thermal diffusivity curves for two single-crystal superalloys during the heating cycle

## CONCLUDING COMMENTS

The thermophysical properties measured are part of those required for the numerical casting simulation models currently in use. Clearly, the melting/solidification process creates a complex set of interrelated properties which must be accurately quantified in order to be effectively utilized in the models. Future work will include density measurements into the fully-molten region for both materials as well as specific heat measurements for the single-crystal material.

The results presented also demonstrate the power of combining bulk thermophysical property data to yield a better understanding of the behavior of multi-component alloys. For example, the complex behavior observed in the results of the IN-718 alloy was clarified, in part, by the cross-correlation of the thermal diffusivity and specific heat data. Finally, the data highlight the ability of both the laser flash apparatus and differential scanning calorimeter to detect and quantify seemingly insignificant events in complex materials.

## REFERENCES

1. Bräuer, G., L. Dusza and B. Schulz, 1992. "New Laser Flash Equipment LFA 427," *Interceram* 41[7/8]:489-492.

2. Henderson, J. B., W.-D. Emmerich and E. Wassmer, 1988. "Measurement of the Specific Heat and Heat of Decomposition of a Polymer Composite to High Temperatures." *Journal of Thermal Analysis* 33:1067-1077.

3. Kaschnitz, E., J. L. McClure and A. Cezairliyan, 1994. "Measurements of Selected Thermophysical Properties of a Nickel-Base Alloy (Inconel 625) at High Temperatures by Pulse Heating Techniques." Presented at the 12th Symposium on Thermophysical Properties, Boulder, CO.

4. Taylor, R. E., H. Groot and J. B. Henderson, 1993. "Thermal Diffusivity and Electrical Resistivity of Molten Materials," *High Temperatures - High Pressures* 25:569-576.

# Thermophysical Property Measurements for Casting Process Simulation

R. A. OVERFELT and R. E. TAYLOR

## ABSTRACT

Solidification models have become very advanced in recent years as software developers have added fluid flow capabilities, stress analysis, and microstructural modeling. Unfortunately, the basic thermophysical and related property database required for simulating the freezing of complex industrial alloys is almost non-existent. Recently Auburn University and Purdue University have formed a teaming relationship to establish reliable measurement techniques for the most critical properties: transformation temperatures, thermal conductivity, specific heat, latent heat, solid fraction evolution, surface tension, and viscosity. Some of the properties in the mushy zone depend critically upon the cooling rate of the casting and thus must be known for representative cooling rates. This paper reviews the application of the various property measurement techniques to molten materials and gives representative data.

## INTRODUCTION

Foundries all over the world are applying advanced computer simulation technology to understand the critical aspects of heat and fluid flow and evolution of defects and metallurgical structures in metal casting processes. The computational tools are enabling the design and production of higher quality, more economical castings. However, the required input data of thermophysical properties necessary for accurate simulations are frequently nonexistent or unreliable for many industrial alloys of interest. Accurate casting models require integrated, self-consistent thermophysical property data sets for reliable simulation of the complex solidification processes.

A large array of thermophysical properties must be available for input before reliable modeling of a casting process can be accomplished. These data must be evaluated from room temperature, through the mushy zone, and to approximately 100°C superheat. All current commercial software require that the thermal conductivity, specific heat, latent heat, solidus and liquidus temperatures, and density be known for the alloy being modelled. More advanced software incorporates mold filling and convection effects and thus need data on the viscosity, wetting angle, and surface tension of the molten alloy. Some of these are more

R.A. Overfelt, Auburn University, 231 Leach Center, Auburn, AL 36849

R.E. Taylor, TPRL Inc., 2595 Yeager Road, West Lafayette, IN 47906 critical than others, depending upon the specific casting process being modeled. For example, if a casting being modeled does not exhibit thin walls, then the surface tension and mold wetting angles would be of minor importance. However, accurate determination of these same parameters will be required to simulate the filling of thin walls in precision aerospace and automotive castings.

Accurate and repeatable thermophysical property measurements are experimentally difficult. Convection effects in samples that are molten often exacerbate the difficulties. However, recently the Thermophysical Properties Research Laboratory (TPRL) at Purdue University and the Space Power Institute (SPI) of Auburn University have established a teaming relationship to develop dedicated capabilities to measure critical thermophysical properties in support of industrial process simulation efforts. This paper presents an overview of the various measurement techniques being utilized and/or under development for these measurements.

## MEASUREMENT TECHNIQUES

### THERMAL DIFFUSIVITY AND THERMAL CONDUCTIVITY

Thermal diffusivity ($\alpha$) is an important property in its own right as it is required for applications involving transient heat flow conditions. In addition, it offers a convenient, economical and accurate method of determining the thermal conductivity ($k$). The relationship between $k$ and $\alpha$ is given by $k = \alpha C_p \rho$ where $C_p$ is the specific heat and $\rho$ is the density. The measurements of these properties are easier and more accurate than the measurement of the parameters (i.e., heat fluxes and temperature gradients) required to determine thermal conductivity directly from steady-state experiments. Thus, the majority of diffusivity measurements are performed in order to obtain thermal conductivity values and indeed, the overwhelming percentage of conductivity values today are obtained from diffusivity results. The accuracy of these thermal conductivity values are at least equal to, and usually substantially exceed, the values obtained from steady-state methods.

The most popular diffusivity technique is the laser flash method (1) which has become an ASTM standard (2). The thermal mass of the system can be made quite small so that it is possible to quickly change temperature and to obtain data at a number of temperatures over a large temperature range rapidly. Thus, a large amount of data can be generated in a short period of time. Another outstanding feature of the method is the on-line comparison of experiment to theory. This greatly aids in detailing the need for and the making of corrections for heat losses, non-uniform heating and finite pulse time.

The flash method was first described in 1960 by Parker, Butler, Jenkins, and Abbott of the U.S. Naval Radiological Defense Laboratory (3). In the standard technique, the front face of a small disk-shaped sample (often about the size of a small coin) is subjected to a very short burst of laser energy. Irradiation times are approximately one millisecond or less. The rear face temperature rise is typically 1 to 2°C and temperature rise times of less than one second are often exhibited. The resulting temperature rise of the rear surface of the sample is measured and thermal diffusivity values are computed from the time for the back surface to reach 50% of its maximum temperature. If the thickness of the sample is L and the 50% rise time is $t_{1/2}$, then the thermal diffusivity is determined simply as

$$\alpha = 0.1388 \, \frac{L^2}{t_{1/2}} \, .  \qquad (1)$$

The details of the standard laser flash technique for experiments on solids are well described in the literature (1,4). The extension of the technique to molten materials involves encapsulating the samples. Figure 1 shows typical thermal diffusivity data measured at Purdue University on four germanium samples (5). These data were taken using an early, horizontally-oriented flash diffusivity device. Data taken upon heating and upon cooling are shown for sample 2. The repeatability of the technique is quite good and the large change in thermal diffusivity after melting is readily apparent.

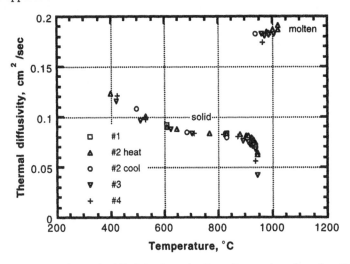

FIGURE 1. Thermal Diffusivity Data for Four Germanium Samples (5).

An instrument dedicated for evaluating molten materials has recently been established in the Thermophysical Properties Research Laboratory at Purdue University. The essential features of the current instrument are described in the literature (6). Samples are contained in a quartz or sapphire cup equipped with a lid. The choice of cup material is dictated by the melting point and reactivity of the material under investigation. The laser beam can pass through the transparent top to impinge on the upper surface of the metal and the resulting rear face temperature rise can be monitored through the transparent cup's bottom by means of an infrared detector. Losses in the transmission through the container are not important as the absolute temperature rise on the back surface is not required. Details of how to know and control the sample thickness are proprietary. Convection effects in the molten samples are minimized by maintaining thin samples with corresponding fast experiments and flash heating on the tops of the samples.

## SPECIFIC HEAT AND LATENT HEAT EVOLUTION

Differential scanning calorimeters have been commercially available from many manufacturers for years and only a summary of their application for studying the solidification of molten materials will be given here. A Netzsch Model 404 differential scanning calorimeter (DSC) is used at TPRL to measure specific heats from 323 to 1673 K (6). Energetics of transformations and reactions can also be determined from 296 to 1773 K. The system is vacuum tight and therefore samples can be tested under inert, reducing or oxidizing atmospheres as well as under vacuum. The device uses a single heating chamber for both the unknown sample and the standard reference material and is carefully calibrated so that the differential thermocouple voltages from the two are directly related to enthalpy differences. Instrument control and data acquisition are accomplished via a standard PC software and electronics package. Data evaluation is accomplished by standard PC software allowing computation of peak and onset temperatures, inflection points, partial area integration, specific heat, energetics, etc.

Specific heat is determined by first running a baseline over the temperature range of interest with the sample and standard crucibles both empty to establish a zero baseline for the instrument. The standard reference material and the sample are then run and the specific heat is calculated by the ratio method. That is:

$$C_{p;s} = C_{p;std} \ \frac{m_{std}}{m_s} \ \frac{\Delta\mu V_s}{\Delta\mu V_{std}} \tag{2}$$

where $C_{p,s}$ = sample specific heat; $C_{p,std}$ = specific heat of the standard; $m_s$ = sample mass; $m_{std}$ = mass of the standard; $\Delta\mu V_s$ = differential microvolt signal between the baseline and the sample and, $\Delta\mu V_{std}$ = differential microvolt signal between the baseline and the standard.

Figure 2 shows typical apparent specific heat data upon cooling for superalloy 718. The sample was cooled through the transformation at a rate of 5°C/min and apparently began freezing at about 1325°C. The small peak in the curve at about 1250°C is associated with the formation of carbides. The unknown latent heat is easily determined by integrating the apparent specific heat curve from the beginning to the end of solidification and subtracting the baseline contribution due to the actual specific heat. In addition, the latent heat of an unknown can also be evaluated from a sample with a known transformation enthalpy as

$$\Delta H_f = \Delta H_{std} \ \frac{C_{p;s} \ A_{p;s}}{C_{p;std} \ A_{p;std}} \tag{3}$$

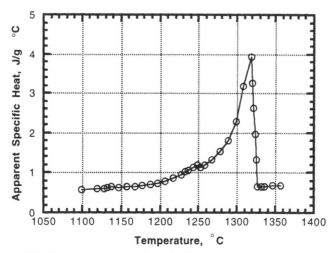

FIGURE 2.  Apparent Specific Heat Data for Superalloy 718.
Sample Cooled at 5°C/min. (7)

where $A_{p;std}$ = peak area of standard transformation, $\Delta H_{std}$ = standard transformation enthalpy, $\Delta H_f$ = sample latent heat, and $A_{p;s}$ = peak area of sample transformation.

## FRACTION SOLID EVOLUTION

The fraction solid - temperature relationship is critical in solidification modeling for accurate simulation of the release of the latent heat over the mushy zone temperature range. This relationship can be established using three techniques: (a) equilibrium melting and quenching followed by metallographic assessment of the quenched in liquid, (b) cooling curve analysis as described by Backerud(8), and (c) sequential integration of the area under the effective specific heat curve. Figure 3 shows the cooling curve and first derivative for a sample of superalloy 718 completely melted at 1460°C and then removed from the furnace to solidify. The cooling rate for this sample just before the start of freezing was approximately 9°C/sec. The sample experienced about 10°C undercooling prior to the beginning of solidification. The sample recalesced to 1340°C and then continued cooling normally. Solidification was complete at 1150°C. The solidification reaction is clearly seen in the cross-hatched region of the first derivative curve, the large initial spike representing the sudden release of latent heat in the undercooled sample and the small final peak due to the terminal γ–γ' eutectic in this alloy.

Sequential integration of the first derivative of the cooling curve from the beginning to the end of solidification represents the total heat generated.(6) Tracking the fraction of heat liberated during the reaction provides a dynamic estimate of the fraction-solid versus temperature relationship. Figure 4 shows fraction-solid vs. temperature data from (a) the sample whose cooling curve is shown in Figure 3 plus an additional similarly processed sample (the solid curves), (b) the sample whose apparent specific heat curve is shown in Figure 3 plus an additional DSC-processed sample cooled at 20°C/min. (the dashed curves), and (c) many samples (solid data points) that were held for long times in the mushy zone and then quenched in water followed by metallographic assessment of the equilibrium fraction-solid (9-11).

In Figure 4, the solid-fraction vs. temperature curves taken from the melted and quenched samples exhibit a smooth monotonic increase in fraction solid with decreasing temperature through the mushy zone. The quenched samples indicated an equilibrium liquidus of 1345°C and an equilibrium solidus of 1240°C. There was no evidence of any terminal eutectic in the quenched samples. The dynamically cooled samples with in-situ thermocouples exhibited typical amounts of undercooling at the beginning of freezing and then recalescence as solidification proceeded. The "tails" at the end of solidification (~1130-1150°C) of the these superalloy 718 samples indicate a small amount of terminal γ–γ' eutectic.

Figure 4 also shows the evolution of fraction-solid calculated by sequential integration of the apparent specific heat curve shown in Figure 2 as well the additional DSC sample continuously cooled at 20°C/min. The evolution of fraction-solid from the DSC samples occurs at significantly lower temperatures compared to both the quenched samples and the samples dynamically frozen with an in-situ thermocouple. The temperature data from the DSC experiments come from a thermocouple located <u>outside</u> of both the sample and its ceramic crucible. Apparently this thermocouple doesn't respond quickly enough to follow the transformation accurately, particularly at the higher cooling rate. The use of metallic pans to hold samples and slow cooling rates minimizes this problem. Additional research into this effect is continuing and will be the subject of a future publication. The best estimate of fraction-solid evolution for modeling dynamic solidification processes is obtained from in-situ thermocouples in samples cooled at rates corresponding to those experienced in practice.

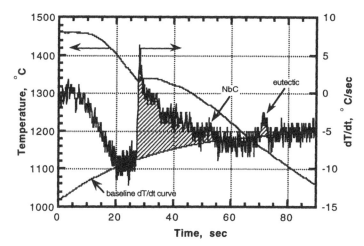

Figure 3. Cooling Curve and First Derivative from a Sample of Superalloy 718. Baseline Curve is an Exponential Curve Fit Representing Cooling Without Any Transformation as Described by Backerud(8). Cross-Hatched Region Represents the Heat of Fusion from the Solidification Reactions.

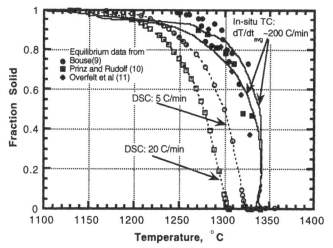

FIGURE 4. Fraction-solid Evolution for Superalloy 718 Determined from DSC Measurements, Cooling Curve Analysis from an In-situ Thermocouple, and Equilibrium Melted and Quenched Samples.

## DENSITY

Although push-rod dilatometers are commercially available from many manufacturers for measuring thermal expansion and contraction, they must be modified for studying solidification phenomena. For materials such as aluminum alloys that do not form carbides, a graphite piston arrangement has been used successfully. For reactive metals, high purity alumina is often used.

For example, the expansion of an aluminum alloy was first measured in the solid region using the standard push-rod dilatometer with a NBS standard as the reference material. A hollow cylinder of graphite 6.352 cm long, 1.055 cm OD and 0.6475 cm ID was machined along with two end plugs 0.811 cm long by 0.6470 cm diameter. A conventional expansion sample was machined from the same graphite and its expansion was measured using the dual push-rod dilatometer. Alloy samples 0.6364 cm diameter by 5.084 cm long were machined to fit inside the graphite cylinder so that the graphite end plugs extended beyond the ends of the graphite cylinder when these plugs were against the alloy sample. For measurements through the mushy zone, a single push-rod head designed to have a relatively large travel under a constant load was substituted for the dual push-rod head. The net expansion of the aluminum alloy and graphite plugs against the silica dilatometer was measured and the expansion of the fused silica used in the dilatometer was known. The contribution of the graphite plugs was then subtracted from the net expansion to yield the expansion of the alloy in the solid region. In the mushy and molten region, the alloy filled the diameter of the graphite cylinder. Net volume changes for the alloy could be calculated from the known radial expansion of the graphite and the net movement of the push-rod. Density values could then be determined from the calculated volume changes and known mass. While the push-rod accurately follows the alloy expansion during heating, problems can develop as the alloy melts. Figure 5 shows typical data from superalloy 718 up to approximately the equilibrium solidus temperature of 1250°C. This data was quite reproducible. However, once into the molten region, liquid alloy squeezed into the gap between the pushrod and the crucible prevented acquiring data in the molten region. Rothrock and Kirby(12) developed an approach using telemicroscopes to make measurements at high temperatures and this technique is being modified to track the position of an alloy's molten surface. These results will be reported in a future publication.

## SURFACE TENSION

Surface tensions of molten metals can be measured by many techniques: sessile-drop, maximum bubble pressure, pendant-drop, capillary-rise, drop weight, and oscillating drop methods(13). The sessile drop technique has been widely utilized because of its many advantages, for example, measurements over a wide range of temperatures. Although the method is inherently straightforward by utilizing a molten drop resting on a horizontal ceramic substrate, great experimental care must be exercised to ensure the absence of contaminants and a level substrate.

The fundamental theory of capillarity was independently published by Young(14) and Laplace(15) in 1805 and relates the pressure difference, $\Delta P$, across an interface to the surface tension, $\gamma$, and the principal radii of curvature, $R_1$ and $R_2$. Mathematically,

$$\Delta P = \gamma \left( \frac{1}{R_1} + \frac{1}{R_2} \right). \tag{4}$$

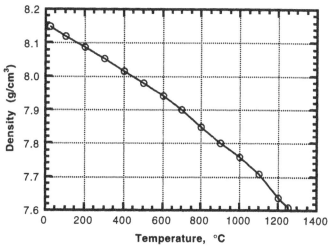

FIGURE 5.  Density of Superalloy 718 Measured During Heating Using the
Push-rod Technique (7).

Bashforth and Adams(16) modified Eqn. (4) for the determination of surface
tension in 1883 and developed a theoretical description of the contour of a
cylindrically symmetrical sessile drop resting on a non-wetting substrate.  The
equilibrium drop shape is characterized by its surface tension and the interfacial
energy between the molten drop and its substrate as shown in Figure 6.  Bashforth
and Adams prepared a set of tables which can be used to fit an experimentally
measured drop shape to the theoretical shape.  The calculations are rather tedious
and Butler and Bloom(17) were prompted to develop an iterative computational
procedure to automate the curve fitting process by minimizing the error between the
theoretical curve and the experimental data.   The many point measurement
technique of Butler and Bloom is superior to techniques that utilize only two or
three measurements of the drop shape (13).

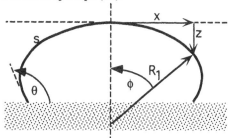

FIGURE 6.  The shape of a typical sessile drop resting on a substrate.

## VISCOSITY

Although measurements of the viscosities of molten  metals have been reported
using several techniques, the dominant technique at moderate to high temperatures
is the oscillating cup technique.  In this method, a molten metal is contained within

a ceramic vessel suspended by a torsional pendulum. Torsional oscillations are then induced and the resulting motion is damped primarily by viscous dissipation within the molten metal under investigation. The logarithmic decrement of the amplitude of torsional oscillation, $\delta$, required to calculate the viscosity must be due solely to the viscous damping of the molten metal. However, small amounts of additional damping arise due to internal friction in the suspension wire and viscous damping due to inert cover gas. These extraneous contributions are determined for each sample/crucible assembly by oscillation experiments after the sample freezes, and then simply subtracted from the decrements measured with the molten sample to yield the decrement solely due to the molten metal's viscosity.

The viscosity of a molten metal can then be determined by measuring the time period and decay of the oscillations. The principal advantages of this technique are it's mechanical simplicity and the ability to measure the time period and amplitude decay with great precision. The motion of a torsional pendulum undergoing damped oscillations can be described by

$$\theta(t) = \theta_0 \exp\left(-\frac{\delta}{\tau} t\right) \cos\left(\frac{2\pi}{\tau} t + \psi\right) \tag{5}$$

where $\theta(t)$ is the time-dependent angular displacement, $\theta_0$ is the initial angular displacement, $\delta$ is the logarithmic decrement of the amplitude of oscillation, $\tau$ is the period of oscillation, t is the time, and $\psi$ is the oscillatory phase shift. Only the logarithmic decrement and the time period need to be measured for calculating the viscosity of the molten sample. These can be obtained by a best-fit of Eq. (5) to the observed motion of the torsional suspension system.

A number of analytical equations have been theoretically developed and experimentally tested to relate the observed time period and decrement of the oscillating assembly to the sample's viscosity. Roscoe's equation(18,19) has been widely used and is considered to provide very accurate values of viscosity (20). In fact, application of Roscoe's formula with a small corrrection factor has been shown to accurately reproduce calibration quality viscosity data for mercury, lead, tin, bismuth, and indium obtained using the well-accepted capillary technique(13). All dimensions and torsional inertias of the torsional assembly must be corrected for thermal expansion effects. An iterative procedure of successive approximation is required to apply Roscoe's technique for calculating an unknown viscosity.

The absolute viscosity of superalloy 718 has been experimentally evaluated over the temperature range of 1350-1475°C and is shown on Figure 7. The oscillating viscometer utilized has been described elsewhere (21). Two samples were utilized and neither showed any evidence of crucible reactions around their circumferences or their bottoms and only minimal oxidation on their free surfaces (tops). Viscosity measurements were made both during the heating cycle (3 measurements at each temperature) and during the cooling cycle (3 measurements at each temperature) for each sample (21). The data set exhibited an Arrhenius behavior with the viscosity decreasing exponentially with temperature. Through a least-squares curve fit procedure, the Arrhenius equation describing this behavior was determined to be

$$\mu = 0.179 \exp\left(\frac{50.2}{R\,T}\right) \quad \pm 3\% \quad \text{cP (or mPa sec)} \tag{6}$$

where R is the universal gas constant (0.008314 kJ/mole/K) and T is the absolute temperature in K.

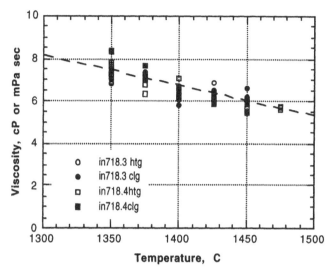

FIGURE 7. Measured Temperature Dependance of Viscosity of Superalloy
718. Dashed Line Represents an Arrhenius Equation Fit (21).

The correlation coefficient of the curve fit to the experimental data is an excellent
0.985. The uncertainty is estimated to be approximately ±3%. The activation
energy for viscous flow was found to be 50.2 kJ/mole for superalloy 718, identical
to that reported for pure nickel (22).

## SUMMARY

Various techniques for the measurement of thermophysical properties have been
developed over the years and recently have begun to be applied to molten materials.
Although accurate and repeatable thermophysical property measurements are
experimentally difficult, measurements to date on semiconductor materials, ferrous
and non-ferrous alloys have shown that the techniques are extremely useful and
reliable when carefully applied.

The same concerns for precision and accuracy on measurements with solids
(e.g., uncorrected heat losses, temperature uniformities, transient effects, signal-to-
noise ratios, etc.) also apply to molten samples. In addition, the measurement of the
transport properties of diffusivity and viscosity in molten samples  must be
performed with minimal convection in the molten sample itself.

Finally, the evolution of latent heat over the mushy zone is very dependent upon
the cooling rate. Solidification models that do not calculate microstructures as part
of their solution must contain the solid-fraction evolution data representative of the
cooling rate of the alloy being modelled. These data are easily obtained from
cooling curve analyses of sample castings.

## ACKNOWLEDGMENTS

The authors gratefully acknowledge the financial support received from NASA's
Office of Space Access and Technology under Grant No. NAGW-1192, from

ARPA under Agreement No. MDA972-93-2-0001, from the Investment Casting Cooperative Arrangement (Howmet Corporation, PCC Airfoils Inc., General Electric Aircraft Engines, United Technologies Corp., and UES Inc.) chaired by Howmet Corporation, and from Ford Motor Company.

## REFERENCES

1. R.E. Taylor and K.D. Maglic, "Pulse Method for Thermal Diffusivity Measurement," Compendium of Thermophysical Property Measurement Methods, Vol. 4, Plenum Publishing Corp., N.Y., N.Y., 1984, pp. 305 - 336.

2. Designation: E 1461 - 92, "Standard Test Method for Thermal Diffusivity of Solids by the Flash Method," 1993 Annual Book of ASTM Standards, Vol. 14.02, General Methods and Instrumentation, pp. 750 - 757.

3. W.J. Parker, R.J. Jenkins, C.P. Butler, and G.L. Abbott, "Flash method of determining thermal difusivity, heat capacity, and thermal conductivity," (Report USNRDL-TR-424, US Naval Radiological Defense Laboratory, 1960).

4. R.E. Taylor, "Critical Evaluation of the Flash Method for Measuring Thermal Diffusivity," Rev. Int. Hautes Temper. et Refract., 1975, pp. 141 - 145.

5. Raymond E. Taylor, Lawrence R. Holland, and Roger K. Crouch "Thermal Diffusivity Measurements on Some Molten Semiconductors," High Temp. - High Press., Vol. 17, 1985, pp. 47 - 52.

6. R.E. Taylor, "Test Methods for Molten Superalloys," (Report TPRL-1347-M, Purdue University Thermophysical Properties Research Laboratory, 1994).

7. R.E. Taylor, H. Groot, and J. Ferrier, "Thermophysical Propoerties of IN718," (Report TPRL-1347-IN718, Purdue University Thermophysical Properties Research Laboratory, 1994).

8. Lennart Backerud, Ella Krol, and Jarmo Tamminen, Solidification Characteristics of Aluminum Alloys, Vol 1: Wrought Alloys, (Oslo, Norway: Skanaluminum, 1986) p 67.

9. G.K. Bouse, "Application of a Modified Phase Diagram to the Production of Cast Alloy 718 Components," Superalloy 718 - Metallurgy and Applications, ed. E.A. Loria (Warrendale, PA: The Minerals, Metals, and Materials Society, 1989) p. 72.

10. B. Prinz and G. Rudolf, "Microporobe Measurements to Determine the Melt Equilibria of High-Alloy Nickel Materials," Mikrochimica Acta (Wien), Suppl. 11 (1985), p. 275-287

11. R.A. Overfelt and Rong-Jiun Su, unpublished research.

12. Rothrock, B.D., and Kirby, R.K., "An Apparatus for Measuring Thermal Expansion at Elevated Temperatures," J. Res. Natl. Bur. Stand., 71C, 1967, pp. 85 - 91.

13. Takamichi Iida and Roderick I.L. Guthrie, The Physical Properties of Liquid Metals, (Oxford: Clarendon Press, 1988) p. 188.

14. T. Young, in Miscellaneous Works, Ed. by G. Peacock, 1855 (London: Murray, 1855) p. 418.

15. P.S. De Laplace, "Mechanique Celeste," Suppl. to Book 10 (Paris, Fr.: Impr. Imperiale,1806); also English version translated by N. Bowditch, Vol. IV, (New York: Chelsea Publishing, 1966) pp. 685-805.

16. F. Bashforth and J.C. Adams, <u>An Attempt to Test the Theories of Capillary Action</u>, (Cambridge: Cambridge University Press, 1883).

17. James N. Butler and Burton H. Bloom, "A Curve-Fitting Method for Calculating Interfacial Tension from the Shape of a Sessile Drop," <u>Surface Science</u>, 4(1966) p. 1-17.

18. R. Roscoe, "Viscosity Determination by the Oscillating-Vessel Method I: Theoretical Considerations," <u>Proc. Phys. Soc.</u>, 72(1958) p. 576-584.

19. R. Roscoe and W. Bainbridge, "Viscosity Determination by the Oscillating-Vessel Method II: The Viscosity of Water at 20C", <u>Proc. Phys. Soc.</u>, 72(1958) p. 585-595.

20. H.R. Thresh, "The Viscosity of Liquid Zinc by Oscillating a Cylindrical Vessel," <u>Trans. Met. Soc. AIME</u>, 233(1965) p. 79-88.

21. Tony Overfelt, Craig Matlock, Dixie Matlock, and Vivek Sahai, " Surface Tension and Viscosity of Superalloy 718," (Auburn University Space Power Institute, 1995).

22. <u>Smithells Metals Reference Book</u>, Eric A. Brandes, Ed. (London: Butterworths, 1983) p. 14-7.

# SESSION 8

## CONTACTS AND JOINTS

SESSION CHAIR
D. L. McElroy

# Effect of Interstitial Materials on Joint Thermal Conductance

## C. MADHUSUDANA and E. VILLANUEVA

### ABSTRACT

A comparison of the effect of some solid interstitial materials  on the thermal conductance of joints indicates that such materials could be used as either  thermal enhancers or as thermal isolators. A new correlation of experimental data on the effectiveness of foils shows that the effectiveness depends strongly on the relative thermal conductivities and the relative hardnesses of the foil and the base materials. The correlation also showed that the effectiveness is also influenced by the ratio of the foil thickness to the surface roughness. The effectiveness, however, appears to be essentially insensitive to the contact pressure variations.

## 1. INTRODUCTION

The actual solid-to-solid contact area, in most mechanical joints, is only a small fraction, less than 1% in most applications, of the apparent area (see Fig. 1). The voids between the actual contact spots are usually occupied by some conducting substance such as air. Other interstitial materials may be deliberately introduced to control, that is, either to enhance or lessen  the thermal contact conductance: examples include foils, powders, wire screens and epoxies. To enhance the conductance, the bare metal surfaces may be coated with metals of higher thermal conductivity by electroplating or vacuum deposition. Greases and other lubricants also provide alternative means of enhancing the thermal conductance of a joint.

In this context, the thermal contact conductance, h, of a joint is defined as the heat flux divided by the additional temperature drop at the interface. It has, therefore, the units of $W/(m^2 K)$.

In this paper, an overview of the effect of interstitial materials on joint conductance  will be first presented. This will be followed by a more detailed discussion of foils as control materials and a correlation will be derived for estimating their effectiveness.

Chakravarti Madhusudana and Eliseo Villanueva, School of Mechanical and Manufacturing Engineering, University of New South Wales, Sydney, 2052, Australia.

**SOLID 1**

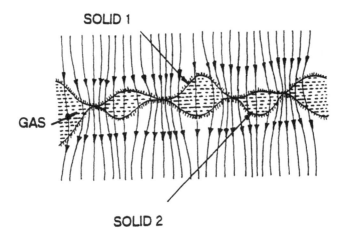

**GAS**

**SOLID 2**

Fig. 1 Schematic Diagram of Heat Flow Through a Joint

## 2. INTERSTITIAL CONTROL MATERIALS - AN OVERVIEW

Filler materials may be used either to increase or to decrease the thermal contact conductance and, therefore, provide a means of thermal control. Furthermore, the thermal condcutance of a joint with an interstitial material is less sensitive to mechanical loading and surface conditions and thus offers predictability of heat transfer behaviour. Comprehensive reviews of the general role of interstitial materials in controlling the contact conductance have been published, see for example, [1,2]. Some reviews have dealt with specific applications, e.g., thermal control material for space craft [3], or thermal enhancement for electronic systems [4].

When interstitial materials are used for the control of thermal conductance, it is desirable to have some means of comparing their effectiveness. A simple definition of effectiveness is

$$e = h_{cm} / h_{bj} \qquad (1)$$

in which h is the conductance and the subscripts cm and bj refer to the control material and the bare junction, respectively. With this definition, an effectiveness greater than 1 indicates enhancement of thermal conductance, whereas control materials yielding an effectiveness of less than 1 will be suitable materials for thermal isolation. It must be noted, however that other parameters such as the relative thickness of the interstitial material, may have a significant effect on the thermal conductance.

In this paper, because of the different ways they control the thermal contact conductance and their use in different applications, metal foils, wire screens, insulation sheets, and coatings will be considered separately.

## 2.1 METALLIC FOILS

In situations where the mechanical load on the joint has to be limited and/or when the joint is in vacuum, metal foils may be sandwiched between the bare metal surfaces. It would be then expected that the foil flows into the gaps between the surfaces thus increasing the actual contact area and enhancing the thermal conductance (Fig. 2).

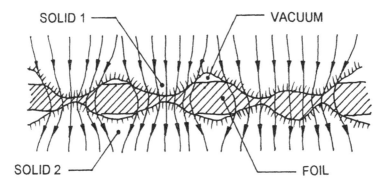

Fig. 2 Schematic Diagram of The Use of Foil to Enhance the Joint Conductance

One of the earliest experimental investigations into the effect of interfacial metal foils on thermal contact conductance was conducted by Fried and Costello [5]. Foils of lead and aluminium were used as interstitial materials between aluminium 2024-T3 surfaces. Cunnington's [6] experiments, on the other hand, compared the thermal conductance of the bare aluminium (6061-T4) junction with that of the joint into which an indium foil had been inserted. In each case, the bare junction thermal conductances were measured in vacuums of less than $10^{-4}$ Torr.

The results of these tests indicated that:

> The thermal contact conductance of the bare joint in vacuum is significantly increased by the insertion of a metallic foil.
> Softer materials such as lead or indium were more effective than aluminium as shims in enhancing the thermal contact conductance.

Cunnington's tests show that the foil was more effective in enhancing the thermal conductance of the rougher surfaces indicating that the optimum thickness depends on the roughness of surfaces. In these tests, it appears that the thickness of the foil used was closer to the optimum required for the rougher of the two pairs of surfaces.

In another early experimental investigation, undertaken by Koh and John [7], foils of copper, aluminium, lead and indium were separately tested as interstitial materials between a pair of mild steel surfaces. Although copper and aluminium have high thermal conductivities, it was found that the insertion of these foils actually reduced the thermal contact conductance, whereas lead and indium foils contributed towards an increase in thermal conductance. It was therefore confirmed that the foil softness was more significant in increasing the thermal conductance than the foil thermal conductivity. In another series of tests, the same authors found that there was an optimum thickness of foil which would result in maximum enhancement of joint thermal conductance. Apparently, thick foils are not pliable enough to fill the voids in the joint, while too thin a foil may not provide sufficient conduction material to fill the gaps in the interface. For the surface roughnesses range 4 μm to 5 μm rms encountered in the tests, the optimum thickness was found to be about 25 μm; at this thickness, the value of the thermal conductance was about three times that for the bare junction. It was also found, for the specimens tested, a foil thickness greater than 100 μm produced no improvement in the thermal contact conductance.

The enhancement of the thermal contact conductance by the use of aluminium foil was also reported by Sauer et al [8], and Sauer [2]. The bare junction was of 2024-T4 aluminium surfaces (surface roughness 0.77 μm) the contact pressure range was 1 to 4 MPa, the foil thickness was 640 μm. The conductance increased by a factor of approximately 3 over the complete range of pressures. However, the bare junction was tested in air rather than in vacuum.

Yovanovich [9] conducted a detailed experimental study of the effect of lead, tin, aluminium and copper foils on the thermal conductance of Armco iron joints. The contact pressure ranged from 2 to 10 MPa and the foil thicknesses from 10 to 500 μm. Unlike the tests of [5] and [6], the tests were conducted in atmospheric conditions. It was found that an optimum thickness, corresponding to minimum joint resistance, existed in all cases. The ratio of the optimum thickness of the foil to the surface roughness (rms) was found to be about 2 for lead and tin, between 0.48 and 0.58 for aluminium and 0.68 for copper. It was proposed that a foil material may be ranked by the ratio of its thermal conductivity to hardness; the larger the ratio the greater will be the increase in thermal contact conductance.

The investigations of O'Callaghan et al [10,11] led them to conclude that, in the absence of macroscopic constrictions, the optimum film thickness should be of approximately the same magnitude as the separation between mean planes of the solid surfaces. Thus, for nominally flat surfaces, a relatively thin foil would be appropriate in enhancing the conductance. For surfaces with macroscopic errors of form or when macroscopic thermal distortions are expected, then the thickness should larger in order to bridge any gaps that would otherwise be formed. This second conclusion confirms the results of Fried and Costello [5] who used a very thick lead foil to maximize the conductance of a pair of surfaces with large flatness deviation.

## 2.2 WIRE SCREENS

Fried and Costello [5] considered that the thermal contact conductance of poorly matching surfaces could be increased the introduction of a copper wire screen which would conform to the (large scale) surface irregularities and, therefore, would assure a large but finite number of contact spots. Their experiments, however, showed that the thermal conductance, in fact decreased due to the introduction of the wire cloth. It was concluded that the increase in thermal resistance caused by the reduced effective contact area more than offset any reductions in individual contact spot resistance. In subsequent literature, therefore, wire screens have also been considered as a means of decreasing the thermal contact conductance, i.e., as thermal isolation materials, especially for those applications, such as cryogenic storage vessels and supports, where mechanical strength is important.

The detailed experimental investigations of Gyorog [12] using stainless steel and titanium wire mesh screens also confirmed their use as thermal isolation materials. It was noted that coarse meshes, because of the fewer contact spots and the large wire diameters, gave rise to much lower thermal conductances than fine meshes did. He also found that the thermal isolation can be increased separating the screen from the aluminium surface with a stainless steel shim.

Experimental studies of stainless steel wire screens as the interstitial material were also carried out by Sauer et al [8]. It was found that the thermal contact conductance increased with increasing mesh number; the finer the mesh, the larger the number of contact points and hence, the larger the thermal conductance. These results thus confirm those of Gyorog.

A theoretical model for the prediction of the contact conductance of a joint containing a wire screen was proposed by Cividino and Yovanovich [13]. Among other things, this model assumed elastic deformation of smooth, clean wires and equal loading at all nodes. The theory was found to consistently overestimate the thermal conductance when a comparison was made with the measured values.

The thermal behaviour of copper wire gauzes inserted between stainless steel surfaces was the subject of an experimental study by O'Callaghan and Probert [11]. They found that the presence of the gauze increased the thermal conductance in vacuum but decreased it in air.

Further experimental and theoretical work on the effect of copper wire gauzes was carried out by Al-Astrabadi et al [14]. They observed that macroscopic constriction effects, due either to thermal distortions or to badly mating surfaces, may be reduced or even eliminated by the insertion of a copper gauze between the surfaces. The resulting reduction in thermal constriction resistance may more than compensate for the greater bulk thermal resistance because of the insert. Thus one might conclude that insertion of wire screens would decrease the thermal resistance if there are large scale surface irregularities; the resistance is likely to increase if the surfaces are flat and conforming. In any case, the direction of change in the thermal contact resistance would depend ultimately on the parent metal / wire material combination. Al-Astrabadi et al also noted that the method of weaving the screens results in the weft being a series of almost straight wires, all in one plane, with the warp interlaced. Contact, therefore, occurs only between the warp and the solid surfaces, i.e., only at every other crossing. As pointed out in [15], this would be one of the reasons why the theory of Cividino and Yovanovich, which assumed that contact occurred at every crossing, overestimated the thermal conductance.

## 2.3 SURFACE COATINGS

One way of reducing the thermal contact resistance of a joint would be to electroplate or coat the surfaces with a material of high thermal conductivity. Whenever coatings are contemplated, the mechanical strength, stability or durability with respect to operating conditions and time and adhesion to the parent surface are also important considerations. Plating would also result in a change in the contact geometry for a given load, since the plating material might have a different surface roughness and a different flow pressure than the bare material. For maximum benefit, both surfaces must be coated; when only one surface is coated, the whole constriction (or spreading) has to still take place in the other uncoated material.

Antonetti and Yovanovich [16] derived the following relationship between the thermal contact conductance and the contact pressure for nominally flat, coated surfaces:

$$h\sigma / (k' \tan \theta ) = 1.25 (P / H')^{0.95} \qquad (2)$$

In this expression k' is an effective thermal conductivity. This is determined by a thermal analysis, taking into account the different thermal conductivities of the coating materials and the substrates and the amount of constriction. H' is the effective hardness determined from a series of microhardness tests. $\sigma$ is the rms surface roughness and $\tan \theta$ is the mean absolute surface slope. The surface slope, measured in radians, is simply the inclination of the surface asperity with respect to the mean (horizontal) plane.

An experimental study of the effect of metallic coatings on the thermal contact conductance of turned surfaces was reported by Kang et al [17]. Tests using coatings of tin, indium and lead, four different thicknesses in each case, on bare aluminium 6061-T6 surfaces indicated that, for each coating material/surface combination, there existed an optimum thickness which yielded the maximum thermal conductance. The optimum thickness was found to be in the range 0.2-0.5 µm for tin, 2 µm for indium and 1.5-2.5 µm for lead. The surface roughness ranged from 0.368 µm to 0.856 µm rms, but in most tests it was about 0.7 µm rms.

It may be recalled that the correlation proposed in [14] was applicable only to flat surfaces and also required experimental and theoretical determination of the effective hardness and effective thermal conductivity. In order to derive a general, usable correlation, Lambert and Fletcher [18] analyzed 654 individual data points obtained from tests on 99 joints by nine different investigators and obtained the relation:

$$(h\sigma / k' \tan \theta) = 0.00977 (P / H')^{0.520} \qquad (3)$$

Since engineering surfaces are typically non-flat and wavy, the data for optically flat surfaces were excluded in a subsequent analysis. The remaining 579 data points, from 85 joints, yielded the correlation:

$$(h\sigma / k' \tan \theta ) = 0.00503 (P / H')^{0.455} \qquad (4)$$

In cases where the combined mean slope, $\tan \theta$, of the surfaces was not furnished in the source investigation, it was estimated from the correlation:

$$\tan \theta = \sqrt{(\sigma / 100)} \qquad (5)$$

In either case, it could be seen that the power to which the relative pressure term has to be raised is significantly smaller than the slope of 0.95 obtained for flat surfaces.

This indicates that non-flat, or wavy surfaces are less sensitive to contact pressure variation than are flat surfaces.

## 2.4 INSULATING INTERSTITIAL MATERIALS

In applications such as isolation of spacecraft equipment, cryogenic storage tanks and, in general, wherever strong insulating supports are required, it is necessary to increase the thermal contact resistance. The use of wire screens as a means of increasing the thermal contact resistance has already been noted. Another method of achieving thermal isolation is by the introduction of low thermal conductivity interstitial materials in the interface.

The works of Fletcher et al [19], Smuda and Gyorog [20], Gyorog [12], Fletcher and Miller [21] and Fletcher et al [22] are just some examples of the experimental investigations into the effect of thermal isolation materials on the thermal contact resistance. Carbon paper, ceramic paper, WR-X-AQ felt and T-30LR laminate were found to be good materials for increasing the thermal resistance. Typically, the thermal contact conductance values for aluminium 2024-T4 junctions reduced from 1000 to between 1 and 10 BTU /hr.ft$^2$.$^{\circ}$F ( from 5680 to between 5.68 and 56.8 W/m$^2$ K) in the pressure range 100 to 300 psi ( 0.69 to 2.07 MPa) and the temperature range - 100 to 200 $^{\circ}$F ( -73 to 93 C). It was noted in ref [19] that Teflon, because of its comparatively high thermal conductivity, produced a relatively small reduction in the thermal contact conductance. Gyorog [12] observed that dusting of surfaces with powders such as manganese oxide or rutile produced noticeable reductions in the conductance but the results were somewhat unpredictable and difficult to control.

The results of [21] for gasket materials for spacecraft joints indicated that elastomers were suitable for thermal control, and produce an order of magnitude (or better) reduction in thermal contact conductance. In particular, the silicone elastomers could be used over a large temperature range: - 130 to 500 $^{\circ}$F (- 90 to 260 C). By contrast, polyethylene materials, although effective in providing isolation, could be used only over a narrow temperature range, typically - 60 to 150 $^{\circ}$F (- 51 to 66 C)

## 2.5 SUMMARY

Table1 provides a comparison of typical values of the effectiveness that may be expected by the use of the interstitial materials discussed in this section. In each case the substrate material is an aluminium alloy, such as 2024-T3, and the results refer to tests conducted in vacuum.

TABLE 1 THE EFFECTIVENESS OF SOME INTERSTITIAL MATERIALS
ON AN ALUMINIUM ALLOY IN VACUUM

| Foil | Conductivity (W/mK) | Hardness (MPa) | Effectiveness |
|---|---|---|---|
| Indium | 24 | 9 | 5 to 8 |
| Lead | 35 | 39 | 2.4 to 4 |
| Copper | 384 | 785 | 1.3 to 2.25 |
| Tin | 60 | 52 | 3 to 5 |
| Aluminium | 204 | 265 | 1.5 to 2.75 |
| **Wire Screen** | **Conductivity (W/mK)** | **Mesh Number** | **Effectiveness** |
| Copper | 384 | 50 | 0.87 |
| **Coating** | **Conductivity (W/mK)** | **Hardness (MPa)** | **Effectiveness** |
| Tin | 60 | 52 | 0.8 to 1.7 |
| Silver | 425 | 390 | 0.8 to 2 |
| Indium | 82 | 9 | 1.9 to 7 |
| **Insulators** | **Conductivity (W/mK)** | **Density (kg / m$^3$)** | **Effectiveness** |
| Mica | 0.363 | 208 | 0.04 to 0.05 |
| Teflon | 2.34 | 160 | 0.15 to 0.25 |
| WRP-AX-AQ Felt | 0.069 | 288 | 0.0006 to 0.0007 |

## 3. CORRELATION FOR THE EFFECTIVENESS OF FOILS

In published literature, there have been very few correlations which relate the effectiveness of foils to the system parameters. One of these, proposed by Couedel et al [23] considers only the hardnesses and the thermal conductivities. As discussed earlier, the thickness of the foil and the contact pressure are likely to affect the thermal contact conductance. These factors were taken into account in the present work. In order to focus the attention on the enhancement of the solid spot conductance of nominally flat surfaces, only tests done in vacuum on surfaces with negligible flatness deviation were considered. These restrictions apply to the correlation proposed in [23] also. The effect of mean junction temperature on the thermal properties, however, was taken into account. The present correlation is based on the experimental results of Peterson and Fletcher [24] for copper, aluminium, tin and lead foils and the results of [6] for indium foils. The ranges of parameters used are shown in Table 2. The bare material is aluminium.

TABLE 2 RANGE OF PARAMETERS USED IN CORRELATION

| $k_f / k_b$ | $t / \sigma$ | $H_f / H_b$ | $P / H_b$ |
|---|---|---|---|
| 0.123 to 2.29 | 20.8 to 75.7 | 0.0122 to 0.721 | 0.00051 to 0.0703 |

In Table 2, k is the thermal conductivity, t is the foil thickness, $\sigma$ is the surface roughness, H is the hardness, P is the contact pressure and the subscripts f and b refer to the foil and the bare materials, respectively.

A multilog regression analysis yielded the following result:

$$(h_f / h_b) = 0.236 \, (k_f / k_b)^{1.88} \, (t / \sigma)^{-0.268} \, (H_f / H_b)^{-1.821} \, (P / H_b)^{-0.043} \qquad (6)$$

with a correlation coefficient of 0.943

This correlation is shown plotted in Fig. 3. It can be seen that the correlation fits the experimental data reasonably well except for two stray points. From Eq (10), it is evident that the relative conductivity and the relative hardness strongly affect the effectiveness. It may also be seen that the relative thickness also has a significant effect on the effectiveness. The effectiveness is seen to decrease with foil thickness. A probable reason for this is that in all of the data considered, the thickness was considerably greater than the optimum thickness. The effectiveness, however, seems to be only a weak function of the contact pressure. This is attributable to the fact that both $h_f$ and $h_b$ increase with pressure. This means that the same degree of enhancement may be expected over the range of contact pressures considered. It may be noted, however, that the above discussion applies to the limited amount of available data. The conclusions need to be confirmed when a large amount of relevant experimental data becomes available.

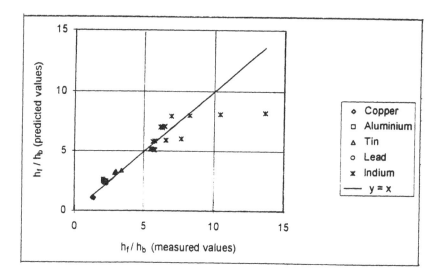

Fig. 3 Correlation for the Effectiveness of Metallic Foils

## 4. CONCLUSIONS AND RECOMMENDATIONS

Interstitial materials may be used to control the thermal contact resistance of pressed joints. Solid interstitial materials may be used to either enhance or diminish the thermal contact conductance. Thermally conductive foils and coatings may be used to enhance the joint thermal conductance by upto a factor of 10. Insulating materials may increase the thermal resistance by a factor of 1000 or greater.

The effectiveness of foils not only depends on their relative thermal conductivity and relative softness, but also on their thickness relative to the surface roughness of the joint surfaces. The effectiveness appears to be essentially insensitive to the contact pressure.

Further experimental work and analysis should deal with thinner foils, the effect of flatness deviations, and the effect of the interstitial gases.

## REFERENCES

1. Snaith, B., P.W. O'Callaghan, and S.D. Probert, 1984 "Interstitial Materials for Controlling Thermal Conductance across Pressed Metallic Contacts", *Applied Energy,* 16: 175-191.

2. Sauer, H.J., Jr., 1992. "Comparative Enhancement Of Thermal Contact Conductance Of Various Classes Of Interstitial Materials", NSF/DITAC Workshop, Melbourne, 103-115.

3. Fletcher, L. S., 1972. "A Review Of Thermal Control Materials For Metallic Junctions" *J Spacecraft and Rockets,* 849-950.

4. Fletcher, L.S., 1990. "A Review Of Thermal Enhancement Techniques for Electronic Systems", *IEEE Trans, Components, Hybrids and Manufacturing Technology,* 13:1012-1021

5. Fried, E. and F.A. Costello, 1962. "Interface Thermal Contact Resistance Problem In Space Vehicles", *ARS Journal,* 237-243

6. Cunnington, G.R., Jr., 1964. "Thermal Conductance Of Filled Aluminium and Magnesium Joints In A Vacuum Environment", ASME Paper 64-WA/HT-40.

7. Koh, B. and J.E. John, 1965. "The Effect Of Foils On Thermal Contact Resistance", Paper 65-HT-44, ASME-AIChE Heat Transfer Conference.

8. Sauer, H.J., Jr., C.R. Remington, W.E. Stewart, Jr. and J.T. Lin, 1971. "Thermal Contact Conductance With Several Interstitial Materials", Internat'l Conference on Thermal Conductivity (11): 22-23 (unpublished).

9. Yovanovich, M.M., 1972. "Effect Of Foils Upon Joint Resistance: Evidence Of Optimum Thickness", Prog In Aeronautics and Astronautics, 31: p 227-245.

10. O'Callaghan, P.W., B. Snaith, and S.D. Probert, 1983. "Prediction Of Interfacial Filler Thickness For Minimum Thermal Contact Resistance", *AIAA Journal,* 21 ( 9):1325-1329.

11. O'Callaghan, P.W. and S.D. Probert, 1988. "Reducing The Thermal Resistance of A Pressed Contact", *Applied Energy,* 30: 53-60.

12. Gyorog D.A., 1970. "Investigation Of Thermal Isolation Materials For Contacting Surfaces", *Prog In Astronautics and Aeronautics,* 24: 310-336

13. Cividino, S., and M.M. Yovanovich, 1975. "A Model For Predicting The Joint Conductance Of A Woven Wire Screen Contacting Two Solids", *Prog In Astro and Aero,* 39:111-128.

14. Al-Astrabadi, F.R., P.W. O'Callaghan, S.D. Probert, and A.M. Jones, 1977 "Thermal Resistances Resulting From Commonly Used Inserts Between Stainless Steel Static Bearing Surfaces", *Wear ,* 40(3): 339-350.

15. Madhusudana, C.V. and L.S. Fletcher, 1986. "Contact Heat Transfer - The Last Decade" *AIAA Journal*, 24 (3): 510-523.

16. Antonetti, V.W. and M.M. Yovanovich, 1985. "Enhancement Of Thermal Contact Conductance By Metallic Coatings: Theory and Experiment", *Trans ASME, Journal Of Heat Transfer*, 107: 513-519.

17. Kang, T.K., G.P. Peterson, and L.S. Fletcher, 1990. "Effect Of Metallic Coatings On Thermal Contact Conductance Of Turned Surfaces", *Trans ASME, J Heat Transfer*, 112: 864-871

18. Lambert, M.A. and L.S. Fletcher, 1993. "A Correlation For The Thermal Contact Conductance Of Metallic Coated Metals", Paper AIAA 93-2778.

19. Fletcher, L.S., P.A. Smuda and D.A. Gyorog, 1969. "Thermal Contact Resistance Of Selected Low-Conductance Interstitial Materials", *AIAA Journal*, 7( 7): 1302-1309.

20. Smuda, P.A. and D.A. Gyorog, 1969. "Thermal Isolation With Low Conductance Interstitial Materials Under Compressive Loads", AIAA Paper 69-25.

21. Fletcher, L.S. and R.G. Miller, 1973. "Thermal Conductance Of Gasket Materials For Spacecraft Joints", *Prog In Astronautics and Aeronautics*, 35: 335-349.

22. Fletcher, L.S., M.R. Cerza, and R.L. Boysen, 1976. "Thermal Conductance and Thermal Conductivity Of Selected Polyethylene Materials", AIAA Paper 75-187.

23. Couedel, D., L.S. Fletcher, and G.P. Peterson, 1994. "A Correlation for the Thermal Contact Conductance of Interstitial Metallic Foils", Proc 10th Int Heat Transfer Conference, 6:337-342.

24. Peterson, G.P. and L.S. Fletcher, 1988. "Thermal Contact Conductance In The Presence Of Thin Metal Foils", AIAA Paper 88-0466.

# Analytical Models for Solid-Solid Contact in Thermal Transient States

A. DEGIOVANNI, S. ANDRÉ and D. MAILLET

## ABSTRACT

In this paper, we present an analytical model for solid-solid contact in thermal transient states. It has already been shown that a thermal contact in steady state can be characterized, for most cases, by an association of five resistances. These correspond to the resistance of the interstitial fluid, the two constriction resistances and the two asperity resistances. In transient regime, it is shown that the preceding resistances are replaced by transfer matrices. The solid-solid contact is modeled then by an association of five transfer matrices.

The "two-pair terminal network" (quadrupole) concept is first presented in the case of a wall under transient thermal stress. The form of the transfer matrix corresponding to the constriction of flux lines is derived and given in both the case of a semi-infinite medium and of a finite medium.

Finally, these results have been used to model two types of solid-solid contact: the plane contact of two surfaces of large dimensions, and the thermal contact between a sample and a thermocouple.

## NOMENCLATURE

a : thermal diffusivity
A,B,C,D: quadrupole coefficients
e : thickness
Ji : Bessel functions of the first kind
p: Laplace variable
q : heat flux density $(J.m^{-2})$
r : space variable in cylindrical coordinates
R: resistance
S: cross-sectional area

T : temperature
z : space variable in cylindrical coordinates
Z: impedance
$\varphi$ : transformed heat flux
$\lambda$ : thermal conductivity
$\theta$: transformed temperature in space

Upperscripts:
-- : refers to a variable in Laplace domain

Alain Degiovanni, Stéphane André, Denis Maillet, LEMTA URA CNRS 875 - INPL - Université de Nancy I, 2 av. de la Forêt de Haye, BP 160, 54504 Vandoeuvre-Lés-Nancy Cédex, France.

## INTRODUCTION

In steady-state conditions, a plane layer reduces to a thermal resistance. In the same manner, in transient state conditions, the layer may be modeled by a thermal quadrupole, which links Laplace transforms of input (subscript 1) and output (subscript 2), temperatures ($\overline{T}$) and heat fluxes ($\overline{q}$) (see figure 1).

We can use, for example, the matrix notation:

$$\begin{bmatrix} \overline{T}_1 \\ \overline{q}_1 \end{bmatrix} = \begin{bmatrix} A & B \\ C & D \end{bmatrix} \begin{bmatrix} \overline{T}_2 \\ \overline{q}_2 \end{bmatrix} \tag{1}$$

where $\begin{bmatrix} A & B \\ C & D \end{bmatrix}$ denotes the inverse transfer matrix and $A = D = \cosh(k.e)$ ,

$C = \lambda k S . \sinh(k.e)$ ; $B = (AD-1)/C$ where $k^2 = p/a$ (with p being the Laplace variable).

In the case of a semi-infinite wall ($e \rightarrow \infty$), one obtains the wall impedance writing that $\overline{q}_2 = 0$. Equation (1) leads to $\overline{T}_1 = Z_b . \overline{q}_1$ with $Z_b = \dfrac{A}{C}$ for $e \rightarrow \infty$, that is

$$Z_b = \frac{1}{\lambda k S} \tag{2}$$

## I-CONSTRICTION IMPEDANCE FOR A SEMI-INFINITE MEDIUM

The problem is classically formulated as shown by figure 2 below (see [1] for example).

The physical description of this heat transfer situation is described by the following set of equations.

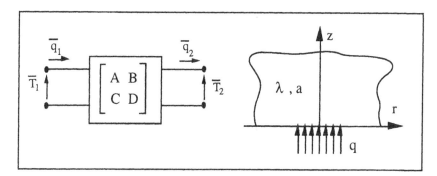

Figure 1 : Quadrupole representation.     Figure 2 : The semi-infinite model.

$$\begin{cases} \dfrac{1}{r}\dfrac{\partial}{\partial r}\left(r\dfrac{\partial T}{\partial r}\right)+\dfrac{\partial^2 T}{\partial z^2}=\dfrac{1}{a}\dfrac{\partial T}{\partial t} \\[2mm] z=0 \quad\ -\lambda\dfrac{\partial T}{\partial z}=q(r,t) \\[2mm] z\to\infty \quad\ T=0 \\[2mm] r\to\infty \quad\ T=0 \\[2mm] r=0 \quad\ \dfrac{\partial T}{\partial r}=0 \\[2mm] t=0 \quad\ T=0 \end{cases} \qquad (3)$$

If we apply the Hankel transform in r and the Laplace transform in time t, the transformed temperature is

$$\theta=\int\limits_0^\infty\int\limits_0^\infty r.J_0(\alpha.r).\exp(-pt).T(r,z,t).dr.dt$$

Equation (3) becomes :

$$\dfrac{\partial^2\theta}{\partial z^2}-\left(\alpha^2+\dfrac{p}{a}\right).\theta=0$$

with

$$z=0 \qquad -\lambda\dfrac{\partial\theta}{\partial z}=\varphi(\alpha,p)$$

$$z\to\infty \qquad \theta=0$$

The solution falls immediately with the impedance form:

$$\theta(\alpha,p,z)=\varphi(\alpha,p)\dfrac{\exp(-\gamma z)}{\gamma\lambda} \qquad (4)$$

with $\gamma^2=\alpha^2+\dfrac{p}{a}$

Equation (4) is formally interesting but one would generally have more interest in "non-local" effects as for example, the average temperature at the contact surface. If q(r,t) is assumed to be equal to 0 for r greater than $r_0$, a relation can be found between the total heat flux crossing surface $\pi.r_0^2$ and the mean temperature upon this surface at z = 0.

The inverse Hankel transform allows the determination of the temperature in the Laplace domain

$$\overline{T}(p,r,z)=\int\limits_0^\infty\dfrac{\alpha\varphi}{\gamma\lambda}J_0(\alpha.r).\exp(-\gamma z)d\alpha \qquad (5)$$

and the mean temperature and heat flux at z = 0

$$\overline{T}_m = \frac{1}{r_0^2} \int_0^{r_0} 2r\overline{T}dr \tag{6}$$

$$\overline{q}_m = \int_0^{r_0} 2\pi r \overline{q}(r,p)dr \tag{7}$$

The constriction impedance $Z_c(p)$ can thus be obtained:

$$Z_c(p) = \frac{\overline{T}_m}{\overline{q}_m} \tag{8}$$

In the classical case where $q(r,t) = q_0(t)$ for $0 < r < r_0$ and $q(r,t) = 0$ for $r > r_0$ then

$$\varphi(\alpha,p) = \varphi_0 \frac{r_0}{\alpha} J_1(\alpha.r_0)$$

with $\varphi_0$, Laplace transform of $q_0(t)$.

Equation (6) becomes:

$$\overline{T}_m = \frac{2\varphi_0}{\lambda} \int_0^\infty \frac{J_1^2(\alpha.r_0)}{\alpha\gamma} d\alpha$$

and the constriction impedance can be written as

$$Z_{c_-} = \frac{\overline{T}_m}{\pi r_0^2 \varphi_0} = \frac{2}{\pi\lambda r_0} \int_0^\infty \frac{J_1^2(\varepsilon)}{\varepsilon\sqrt{\varepsilon^2 + p^*}} d\varepsilon \qquad \text{with } p^* = \frac{pr_0^2}{a} \tag{9}$$

Practically, a very good approximate of integral (9) can be found, that can be represented by the constriction resistance (obtained in the steady-state regime) connected in parallel with the impedance of a rod of radius $r_0$ (see figure 3) and obtained as follows:

for $p^* \to 0$ $\qquad Z_{c_-} \to R_{c_-} = \dfrac{8}{3\pi^2 \lambda r_0}$

for $p^* \to \infty$ $\qquad Z_{c_-} \to Z_b = \dfrac{1}{\sqrt{\lambda\rho C}\sqrt{p}\ \pi r_0^2}$ $\qquad$ and the approximation is

$$Z_{c_-} \approx \frac{1}{\dfrac{1}{R_{c_-}} + \dfrac{1}{Z_b}} = \frac{8}{3\pi^2 \lambda r_0 \left(1 + \dfrac{8}{3\pi}\sqrt{\dfrac{pr_0^2}{a}}\right)} \tag{10}$$

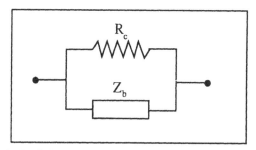

Figure 3 : Resistive schema of the thermal contact in steady-state.

Figure 4 compares Equation (9) and (10) in Bode's diagram. One can notice that the impedance obtained in (4) is independant of the boundary conditions at $z = 0$ $(\exp(-\gamma z)/\gamma\lambda)$. On the other hand, impedance $Z_{c_-}$ depends on the spatial distribution of the heat flux density on the circular surface of radius $r_0$.

Nevertheless, for almost all applications, the approximation of a constant heat flux density is fair. The other limiting case is obtained for a flux distribution that corresponds to a uniform temperature (see [2] and [3]), the expression of the impedance becoming:

$$Z_{c_-}^* = \frac{T_0}{\overline{q}_m} = \frac{1}{4\lambda r_0 \int\limits_0^\infty \frac{\sqrt{\varepsilon^2 + p^*}}{\varepsilon^2} J_1(\varepsilon)\sin\varepsilon d\varepsilon} \tag{11}$$

and in a similar manner as before $R_{c_-}^* = \dfrac{1}{4\lambda r_0}$ $\qquad$ $(Z_{c_-}^*$ for $p^* \to 0)$

and $\qquad\qquad$ $Z_b = \dfrac{1}{\sqrt{\lambda\rho C}\sqrt{p}\ \pi r_0^2}$ $\qquad$ $(Z_{c_-}^*$ for $p^* \to \infty)$

thus leading to $\qquad$ $Z_{c_-}^* \approx \dfrac{1}{4\lambda r_0\left(1 + \dfrac{\pi}{4}\sqrt{\dfrac{pr_0^2}{a}}\right)}$ $\tag{12}$

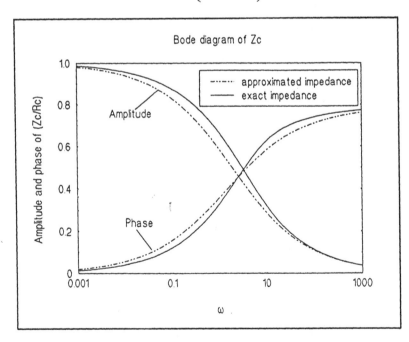

Figure 4 : Bode's diagram of the approached and exact constriction impedances.

At short times, this impedance is identical to the one obtained for a uniform heat flux density (equation (9)).

## II-THE FINITE MEDIUM

This problem is illustrated in figure 5: A cylinder of circular base (radius R) is perturbed on one face. If the boundary condition at r = R is prescribed according to the following general form:

$$\frac{\partial T}{\partial r} + k_0 T = 0 \qquad \text{at } r = R$$

then the problem is strictly equivalent to the previous one as far as the integral transform in r is changed, according to [4], and becomes in that case the Hankel transform:

$$v(\alpha_n) = \int_0^R r J_0(\alpha_n r) . u(r) . dr$$

with $\alpha_n$ solution of $\alpha . J_1(\alpha R) + k_0 . J_0(\alpha R) = 0$ where $v(\alpha_n)$ is the transform of u(r) and the inverse transform is given by:

$$u(r) = \sum_{n=0}^{\infty} \frac{J_0(\alpha_n r) . v(\alpha_n)}{N(\alpha_n)} \qquad \text{with } N(\alpha_n) = \int_0^R r J_0^2(\alpha_n r) . dr$$

Therefore, the transformed temperature is

$$\theta = \int_0^R \int_0^\infty r . J_0(\alpha_n . r) . \exp(-pt) . T(r, z, t) . dr . dt$$

and applying the transform on equation (3) leads to:

$$\frac{\partial^2 \theta}{\partial z^2} - \left(\alpha^2 + \frac{p}{a}\right) . \theta = 0$$

The solution is obtained directly using the transfer matrix:

$$\begin{bmatrix} \theta(\alpha, p, 0) \\ \varphi(\alpha, p, 0) \end{bmatrix} = \begin{bmatrix} A(\alpha, p) & B(\alpha, p) \\ C(\alpha, p) & D(\alpha, p) \end{bmatrix} \begin{bmatrix} \theta(\alpha, p, e) \\ \varphi(\alpha, p, e) \end{bmatrix} \qquad (13)$$

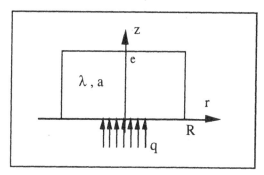

Figure 5 : The finite model.

$$A = D = \cosh(\gamma e)$$

with $\quad B = \sinh(\gamma e)/\lambda\gamma \quad$ with $\quad \gamma^2 = \alpha^2 + p/a \quad$ and with $\quad \varphi = -\lambda\dfrac{\partial\theta}{\partial z}$

$$C = \lambda\gamma\sinh(\gamma e)$$

If one is interested only in the macroscopic effect, the solution (13) is not convenient. Let us study the following case:

$$\text{at } z = 0 \quad -\lambda\frac{\partial T}{\partial z} = q(r,t) \quad 0 < r < r_0$$
$$= 0 \quad R < r < r_0$$

at $z = e \quad$ a uniform boundary condition (independant of r)

$$\text{at } r = R \quad -\lambda\frac{\partial T}{\partial r} = 0$$

The following macroscopic variables are then introduced:

$$\overline{T}_m(0) = \frac{1}{r_0^2}\int_0^{r_0} 2r\,\overline{T}(0)\,dr$$

$$\overline{T}_m(e) = \frac{1}{R^2}\int_0^R 2r\,\overline{T}(e)\,dr$$

$$\overline{q}_m(0) = \int_0^{r_0} 2\pi r\,\overline{q}(0)\,dr$$

$$\overline{q}_m(e) = \int_0^R 2\pi r\,\overline{q}(e)\,dr$$

with $\quad \begin{array}{l}\overline{T}(0) \text{ and } \overline{T}(e) \text{ the Laplace transforms of the temperature at } z = 0 \text{ and } z = e. \\[4pt] \overline{q}(0) \text{ and } \overline{q}(e) \text{ the Laplace transforms of the heat flux density at } z = 0 \text{ and } z = e \left(\overline{q} = -\lambda\dfrac{\partial\overline{T}}{\partial z}\right).\end{array}$

A transfer matrix similar to (13) cannot be rigorously found; on one hand, this matrix depends upon the flux density distribution at $z = 0$ and on the other hand upon the uniform condition at $z = e$ (prescribed temperature, prescribed flux, or heat exchange coefficient).

Concerning the condition at $z = 0$, It has been shown previously (section I) that the discrepancy between the borderline cases is maximum in the steady state regime and less than 8% (from 1/4 to $8/3\pi^2$). We then choose $q(r,t) = q_0(t)$.

Concerning the condition at $z = e$, the difference between uniform temperature or uniform heat flux conditions can be very large when e is small compared to R (see [5]); in that case, the discrepancy is still maximum for the steady-state regime. It has been shown that for $e/R > 1$, the solutions are identical within 1%. Therefore, it is possible to write [6]:

$$\begin{bmatrix}\overline{T}_m(0) \\ \overline{q}_m(0)\end{bmatrix} = \begin{bmatrix} A(p) & B(p) \\ C(p) & D(p)\end{bmatrix}\begin{bmatrix}\overline{T}_m(e) \\ \overline{q}_m(e)\end{bmatrix} \tag{14}$$

with $\quad \begin{array}{l} A(p) = \cosh(ke) \\[4pt] B(p) = \sinh(ke)/\pi R^2 k\lambda + \cosh(ke)\displaystyle\sum_{n=1}^{\infty} F_n\end{array}$

$$\left|\begin{array}{l} C(p) = \pi R^2 k\lambda \ \sinh(ke) \\[2mm] D(p) = \cosh(ke) + \pi R^2 k\lambda \ \sinh(ke) \sum_{n=1}^{\infty} F_n \end{array}\right.$$

with $k^2 = p/a$ and

$$F_n = \frac{4J_1^2(\alpha_n r_0)}{\pi\lambda\alpha_n^2 \ R^2 \gamma_n J_0^2(\alpha_n R)r_0^2} \tag{15}$$

The transfer matrix (14) can be expressed as a product of two matrices; the first one represents a "wall" of thickness e (equation (1)), and the second characterizes the "constriction" of the current lines. This is mathematically expressed as follows:

$$\begin{bmatrix} A(p) & B(p) \\ C(p) & D(p) \end{bmatrix} = \begin{bmatrix} A_1 & B_1 \\ C_1 & D_1 \end{bmatrix}\begin{bmatrix} A_2 & B_2 \\ C_2 & D_2 \end{bmatrix}$$

$$A_1 = D_1 = \cosh(ke)$$

with $\quad B_1 = \sinh(ke)/\pi R^2 k\lambda \qquad (16) \qquad$ and $\quad B_2 = \sum_{n=1}^{\infty} F_n = Z_c \qquad (17)$

$$C_1 = \pi R^2 k\lambda \ \sinh(ke) \qquad\qquad\qquad\qquad C_2 = 0$$

$$A_2 = D_2 = 1$$

One can verify numerically that $\sum_{n=1}^{\infty} F_n \to Z_{c_\infty}$ when $r_0/R \to 0$.

It is obvious that the problem is identical for every kind of geometry, for example in the case of cylinders having a rectangular base [2]; only the expression of $Z_c$ is changed.

## III-APPLICATION TO THE MODELING OF A SOLID-SOLID CONTACT

The solid-solid contact is modeled by a great number of cylindrical unit cells (see fig. 6).

### III-1 The simplified model:

If the heat flux that is transfered by the interstitial fluid is neglected, the modeling of the problem is greatly simplified if results of section II are used. We can actually write

$$\begin{bmatrix} \overline{T}_m(-L_1) \\ \overline{q}_m(-L_1) \end{bmatrix} = \begin{bmatrix} A_1 & B_1 \\ C_1 & A_1 \end{bmatrix}\begin{bmatrix} 1 & Z_{c_1} \\ 0 & 1 \end{bmatrix}\begin{bmatrix} \overline{T}_m(-\delta_1) \\ \overline{q}_m(-\delta_1) \end{bmatrix}$$

$$\begin{bmatrix} \overline{T}_m(-\delta_1) \\ \overline{q}_m(-\delta_1) \end{bmatrix} = \begin{bmatrix} A_{1'} & B_{1'} \\ C_{1'} & A_{1'} \end{bmatrix}\begin{bmatrix} \overline{T}_m(0) \\ \overline{q}_m(0) \end{bmatrix}$$

$$\begin{bmatrix} \overline{T}_m(0) \\ \overline{q}_m(0) \end{bmatrix} = \begin{bmatrix} A_{2'} & B_{2'} \\ C_{2'} & A_{2'} \end{bmatrix} \begin{bmatrix} \overline{T}_m(\delta_2) \\ \overline{q}_m(\delta_2) \end{bmatrix}$$

$$\begin{bmatrix} \overline{T}_m(\delta_2) \\ \overline{q}_m(\delta_2) \end{bmatrix} = \begin{bmatrix} 1 & Z_{c_2} \\ 0 & 1 \end{bmatrix} \begin{bmatrix} A_2 & B_2 \\ C_2 & A_2 \end{bmatrix} \begin{bmatrix} \overline{T}_m(L_2) \\ \overline{q}_m(L_2) \end{bmatrix}$$

where the $X_i$'s, $X_{i'}$'s stands for the coefficients of a "wall" quadrupole given by equations (16) and the $Z_{ci}$'s are the impedances given by equations (17).

If we associate the chain of quadrupoles, we have finally

$$\begin{bmatrix} \overline{T}_m(-L_1) \\ \overline{q}_m(-L_1) \end{bmatrix} = \begin{bmatrix} \mathcal{A} & \mathcal{B} \\ C & \mathcal{D} \end{bmatrix} \begin{bmatrix} \overline{T}_m(L_2) \\ \overline{q}_m(L_2) \end{bmatrix} \qquad (18)$$

which is represented with the scheme of figure 7.

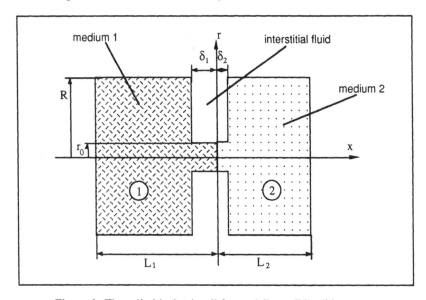

Figure 6 : The cylindrical unit cell for modeling solid-solid contact.

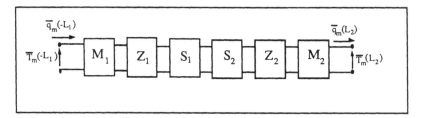

Figure 7 : Quadrupole schema of the unit cell under transient regime.

### III-2 The complete model:

If the heat flux transfered through the interstitial fluid is taken into account, it becomes very difficult to directly obtain the transfer matrix of the system in the same way as for equation (18). Therefore, another step has been employed.

This problem has already been examined under steady-state conditions [2]. Beginning with an exact model, a simplified scheme, which is valid for realistic contact conditions, has been built and is shown in figure 8. The only limitations are that the ratio of the cross section of the contact spot (s) to the cross section of the unit cell (S) is small compared to 1, and that the thermal conductivity of the fluid $\lambda_f$ is lower or equal to the solid conductivities $\lambda_1$ and $\lambda_2$. This simplified model shows that the fluid resistance $r_f$ is in parallel with the resistance of the contact spot $(r_{S_1} + r_{S_2})$ which is in series with the constriction resistance $\left(r_{C_1} + r_{C_2}\right)$.

The expressions of the resistances are:

- unperturbed media 1 and 2:   $r_1 = \dfrac{L_1}{\lambda_1 S}$   and   $r_2 = \dfrac{L_2}{\lambda_2 S}$

- fluid resistance:   $r_f = \dfrac{\delta_1 + \delta_2}{\lambda_f S}$

- contact spot:   $r_{S_1} = \dfrac{\delta_1}{\lambda_1 s}$   and   $r_{S_2} = \dfrac{\delta_2}{\lambda_2 s}$

- constriction resistance:   $r_{C_1} = \dfrac{K}{\lambda_1 \sqrt{s}}$   and   $r_{C_2} = \dfrac{K}{\lambda_2 \sqrt{s}}$

where K is a function of the geometry of the unit cell (especially of the ratio s/S) and of the shape of the contact spot. For example, in the case where s/S --> 0 and for a cell with circular base, K = 0.48 (see equation (10)).

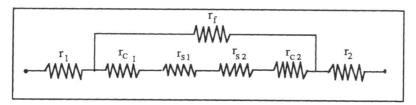

Figure 8 : Resistive schema of the unit cell according to the complete model.

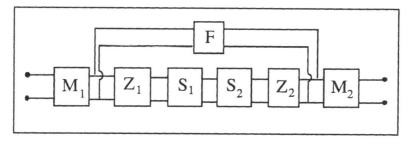

Figure 9 : Quadrupole schema of the unit cell according to the complete model.

Using an analogy with the steady-state approach, the sketch shown in figure 9 is proposed for the transient state:
- the "wall" resistances are replaced by "wall" quadrupoles
- the "constriction" resistances are replaced by "constriction" quadrupoles.
The quadrupoles $M_1$, $M_2$, $Z_1$, $Z_2$, $S_1$, $S_2$ are identical to those of figure 7

and F is given by $\quad F = \begin{bmatrix} A & B \\ C & D \end{bmatrix} \quad$ with $\quad$ A,B,C,D being the wall quadrupole

coefficients based on the thermophysical properties of the fluid.

### IV-APPLICATION TO THE SAMPLE-THERMOCOUPLE CONTACT

The previous result is now applied to calculate the error occuring for surface temperature measurements using a thermocouple.
The problem is illustrated on figure 10. A homogeneous sample, cylindrical in shape (radius R) and of thickness e, is considered. A thermocouple is applied on this sample and will be modeled by an infinite fin of radius $r_0$ . Thesample-thermocouple contact introduces a contact resistance $R_c$. A heat flux is applied on face z = 0 of the sample (dirac distribution: $\bar{q} = 1$); the problem consists in evaluating the error induced by the thermocouple on face z = e. The equivalent scheme that corresponds to the sample alone is shown on figure 11 (h is the surface heat exchange coefficient); the scheme corresponding to the complete system is presented in figure 12. In the transformed domain, the measured temperature can be readily found to be

$$\bar{T}_m = \frac{Z_d}{A(1 + 2hS.Z_u) + C.Z_u + BhS(1 + hS.Z_u)} \tag{19}$$

with $\quad Z_u = Z_c + R_c + Z_d$, $\quad A = \cosh(ke)$, $\quad B = \sinh(ke)/\lambda kS$, $\quad C = \lambda kS.\sinh(ke)$,

$\quad Z_c \approx 8/3\pi^2 \lambda r_0 (1 + 8kr_0/3\pi) \quad$ and $\quad Z_d = 1/\pi r_0 \sqrt{\rho C_t \lambda_t r_0^2 p + 2h_2 \lambda_t r_0}$

$Z_d$ is the thermal impedance of a thin of radius $r_0$ and its expression has been calculated in [7].

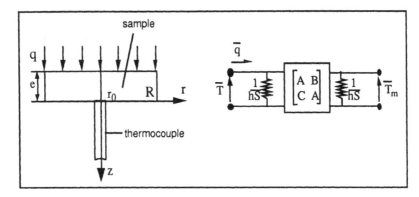

Figure 10 : Model of the
sample-thermocouple contact.

Figure 11 : Quadrupole schema
for the sample alone.

This temperature must be compared to the temperature taken in the absence of the thermocouple

$$\overline{T}_{m_0} = \frac{1}{C + 2hSA + Bh^2S^2} \qquad (20)$$

Comparisons with an exact solution (obtained with the separation of variables method) have revealed an error of less than 1% in all the cases tested. Using (19) also reduces calculation times by a factor of about 100.

## CONCLUSION

This work has pointed out the "constriction" impedance when thermal contacts are studied in the transient state. This allows fast calculations of complex systems where the heat transfer is not strictly one dimensional.

## REFERENCES

1. Beck J.V., Cole K.D., Haji-Sheikh A., Litkouhi B., 1990. "Heat Conduction Using Green's Functions." Series in Computational and Physical Processes in Mechanics and Thermal Sciences, Hemisphere Publishing corporation.
2. Degiovanni A., Moyne C. ,1989. "Résistance thermique de contact en régime permanent. Influence de la géométrie du contact". Rev.Gén.Thermique, Fr., n°334, pp 1-8.
3. Carslaw H.S., Jaeger J.C. ,1959. "Conduction of Heat in Solids" Oxford at Clarendon Press.
4. Özisik M. N., 1980. "Heat conduction." Wiley-Interscience Publication, John Wiley and Sons.
5. Degiovanni A., Sinicki G. , Géry A., Laurent M., 1984. "Un modèle de résistance thermique de contact en régime permanent." Rev.Gén.Thermique, Fr., n°267.
6. Degiovanni A., Lamine A. S. , Moyne C., 1992. "Thermal Contacts in Transient States: a New Model and Two Experiments" Journal of Thermophysics and Heat Transfer, vol. 6, n°2, pp 356-363.
7. Lachi M., Degiovanni A., 1992. "Influence de l'erreur de mesure de température de surface par thermocouples de contact sur la détermination de la diffusivité thermique par méthode flash" J. Phys. III, France **2** , pp 2247-2265.

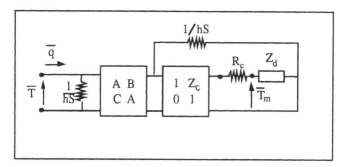

Figure 12 : Quadrupole schema of the sample-thermocouple contact in transient state.

# CERAMICS

**SESSION CHAIRS**
R. E. Taylor
D. G. Cahill

# A Model for Thermal Conductivity of Irradiated UO$_2$ Fuel

P. G. LUCUTA,[1] H. J. MATZKE,[2] I. J. HASTINGS[1]
and P. G. KLEMENS[3]

## ABSTRACT

The thermal conductivity of irradiated UO$_2$ fuel is discussed considering the effects of burnup (dissolved and precipitated solid fission products), porosity and fission-gas bubbles, deviation from stoichiometry and radiation damage. Model parameters are selected based on previously published results from SIMFUEL (simulated high burnup UO$_2$ fuel). A model developed to describe the thermal conductivity of irradiated UO$_2$ accounts for the above effects; it reflects the knowledge available today, and is recommended for use in fuel performance codes. The model is validated against available published data on thermal conductivity of irradiated fuel. This model can be incorporated into fuel performance codes to improve calculations of operating temperatures and predictions of behaviour of irradiated fuel under normal and accident conditions, including the extended burnup.

## INTRODUCTION

The thermal conductivity of UO$_2$ is one of the most important properties as it determines the fuel operating temperature, affecting directly fuel performance and behaviour. There are several previous reviews [1-5] of UO$_2$ thermal conductivity, analyzing the experimental published data and the effects of different factors such as porosity and hyperstoichiometry. It is generally agreed [5] that thermal conductivity of irradiated UO$_2$ is affected by the changes that take place in the fuel during irradiation: solid fission-product build-up, both in solution and as precipitates, fission gas-bubble formation, possible deviation from stoichiometry and radiation damage.

There are only a few available published data on the thermal conductivity of irradiated UO$_2$ fuel [6-11]. The data of Daniel and Cohen [7] published about 30 years ago, analyzed by Marchandise [12], were considered the most reliable.

[1] AECL, Chalk River Laboratories, Chalk River, Ontario, Canada, K0J 1J0
[2] Institute for Transuranium Elements, Joint Research Centre, European Commission, Postfach 2340, D-76125 Karlsruhe, Federal Republic of Germany
[3] University of Connecticut, Department of Physics, Storrs, CT 0629-3046

Despite the need for data on thermal conductivity of irradiated fuel, there are still difficulties in the interpretation of these results, because the effects from all the changes induced by irradiation are present, and overlap substantially, masking the effect of individual changes.

A useful way to quantify the changes in the fuel affecting thermal conductivity has been the use of SIMFUEL (SIMulated high burnup FUEL) [13-15]. SIMFUEL replicates the composition and the microstructure (*without fission-gas bubbles*) of irradiated fuel by adding stable additives to the $UO_2$. Some additives are soluble in the fluorite-type lattice of the fuel and others precipitate as a second phase [13-15]. The thermal conductivity was determined for different simulated burnups (1.5, 3, 6 and 8 at%[1] burnup) from thermal diffusivity, specific heat and density measurements [16-21]. The use of SIMFUEL permitted the assessment and analysis of single effects on thermal conductivity: the effect of fission products [22-24] and the effect of deviation from stoichiometry [21] were quantified from the results in the form of factors applied to thermal conductivity of unirradiated $UO_2$.

Radiation damage creates additional defects in the fluorite lattice especially in the "cold" regions of the fuel, temperatures below 973 K (at higher temperatures the defects from radiation damage are annealed) [25]. Earlier work at Chalk River [25-27], based on in-reactor measurements of fuel central temperatures, determined the effect of radiation damage (thermal neutron flux) on fuel thermal conductivity. The effect of alpha decay-damage on thermal conductivity of plutonium and americium oxides has also been investigated for various temperatures and numbers of alpha disintegrations [28]. Presently, fuel performance codes account by a cut-off value for the effect of radiation damage on fuel thermal conductivity for temperatures below 800 K.

This paper provides an analytical expression for thermal conductivity of irradiated $UO_2$ fuel as a function of temperature and burn up that accounts for all of the changes taking place during irradiation: solid fission-product build-up (dissolved and precipitated), pores and fission-gas-bubble formation, radiation damage and changes in stoichiometry.

## MODEL FOR THERMAL CONDUCTIVITY OF IRRADIATED $UO_2$

The thermal conductivity of irradiated $UO_2$ fuel [5] is affected by:
- solid fission products
  - dissolved, and
  - precipitated
- pores and fission-gas bubbles
- deviation from stoichiometry
- radiation damage.

---

1 at% ≈ 225 MW h/kgU ≈ 9375 MW d/teU

In this paper, the effects from three (fission products, deviation from stoichiometry and radiation damage) of four factors limiting thermal conductivity are individually quantified and described as function of burnup and temperature (between 600 and 1800 K). The uncertainties related to pore/fission-gas bubbles are only briefly discussed.

The model for the parametric dependence of irradiated $UO_2$ thermal conductivity, $\lambda$, can be provided in a form of contributing factors for each individual effect to unirradiated fuel conductivity:

$$\lambda = \lambda_o(T)\kappa_1(\beta,T)\kappa_2(p)\kappa_3(x)\kappa_4(T) \qquad (W/mK) \qquad (1)$$

where $\lambda_o$ is the analytical expression for thermal conductivity of unirradiated $UO_2$, $\beta$ denotes the burn up in at%, T is the temperature in Kelvin, $\kappa_1(\beta)$ is the burnup dependence factor, $\kappa_2(p)$ is the porosity/bubbles (p) contribution, $\kappa_3(x)$ describes the effect of deviation from stoichiometry (x) and $\kappa_4(T)$ refers to the radiation damage.

The thermal conductivity of unirradiated $UO_2$ has two dominant contributions: conduction through lattice vibrations and conduction by electronic processes (conduction by radiation through the lattice can be neglected [3]). The former plays a major role for temperatures below 1800 K and the latter is responsible for the observed increase in conductivity above 1800K. Harding's expression for the thermal conductivity of unirradiated $UO_2$ [3] is the one considered for $\lambda_o$:

$$\lambda_o = \frac{1}{0.0375 + 2.16510^{-4}T} + \frac{4.715 \cdot 10^9}{T^2} \exp(-\frac{16361}{T}) \qquad (2)$$

The burnup factor $\kappa_1(\beta)$ describes the effect of the fission products build up and it consists of two contributions:

$$\kappa_1(\beta,T) = \kappa_{1d}(\beta,T)\kappa_{1p}(\beta,T) \qquad (3)$$

where $\kappa_{1d}$ describes the effect of the dissolved fission products and $\kappa_{1p}$ accounts for the effect of the precipitated fission products. Based on SIMFUEL experimental data and analysis of the results [5], the effect of the dissolved fission products is given as a function of burnup and temperature by:

$$\kappa_{1d}(\beta,T) = (\frac{1.09}{\beta^{3.265}} + \frac{0.0643}{\sqrt{\beta}}\sqrt{T})arctan(\frac{1}{1.09/\beta^{3.265} + 0.0643\sqrt{T}/\sqrt{\beta}}) \qquad (3a)$$

The effect of the precipitated fission products are quantified by a Maxwell's-type factor for a precipitated second phase [5,22]; a temperature distribution function was added to delineate the fission-product-precipitates formation above 1200K:

$$\kappa_{1p}(\beta,T) = 1 + \frac{0.019\beta}{(3-0.019\beta)} \frac{1}{1+\exp(-(T-1200)/100)} \qquad (3b)$$

The burn up, $\beta$, expressed in at.% is related to the volume fraction of the precipitates [5] by $q=0.0038\beta$.

The pore/bubbles factor, $\kappa_2$ is dependent on the volume fraction and the shape of the pores/ bubbles. For small spherical pores/bubbles, uniformly dispersed, a modified Loeb expression [16], which includes the temperature dependence can be used:

$$\kappa_2(p,T) = 1 - (2.58 - 0.58 \cdot 10^{-3} T)p \qquad (4a)$$

where p is the pore/bubble volume fraction up to 9 vol%. The Loeb expression is clearly inadequate for high porosities; it thus can not be generally used for irradiated fuel, despite the fact that is often applied for unirradiated $UO_2$.

The Maxwell-Eucken formula [29] includes a pore shape factor, and can be applied for a higher porosity amount (up to 20 vol%):

$$\kappa_2(p) = \frac{1-p}{1+(\sigma-1)p} \qquad (4b)$$

where $\sigma$ is a porosity shape factor, equal to 1.5 for spherical pores, larger for flatter pores and smaller for tubular porosities; many different measurements lead to different values for $\sigma$. The Maxwell equation is considered in our model.

Analyzing the effect of deviation from stoichiometry on irradiated fuel conductivity there are two distinct regimes that would apply:

• normal operating conditions (NOC), and
• accident conditions and fuel defects.

Recent results from irradiated fuel [30,31] show no deviation from stoichiometry under NOC; so $\kappa_3(o) = 1$, the case of the present analysis.

Radiation damage from neutrons, $\alpha$-decay and fission fragments increases the number of lattice defects and consequently reduces the thermal conductivity of the fuel. The radiation-induced decrease in $\lambda$ occurs quite rapidly in a matter of days. Oxygen defects are known to anneal at around 500 K (hence below fuel operating temperatures) and uranium defects largely anneal at temperatures around 1000 K. This explains why most changes are seen below 1000 K. The factor describing the radiation damage can be expressed as a function of temperature [5] by:

$$\kappa_4 = 1 - \frac{0.2}{1+\exp(T-900)/80} \qquad (5)$$

Using the above model, the predicted $UO_2$ thermal conductivity is shown in Figure 1as a function of temperature for six different burnups.

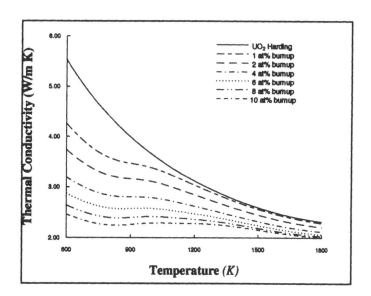

Figure 1.        Thermal conductivity of irradiated fuel predicted by the model for
six different burnups (5 vol% pore/bubble fraction) as a function of
temperature.

The reduction in thermal conductivity is significant especially in the low
temperature region of the fuel where radiation damage occurs. There is a large
reduction predicted at extended burnup in the 600 to 700 K range corresponding to
fuel periphery operating temperatures; this could be related to the Rim effect [32]
observed in light water reactor fuel pellets at burnups over 6 at.%.

## VALIDATION BASED ON IRRADIATED FUEL DATA

There are two sources [6-7] from which the data on thermal conductivity of
irradiated fuel are taken in this section. To validate the recommended model for
thermal conductivity of irradiated fuel, the available data will be briefly reviewed;
the fit of the data and the inconsistencies encountered are also discussed. The
measurements of thermal conductivity of irradiated UO₂ fuel are subject to error
sources. In-pile measurements usually give the thermal conductivity integral from
centre-line temperature measurements and are affected by uncertainty in the gap
conductance. The uncertainty is about ±15% for these results. Out-of-pile
measurements are done usually after some cooling time of the fuel (one or more
years) and the radiation damage factor is different. Furthermore, post-irradiation
measurements are affected by fission- product precipitation during the laser flash
measurements.

### *Chalk River Laboratories (CRL) Data*

In 1962, results from instrumented in-reactor experiments done at Chalk River were published [6]. The experiments concerned primarily measurements of fission gas release from irradiated fuel. However, thermal conductivity was also deduced from the centre-temperature measurements of the irradiated fuel pellets.

The irradiated fuel thermal conductivity results were obtained over a large temperature range, from 500 K to 2700 K and showed as well that the thermal conductivity of irradiated fuel is reduced compared to that of fresh $UO_2$. The results were discussed and compared with those from Bettis and Hanford [7]. The Chalk River results are plotted in Figure 2 along with those predicted by us and the unirradiated $UO_2$ data. However, the burnup of these fuel pellets is small (a burnup of 0.9 at.% has been reported), so, practically, the degradation occurred due to radiation damage in the low temperature region. The CRL experimental data were lower than those from unirradiated $UO_2$ below 1400 K. The thermal conductivity values decrease with the temperature to about 1200 K and above they level and increase above 1500 K. Taking into account the large experimental errors in the measurements (±15%), a reasonable prediction of these results is obtained using the proposed analytical expression for the burnup of 1 at.% and a porosity of 7 vol%.

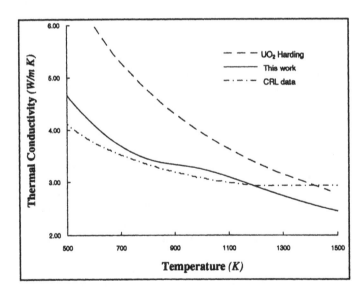

Figure 2.    Thermal conductivity results measured at Chalk River [6] on irradiated $UO_2$ (0.9 at% burnup) compared to those measured on fresh $UO_2$.

### Bettis Data

In 1964, Daniel and Cohen [7] reported in-pile measurements from Bettis Atomic Power Laboratories on irradiated fuel. They reported results from two experiments. The average values of effective thermal conductivity were obtained from three pairs of thermocouples. The data were obtained for centre temperatures up to 1050 K at various burnups to 10 at.%. Despite the fact that the data are plotted and accurately tabulated, they are difficult to analyze. Their effective conductivity consists of two effects: fuel thermal conductivity and fuel-clad gap conductance. Most of the data span the temperature range for which the radiation damage has a significant impact. This fact could explain the relative large reduction in fuel thermal conductivity observed at low burnup (< 1 at.%).

However, their data show lower thermal conductivity for irradiated fuel and the degradation is higher as the burnup increases. The authors noticed a reduction in the effective thermal conductivity of $UO_2$ from 3.5 W/m K at 0.02 at.% burnup to 1.7 W/m K at 11 at.% burnup ($28 \times 10^{20}$ fission/cm³). These values are not representative of constant operation since the fuel temperatures decreased significantly during irradiation.

The Bettis data reported by Daniel and Cohen were carefully analyzed by Marchandise [12], who proposed an analytical expression as a function of the burnup and temperature to fit Daniel&Cohen's data:

$$\lambda = \frac{1}{0.056 + 0.00021 T + \dfrac{18}{T} + \beta \dfrac{6.184}{T}} \qquad (W/mK) \qquad (6)$$

where $\beta$ is the burnup expressed as fission/cm³ and T is the temperature in Kelvin. This expression was plotted by Marchandise for six different burnups between 0.2 and 3.2 at.% and it was extrapolated to temperatures up to 1900 K. This expression and our model for the thermal conductivity of irradiated fuel are plotted comparatively in Figures 3a, 3b and 3c for burnups of 1.5, 3 and 8 at.%, respectively. The unirradiated $UO_2$ and SIMFUEL (of equivalent burnup) data are also shown in the Figures; Daniel and Cohen's measurements, corrected for gap conductance, are displayed for each burnup by open diamond symbols.

Clearly the measured SIMFUEL data for the 7 vol% porosity represent the upper limit for the fuel thermal conductivity, especially for temperatures below 1200 K where the effect of the precipitated fission products is nil. Our proposed analytical expression and Marchandise's relation match well for each of the three burnups represented and both are in good agreement with the experimental data of Daniel and Cohen corrected for the gap conductance.

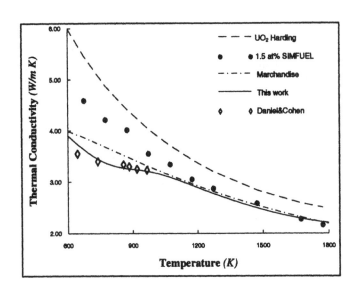

Figure 3a.    Irradiated fuel thermal conductivity (1.5 at.% burnup) as a function of temperature compared to the Marchandise expression (Eqn.6) and irradiated fuel data from Daniel and Cohen [7].    The 1.5 at% SIMFUEL and $UO_2$ data are also plotted.

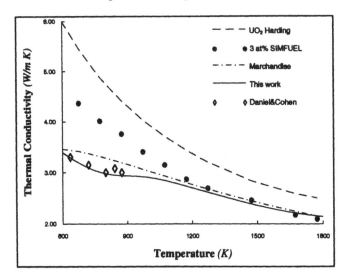

Figure 3b.    The variation of irradiated fuel (3 at.% burnup) thermal conductivity compared to Marchandise expression and irradiated fuel data from Daniel and Cohen.

586

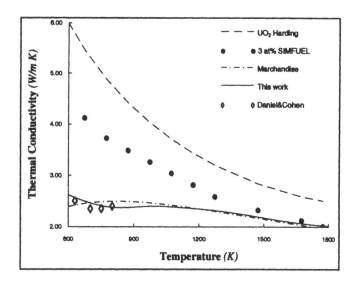

Figure 3c.     The variation of irradiated fuel (8 at.% burnup) thermal conductivity compared to Marchandise expression and irradiated fuel data from Daniel and Cohen. The SIMFUEL and $UO_2$ measured data are also plotted.

## CONCLUSIONS

An analytical expression for thermal conductivity of irradiated $UO_2$ valid to 1800 K was derived taking into account the effects of solid dissolved and precipitated fission products; porosity/fission-gas-bubbles, deviation from stoichiometry and radiation damage on intrinsic thermal conductivity of $UO_2$. The parameters involved with each thermal conductivity limiting mechanism in the model were first determined:

o     The effects of solid-fission-product build-up and deviation from stoichiometry were assessed and quantified based on measured values and modelling of SIMFUEL results.

o     The porosity factor suggested by Maxwell was used throughout the model validation, emphasising the need for a better approach to quantify the pore/bubble factor based on experimental data.

o     The radiation damage effect was quantified based on earlier measurements of irradiated fuel.

To validate the model, previous in-pile experimental results were used:

o    After determining the model's porosity parameter from the CRL data [6], there is a consistent agreement over the temperature range (500 to 1300 K).

o    The results from Bettis [7], in the manner analyzed by Marchandise [12] were well fitted by the analytical function proposed for irradiated fuel.

## ACKNOWLEDGEMENTS

The financial support  for SIMFUEL work over the last five years from the CANDU Owners Group (COG) that made these results possible is acknowledged. Also, we thank many colleagues: M. Tayal (AECL),  D. Olander (University of California at Berkeley) and  W. Wiesenack (Halden)  for their useful comments and suggestions on the manuscript.

## REFERENCES

1.    Martin D.G.,  J. Nucl. Mater. **110** (1982) 73.
2.    Hyland G.J.,  J. Nucl. Mater. **113**  (1983) 125.
3.    Harding J.H. and Martin D.G.,  J. Nucl. Mater. **166**  (1989) 166.
4.    Washington A.B.G., UKAEA report TRG 2236 (D) (1973).
5.    MATPRO - A Handbook of Materials Properties for Use in the Analysis of Light Water Reactor Fuel Rod Behaviour, TREE-NUREG-1005, EG&G Idaho, Inc. (1989).
6.    Robertson J.A.L., Ross A.M., Notley M.J.F. and MacEwan J.R., J. Nucl. Mater., **7**, 3 (1962) 223.
7.    Daniel R.C. and Cohen I., Bettis Atomic Power Report WAPD-246 (1964).
8.    Shibab  S., Belgonucleaire High Burnup Chemistry Club (HBC) topical report, HBC 91/34 (1991) - private communication.
9.    Wiesenack W., IAEA Technical Committee Meeting on Fission Gas Release and Fuel Rod Chemistry Related to Extended Burnup, 1992 April 27-May 1, Pembroke, Canada, IAEA-TECDOC-697, 1993 April.
10.   Nakamura J., Uetsuka H., Kohno N., Oheda E., Sukegawa T. and Furuta T., Enlarged Halden Programme Group Meeting, Bolkesjø, Norway, 30 October - 4 November 1994.
11.   Carrol J.C., Gomme R.A. and Leech N.A., Enlarged Halden Programme Group Meeting, Bolkesjø, Norway, 30 October - 4 November 1994.
12.   Marchandise H.,  CEC report EUR - 4568f (1970).
13.   Lucuta P.G., Palmer B.J., Matzke Hj. and Hartwig D.S., Proc. Second Int. Conf. CANDU Fuel, Ed. I.J. Hastings, CNS, Toronto (1989) 132.

14. Lucuta P.G., Verrall R.A., Matzke Hj. and Palmer B.J., J. Nucl. Mater. **178** (1991) 48.

15. Matzke Hj., Lucuta P.G. and Verrall R.A., J. Nucl. Mater. **185** (1991) 292.

16. Lucuta P.G., Matzke Hj., Verrall R.A. and Tasmann H.A., J. Nucl. Mater. **188** (1992) 198.

17. Matzke Hj., Lucuta P.G., Verrall R.A., and Hiernaut J.P., 22$^{nd}$ Int. Conf. on Thermal Conductivity , 1993 November 7-10, Tempe, Arizona, Proc. Thermal Conductivity 22, Ed. T. Tong, Technomic (1994), 904.

18. Lucuta P.G., Verrall R.A., Matzke Hj. and Hastings I.J., IAEA Technical Committee Meeting on Fission Gas Release and Fuel Rod Chemistry Related to Extended Burnup, 1992 April 27-May 1, Pembroke, Ontario, Canada, IAEA-TECDOC-697, !993 April.

19. Lucuta P.G., Verrall R.A., Matzke Hj. and Hastings I.J., Proc. Third Int. Conf. CANDU Fuel, Ed. P.G. Boczar, CNS, Toronto (1992) 2-61.

20. Lucuta P.G., Matzke Hj. and Verrall R.A., IAEA Technical Committee Meeting on Water Reactor Fuel Element Modelling at High Burnup and Experimental Support, Bowness-on- Windermere, UK, 18-23 September 1994.

21. Lucuta P.G., Matzke Hj. and Verrall R.A., J.Nucl. Mater. **223**, (1995), 51.

22. Lucuta P.G., Matzke Hj. and Verrall R.A., J.Nucl. Mater. **217**, (1994), 279.

23. Lucuta P.G., Matzke Hj., Verrall R.A., and Klemens P.G., 22$^{nd}$ Int. Conf. on Thermal Conductivity, 1993 November 7-10, Tempe, Arizona, Proc. Thermal Conductivity 22, Ed. T. Tong, Technomic (1994), 894.

24. Klemens P.G. and Lucuta P.G., - 23$^{d}$ Thermal Conductivity International Conference, Oct 30-Nov. 1 1995, Nashville, Tennessee - in these proceedings.

25. Ross A.M., Atomic Energy of Canada Limited report, AECL-1096 (1960).

26. Hawkings R.C. and Robertson J.A..L., Atomic Energy of Canada Limited report AECL-1733 (1963).

27. Hawkings R.C. and Bain A.S., Atomic Energy of Canada Limited report AECL-1790 (1963).

28. Schmidt E.H., Richter J., Matzke Hj. and Van Geel J., 22$^{nd}$ Int. Conference on Thermal Conductivity, 1993 November 7-10, Tempe, Arizona, Proc. Thermal Conductivity 22, Ed. T. Tong, Technomic (1994), 920.

29 Winter P.W. and MacInnes D.A., IAEA Technical Committee Meeting on Water Reactor Fuel Element Computer Modelling in Steady-State, Transient and Accident Conditions, Preston, UK, 1988.

30. Matzke Hj., J.Nucl. Mater., **223**, (1995), 1.

31. Matzke Hj., J.Nucl. Mater., **208**, (1994), 18.

32. Cunningam M. E., Freshley M.D. and Lanning D.D., J. Nucl. Mater. **188**, (1992) 19.

# Phonon Scattering in SIMFUEL—Simulated High Burnup UO$_2$ Fuel

P. G. KLEMENS[1] and P. G. LUCUTA[2]

## ABSTRACT

Reduction in UO$_2$ thermal conductivity with burnup has been experimentally observed using simulated high burnup fuel - termed SIMFUEL. (Thermal conductivity was obtained from thermal diffusivity and specific heat measurements). This reduction could be explained by the mass difference between the fission-product solute atoms in the fluorite lattice and U atoms alone, without a need for a strain term.

The calculation of the misfit in ionic radius shows that the largest distortion effect is obtained for simulated fission products of Mo, Pd, Ru, and Rh. Nevertheless these atoms precipitate outside of the fluorite lattice due to the large radial misfit that limits their solubility in the matrix. This explains well the agreement obtained for SIMFUEL between experimental phonon scattering values and those calculated from the mass differences alone. Note that if all the Mo, Rh, Ru cations would be in solution, the phonon scattering would be enhanced by a factor of three and the thermal conductivity of SIMFUEL reduced further by roughly a factor of 0.6.

## INTRODUCTION

Thermal conductivity is an important property of UO$_2$, because it controls fuel operating temperatures and therefore influences almost all processes related to fuel performance and behaviour. Changes in fuel thermal conductivity occur during irradiation because of fission-gas bubble formation, pores, cracks, fission-product build-up (dissolved and precipitated), and possibly, changes in the oxygen to uranium ratio (O/U).

The thermal conductivity of UO$_2$ has been critically analyzed and reviewed

---

[1]  Dept. of Physics, University of Connecticut, Storrs, CT 06269-3046 USA
[2]  AECL, Chalk River Laboratories, Chalk River, Ontario, Canada, K0J 1K0

theoretically by Marchandise [1], Martin [2], Hyland [3], Harding and Martin [4], Klemens [5] and Lucuta et al. [6]. The thermal conductivity of irradiated $UO_2$ depends on the deviation from stoichiometry, the burnup and the fractional porosity, as well as the temperature.

The influence of the various parameters on fuel thermal conductivity is important to assess fuel performance and behaviour. Thermal conductivity measurements of SIMFUEL provide a useful way to determine the effects of burnup and deviation from stoichiometry on fuel thermal conductivity [7-10].

SIMFUEL replicates the chemical state and microstructure of irradiated fuel by adding non-radioactive elements to $UO_2$ powder in amounts appropriate for the chosen burnups [11]. The preparation route features high-energy grinding and spray drying to achieve homogeneous dispersion, and sintering to provide atomic-scale mixing, so the phase structure is representative of high-burnup fuels that operated at high temperatures–without fission-gas bubbles (gases and volatiles are not added because they would not be retained during sintering). Extensive characterization has shown fission-product atoms dissolved in the $UO_2$ oxide fluorite matrix, and two types of precipitated "fission products": spherical metallic Mo-Ru-Pd-Rh compounds and ceramic phases [11-13].

Thermal conductivities of stoichiometric and hyperstoichiometric SIMFUEL with an equivalent burnup of 1.5, 3, 6 and 8 at.% were earlier determined from thermal diffusivity and specific heat measurements, and were found to be lower than those of the $UO_2$ conductivity [7-10]. Thermal resistivity, the inverse of conductivity, increased linearly with the temperature and burnup [7-10]. The reduction in SIMFUEL thermal conductivity, corrected for porosity, reflects the dependence on burnup due to the contributions from solid precipitated phases and dissolved oxides in the matrix. The solid precipitated phases, mainly metal particles, have a small positive contribution to the thermal conductivity that is considered using a correction factor [14,15].

This paper analyses the reduction in thermal conductivity of SIMFUEL matrix using the interference between mass difference and distortion in the scattering amplitudes of the solute fission product and uranium host atoms in the SIMFUEL fluorite $UO_2$ lattice.

## BACKGROUND

The degradation in the thermal conductivity of SIMFUEL fluorite matrix was analyzed in terms of the phonon scattering by point defects and yielded an empirical value for the scattering cross-sections [14,15]. The value was compared to theoretical scattering cross-sections, under the following assumptions.
- all metal ions are in solid solution, and
- the solute atoms scatter the phonons in virtue of their difference in mass from that of uranium and contributions to the scattering from size misfit can be neglected.

Although point defect scattering parameters deduced earlier from the thermal conductivity agreed with the calculated cross-sections, [14,15] the assumptions are not completely valid. Some of the metal ions have very limited solubility, and therefore metal precipitates are present in the microstructure [11]. Furthermore there is a difference in the ionic radii between solvent and solute metal ions, and an empirical distortion term had to be invoked by Fukujima et al. [16,17] to explain thermal conductivity reduction in single additive doped $UO_2$. For these reasons the phonon scattering calculations for SIMFUEL matrix have been repeated with these factors in mind.

The present analysis, based on Ackerman's expression [18] for phonon scattering by solute atoms, differs from that used by Fukujima et al. [16,17]. This analysis considers interference between mass difference and distortion in the scattering amplitudes, while Fukujima's formula [16] just adds scattering probabilities, so that agreement was obtained by using an empirical distortion term. In the present case the interference is mainly constructive (light masses are accompanied by small ionic radii and heavy metal atoms have larger ionic radii). It will be shown that calculated scattering is much stronger than that observed experimentally if all fission product atoms are dissolved within the $UO_2$ fluorite lattice. However, the ions of the largest misfit (Pd, Ru, Rh, and Mo) are known to have very limited solubility in the $UO_2$ lattice and they precipitate outside as has been observed in SIMFUEL [11] and irradiated fuel [19-21]. If their contribution to the phonon scattering in the fluorite matrix is absent, the calculated scattering from the other solute fission product atoms agree with the experimentally observed point defect scattering.

## POINT DEFECT SCATTERING OF PHONONS

For substitutional solute atoms which differ from the host lattice atoms only in respect to the atomic mass, the phonon reciprocal mean free path is [5]:

$$1/l_p(\omega) = c(\Delta M/M)^2 a^3 (4\pi v^4)^{-1}\omega^4 \qquad (1)$$

where $a^3$ is the volume per atom, $v$ the phonon velocity, $M$ the atomic mass, $\Delta M$ the mass difference between the solute fission product and uranium atoms and c the solute concentration per atom. For a compound such as $UO_2$/SIMFUEL, $M$ is the average atomic mass.

If there is also lattice distortion from the solute atoms because of the size misfit, the fractional mass difference $\Delta M/M$ is replaced by [18]:

$$(\Delta M/M)_{eff} = \Delta M/M + 2g\Delta V/V \qquad (2)$$

where $g$ is a Grueneisen anharmonicity parameter and $\Delta V/V$ is the fractional volume difference between the solute cation and the solvent uranium ion in the fluorite lattice.

In our previous analysis [14,15], after correcting for pores and metallic inclusions to obtain the thermal conductivity of SIMFUEL matrix $\lambda_m$, this was compared to $\lambda_o$, the thermal conductivity of fully dense pure $UO_2$, by means of:

$$\lambda_m/\lambda_o = (\omega'/\omega_D)\,\text{arctan}\,(\omega_D/\omega') \tag{3}$$

Here, $\omega_D$ is the Debye frequency and $\omega'$ is the frequency at which the point defect mean free path $l_p(\omega)$ equals the intrinsic mean free path $l_o(\omega,T)$.

Now $l_o(\omega,T)$ and $\lambda_o$ are related by:

$$l_o(\omega,T) = \lambda_o(T)\,(\omega_D/\omega)^2 a^3\,(3kv)^{-1} \tag{4}$$

where $k$ is the Boltzmann constant. Thus the equality $l_p(\omega')=l_o(\omega',T)$ and Equation (1) lead to:

$$(\omega'/\omega_D)^2 = 2kv\,(3\pi\lambda_o a^2)^{-1}\,(6\pi^2)^{-\frac{1}{3}}\,[c\,(\Delta M/M)_{eff}]^{-1} \tag{5}$$

*Note: In Equation (11) of our previous paper [14], $\lambda_o$ was mistakenly written as $\lambda_m$.*

From Equation (3), observed values of $\lambda_m/\lambda_o$ yield the scattering parameter, $\omega'/\omega_D$, for each temperature and burnup. They have been determined as graphic solutions as it is shown in Figure 1.

The experimental thermal conductivity data of UO₂ and SIMFUEL reported earlier [7,10] were used to determine the scattering parameter. The values of $\omega'/\omega_D$ are temperature and burnup dependent. As burnup increases, the concentration of dissolved fission products increases, and the contribution by dissolved atoms to the scattering of the phonons increases. (Note that for increasing values of the scattering parameter, the scattering from the dissolved fission product atoms decreases). The results also show that as the temperature increases the contribution of the dissolved atoms to phonon scattering decreases.

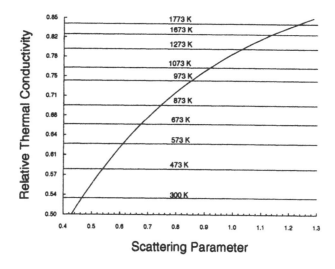

Figure 1. Scattering parameter, $\omega'/\omega_D$, determined from relative thermal conductivities of 8 at% burnup SIMFUEL for various temperatures.

The variation of the calculated scattering parameter with temperature for simulated burnups of 1.5, 3 and 8 at% are shown in Figure 2.

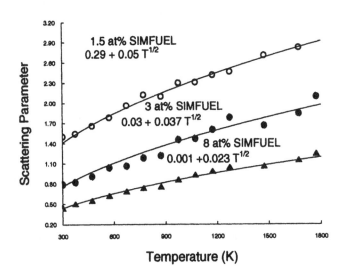

Figure 2.    Scattering by the dissolved fission products in 1.5, 3 and 8 at% SIMFUEL as a function of the temperature. The lines represent the best fit to a square root temperature dependence described by the expression given for each burnup. Data points are from [15].

Since $\lambda_o \propto T^{-1}$, $\omega'/\omega_D$ should vary as $T^{1/2}$ provided the solute concentrations do not vary with temperature. This variation is plotted against the observed values for each burnup in Figure 2. This is indeed found to be true for the 1.5, 3 and 8 at% simulated burnup SIMFUEL. The temperature dependence of the scattering parameter described by the expressions shown in Figure 2 allow one to deduce the effect of dissolved fission product on the thermal conductivity of irradiated fuel and were used in another paper of these proceedings [22].

The values of $c(\Delta M/M)^2$ could be obtained from Equation (5); they were found to be:

$c(\Delta M/M)^2 = 4.8 \times 10^{-3}$ for 3 at% burnup SIMFUEL, and

$c(\Delta M/M)^2 = 15 \times 10^{-3}$ for 8 at% burnup SIMFUEL

These values can be compared to those calculated from the mass difference alone, assuming all the additives from SIMFUEL compositions to be in solution. For the 3 and 8 at% burnup compositions, these are $7 \times 10^{-3}$ and $18 \times 10^{-3}$ respectively.

Thus, there seems to be a reasonable agreement between the theory and experiment by using the mass difference alone. However, it did seem that scattering increased more than linearly with the solute atoms content, suggesting that at 8 at% burnup not all the metal ions are in solution.

Normal three phonon processes [23], i.e. processes which conserve phonon momentum, should enhance the effect of point defects at higher concentrations. This may make the scaling discrepancy larger, but in absence of information about the strength of these processes, no quantitative conclusions are possible at this time.

## ESTIMATED EFFECT OF SIZE MISFIT

A substitutional atom, of volume $V+\Delta V$, replacing an atom of volume $V$, causes a radial displacement which is non-additional and acts, for purposes of phonon scattering, like an expansion $\Delta V$ localised at the centre, and changes the local phonon velocity $\upsilon$ by: $-g\upsilon(\Delta V/V)$. Since an increase in mass $\Delta M$ reduces the local velocity by $-(\Delta M/2M)\upsilon$, one obtains the effective value $(\Delta M/M)_{\text{eff}}$ given by Equation (2). Using ionic radii for appropriate valence states of the substitutional fission product ions [24], and with $\Delta V/V = 3\,\Delta r/r$, values for $(\Delta M/M)_{\text{eff}}$ are obtained. Table 1 gives $c\,(\Delta M/M)^2$ for each solute additive in the 3 at% burnup SIMFUEL composition. Note that $c$ is the concentration of cations per atom, which is 1/3 of the concentration per mole of UO₂.

If all solute cations are assumed to be in solution:

$$\sum c_i (\Delta M/M)_{\text{eff}}^2 = 23 \cdot 10^{-3}$$

which is considerably larger than the value of $4.8\ 10^{-3}$ which is obtained from $\omega'/\omega_D$ values which fit the thermal conductivity. However if the contribution from Mo, Pd, Rh and Ru are completely excluded:

$$\sum c_i (\Delta M/M)_{\text{eff}}^2 = 6.01 \times 10^{-3}$$

which is in reasonable agreement with the experimental data (slightly larger).

The contribution from Ba and Zr can be diminished as these fission products are only partially soluble in the fluorite lattice and they also form BaZrO₃-type ceramic precipitates. If about half of the Ba and Zr atoms are dissolved in the lattice [11]:

$$\sum c_i (\Delta M/M)_{\text{eff}}^2 = 5.1 \times 10^{-3}$$

which is in good agreement with the experimental data

Thus if all simulated fission products were in solution and contribute to the phonon scattering, the process would be stronger by a factor of 4 and the thermal conductivity lower by a factor of 2; if the contributions from Mo, Pd, Ru and Rh are not considered and the contribution of the Ba and Zr decreased to their solubility level good agreement between theory and experiment is obtained.

**Table 1.**    Effective scattering parameter for 3 at% burnup composition. The last four additives precipitate outside the matrix. Clearly the last four additives have the largest misfit.

| FP | wt% | mol% | $\Delta M/M$ | $4/3 \Delta V/V$ | $(\Delta M/M)_{eff}$ | $c_i(\Delta M/M)_{eff}^2$ |
|---|---|---|---|---|---|---|
| BaO | 0.147 | 0.200 | -0.374 | +1.140 | 0.77 | 0.00119 |
| CeO$_2$ | 0.285 | 0.447 | -0.363 | -0.150 | -0.51 | 0.00118 |
| LaO$_3$ | 0.106 | 0.153 | -0.366 | +0.150 | -0.22 | 0.00007 |
| Nd$_2$O$_3$ | 0.460 | 0.139 | -0.348 | +0.078 | -0.27 | 0.00054 |
| SrO | 0.072 | 0.187 | -0.556 | +0.245 | -0.106 | 0.00002 |
| Y$_2$O$_3$ | 0.041 | 0.099 | -0.554 | -0.186 | -0.791 | 0.00062 |
| ZrO$_2$ | 0.339 | 0.744 | -0.544 | -0.186 | -0.730 | 0.00396 |
| *MoO$_3$* | 0.359 | 0.673 | -0.526 | -1.080 | -1.606 | 0.0174 |
| *PdO* | 0.149 | 0.330 | -0.489 | -0.525 | -1.014 | 0.00339 |
| *Rh$_2$O$_3$* | 0.028 | 0.060 | -0.500 | -0.897 | -1.397 | 0.00117 |
| *RuO$_2$* | 0.346 | 0.442 | -0.507 | -0.927 | -1.434 | 0.00909 |

*Note: $(\Delta M/M)_{eff} = \Delta M/M + (4/3) \Delta V/V$ to convert concentration per atom basis to agree with M which is defined per average atom.*

## CONCLUSIONS

- The reduction in thermal conductivity of SIMFUEL matrix is well explained by the phonon scattering due to the combined effect of the mass difference and distortion in the scattering amplitudes with the largest contribution from the mass difference.
- The precipitated fission product in the metal particles have the largest misfit into the fluorite lattice. This elastic misfit energy may explain, at least in part, their very limited solubility in the fluorite lattice.

## ACKNOWLEDGEMENTS

The authors are grateful to D. Olander (Berkeley University) for useful comments and suggestions and Hj. Matzke (Institute for Transuranium Elements) for reviewing the manuscript. The financial support of CANDU Owners Group (COG) for the work on SIMFUEL thermal properties is also acknowledged.

## REFERENCES

1. Marchandise H., CEC report EUR - 4568f (1970).
2. Martin D.G., J. Nucl. Mater. **110** (1982) 73.
3. Hyland G.J., J. Nucl. Mater. **113** (1983) 125.
4. Harding J.H. and Martin D.G., J. Nucl. Mater. **166** (1989) 166.
5. Klemens P.G., High Temps.-High Pressures **17**, (1985) 41.
6. Lucuta P.G., Matzke Hj. and Hastings I.J.- J. Nucl. Mater., - to be published.
7. Lucuta P.G., Matzke Hj. Verrall, R.A. and Tasman H.A., J. Nucl. Mater. **188** (1992) 198.
8. Lucuta P.G., Verrall R.A., Matzke Hj. and Hastings I.J., IAEA Technical Committee Meeting on Fission Gas Release and Fuel Rod Chemistry Related to Extended Burnup, 1992 April 27-May 1, Pembroke, Ontario, Canada.
9. Matzke Hj., Lucuta P.G., Verrall R.A., and Hiernaut J.P., 22nd Int. Conf. on Thermal Conductivity , 1993 November 7-10, Tempe, Arizona, Thermal Conductivity 22, Technomic, Ed. T. Tong, (1994).
10. Lucuta P.G., Matzke Hj. and Verrall, R.A.. J. Nucl. Mater. **223** (1995) 51.
11. Lucuta P.G., Verrall R.A., Matzke Hj. and Palmer B.J., J. Nucl. Mater. **178** (1991) 256.
12. Lucuta P.G., Palmer B.J., Matzke Hj. and Hartwig D.S., Proc. Second Int. Conf. CANDU Fuel, Ed. I.J. Hastings, CNS, Toronto (1989) 132.
13. Matzke Hj., Lucuta P.G. and Verrall R.A., J. Nucl. Mater. **185** (1991) 292.
14  Lucuta P.G., Matzke Hj. and Verrall, R.A. and Klemens P.G., 22nd Int. Conf. on Thermal Conductivity, 1993 November 7-10, Tempe, Arizona, Thermal Conductivity 22, Technomic, Ed. T. Tong, (1994).
15. Lucuta P.G., Matzke Hj. and Verrall, R.A.., report AECL-11235 (1994).
16. Fukushima S., Ohmichi T., Maeda A. and Watanabe H., J. Nucl. Mater. **102**, (1981) 30.
17. Fukushima S., Ohmichi T., Maeda A. and Handa M., J. Nucl. Mater. **114** (1983) 312.
18. Ackerman M.W. and Klemens P.G., J. Appl. Phys. **42**, (1971), 968.
19. Bramman J.I., Sharpe R.M., Thom D. and Yates G., J. Nucl. Mater., **25**, (1968), 201.
20. Ewart F.T., Taylor R.G., Horspool J.M. and James G., J. Nucl. Mater., **61**, (1976), 254.
21. Kleykamp H., Paschoal J.O., Pesja R. and Thummler F., J. Nucl. Mater., **130**, (1985), 426.
22. Lucuta P.G., Matzke Hj., Hastings I.J. and Klemens P.G., - these proceedings.
23. Carruthers P., Rev. Mod. Phys. **33**, (1961), 92.
24. "Crystal Ionic Radii of the Elements", Handbook of Chemistry and Physics, 63d ed. CRC Press, BocaRaton, FA, (1983), F-179.

# The Thermal Conductivity of "Ceramic Powders"

T. ASHWORTH and E. ASHWORTH

## ABSTRACT

A new specimen cell, used in conjunction with a previously described dual heat-flux metered system, has been fabricated to allow measurement at different densities of the thermal conductivity of powders used as the components of ceramic materials. The design and fabrication of the cell is discussed in this paper. Thermal conductivity values as a function of density for two different sizes of aluminium powders (18 micron and 5 micron) at room temperature are presented. Values of about 0.2 W/(m-K) were found at approximately 1000 Kg/m$^3$, increasing 5- to 10-fold for the powder compressed to 1500 Kg/m$^3$. Theories of thermal resistance of multiple materials are discussed. Luikov et al's model can be made to fit the data at the low densities by adjusting parameters. At higher densities, metal-to-metal contact resistance is an important component of the thermal conductivity. The results show that the thermal conductivity of the powder is relatively insensitive to density until the density increases beyond a critical point when the thermal conductivity notably increases with increasing density. This is of significant importance in the design of the ceramic combustion synthesis when control of the conversion zone determines the quality of the produced material.

## INTRODUCTION

In the combustion synthesis of ceramics and intermetallic materials (self-propagating high temperature synthesis), a mixture of powders is ignited at one end and a combustion front advances through the material. As it does so, the powders are converted into a solid material in the small conversion zone. The properties of the material produced depend both on the starting materials and the manner, particularly the velocity, in which the conversion zone progresses. Propagation

T. Ashworth, Physics Department, South Dakota School of Mines and Technology, 501 East St. Joseph Street, Rapid City, SD 57701-3995, USA

E. Ashworth, Mining Engineering Department, South Dakota School of Mines and Technology, 501 East St. Joseph Street, Rapid City, SD 57701-3995, USA

velocity depends on several parameters [1]. The thermal conductivity of the powder mix is of obvious importance, since it determines the rate at which energy is propagated forwards; that is, the size of the heating zone. Also, the density of the powder mixture can be changed by compression to change the volumetric combustion energy of the powder. Unfortunately, this simultaneously changes the thermal conductivity and thermal diffusivity of the material. For this reason, data for the thermal conductivity of powder mixtures as a function of density are important. Typical powder mixtures used are Titanium-Carbon and Nickel-Aluminium.

Over the years, a dual heat-flux metered system (Figure 1 and [2]) has been used by the authors for numerous measurements on rocks and concretes [3], mainly to obtain thermal conductivity of these materials as a function of moisture content. It has been found that the thermal resistance of an interface between a metallic surface and another surface separated by Teflon, or by a plastic bag, can be calibrated, as shown in Figure 2. The thermal resistance of an interface, which uses Teflon as the contact medium, changes with time, with the type/thickness of the Teflon, and with the roughness of the surfaces, but it is well-defined and consistent [4]. A plastic bag used as the contact medium produces a very stable thermal resistance, which is essentially invariant with time; it is, however, sensitive to temperature. Each plastic bag has a different thermal resistance, so that calibration of each bag is needed. It is the use of plastic bags which has allowed different moisture conditions to be maintained within a specimen [3, 5], which in turn allowed thermal conductivity measurements to be obtained in both thermal and moisture (dynamic) equilibrium. The heat-flux metered system is housed in a temperature controlled environment; the system has been shown to be very reproducible [4]. Thus, the system is able to sensitively monitor changes in thermal conductivity.

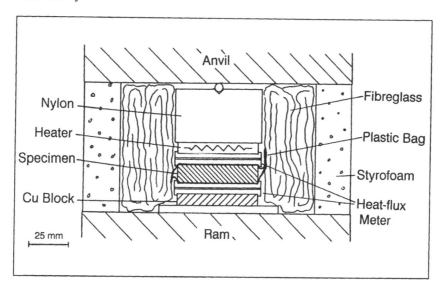

Figure 1. Schematic diagram of the dual heat-flux meter thermal conductivity system.

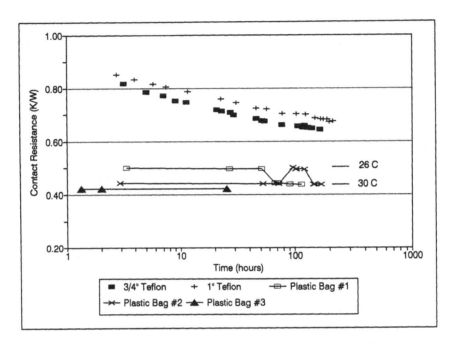

Figure 2. Thermal resistance of Teflon and plastic bags used as interfacial materials.

Using this apparatus, preliminary measurements were carried out some time ago on a titanium-carbon powder mixture. These measurements indicated that the thermal conductivity was sensitive to the design of the cell which contained the powder mix; they also suggested that the variation of conductivity was not as sensitive to density as was expected. Because of the importance of the combustion synthesis process and because of our interest in porous materials and in boundary resistance in various material systems, we have undertaken the development of a system which was capable of determining the thermal conductivity of powder mixtures as a function of density.

## DESIGN OF SPECIMEN CELL

From the preliminary measurements it was anticipated that the materials would have thermal conductivities in the range 0.1 to 1 W/(m-K). The current apparatus accommodates specimens or specimen cells of slightly less than 50 mm diameter and 10 mm height. Taller specimens lead to larger heat losses which would not be as well compensated for by the dual heat flux measurements. In order that the containment cell would not dominate the heat flow for low density specimens, the cell needed to have a thermal resistance of 300 K/W or higher. Also, there had to be a compromise to give the cell adequate load-bearing capability. As indicated, it was desired to have the height of the cell constant for

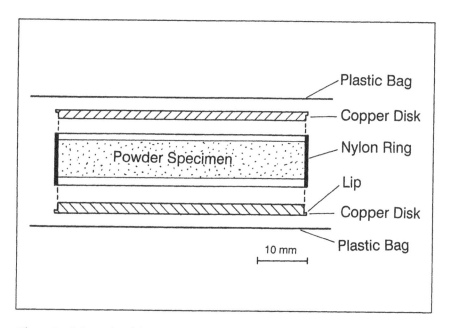

Figure 3. Schematic of the specimen cell used for powders.

the varying powder densities to make heat losses more uniform. The interfacial configuration needed to remain the same for each powder density.

The configuration developed to accommodate these conflicting requirements has copper plates at each end of a thin-walled (0.40 mm) nylon ring (R) of height 9.6 mm. (Figure 3). Several copper plates were fabricated of different thickness but with approximately the same lip thickness. Thus, the gap between the copper plates (see Table I), which contained the powder could be varied thereby

TABLE I - CONFIGURATION OF COPPER PLATES AND CHANGES IN
GAP OF THE SPECIMEN CELL

| Number | Configuration | Overall Thickness | Thickness of disks | Gap (inches) | Gap (mm) |
|--------|---------------|-------------------|--------------------|--------------|----------|
| 1 | "1+R+2" | 0.4085 | 0.0881 | 0.3204 | 8.14 |
| 2 | "1+R+3" | 0.4170 | 0.1146 | 0.3024 | 7.68 |
| 3 | "1+R+4" | 0.4136 | 0.1322 | 0.2814 | 7.15 |
| 4 | "1+R+5" | 0.4130 | 0.1421 | 0.2709 | 6.88 |
| 5 | "7+R+5" | 0.4164 | 0.1565 | 0.2599 | 6.60 |
| 6 | "8+R+5" | 0.4140 | 0.1694 | 0.2446 | 6.21 |
| 7 | "4+R+5" | 0.4240 | 0.1811 | 0.2429 | 6.17 |
| 8 | "9+R+5" | 0.4175 | 0.1860 | 0.2315 | 5.88 |
| 9 | "6+R+5" | 0.4254 | 0.2084 | 0.2170 | 5.51 |

producing changes in the density of a specific powder specimen. The outer surfaces of each copper disk were lapped with a range of wet/dry emery down to 600 grit followed by 1.0, 0.3 and 0.05 micron alumina slurry.

## MEASUREMENTS

Contact resistance between the copper surfaces of the heat-flux meters and a surface representing the powder specimen cell was determined by placing a disk of copper inside a bag and performing a thermal resistance measurement. The thermal resistance of the copper disk was very small; it was calculated and then subtracted from the total thermal resistance to give the interfacial resistance. This was repeated for all bags used in the powder measurements; the same bag was used for all measurements on one powder.

At the beginning of each sequence of measurements, the cell was loaded with enough "fluffy" powder to obtain physical contact with both inner copper surfaces using the two thinnest disks - configuration "1+R+2". After each thermal conductivity measurement, one of the disks was sequential changed to decrease the gap (Table 1) and hence increase the density of the powder. The disks were sized in such a way that at only one disk had to be changed at a time. With a little practice, a copper disk could be changed without any significant loss of powder, and even small losses could be measured if they occurred. Thus, the powder had minimal disturbance between measurements, and the entire sequence of measurements was made on the same specimen of powder.

All measurements were carried out with the cell environment set at 303 K. For the results presented, the heater input was not changed, so that the temperature differences across the specimen cell did change with the cell's thermal resistance. Typically, temperature differences across the specimen cell ranged between 1 and 5°C and average temperatures of the powder specimen ranged from 30.1 to 31.8°C. Temperature effects on thermal conductivity for these small changes in average temperature have not been considered at this time.

Raw data simply comprises the resistance of the thermistors in the heat-flux meters, together with the physical parameters of the specimen cell. Whilst the system is coming to equilibrium, the difference between the resistance of the pairs of thermistors is monitored by the output of Wheatstone Bridge circuits on a strip chart recorder. When equilibrium has been established, usually in 2 to 6 hours, the resistance of five thermistors is measured directly using a multimeter with 4½ digit resolution. Thermistor calibrations, thermal parameters of the materials involved, and interfacial thermal resistance data are stored in a spreadsheet. Values of the total thermal resistance, thermal resistance of the cell wall, and the thermal conductivity of the specimen are calculated within this spreadsheet.

## RESULTS

Measurements on two aluminium powders are reported at a nominal average temperature of 303 K. Specimen #1 was an 18 micron Al powder

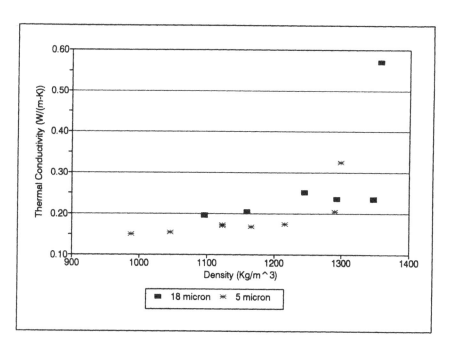

Figure 4. Thermal Conductivity of Aluminium powder -- low density comparison

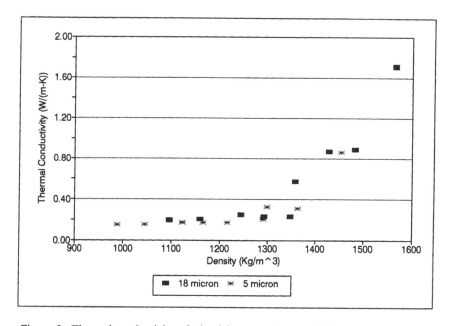

Figure 5. Thermal conductivity of Aluminium powder -- to limiting density.

manufactured by Alcoa; specimen #2 was a 5 micron Al powder manufactured by Reynolds. Figures 4 and 5 show the results obtained.

Figure 4 shows the values of thermal conductivity for both powders at low density. The reproducibility of the results can be seen; each point shown on the graphs is a double or triple data point for each density. It should be noted that the data for the 5 micron powder at ~1120 Kg/m$^3$ is a quadruple point. Two data points were taken with the fibre-glass insulation around the conductivity stack, and two more were taken with absence of the insulation. This shows the efficacy of the dual heat-flux system to automatically correct for heat losses.

Figure 5 shows the full scope of the data varying from the "fluffy" powder to powder which was compressed to the maximum extent possible by simple means. In order to achieve the maximum compression the copper plates were pressed together using a vice to the point that the powder became close to a competent solid.

## DISCUSSION

There appears to be very limited data for powdered samples as a function of powder density. For example, Gilcrist and Preston [6] report on the thermal conductivity of stainless steel powders as a function of different fluids in the pore space. But measurements were only at one powder density for each of the two powders, 13 μm and 42 μm in size. The reported thermal conductivities in air were 0.19 W/(m-K) and 0.23 W/(m-K), respectively, which are comparable with the data presented here. In both cases, the powders were hand tapped into the containers giving values of ε (porosity of continuous phase in volume %) of 0.44 and 0.40, respectively.

Whilst Luikov et al [7] discuss the effects of external pressure on the inter-granular contact resistance in their important review of conduction in porous systems, data they present from their own work and from a large number of references do not include the dependence of apparent thermal conductivity on external pressure or density.

There are a large number of models which could be considered. Pratt [8] summarizes some common models. Ashworth [3] and Woodside and Messmer [9] discuss some of the earlier models in more detail.

By assuming that the thermal resistance of all the aluminium particles into one series resistance and the air voids into another series resistance, the overall resistance can be calculated [3,9]. Assuming typical values for the thermal conductivity of air (0.024 W/(m-K)) and aluminium (237 W/(m-K)), the thermal conductivity for any porosity can be found. The thermal conductivity values are shown in Figure 6 and represent the theoretical lower bound.

The thermal conductivity for a parallel model assumes the air resistance and aluminium resistance are in parallel. Values are graphed in Figure 6; the curve is shown as quickly rising to a high conductivity, even for a small value of n. As expected, this model gives the minimum thermal resistance and hence the upper bound for the thermal conductivity. The model appears to be applicable in some cases where the thermal conductivity values of the components are of the same order of magnitude. The model was used, for example, by Whittaker et al [10] in calculating the thermal conductivity of ceramic fibers in epoxy. Schneider et al [11] uses an extension of the model to a three component system (polyurethane foam at a range of moisture levels). The model is attributed to Schuhmeister [12] and provides a very good fit for Schneider's experimental data.

A third model is the geometric mean. This model has no direct analogy with lumped parameter combinations but represents a random distribution of the phases. The values of thermal conductivity rise quickly with n giving a curve just below that for the parallel model (not shown). Woodside and Messmer [9] concluded that this model over estimates the thermal conductivity when the conductivity ratio of the two phases is greater than 20, which is certainly the case for the data presented in this paper.

Additional models are presented in Gilcrist and Preston [6], as they compare their data with six models that do not consider contact resistance. Assumptions are perfectly spherical particles of one size, no convective contribution to heat flow, gas at atmospheric pressure and gas was the continuous medium.

Two of these models (Swift [13] and Deissler and Eian [14,15]) include only the variable of the ratio of the thermal conductivities of the solid-powder state to that of the gas-liquid phase. Thus, changes in density (porosity) do not affect the calculated conductivities. Swift's model gives a value of 0.36 W/(m-K); the Deissler and Eian model gives a value of 0.31 W/(m-K). Both of these values are not inconsistent with the experimental value obtained here for about 50% aluminium and both of these models give the best fit with the Gilcrist and Preston data. However, these two models are inappropriate to explain the variation of thermal conductivity with density.

The Kampf-Karsten model [16] gives

$$\lambda_{pow} = [\ 1\ -\ \{\ n\ (\lambda_{air}\ /\ \lambda_{al}\ -\ 1)\ /\ (1 + n^{1/3}\ [\lambda_{air}\ /\ \lambda_{al}\ -\ 1])\}\ ]\ *\ \lambda_{air} \qquad (1)$$

where $\lambda_{pow}$ = thermal conductivity of the powder,

$\lambda_{air}$ = thermal conductivity of still air,

$\lambda_{Al}$ = thermal conductivity of aluminium,

and n = percent of aluminium in the mixture (by volume).

The Eucken model [17], which is an extension of Maxwell's model, gives the comparable formula:

$$\lambda_{pow} = \frac{[\ 1\ +\ 2\ n\ (\lambda_{air}\ /\ \lambda_{al}\ -\ 1)\ /\ (2 + \lambda_{air}\ /\ \lambda_{al})\ ]\ *\ \lambda_{air}}{[\ 1\ -\ n\ (\lambda_{air}\ /\ \lambda_{al}\ -\ 1)\ /\ (2 + \lambda_{air}\ /\ \lambda_{al})\ ]} \qquad (2)$$

Even though these models have different formulations, they give very similar curves as functions of percent of aluminium; only the Eucken curve is shown in Figure 6.

Russell [18] includes the variable $n^{2/3}$ together with the ratio $\lambda_{air}/\lambda_{al}$ in his formula. A plot of the thermal conductivity (not shown) as a function of n gives basically the same curve as Eucken and Kampf-Karsten. At n = 60% (porosity of 40%), the three models predict thermal conductivities of 0.12, 0.13, and 0.15 W/(m-K), respectively.

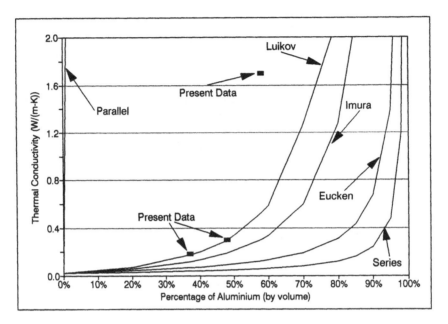

Figure 6. Thermal conductivity as a function of density for five models and experimental data.

The best of these six models for the current data is Imura and Takegoshi [19]. Gilcrist and Preston indicate that this model gives the highest values; the same has been found here as can be seen in Figure 6. This model was a semi-analytical derivation for $\varepsilon$ (or $1 - n$) in the range 0.39 to 0.21. However, even though the model seems appropriate for a wider range of n and produces higher values than the other models, the thermal conductivities do not rise quickly enough. Thus, another component to the thermal resistance needs to be considered; this is missing from all of the models considered so far.

The remaining model considered by Gilcrist and Preston was that of Flinta [20]. This model gives inappropriate values for both low and high values of n; the contact resistance is given as an equivalent thermal conductivity, which, when increased to represent increasing density, has insignificant effect on the calculated powder conductivity.

Luikov et al's model [7] includes the effects at the contact of both the conduction through the points of physical contact and by the gas and radiation in the region of the physical contact and in the spaces between the microscopic contact points. To obtain the fit shown in Figure 6, the value of the product of the surface micro-roughness and coefficient of particle adhesion had to be increased by a factor of 4 from the values they recommend for metals. For this curve, the conduction through physical contact was set at zero. Formulae are included in Luikov et al's paper for the physical contact conduction as a function of intergranular pressure; it is likely that this missing component which is likely to be the cause of deviation above the 50% aluminium content. Since the applied pressure within the cell cannot presently be measured, we are unable to estimate its dependence on specimen density.

## CONCLUSIONS

It is concluded that models which simply consider the effects of a two phase system are inappropriate for the description of the thermal conductivity of aluminium powder. At the lower densities involved in the present measurements (less than 1300 Kg/m$^3$), there is a slow increase of conductivity with density. Even in this region, the two phase models are inadequate. In order to achieve agreement with the data, a term representing the non-direct component of the contact resistance must be included, as in Luikov et al's model. At higher densities where the intergranular pressure increases, the additional component of the direct contact resistance appears to be required.

In a private communication, Tye [21] indicated that he had observed similar behaviour as a function of applied load for metallic fibers.

For the control of the combustion synthesis process, the nearly constant thermal conductivity at lower densities allows the combustion energy per unit volume to be varied without deleterious effects. However when a certain density is reached, where the direct particle-particle boundary resistance begins to be significantly reduced, the thermal conductivity increases sharply. Using powders beyond this density would greatly reduce the effectiveness of the process, by increasing the size of the heating and conversion zones, thereby reducing the process temperature.

## ACKNOWLEDGMENTS

The authors wish to thank Dr. J. A. Puszynski of the Chemical Engineering Department, SDSM&T, for his advice regarding the combustion synthesis process, and for providing the samples of powder.

## REFERENCES

1. Puszynski, J. A., 1995. Chemical Engineering Department, SDSM&T, Personal Communication.

2. Ashworth, T., E. Ashworth, and S. F. Ashworth, 1991. "A New Apparatus for Materials of Intermediate Conductivity," Thermal Conductivity 21, C. Cremers and A. Fine, Editors, Plenum Press, NY, pp. 51-66.

3. Ashworth, E., 1992. "Heat Flow into Underground Openings -- Significant Factors," Ph.D. Dissertation, University of Arizona, 478 p.

4. Ashworth, T. and E. Ashworth, 1993. "Thermal Resistance of Teflon as an Interfacial Medium," presented at 22nd International Thermal Conductivity Conference, Arizona State University, Tempe, November.

5. Ashworth, E., 1992. "The Thermal Conductivity of Tuff - Moisture Effects," Proceedings of 33rd U. S. Symposium on Rock Mechanics, Tillerson, J. R. and W. R. Wawersik, Editors, A. A. Balkema Publishers, Rotterdam, pp. 859-868.

6. Gilcrist, K. E. and S. D. Preston, 1985. "The Thermal Diffusivity/Conductivity of 316 Stainless Steel Powders in Air and Nitric Acid Mixtures at Room Temperature," Thermal Conductivity 18, T. Ashworth and D. R. Smith, Editors, Plenum Press, NY, pp. 713-722.

7. Luikov, A.V., A.G. Shashkov, L.L. Vasiliev, and Yu. E. Fraiman, 1968. "Thermal Conductivity of Porous Systems," Int. J. Heat Mass Transfer, Vol. 11, pp. 117-140.

8. Pratt, A.W., 1969. "Heat Transmission in Low Conductivity Materials --III Thermal Conductivity in Granular Materials" in Thermal Conductivity, Volume I, edited by R.P. Tye, Academic Press, London, pp. 319-326.

9. Woodside, W. and J. H. Messmer, 1961. "Thermal Conductivity of Porous Media, I and II," J. Appl. Phys., Vol. 32, pp. 1688-1706.

10. Whittaker, A.J., M.L. Allitt, D.G. Onn, and J.D. Bolt, Thermal Conductivity 21, C. Cremers and A. Fine, Editors, Plenum Press, NY, pp. 187-198.

11. Schneider, A.M., H.J. Goldsmid, and B.N. Hoschke, 1989. "Heat Transfer through Porous Media," Thermal Conductivity 20, D.P.H. Hasselman and J.R. Thomas, Jr., Plenum Press, NY, pp. 415-421.

12. Schuhmeister, J., 1877. Ber. K. Akad. Wiss., Wien. Math.-Naturwiss. Klasse, 76:283.

13. Swift, D.L., 1966. "Thermal Conductivity of Spherical Metal Powders including the Effect of an Oxide Coating," Int. J. Heat Mass Transfer, Vol. 9, p. 1061.

14. Deissler, R.G. and C.S. Eian, 1952. "Investigation of Effective Thermal Conductivities," NACA RM E52 CO5.

15. Eian, C.S. and R.G. Deissler, 1953. "Effective Thermal Conductivities of MgO, Stainless Steel and Uranium Oxides in Various Gases," NACA RM E53 GO3

16. Kampf, H. and G. Karsten, 1970. "Effects of Different Types of Void Volumes on the Radial Temperature Distribution of Fuel Pins," Nucl. Appl. Technol., Vol. 9, p. 288.

17. Euken, A., 1932. "The Thermal Conductivity of Ceramic Refractory Materials," VDI Forschung, Vol. 353.

18. Russell, H.W., 1935. "Principles of Heat Flow in Porous Insulators," J. Am. Ceram. Soc. <u>Vol. 18</u>, p. 1.

19. Imura, S. and E. Takegoshi, 1974. "Effect of Gas Pressure on the Effective Conductivity of Packed Beds," Heat Transfer-Japanese Research, <u>Vol. 3</u>, No 4, pp. 13-26.

20. Flinta, J.E., 1958. "Thermal Conductivity of Uranium Oxide," Fuel Elements Conference, Paris, TID-7546, pp. 516-525.

21. Tye, R. P., 1995. Private Communication about unpublished work.

**SESSION 10**

---

# ORGANICS

SESSION CHAIR
J. R. Booth

# Organic Compounds: Correlations and Estimation Methods for Thermal Conductivity

G. LATINI, G. PASSERINI, F. POLONARA and G. VITALI

## ABSTRACT

Design of industrial devices frequently requires a wide and precise knowledge of the transport properties of all the fluids involved since efforts to reduce environmental impact often concentrate in the optimisation of their employment and in the use of new ones. The thermal conductivity is one of the most critical properties because organic compounds are widely required both for the production and the operation of the greater part of appliances.

In this paper an extension of a previously determined predictive method to new families of organic compounds is introduced: a simple equation relates the liquid thermal conductivity to the reduced temperature. It is effective in the range from the normal freezing point nearly to the critical point, though it does not take into account critical enhancement. Parameters are extracted from available thermophysical properties and evaluated for each family. Mean deviations are usually below 5% while typical maximum deviations are below 10% in the reduced temperature range 0.3 to 0.95.

## INTRODUCTION

This work marks the third step of a research project started ten years ago. The first part of the research centered on the evaluation of estimation methods for the liquid thermal conductivity of organic fluids in the saturated state. As a first result a formula was able to evaluate such property for almost all the organic compounds. The formula was effective in a rather wide temperature range and required only a single, reliable experimental datum.

A second step led to a new formula to evaluate from physical properties the parameter once calculated by means of the experimental datum, thus making the method predictive. This formula was specifically developed for refrigerants. While analysing these compounds a new form for the main equation was reached.

Giovanni Latini, Giorgio Passerini, Fabio Polonara, Giuliano Vitali, Dipartimento di Energetica, Università di Ancona, via Brecce Bianche, 60100 Ancona, Italy

This expression possibly represented the simplest estimation method: a single straight line with just one parameter was able to predict the liquid thermal conductivity of refrigerants in the saturated state from the normal freezing point near to the critical point.

Aim of this work is to demonstrate that this new formula is effective also for other organic compounds to prove again that the method is "general". A new organization of data is proposed within this paper. In previous works organic fluids were grouped according to chemical families. Now alkanes and aromatic compounds are still organised as families but we grouped other fluids within sets, consisting of all the compounds whose moleculas contain the same number of carbon atoms. Results are presented only for the first two sets: compounds of methane derivatives and of ethane derivatives.

Table 1 shows the analysed compounds. Mean deviations between predicted results and experimental data available in literature are usually below 5% while typical maximum deviations are below 10% in the reduced temperature range 0.3 to 0.95.

**PREVIOUS WORK ON THE METHOD**

This work follows the introduction of a simple and reliable formula originally conceived as an estimation tool and then modified to obtain reliable prediction capabilities for carbon and halocarbon refrigerants. The development of the equations proceeded from an equation proposed by one of the present authors [1, 2]. This equation was established for pure fluids and had the form:

$$\lambda = A \frac{(1 - T_r)^{0.38}}{T_r^{1/6}} \tag{1}$$

where the constant A was determined from experimental data for many organic liquids along the saturation line or near saturation. A first attempt to link factor A with one or more thermophysical properties was fulfilled for the refrigerant families and led to the expression:

$$A = A^* \frac{T_c^{1/6}}{M^{1/2}} \tag{2}$$

where A* was a constant which assumed two different values, one for the fully halogenated methane and ethane refrigerants, another one for the methane based refrigerants having a single hydrogen.

The method was improved by present authors [3, 4], extending the reduced temperature range of validity up to 0.95. A new equation was developed to describe saturated liquid thermal conductivity as a function of reduced temperature:

$$\lambda = A(1/T_r - 1)^{1/3} \tag{3}$$

A further step was possible by writing a linear expression:

$$\lambda = 4^{1/3} \cdot A \cdot \left(1 - \frac{3}{4}T_r\right) = B \cdot \left(1 - \frac{3}{4}T_r\right) \qquad (4)$$

This equation represents the tangent of the function given by Eq. (3) at the inflection point $T_r = 2/3$. It describes a straight line and it is well known that the liquid thermal conductivity is approximately a linear function of temperature. The constant B had the same characteristics of the constant A. It was linked with some thermophysical properties in the form:

$$B = B^* T_c^\alpha P_c^\beta M^\gamma \qquad (5)$$

where $B^*$, $\alpha$, $\beta$, and $\gamma$ are numerical constants whose values were different for the methane series refrigerants and the ethane series refrigerants.

At this point the method could be regarded as a predictive one since no experimental datum was required for the evaluation of the liquid thermal conductivity.

## EXTENSION OF THE METHOD TO NEW ORGANIC COMPOUNDS

The experimental thermal conductivity values used for the new analysis were taken from several papers but most of them appear in ref. [7]. It must be pointed out that some of the data do not refer to the saturation state. They are usually the result of experiments performed at low pressures with different techniques. A preliminary investigation was made about their reliability, both by comparing values belonging to different data sets and evaluating their dependence on the pressure.

After data validation, a first work of comparison was performed in order to check the capability of the Eq.(4) to evaluate liquid thermal conductivity of alkanes, aromatic compounds and other substances derivatives of methane and ethane series. It was found that Eq.(4) is able to predict their behaviour with almost the same accuracy it showed when evaluating refrigerants. Thus we concentrated our efforts on the factor B which had to be linked with one or more thermophysical properties.

We started to try to link B with properties related to the critical point while the use of the normal boiling point $T_b$ and the normal freezing point $T_f$ was avoided since their values depend on the normal pressure. Unfortunately we found that values of the critical volume $V_c$ are not easily available in literature so we decided to relate the factor B with the critical pressure $P_c$, the critical temperature $T_c$ and the molecular mass M. Thus we achieved the same form of Eq.(5). The new values for $B^*$, $\alpha$, $\beta$, and $\gamma$ are reported in Table 2. A zero value for an exponent means no manifest dependence of B from the related property. A special case is that of aromatic compounds: a single value for the factor B is required to analyse all the compounds belonging to the family. This is due to the comparable molecular structures of these substances.

Table 1 - Investigated compounds and some of their properties [5,6]

| Compound | Reduced Formula | M | $T_f$ [K] | $T_b$ [K] | $T_c$ [K] | $P_c$ [bar] | $V_c$ [m$^3$·kmol$^{-1}$·10$^3$] |
|---|---|---|---|---|---|---|---|
| Alkanes | | | | | | | |
| Methane | CH$_4$ | 16.043 | 90.7 | 111.7 | 190.6 | 46 | 0.0992 |
| Ethane | C$_2$H$_6$ | 30.07 | 89.9 | 184.5 | 305.4 | 48.8 | 0.1483 |
| Propane | C$_3$H$_8$ | 44.094 | 85.5 | 231.1 | 369.8 | 42.5 | 0.203 |
| n-Butane | C$_4$H$_{10}$ | 58.124 | 134.8 | 272.7 | 425.2 | 38 | 0.255 |
| n-Pentane | C$_5$H$_{12}$ | 72.151 | 143.4 | 309.2 | 469.7 | 33.7 | 0.304 |
| n-Hexane | C$_6$H$_{14}$ | 86.178 | 155 | 341.9 | 507.5 | 30.1 | 0.37 |
| n-Heptane | C$_7$H$_{16}$ | 100.205 | 182.6 | 371.6 | 540.3 | 27.4 | 0.432 |
| n-Octane | C$_8$H$_{18}$ | 114.232 | 216.4 | 398.8 | 568.8 | 24.9 | 0.492 |
| n-Nonane | C$_9$H$_{20}$ | 128.259 | 219.7 | 424 | 594.6 | 22.9 | 0.548 |
| n-Decane | C$_{10}$H$_{22}$ | 142.286 | 243.5 | 447.3 | 617.7 | 21.2 | 0.603 |
| n-Undecane | C$_{11}$H$_{24}$ | 156.313 | 247.6 | 469.1 | 638.8 | 19.7 | 0.66 |
| n-Dodecane | C$_{12}$H$_{26}$ | 170.34 | 263.6 | 489.5 | 658.2 | 18.2 | 0.713 |
| n-Tridecane | C$_{13}$H$_{28}$ | 184.367 | 267.8 | 508.6 | 676 | 17.2 | 0.78 |
| n-Tetracane | C$_{14}$H$_{30}$ | 198.394 | 279 | 526.7 | 693 | 14.4 | 0.83 |
| n-Pentadecane | C$_{15}$H$_{32}$ | 212.421 | 283 | 543.8 | 707 | 15.2 | 0.88 |
| n-Hexadecane | C$_{16}$H$_{34}$ | 226.448 | 291 | 560 | 722 | 14.1 | 0.88 |
| n-Heptadecane | C$_{17}$H$_{36}$ | 240.475 | 295 | 575.2 | 733 | 13 | - |
| n-Octadecane | C$_{18}$H$_{38}$ | 254.504 | 301.3 | 589.5 | 748 | 12 | - |
| n-Nonadecane | C$_{19}$H$_{40}$ | 268.529 | 305 | 603.1 | 756 | 11.1 | - |
| n-Eicosane | C$_{20}$H$_{42}$ | 282.556 | 310 | 617 | 767 | 11.1 | - |
| Methane derivatives | | | | | | | |
| R10 | CCl$_4$ | 153.823 | 250 | 349.9 | 556.3 | 44.13 | 0.276 |
| R11 | CCl$_3$F | 137.37 | 162.15 | 296.9 | 471.15 | 44.06 | 0.247 |
| R12 | CCl$_2$F$_2$ | 120.91 | 115.15 | 243.4 | 384.95 | 41.13 | 0.247 |
| R13 | CClF$_3$ | 104.46 | 92.15 | 191.7 | 302 | 38.65 | 0.217 |
| R13B1 | CBrF$_3$ | 148.91 | 105.15 | 215.5 | 340.2 | 39.62 | 0.2 |
| R20 | CHCl$_3$ | 119.378 | 209.6 | 334.3 | 536.4 | 53.84 | 0.239 |
| R21 | CHCl$_2$F | 102.93 | 138 | 282.1 | 451.65 | 51.7 | 0.1964 |
| R22 | CHClF$_2$ | 86.47 | 113.15 | 232.3 | 369.33 | 49.9 | 0.169 |
| R23 | CHF$_3$ | 70.01 | 118.15 | 191 | 299.1 | 48.33 | 0.133 |
| Bromoform | CHBr$_3$ | 252.77 | 281.05 | 423.65 | 685.1 | 56.5 | - |
| R30 | CH$_2$Cl$_2$ | 84.933 | 178.1 | 313 | 510 | 63 | - |
| R31 | CH$_2$FCl | 68.478 | 178 | 263.9 | 430 | 63 | - |
| R32 | CH$_2$F$_2$ | 52.024 | 137 | 221.6 | 351.36 | 58.14 | 0.121 |
| Dibromomethane | CH$_2$Br$_2$ | 173.835 | 220.6 | 370 | 583 | 71 | - |
| Nitromethane | CH$_3$NO$_2$ | 61.041 | 244.6 | 374.3 | 588 | 63.1 | 0.1732 |

Table 1(cont.) - Investigated compounds and some of their properties [5, 6]

| Compound | Reduced Formula | M | $T_f$ [K] | $T_b$ [K] | $T_c$ [K] | $P_c$ [bar] | $V_c$ [m$^3$·kmol$^{-1}$·10$^3$] |
|---|---|---|---|---|---|---|---|
| Ethane derivatives | | | | | | | |
| R112 | $C_2F_2Cl_4$ | 203.830 | 298 | 366 | 551 | 38.7 | 0.2711 |
| R113 | $C_2Cl_3F_3$ | 187.38 | 238.15 | 320.7 | 487.5 | 34.37 | 0.329 |
| R114 | $C_2Cl_2F_4$ | 170.92 | 179.15 | 276.8 | 418.8 | 32.48 | 0.307 |
| R114B2 | $C_2Br_2F_4$ | 259.824 | 162.65 | 320.15 | 487.65 | 55.55 | 0.341 |
| R115 | $C_2ClF_5$ | 154.467 | 167 | 235.2 | 353.2 | 32.3 | 0.2518 |
| R116 | $C_2F_6$ | 138.012 | 172.4 | 194.9 | 293 | 30.6 | 0.222 |
| R123 | $C_2HCl_2F_3$ | 152.93 | 165.95 | 301.85 | 456.94 | 36.68 | 0.2725 |
| R123a | $C_2HCl_2F_3$ | 152.93 | 195.15 | 301.15 | 461.1 | 44.7 | 0.2783 |
| R124 | $C_2HClF_4$ | 136.475 | 74 | 260 | 395.6 | 36.34 | 0.244 |
| R125 | $C_2HF_5$ | 120.02 | 170 | 224.7 | 339.33 | 36.39 | 0.21 |
| Trichloroethylene | $C_2HCl_3$ | 131.389 | 186.8 | 360.4 | 572 | 50.5 | 0.256 |
| R132b | $C_2H_2Cl_2F_2$ | 134.941 | 171.95 | 319.95 | 493.15 | 49 | 0.2267 |
| R133a | $C_2H_2ClF_3$ | 118.486 | 145.91 | 279.15 | 432.02 | 49.4 | 0.2395 |
| R134a | $C_2H_2F_4$ | 102.03 | 172.15 | 247.1 | 374.3 | 40.56 | 0.199 |
| Tetrachl.ethylene | $C_2H_2Cl_4$ | 165.834 | 251 | 394.4 | 620.2 | 47.6 | 0.2896 |
| 1122-tetrachl.ethane | $C_2H_2Cl_4$ | 167.85 | 237 | 419.4 | 661.2 | 58.4 | - |
| R141b | $C_2H_3Cl_2F$ | 116.95 | 169.85 | 304.9 | 477.3 | 41.2 | 0.2521 |
| R142b | $C_2H_3ClF_2$ | 100.49 | 142.15 | 263.4 | 409.6 | 43.3 | 0.231 |
| R152a | $C_2H_4F_2$ | 66.05 | 156.15 | 248.5 | 386.7 | 44.92 | 0.18 |
| 1,2-dibromoethane | $C_2H_4Br_2$ | 187.862 | 283.3 | 404.7 | 646 | 53.5 | 0.3 |
| 1,2-dichloroethane | $C_2H_4Cl_2$ | 98.96 | 237.5 | 356.7 | 566 | 53.7 | 0.225 |
| Acetic acid | $C_2H_4O_2$ | 60.052 | 289.8 | 391.1 | 592.7 | 57.9 | 0.171 |
| Ethylbromide | $C_2H_5Br$ | 108.966 | 154.6 | 311.5 | 503.9 | 62.3 | 0.215 |
| Ethyliodide | $C_2H_5I$ | 155.967 | 165 | 345.6 | 554 | 47 | 0.255 |
| Ethanol | $C_2H_5Br$ | 46.069 | 159.1 | 351.4 | 513.9 | 61.4 | 0.1671 |
| Aromatics | | | | | | | |
| Benzene | $C_6H_6$ | 78.114 | 278.7 | 353.2 | 562.2 | 48.9 | 0.259 |
| Toluene | $C_7H_8$ | 92.141 | 178 | 383.8 | 591.8 | 41 | 0.316 |
| Naphtalene | $C_{10}H_8$ | 128.174 | 353.5 | 491.1 | 748.4 | 40.5 | 0.413 |
| Ethylbenzene | $C_8H_{10}$ | 106.168 | 178.2 | 409.3 | 617.2 | 36 | 0.374 |
| m-Xylene | $C_8H_{10}$ | 106.168 | 225.3 | 412.3 | 617.1 | 35.4 | 0.376 |
| o-Xylene | $C_8H_{10}$ | 106.168 | 248 | 417.6 | 630.3 | 37.3 | 0.369 |
| n-Xylene | $C_8H_{10}$ | 106.168 | 286.4 | 411.5 | 616.2 | 35.1 | 0.379 |

Table 2 - Coefficients of Eq. (5)

| Families | B* | $\alpha$ | $\beta$ | $\gamma$ |
|---|---|---|---|---|
| Alkanes | 0.53 | 0.92 | -0.7 | -0.94 |
| Methane derivatives | 0.28 | 0.18 | 0.36 | -0.61 |
| Ethane derivatives | 0.65 | 0.23 | 0 | -0.57 |
| Aromatics | 0.22 | 0 | 0 | 0 |

## RESULTS

The results are shown in Table 3 in form of Average Absolute Deviation (AAD) and Maximum Absolute Deviation (MAD) in the reduced temperature range 0.35 to 0.95

The method gives average absolute deviations between estimated and experimental thermal conductivity usually less than 5% and maximum absolute deviations usually less than 10%. Figures 1 through 4 show the same information in more details for most of compounds.

In Table 3 some compounds show deviations higher than the typical ones. As a general rule, when high deviations are comparable in their average and maximum values, it can be assumed that errors are mostly due to the factor B, thus to Eq. (5). On the other hand when maximum errors are much higher than average ones this weakness is mostly due the Eq. (4).

According to this explication errors in methane, ethane, toluene, m-xylene, n-xylene and R125 estimations are due to a weakness of Eq. (4). As far as methane and ethane are concerned one could expect a different behaviour for both of them due to their different molecular structure and their low molecular mass. We found that errors could be reduced by replacing the factor 3/4 in Eq. (4) with the new value 0.77 but this would have increased errors related to almost all the other compounds.

The remarkable errors related to bromoform, R32 and R134a. are due to the factor B. The knowledge that errors are only due to the factor B is very important since we expect that deviations should not increase outside the actually investigated temperature range.

## CONCLUSIONS

A prediction method, is proposed for thermal conductivity of 68 organic fluids. It evaluates thermal conductivity in the liquid state along the saturation line and in the subcooled region at pressures near saturation. The method determines thermal conductivity as a function of the reduced temperature. It requires the knowledge of a single parameter which is related to easily available physical constants characteristic of each substance: the critical temperature, critical pressure and molecular weight. The

Table 3 - Comparison between experimental and estimated thermal conductivities

| Compound | Source of experimental data | Const.B of Eq. (4) estimated with Eq. (5) | Const.B of Eq. (4) derived from exper. data | AAD % | MAD % |
|---|---|---|---|---|---|
| Alkanes | | | | | |
| Methane | [7-9] | 0.3350 | 0.3150 | 6.15 | 14.85 |
| Ethane | [7], [8], [10] | 0.2748 | 0.2954 | 5.78 | 15.62 |
| Propane | [7], [11], [12] | 0.2519 | 0.2526 | 2.08 | 4.10 |
| n-Butane | [7] | 0.2389 | 0.2309 | 3.94 | 7.87 |
| n-Pentane | [7], [13] | 0.2324 | 0.2197 | 5.86 | 8.25 |
| n-Hexane | [7], [14] | 0.2286 | 0.2131 | 7.27 | 9.98 |
| n-Heptane | [7], [14] | 0.2244 | 0.2126 | 5.55 | 6.65 |
| n-Octane | [7], [14] | 0.2224 | 0.2111 | 5.34 | 6.97 |
| n-Nonane | [7], [14] | 0.2203 | 0.2064 | 6.75 | 8.08 |
| n-Decane | [7], [14] | 0.2185 | 0.2327 | 6.90 | 6.93 |
| n-Undecane | [7], [14] | 0.2171 | 0.2082 | 4.27 | 5.49 |
| n-Dodecane | [7], [14] | 0.2176 | 0.2071 | 5.07 | 6.43 |
| n-Tridecane | [7] | 0.2154 | 0.2008 | 7.24 | 8.51 |
| n-Tetracane | [7] | 0.2145 | 0.2091 | 2.56 | 4.21 |
| n-Pentadecane | [7] | 0.2142 | 0.2109 | 2.21 | 3.45 |
| n-Hexadecane | [7] | 0.2168 | 0.2142 | 2.25 | 3.50 |
| n-Heptadecane | [7] | 0.2199 | 0.2180 | 2.30 | 4.47 |
| n-Octadecane | [7] | 0.2246 | 0.2220 | 3.16 | 5.28 |
| n-Nonadecane | [7] | 0.2278 | 0.2259 | 3.39 | 7.00 |
| n-Eicosane | [7] | 0.2201 | 0.2150 | 2.31 | 2.92 |
| Methane derivatives | | | | | |
| R10 | [7] | 0.1582 | 0.1739 | 8.98 | 10.95 |
| R11 | [7], [15-17] | 0.1645 | 0.1661 | 1.99 | 4.04 |
| R12 | [7], [15-17] | 0.1672 | 0.1640 | 2.41 | 8.21 |
| R13 | [7], [15] | 0.1711 | 0.1705 | 2.65 | 5.99 |
| R13B1 | [7], [15] | 0.1421 | 0.1459 | 2.67 | 6.57 |
| R20 | [7] | 0.1971 | 0.2008 | 2.31 | 5.13 |
| R21 | [7] | 0.2062 | 0.1995 | 3.59 | 5.33 |
| R22 | [7], [15-18] | 0.2183 | 0.2154 | 1.79 | 6.49 |
| R23 | [7], [19] | 0.2364 | 0.2536 | 6.73 | 10.07 |
| Bromoform | [7] | 0.1326 | 0.1510 | 12.11 | 14.96 |
| R30 | [7] | 0.2544 | 0.2472 | 2.86 | 4.54 |
| R31 | [7] | 0.2814 | 0.2834 | 1.33 | 2.56 |
| R32 | [20], [21] | 0.3117 | 0.3599 | 13.39 | 14.88 |
| Dibromomethane | [7] | 0.1758 | 0.1758 | 0.24 | 0.56 |
| Nitromethane | [7] | 0.3195 | 0.3279 | 2.58 | 2.94 |

AAD(%)=abs{[$\Sigma(\lambda_{calc}/\lambda_{exp}-1)$]/n}100; MAX(%)=max of [abs($\lambda_{calc}/\lambda_{exp}-1$)/]100; $\lambda_{exp}$= experimental thermal conductivity value; $\lambda_{calc}$= estimated thermal conductivity value; n= number of experimental points.

Table 3 (cont) - Comparison between experimental and estimated thermal conductivities

| Compound | Source of experimental data | Const.B of Eq. (4) estimated with Eq. (5) | Const.B of Eq. (4) derived from exper. data | AAD % | MAD % |
|---|---|---|---|---|---|
| Ethane derivatives | | | | | |
| R112 | [20] | 0.1344 | 0.1395 | 3.64 | 4.77 |
| R113 | [7], [15] | 0.1371 | 0.1363 | 2.90 | 7.64 |
| R114 | [7], [15], [16] | 0.1396 | 0.1346 | 3.68 | 9.56 |
| R114B2 | [15] | 0.1139 | 0.1141 | 0.73 | 2.13 |
| R115 | [7], [15], [22] | 0.1422 | 0.1390 | 2.42 | 4.84 |
| R116 | [20] | 0.1452 | 0.1545 | 6.33 | 6.77 |
| R123 | [16], [23-27] | 0.1516 | 0.1520 | 1.53 | 2.16 |
| R123a | [16] | 0.1519 | 0.1487 | 2.17 | 3.28 |
| R124 | [7],[15],[16],[28] | 0.1565 | 0.1599 | 2.32 | 4.60 |
| Trichloroethylene | [7] | 0.1739 | 0.1880 | 7.45 | 9.01 |
| R125 | [16],[21],[27],[29] | 0.1625 | 0.1825 | 4.71 | 18.53 |
| R132b | [20] | 0.1656 | 0.1740 | 4.76 | 7.59 |
| R133a | [20] | 0.1730 | 0.1875 | 7.73 | 9.67 |
| R134a | [23-26],[16],[30-35] | 0.1822 | 0.2026 | 11.38 | 14.72 |
| Tetrachloroethylene | [7] | 0.1553 | 0.1699 | 8.62 | 10.71 |
| 1122-tetrachl.ethane | [7] | 0.1565 | 0.1644 | 4.76 | 7.96 |
| R141b | [28], [33] | 0.1783 | 0.1726 | 3.77 | 13.17 |
| R142b | [28], [34] | 0.1876 | 0.1795 | 4.83 | 7.69 |
| R152a | [25], [28], [18] | 0.2349 | 0.2464 | 4.74 | 7.75 |
| 1,2-dibromoethane | [7] | 0.1460 | 0.1542 | 5.32 | 6.42 |
| 1,2-dichloroethane | [7] | 0.2038 | 0.2207 | 7.65 | 10.44 |
| Acetic acid | [7] | 0.2733 | 0.2722 | 2.63 | 4.87 |
| Ethylbromide | [7] | 0.1879 | 0.1716 | 9.71 | 17.82 |
| Ethyliodide | [7] | 0.1567 | 0.1465 | 6.93 | 8.19 |
| Ethanol | [7] | 0.3075 | 0.3012 | 7.96 | 14.99 |
| Aromatics | | | | | |
| Benzene | [7] | 0.2200 | 0.2426 | 9.29 | 11.18 |
| Toluene | [7], [36] | 0.2200 | 0.2145 | 4.48 | 12.30 |
| Naphtalene | [7] | 0.2200 | 0.2135 | 3.08 | 4.87 |
| Ethylbenzene | [7] | 0.2200 | 0.2110 | 7.62 | 13.98 |
| m-Xylene | [7] | 0.2200 | 0.2157 | 6.91 | 14.80 |
| o-Xylene | [7] | 0.2200 | 0.2118 | 7.10 | 9.80 |
| n-Xylene | [7] | 0.2200 | 0.2145 | 7.28 | 14.88 |

AAD(%)=abs{[$\Sigma(\lambda_{calc}/\lambda_{exp}-1)$]/n}100; MAX(%)=max of [abs($\lambda_{calc}/\lambda_{exp}-1$)/]100; $\lambda_{exp}$= experimental thermal conductivity value; $\lambda_{calc}$= estimated thermal conductivity value; n= number of experimental points.

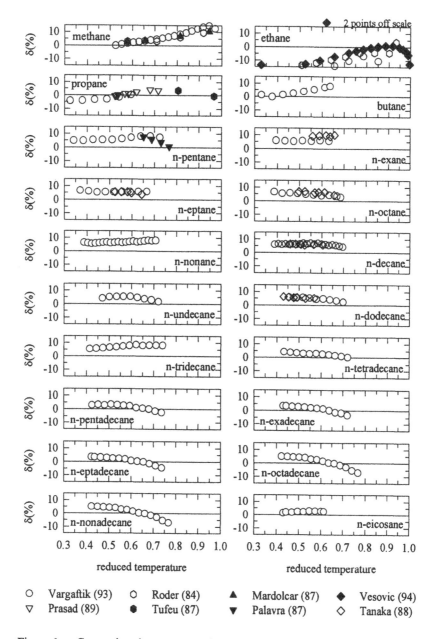

Figure 1 - Comparison between experimental and estimated thermal conductivity for alkanes. $\delta(\%)=100(\lambda_{calc}/\lambda_{exp}-1)$

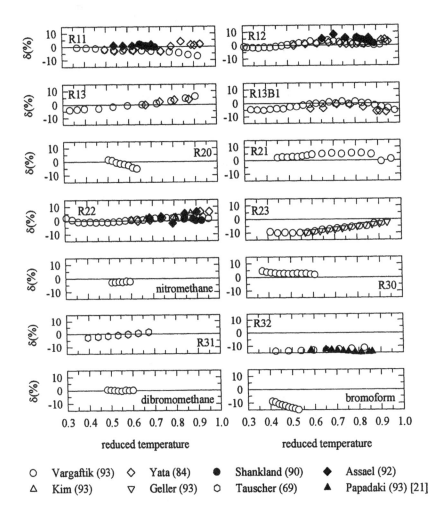

Figure 2 - Comparison between experimental and estimated thermal conductivity for methane derivatives. $\delta(\%)=100(\lambda_{calc}/\lambda_{exp}-1)$

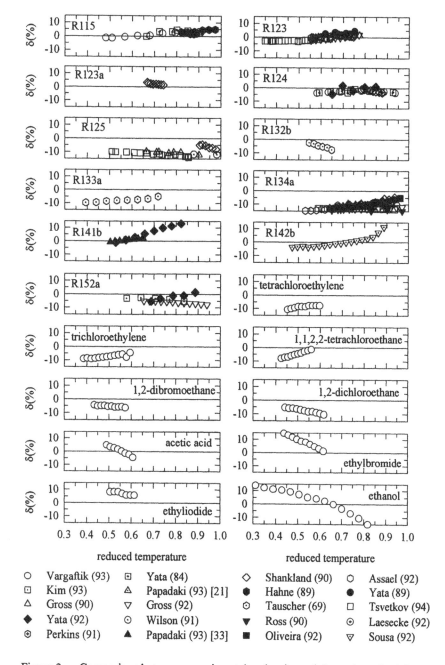

Figure 3 - Comparison between experimental and estimated thermal conductivity for ethane derivatives. $\delta(\%)=100(\lambda_{calc}/\lambda_{exp}-1)$

numerical constants to be used in the monomial parameter are given for all the investigated compounds.

The method, shows average absolute deviations which are generally less than 5%, with maximum absolute deviations usually less than 10% in the reduced temperature range 0.35 to 0.95. Hence it can be useful for engineering purposes. An extension of the method to other organic compounds is to be presented in a future paper.

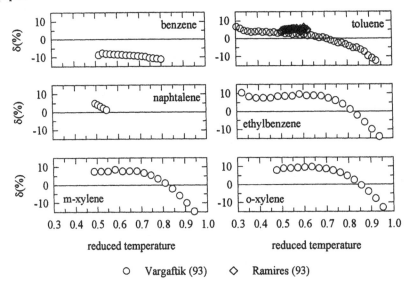

Figure 4 -  Comparison between experimental and estimated thermal conductivity for aromatic compounds.  $\delta(\%)=100(\lambda_{calc}/\lambda_{exp}-1)$

## NOMENCLATURE

| | | |
|---|---|---|
| A | parameter in Eq. (1) | $[W \cdot m^{-1} \cdot K^{-1}]$ |
| A* | parameter in Eq. (2) | |
| AAD | average absolute deviation | (%) |
| B | parameter in Eq. (4) | $[W \cdot m^{-1} \cdot K^{-1}]$ |
| B* | parameter in Eq. (5) | |
| M | molecular weight | $[kg \cdot kmol^{-1}]$ |
| MAD | maximum absolute deviation | (%) |
| $P_c$ | critical pressure | [bar] |
| $T_c$ | critical temperature | [K] |
| $T_r$ | reduced temperature | |
| $V_c$ | critical volume | $[m^3 \cdot kmol^{-1}]$ |
| $\lambda$ | thermal conductivity | $[W \cdot m^{-1} \cdot K^{-1}]$ |
| $\alpha$ | exponent in Eq. (5) | |
| $\beta$ | exponent in Eq. (5) | |

δ      deviation between experimental and
        estimated thermal conductivity    (%)
γ      exponent in Eq. (5)

## ACKNOWLEDGMENTS

This work has been supported by the Ministero dell'Universita' e della Ricerca Scientifica e Tecnologica of Italy.

## REFERENCES

1. C. Baroncini, P. Di Filippo, G. Latini, M. Pacetti, *Int. J. Thermophys.* Vol.2, Pag. 21, (1981).
2. C. Baroncini, G. Latini, P. Pierpaoli, *Int. J. Thermophys.*, Vol. 5 Pag. 387, (1984)
3. G. Latini, P. Pierpaoli F. Polonara, *Proc. 1992 International Refrigeration Conference*, Purdue University, U.S.A. Vol. II, Pag. 489, (1992).
4. G. Latini, G. Passerini F. Polonara, *Int. J. Thermophys.*, Vol. 17 Pag. 85, (1996)
5. R. C. Reid, J. M. Prausnitz, B. E. Poling, The Properties of Gases & Liquids, 4th Edition, McGraw-Hill, New York, (1987).
6. C. F. Beaton, G. F. Hewitt, Physical Property Data for the Desigh Engineer, Hemisphere Publishing Co., New York, (1989).
7. N. B. Vargaftik, L. P. Filippov, A. A. Tarzimanov, E. E. Totskii, Handbook of Thermal Conductivity of Liquids and Gases, CRC Press, Boca Raton, Fl, (1993).
8. H. M. Roder, Experimental Thermal Conductivity Values for Hidrogen, Methane, Ethane and Propane, NBS, USA, (1984).
9. U. V. Mardolcar, C. A. Nieto de Castro, *Berichte der Bunsen-Gesellschaft für Physikalische Chemie - Weinheim*, Vol.91, Pag.152, (1987).
10. V. Vesovic, W. A. Wakeham, J. Luettmer-Strathman, J. V. Sengers, J. Millat, E. Vogel, M. J. Assael, *Int. J. Thermophys.*, Vol. 15, Pag.33, (1994).
11. C. Prasad, G. Wang, J. S. Venart, *Int. J. Thermophys.*, Vol. 10, Pag. 1013, (1989).
12. R. Tufeu, B. Le Neidre, *Int. J. Thermophys.*, Vol. 8 Pag.27, (1987).
13. A. M. F. Palavra, W. A. Wakeham, M. Zalaf, *Int. J. Thermophys.*, Vol.8, Pag. 305, (1987).
14. Y. Tanaka, Y. Itani, H. Kubota, T. Makita, *Int. J. Thermophys.*, Vol.9, Pag. 331, (1988).
15. J. Yata, T. Minamiyama, S. Tanaka, *Int. J. Thermophys.*, *Vol.* 5, Pag. 209, (1984).
16. I. R. Shankland, *AIChE Spring Nat. Meeting, Orlando, Florida* (1990).
17. M. J. Assael, E. Karagiannidis, W. A. Wakeham, *Int. J. Thermophys.*, Vol. 13, Pag. 735, (1992).
18. S. H. Kim, D. S. Kim, M. S. Kim, S. T. Ro, *Int. J. Thermophys.*, Vol. 14, Pag. 937, (1993).
19. V. Geller, M. Paulaitis, D. Bivens, A. Yokozeki, *Proc. IIF/IIR Meeting Comm. B1/B2, Ghent, Belgium* p. 227 (1993).
20. W. H. Tauscher, *Ashrae J.*, Vol. 11, Pag. 97, (1969).
21. M. Papadaki, W. A. Wakeham, *Int. J. Thermophys.*, Vol. 14, Pag. 1215, (1993).
22. E. Hahne, U. Gross, Y. W. Song, *Int. J. Thermophys.*, Vol. 10, Pag. 687, (1989).

23. J. Yata, C. Kawashima, M. Hori, T. Minamiyama, *Proc. 2nd Asian Thermophysical Prop. Conf., Sapporo, Japan* p. 201 (1989).
24. U. Gross, Y. W. Song, J. Kallweit, E. Hahne, *Proc. IIF/IIR Meeting Comm. B1, Herzlia, Israel* p. 103 (1990).
25. U. Gross, Y. W. Song, E. Hahne, *Int. J. Thermophys.*, Vol. 13, Pag. 957, (1992).
26. M. J. Assael, and E. Karagiannidis, *Int. J. Thermophys.*, Vol. 14, Pag. 183, (1993).
27. O.B.Tsvetkov, Y.A.Laptev, A.G.Asambaev, *Int. J. Thermophys.*, Vol.15, Pag.203-214 (1994).
28. J. Yata, M. Hori, T. Kurahashi T. Minamiyama, *Fluid Phase Equil.*, Vol. 80, Pag. 287, (1992).
29. L. C. Wilson, W. V. Wilding, G. M. Wilson, R. L. Rowley, V. M. Felix, and T. Chisolm-Carter, *Fluid Phase Equil.*, Vol. 80, Pag. 167 (1992).
30. M. Ross, J. P. M. Trusler, W. A. Wakeham, and M. Zalaf, *Proc. IIF/IIR Meeting Comm. B1, Herzlia, Israel* p. 89 (1990).
31. A. Laesecke, R. A. Perkins, C. A. Nieto de Castro, *Fluid Phase Equil..*, Vol. 80, Pag. 263 (1992).
32. R. A. Perkins, A. Laesecke, and C. A. Nieto de Castro, *Fluid Phase Equil.*, Vol. 80, Pag. 275 (1992).
33. M. Papadaki, M. Schmitt, A. Seiz, K. Stephan, B. Taxis, W. A. Wakeham, *Int. J. Thermophys.*, Vol. 14, Pag. 173, (1993).
34. A. T. Sousa, P. S. Fialho, C. A. Nieto de Castro, R. Tufeu, B. Le Neindre, *Int. J. Thermophys.*, Vol. 13, Pag. 383, (1992).
35. C. Oliveira, M. Papadaki, W. A. Wakeham, *Proc. 3nd Asian Thermophysical Prop. Conf., Beijing, China* p. 32 (1992).
36. M. L. V. Ramires, J. M. N. A. Fareleira, C. A. Nieto de Castro, M. Dix, W. A. Wakeham, *Int. J. Thermophys.*, Vol. 14, Pag. 1131, (1993).

# A Thermistor Based Method for Measuring Thermal Conductivity and Thermal Diffusivity of Moist Food Materials at High Temperatures

M. F. van GELDER[1] and K. C. DIEHL[2]

## ABSTRACT

A method, built around a small glass encapsulated thermistor was used to measure the thermal conductivity and diffusivity of food materials. High temperature measurements were performed on tomato sauce and tomato paste. Thermal conductivity and thermal diffusivity were measured at 100, 130 and 150 °C and compared to literature values. In general, the values measured in this research were higher than those found in literature. The thermal property values (k, α) for tomato puree were in good agreement. Thermal conductivity of the tomato paste agreed quite well with the reported values. However, for the thermal diffusivity, the values were much larger than reported and variability between the different samples was substantial. This is thought to be caused by alterations in the tomato paste caused by prolonged heating at high temperatures. Measurements of thermal conductivity of liquid nutritional supplements at 95 and 150 °C were also performed. The results showed good replication and were consistent with values of water considering their composition.

## NOMENCLATURE

| | | |
|---|---|---|
| a | = radius of thermistor bead | [m] |
| k | = thermal conductivity | [W/m·°K] |
| q | = heat generation | [W/m$^3$] |
| r | = radial coordinate | [m] |
| R | = resistance | [Ω] |
| t | = time | [s] |
| T | = temperature | [°C] |
| V | = voltage drop across thermistor | [V] |

[1]    PhD Student, Biological Systems Engineering Department, VPI&SU, Blacksburg, VA 24061-0303; to whom correspondence should be addressed.

[2]    Associate Professor, Biological Systems Engineering Department, VPI&SU, Blacksburg, VA 24061-0303.

| | | |
|---|---|---|
| $\alpha$ | = thermal diffusivity | $[m^2/s]$ |
| $\beta$ | = slope of heat generation function | $[V \cdot s^{1/2}]$ |
| $\Gamma$ | = steady state power dissipation of bead per unit volume | $[W/m^3]$ |

Subscripts

| | | |
|---|---|---|
| 1 | = thermal conductivity calibration (e.g. $a_1$) | |
| 2 | = thermal diffusivity calibration (e.g. $a_2$) | |
| b | = bead | |
| f | = final | |
| i | = initial | |
| m | = medium | |
| ss | = steady state | |

## INTRODUCTION

Heat transfer is used in the analysis of food processing operations such as sterilization processes. These processes require temperatures up to 150 °C. At present there is a lack of thermal property data, especially at high temperatures (100 to 150 °C). Few researchers have measured thermal properties at temperatures above 100 °C. Wallapan measured the thermal conductivity, specific heat and density of defatted soy flour at 130 °C, at different moisture contents and densities [1]. Choi and Okos measured the thermal conductivity and thermal diffusivity of tomato concentrate from 20 to 150 °C [2]. Tomato concentrate was measured at a solids content ranging from 4.8 to 80 % (wb). Equations for the properties were fitted to the data. The thermal conductivity of fluid milk products of various fat contents was measured by Kravets at temperatures from 25 to 125 °C [3]. Of the above mentioned studies only Choi and Okos measured thermal diffusivity directly. Only a few methods directly provide values for thermal diffusivity.

Different methods have been used for measuring the thermal conductivity and/or diffusivity of food materials. They can be divided in steady state and transient methods. Among the steady state methods, the guarded hot plate is considered the most accurate and most widely used method for the measurement of thermal conductivity of poor conductors of heat [4]. It is most suitable for dry homogeneous samples that can be formed into a slab. Since steady state conditions may take several hours to develop, this method is unsuitable for use with material in which moisture migration may take place. The method has been used for measuring the thermal conductivity dried or frozen foods.

Food subject to thermal processing generally contains some moisture. Moisture migration would bias a steady-state measurement. For this reason, transient methods are preferred for measuring the thermal conductivity of moist food materials. They have the advantage of being rapid.

The line heat source method is one of the most commonly used transient methods in particular with granular materials [4]. The theory is based on a linear heat source of infinite length and infinitesimal diameter. An implementation of the line heat source method is the line heat source probe, also known as thermal conductivity probe. It is widely used for measurement of thermal conductivity in food. The heat source and temperature sensor are placed in a narrow tube [5]. Its length to diameter ratio is large to approximate a line heat source. The probe is embedded in the material whose thermal conductivity is to be measured. From thermal equilibrium, the heat source is energized and heats the medium with constant power. The temperature response of the medium is a function of its thermal properties. The thermal conductivity is found from the temperature rise measured at the heat source. The line heat source probe in its basic form obtains only a value for thermal conductivity. The probe used by Choi and Okos was equipped with an additional temperature sensor for direct measurement of thermal diffusivity [2].

The thermistor based method can simultaneously yield thermal conductivity and thermal diffusivity. The use of a thermistor for measurement of food thermal properties is limited to the study by Kravets [3]. He utilized a thermistor to obtain thermal conductivity at temperatures up to 125 °C. The objective of the research presented here is to examine the suitability of the bead thermistor method for measuring thermal properties at higher temperatures, which from the perspective of food processing are up to 150 °C.

## THERMISTOR PROBE TECHNIQUE

A thermistor is a resistor with a temperature dependent resistance. Those used for thermal property measurement have a negative temperature coefficient of resistance, meaning that their resistance decreases when their temperature is increased. This property is exploited in the thermistor based method. The thermistor is used both as a temperature sensor and as a point heating source.

Early use of a thermistor probe for measurement of thermal conductivity dates back to 1968 when Chato, a researcher in the biomedical field, utilized a miniature bead thermistor for the measurement of thermal conductivity of tissue [6]. Chato treated the thermistor as a lumped thermal mass. Assuming an instantaneous step change in the surface temperature of the bead he solved the heat transfer problem for an infinite, homogeneous and isotropic solid. Experimentally, the bead surface temperature was measured from the resistance of the thermistor. Researchers following Chato treated the thermistor as a distributed thermal mass and solved the coupled heat transfer problem for thermistor and medium [7, 8]

## HEAT TRANSFER MODEL

The model describing the temperature function in the thermistor and surrounding medium is based in the assumptions that:
- the thermistor is of spherical shape,
- the thermistor is made of isotropic and homogeneous material,
- the surrounding medium is isotropic, homogeneous and infinite in extent,
- heat is being generated uniformly in the thermistor,

- no contact resistance exists between thermistor and medium,
- the mode of heat transfer is pure conduction.

In practice, the last assumption is challenged when measuring the thermal properties of low viscosity fluids. Small beads and low heating power are used in an attempt to delay the onset and minimize the effect of natural convection. In addition, flow inhibitors (e.g. thickeners, glass fiber material) may be used.

Using the assumptions the heat transfer problem can be described by the following one-dimensional heat diffusion equations:

$$\frac{1}{r^2}\frac{\delta}{\delta r}\left[r^2\frac{\delta T_b}{\delta r}\right]+\frac{q(t)}{k_b} \quad =\frac{1}{\alpha_b}\frac{\delta T_b}{\delta t} \qquad 0\le r \le a \qquad (1)$$

$$\frac{1}{r^2}\frac{\delta}{\delta r}\left[r^2\frac{\delta T_m}{\delta r}\right] \quad =\frac{1}{\alpha_m}\frac{\delta T_m}{\delta t} \qquad r \ge a \qquad (2)$$

Symmetry dictates the boundary condition at the center of the thermistor to be:

$$\frac{\delta T_b}{\delta r}=0 \qquad\qquad r=0, \quad t\ge 0 \qquad (3)$$

The assumptions of pure conduction, absence of surface contact resistance and infinity of the medium give as boundary conditions:

$$T_b=T_m \qquad\qquad r=a, \quad t>0 \qquad (4)$$

$$k_b\frac{\delta T_b}{\delta r}=k_m\frac{\delta T_m}{\delta r} \qquad\qquad r=a, \quad t>0 \qquad (5)$$

$$T_m=T_i \qquad\qquad r\rightarrow\infty, \quad t>0 \qquad (6)$$

The initial condition is given by the requirement that the thermistor and the surrounding medium should be at a thermal equilibrium at the start of a test.

$$T_b=T_m=T_i \qquad\qquad t=0 \qquad (7)$$

## SOLUTION

The solution to this coupled heat transfer problem is taken from Valvano [8]. In order to obtain the solution, the form of the heat generation function, q(t), has to be known. A heat generation function of the form $q(t) = \Gamma + \beta \cdot f(t)$ is assumed. $\Gamma$ is the steady state term, f(t) is the transient term which will approach zero for long times. The mathematics involved in solving this heat transfer problem can be found in Valvano [8]. The resulting two equations for the temperature distribution in the thermistor bead and the medium are:

$$T_b(r,t) - T_i = \frac{a^2}{k_b}\left[\frac{k_b}{3k_m} + \frac{1}{6}\left(1 - \left(\frac{r}{a}\right)^2\right)\right](\Gamma + \beta f(t)) - \frac{\Gamma a^3 f(t)}{3k_m\sqrt{\alpha_m\pi}} \quad (8)$$

$$T_m(r,t) - T_i = \frac{a^3\Gamma}{3k_m r}\left[\frac{(a-r)f(t)}{\sqrt{\alpha_m\pi}}\right] + \frac{\beta a^3 f(t)}{3k_m r} - \frac{\Gamma a^4 f(t)}{3k_m r\sqrt{\alpha_m\pi}} \quad (9)$$

Equation (8) describes the temperature distribution inside the bead and is used to derive equations that form the basis for the thermistor based measurement technique.

## MEASUREMENT OF THERMAL CONDUCTIVITY

The test procedure consists of controlling the temperature of the thermistor bead at a specified increment above the initial temperature. The experimentally measured temperature of the bead is a spatial average over its volume. When the limit is taken of the expression for the temperature distribution in the thermistor bead when the time variable goes to infinity and the resulting expression is averaged over the bead's volume, an expression for the average steady state temperature of the thermistor bead is obtained:

$$T_{b_{ss}} - T_i = \frac{\Gamma a^2}{3k_b}\left[\frac{k_b}{k_m} + 0.2\right] \quad (10)$$

The constant internal heat generation in the bead is:

$$\Gamma = \frac{\text{steady state power}}{\text{volume of bead}} = \frac{V_{ss}^2/R_f}{(4/3)a^3} \quad (11)$$

Substituting this expression into equation (11) gives, after rearranging:

$$\frac{1}{k_m} = \frac{4\pi a R_f \Delta T_b}{V_{ss}^2} - \frac{1}{5k_b} \quad (12)$$

This expression shows that when the bead parameters, a and $k_b$, are known, the medium thermal conductivity can be calculated from the experimentally obtained values for $R_f$, T, and $V_{ss}$. The thermistor probe requires calibration with media of known thermal conductivity to estimate values for the bead parameters, a and $k_b$.

## MEASUREMENT OF THERMAL DIFFUSIVITY.

The temperature of the thermistor bead as observed from its resistance, reaches a constant value shortly after energizing. As shown above, this temperature is related to the steady state heat generation term. The volume average contribution of the transient terms in equation (8) (all those involving the function f(t)) must therefore sum to zero. Doing so leads to an expression for the medium thermal diffusivity:

$$\alpha_m = \left[ \frac{a}{\sqrt{\pi} \, (\beta/\Gamma)\left(1 + \frac{k_m}{5k_b}\right)} \right]^2 \qquad (13)$$

The values for $\beta$ and $\Gamma$ of the heat generation function are found from the experimentally obtained power response of the thermistor. In the process of solving the heat transfer problem, Valvano showed that the time-function had to be of the form $t^{-1/2}$. The heat generation function can thus be described as: $q(t) = \Gamma + \beta \cdot t^{-1/2}$. Values for $\Gamma$ and $\beta$ are found through linear regression. The values for the probe radius and thermal conductivity in equation (13) are not identical as those used for estimation of thermal conductivity and have to be determined through calibration with media of known thermal diffusivity.

## MATERIALS AND METHODS

### THERMISTOR PROBE

An instrument was constructed around a glass encapsulated thermistor probe (P60, Thermometrics, Edison, NJ). The probe is shown in figure 1. Teflon coated wires were soldered to the probe lead wires and the junctions were insulated with heatshrink tubing. The glass probe was bonded to a stainless steel holder with an epoxy rated for high temperature use (Omegabond 200, Omega Engineering, Inc., Stamford, CT). The bead extended 6.5 mm from the holder. This distance was large enough to satisfy the infinite boundary condition. Kravets had shown that for a probe with the dimensions as used in this research, a spherical region around the bead with a diameter of 5 mm will suffice [3]. The stainless steel holder was screwed into an aluminum lid where an o-ring provided a seal. The sample chamber was made of brass and had an internal diameter and depth of 21 mm and 44.5 mm, respectively. An o-ring mounted in the lid ensured a hermetic seal.

The thermistor was used in a fixed temperature step mode as mentioned in the section detailing the heat transfer model. This requires that the temperature of the thermistor bead is raised almost instantaneously to a temperature above the initial equilibrium temperature and kept at this elevated temperature for the duration of the test. An electronic circuit, designed by Balasubramaniam, was used [7]. It was adapted for computer interfacing.

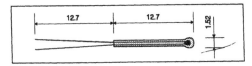

Figure 1. Typical thermistor as used in experiments; dimensions in [mm]

Measurement control and data acquisition took place through software written in Turbo Pascal 4.0 (Borland International, Scotts Valley, CA), running on a personal computer. The circuitry was interfaced to the PC with an I/O-board (PIO-102, MetraByte Corporation, Taunton, MA) and an ADC (DAS-20, MetraByte Corporation, Taunton, MA).

Experiments were performed at high temperatures. A constant temperature bath (Exacal EX-251 HT, NESLAB Instruments, Inc., Newington, NH) was used for temperature control of the measurements.

Probe Calibration

Thermistors were calibrated against a thermal reference (Guildline 9540 Digital Platinum Resistance Thermometer, Smith Falls, Ontario) traceable to NIST. The bead parameters $a_1$, $k_{b1}$, $a_2$ and $k_{b2}$ were estimated from measurements in three reference fluids that cover the range of thermal conductivity and diffusivity found in foods: distilled water, glycerol (Fisher Scientific, Pittsburgh, PA) and HTF 500 (Union Carbide, Danbury, CT). Property data for water and glycerol were obtained from literature [9, 10, 11], except for the density of glycerol for which reliable data were unavailable. It was therefore measured. Property data for HTF 500 were supplied by the manufacturer as graphs [12]. Values for thermal diffusivity were found by calculation from thermal conductivity, density and specific heat. Calibration was performed at 100, 130 and 150 °C. Fluids were tested in triplicate and each sample was used over the entire temperature range.

The test procedure was: three repeat measurements were performed on each sample at each temperature; a temperature step of 2.5 °C was used; data were sampled at a rate of 100 Hz for a duration of 30 seconds; repeats were spaced eight minutes apart to allow the system to return to equilibrium.

A linear regression fit of the observed voltage response to the heat generation function yielded the values of the intercept and slope needed in equations (12) and (13).

TOMATO CONCENTRATE

Thermal conductivity and thermal diffusivity of the tomato puree and tomato paste were determined with the thermistor probe. The tomato concentrate was supplied by Van Den Bergh Foods Co. (Roxbury, CT) in size 10 cans. Three temperatures were chosen for thermal property evaluation, 100, 130 and 150 °C. Two cans per material (puree, paste) were used and from each can two samples were taken for measurement of thermal conductivity and thermal diffusivity. Each sample was tested at all temperatures. A sample of the tomato product was carefully placed in the sample chamber. Care was taken not to enclose air pockets which could alter the thermal properties. The sample chamber was hermetically sealed with the lid with the probe mounted to it. The sample chamber was then submerged in the constant temperature bath. Bath and sample were heated to 100 °C. Measurements were performed with the procedure as outlined in 'Probe Calibration'. The sample was then heated to 130 °C for measurement and subsequently to 150 °C. The sample was allowed to cool to room temperature and visually inspected for heating induced changes.


## NUTRITIONAL SUPPLEMENT

A fluid nutritional supplement was supplied for testing by Abbott Labs/Ross Products Division (Columbus, OH). Products of different solids content and different composition (e.g. fat, protein, carbohydrates) were tested. Some of these products were viscous enough that natural convection was deemed unlikely. For the others, glass fiber pipe insulation was used at 1 - 1.5% (w/v) as a convection inhibiter/reducer. These low concentrations were assumed to have no influence on the thermal properties. Care was taken in the sample preparation to limit air inclusion in the samples as much as possible. Four different products were tested, designated at "TC1", "TC2", "TC3" and "TC4". Each product was supplied in various solids contents. The products were tested at two temperatures, 95 and 150 °C. Duplicate samples were used for all but product "TC2". The combinations of product, solids content and temperature that were tested were specified by the supplier and can be found in Table V in the results. The samples were submerged in the bath, already at measurement temperature. Doing so minimized the exposure of the food material to high temperatures and their potential for altering the properties. Approximately 20 minutes were required, after submersion, for temperature equilibration to take place. Property measurement was performed as described in 'Probe Calibration'.


## RESULTS AND DISCUSSION

### TOMATO CONCENTRATE

Of the four samples of tomato puree taken for measurement, three gave usable data; the power response for one sample turned out to be erratic. This response could have been the result of inhomogeneities (e.g. air pockets) near the probe. This sample was excluded from further analysis. Thermal conductivity data are calculated at each temperature for each of the three samples that gave acceptable data and are presented in Table I for thermal conductivity and Table II for thermal diffusivity. Included in the table are values calculated with the equations that were fitted by Choi and Okos [2]. A solids content of 16% (manufacturer data) was used. The thermal conductivity values found in this study are higher than those reported by Choi and Okos. The discrepancy decreases from  +7.5% at 100 °C to +4.5% at 150 °C. Replication is good but deteriorates with temperature.

Thermal diffusivity seems to agree better with the published data and slightly overestimates them. However the repeatability appears worse with sample #3 yielding much larger values than the other two samples.

TABLE I.   Thermal Conductivity of Tomato Puree (solids content is 16% [w/w]),   [W/m·°C]

|  | Temperature | | |
|---|---|---|---|
|  | 100 °C | 130 °C | 150 °C |
| 1 | 0.6566 | 0.6659 | 0.6378 |
| 2 | 0.6566 | 0.6668 | 0.6591 |
| 3 | 0.6507 | 0.6562 | 0.6677 |
| average | 0.6546 | 0.6630 | 0.6549 |
| Choi & Okos | 0.6087 | 0.6227 | 0.6266 |
| difference [%] | +7.5 | +6.5 | +4.5 |

TABLE II.   Thermal Diffusivity of Tomato Puree (solids content is 16% [w/w]),   [m²/s]

|  | Temperature | | |
|---|---|---|---|
|  | 100 °C | 130 °C | 150 °C |
| 1 | 1.51 | 1.50 | 1.56 |
| 2 | 1.55 | 1.53 | 1.59 |
| 3 | 1.71 | 1.76 | 1.74 |
| average | 1.59 | 1.60 | 1.63 |
| Choi & Okos | 1.50 | 1.56 | 1.59 |
| difference [%] | +6.0 | +2.6 | +2.5 |

A total of six samples of tomato paste were used in the tests. The reason for sampling beyond the initially planned total of four was the inconsistent power response curves obtained with some of the samples. Two samples yielded unrepresentative data and were removed from the analysis at all temperatures. Other samples gave acceptable results at 100 and 130 °C but not at 150 °C. The most likely reason for the results at the higher temperatures was temperature induced changes. It was noticed that the paste had browned and turned into a brittle solid. The sensing part of the thermistor is 1.52 mm in diameter. It measures the effective properties of a very small region around the probe and is therefore sensitive to inhomogeneities. The line heat source probe as used by Choi and Okos has a much larger measurement volume. A second explanation may be the origin of products tested. The products in this study were commercially formulated. Choi and Okos prepared their test products from concentrated and dehydrated fresh tomatoes. The results of the tests are given in Table III for thermal conductivity and Table IV for thermal diffusivity.

## NUTRITIONAL SUPPLEMENT

Thermal conductivity values for the fluid nutritional supplement as related to solids content and temperature are given in Table V. Reference values for water are given for comparative purposes. The thermal conductivity of moist foods is strongly

TABLE III. Thermal Conductivity of Tomato Paste (solids content is 26% [w/w]), [W/m·°C]

|  | Temperature | | |
| --- | --- | --- | --- |
|  | 100 °C | 130 °C | 150 °C |
| 1 | 0.5979 | 0.5978 | 0.5808 |
| 2 | 0.5952 | 0.5792 | |
| 3 | 0.6000 | | |
| 4 | 0.6037 | 0.6179 | |
| average | 0.5992 | 0.5983 | 0.5808 |
| Choi & Okos | 0.5643 | 0.5836 | 0.5920 |
| difference [%] | +6.2 | +2.5 | -1.9 |

TABLE IV. Thermal Diffusivity of Tomato Paste (solids content is 26% [w/w]), [m²/s]

|  | Temperature | | |
| --- | --- | --- | --- |
|  | 100 °C | 130 °C | 150 °C |
| 1 | 1.62 | 1.63 | 1.74 |
| 2 | 1.68 | 2.11 | |
| 3 | 1.52 | | |
| 4 | 1.55 | 1.71 | |
| average | 1.59 | 1.82 | 1.74 |
| Choi & Okos | 1.38 | 1.46 | 1.50 |
| difference [%] | +15.2 | +24.7 | +16.0 |

influenced by their water content. Tests on a number of samples were performed with an uncalibrated probe due to: (1) catastrophic failure of the calibrated probe, (2) time constraints, dictated by the limited shelf life of the products, which prohibited calibration of another probe. Its parameters were to be determined after completion of the measurements with the nutritional supplement. Unfortunately, it also failed before it could be calibrated. Its parameters at 150 °C were estimated using the results of two products that were measured with both the calibrated and uncalibrated probe ("TC3", 40% solids, 150 °C and "TC4", 30% solids, 150 °C). Probe parameters of previously used probes were tried for the uncalibrated probe. Those that yielded thermal conductivity values that best matched those found with the calibrated probe were used to calculate the thermal conductivity at 150 °C for all the samples that were measured with the uncalibrated probe. The probe whose calibrated parameters at 150 °C were substituted for those of the uncalibrated probe, also supplied the probe parameters for the uncalibrated probe at 95 °C. These parameters were however obtained through calibration at 100 °C. Data gathered with the uncalibrated probe are indicated with a superscript. Because of the assumed probe parameters for the uncalibrated probe, the thermal conductivity values obtained with it are less reliable than those found with the calibrated probe.

An increase in solids content appears to decrease the thermal conductivity of the products. This finding is consistent with the commonly found positive relationship between thermal conductivity and moisture content [13]. The

TABLE V. Thermal Conductivity of "TC1", "TC2", "TC3" and "TC4" at High Temperatures, [W/m°C]

| Solids content [% wb] | TC1 | | TC2 | | TC3 | | TC4 | |
|---|---|---|---|---|---|---|---|---|
| | 95 [°C] | 150 [°C] | 95 [°C] | 150 [°C] | 95 [°C] | 150 [°C] | 95 [°C] | 150 [°C] |
| 15 | | 0.6318 0.6273 | | 0.6251 | | | | 0.6113 0.6110 |
| 25 | | | | | | 0.5620 0.5595 | | |
| 30 | | 0.5545 0.5018[c] | 0.5596 | 0.5513 | | | | 0.5230 0.5114[b] |
| 35 | | | | 0.5334 | | | | |
| 40 | 0.4898[a] 0.4895[a] | 0.4949[b] 0.4818[b] | 0.5084 | 0.5069 | | 0.4589 0.4535[b] | 0.4462[a] 0.4451[a] | 0.4497[b] |
| 50 | | | | | 0.4023[a] 0.4013[a] | 0.4186[b] 0.4049[b] | | |
| water, literature value | 0.6762 | 0.6820 | 0.6762 | 0.6820 | 0.6762 | 0.6820 | 0.6762 | 0.6820 |

[a] Measured with TT5, probe parameters taken from H3 at 100 °C,
[b] Measured with TT5, probe parameters taken from H3 at 150 °C,
[c] Unreliable.

replication is excellent (within 1% different) for the measurements at 95 °C at all solids levels and for those at 150 °C at solids levels of 25% and below. The replication at 150 °C for the higher solids levels is worse than those at lower solid levels and temperature. This is probably due to heat induced changes in the product which are more pronounced at higher solids levels. A similar effect was observed with the tomato concentrate. Tomato paste yielded less accurate data (as compared to the literature data) and displayed worse replication than tomato puree.

## CONCLUSIONS

The thermistor probe has the advantage of simultaneous yielding values for thermal conductivity and thermal diffusivity. The small size makes it suitable for measuring very small samples. A problem encountered in this research is providing a durable seal at the probe's surface. The epoxy, though rated for high temperatures, deteriorated over time. Moisture would be pushed into the gap between epoxy and thermistor, reach its leads, and cause probe failure.

The thermistor based method appears to give satisfactory results when used for the measurement of thermal conductivity and diffusivity of tomato puree. With tomato paste the method does not yield data with acceptable accuracy or repeatability probably due to heat induced changes in the product. As most foods heat slowly, nearly all methods for measuring high temperature thermal properties of food will be confronted by this.

For the nutritional supplement, no published data were available. The accuracy of the obtained values can thus not directly be assessed. The values appear to be reasonable. They follow an expected trend with respect to the effect of the solids content. With the exception of the high solids content samples at 150 °C, the replication is within one (1) percent.

## ACKNOWLEDGMENTS

The authors wish to acknowledge the Center for Aseptic Processing and Packaging Studies, North Carolina State University (Raleigh, NC), whose financial support made this research possible. We wish to also thank Van Den Bergh Foods Co. and Abbott Labs/Ross Products Division for supplying samples.

## REFERENCES

1. Wallapapan, K., Sweat, V.E., Arce, J.A. and P.F. Dahm, 1984. "Thermal diffusivity and conductivity of defatted soy flour," *Transactions of the ASAE,* 27(5):1610-1613.
2. Choi, Y. and M.R. Okos, 1983. "The Thermal Properties of Tomato Juice Concentrates." *Transactions of the ASAE,* 26(1):305-311.
3. Kravets, R.R., 1988. *Determination of thermal conductivity of food materials using a bead thermistor.* Ph.D. thesis in Food Science and Technology, Virginia Polytechnic Institute and State University.
4. Mohsenin, N.N., 1980. *Thermal Properties of Foods and Agricultural Materials.* New York, Gordon and Breach Science Publishers, pp. 83-111.
5. Sweat, V.E. and C.G. Haugh, 1974. "A thermal conductivity probe for small food samples." *Transactions of the ASAE,* 17(1):56-58.
6. Chato, J.C., 1968. "A method for the measurement of the thermal properties of biological materials," in *Thermal Problems in Biotechnology,* ASME, NY, pp. 16-25.
7. Balasubramaniam, T.A., 1975. *Thermal conductivity and thermal diffusivity of biomaterials: a simultaneous measurement technique.* Ph.D. dissertation, Northeastern University, Boston MA.
8. Valvano, J.W., 1981. *The use of thermal diffusivity to quantify tissue perfusion.* Ph.D. thesis in Medical Engineering, Harvard University - MIT Division of Health Sciences and Technology.
9. Incropera, F.P. and D.P. DeWitt, 1985. *Fundamentals of Heat and Mass Transfer.* 2nd ed., John Wiley & Sons, New York.
10. Touloukian, Y.S., Liley, P.E. and S.C. Saxena, 1970. *Thermal Conductivity: Nonmetallic Liquids and Gases.* The TPRC data series, Vol. 3, IFI/Plenum Press, New York.
11. Touloukian, Y.S., 1970. *Specific Heat: Nonmetallic Liquids and Gases.* The TPRC data series, Vol. 6 Suppl., IFI/Plenum Press, New York.
12. *UCON Heat Transfer Fluid 500,* 1989. Union Carbide Chemicals and Plastics Company Inc., Specialty Chemicals Division, Danbury, CT.
13. Sweat, V.E., 1986. "Thermal Properties of Foods," in *Engineering Properties of Foods,* M.A. Rao and S.S.H. Rizvi, eds. New York, Marcel Dekker, Inc., pp. 49-87.

# Measuring Thermal Properties of Elastomers Subject to Finite Strain

N. T. WRIGHT, M. G. DA SILVA, D. J. DOSS
and J. D. HUMPHREY

## ABSTRACT

General thermomechanical analyses of elastomers require both thermoelastic and thermophysical properties over a whole range of temperatures and finite deformations. In this paper, a device is described that can measure three components of thermal diffusivity of elastomers that are subject to the biaxial loads, causing finite strains of up to 100 percent, and temperatures from 20 to 100°C. Data analysis is accomplished by a Marquardt parameter estimation using a finite difference solution of the diffusion equation. Reported here are preliminary measurements of the out-of-plane spatial thermal diffusivity in natural gum rubber specimens subjected to uniaxial and equal biaxial loading.

## INTRODUCTION

Elastomers, such as rubber and soft biological tissues, often experience large strains and significant temperature gradients. For example, arteries are often stretched more than 50 percent *in vivo* and may be subject to temperature gradients of tens of kelvins per mm during heat-based clinical procedures. Elastomeric composites, an interesting new class of materials that may become of great technological importance, are similarly subjected to large strain and temperature gradients during typical service conditions. A general thermo-mechanical analysis of elastomers is essential for optimizing designs, especially those customized for a singular purpose and those directly affecting human safety. The mechanical and thermal properties of these materials may change with deformation, temperature, or both. The extent of these changes is largely unknown. Thus, a method is needed for measuring the influence of large deformations on the material properties.

Although data are scant, a general theory for quantifying thermo-mechanical behavior under finite strains has long been available [1]. In particular, the response to thermal or mechanical loads must satisfy the general balance relations of conservation of mass, energy, and linear momentum. Additionally, the second law of thermodynamics and conservation of angular momentum provide restrictions on the constitutive relations. For nonlinear thermoelastic

Neil T. Wright, Mark G. da Silva, David J. Doss, Jay D. Humphrey, Department of Mechanical Engineering, University of Maryland Baltimore County, Baltimore, MD 21228-5398

solids, such an analysis requires two independent constitutive relations for the material:

$$\psi = \hat{\psi}(\mathbf{C}, T) \tag{1_a}$$

$$\mathbf{q}_0 = \hat{\mathbf{q}}_0(\mathbf{C}, T, \nabla_0 T) \tag{1_b}$$

where $\psi$ is the Helmholtz free energy, $\mathbf{C}$ the right Cauchy-Green deformation tensor, $T$ the temperature, $\mathbf{q}_0$ the referential heat flux vector, $\nabla_0 T (= \partial T / \partial \mathbf{X})$ the referential temperature gradient, and $\mathbf{X}$ the position of a material particle in a reference state. Equation $1_a$ defines the thermoelastic behavior (stress-strain-temperature) whereas equation $1_b$ defines the heat transfer properties. Herein, the focus is the measurement of the heat transfer properties and thus, equation $1_b$. The thermoelastic behavior is the subject of a parallel study.

Consider a generalized form of Fourier's Law,

$$\mathbf{q}(\mathbf{x}, t) = \frac{1}{J}\mathbf{F} \cdot \mathbf{q}_0 = \frac{1}{J}\mathbf{F} \cdot \left( -\mathbf{K}(\mathbf{C}, T) \cdot \nabla_0 T(\mathbf{X}, t) \right) \tag{2}$$

where $\mathbf{q}(\mathbf{x}, T)$ is the spatial heat flux vector, $\mathbf{K}(\mathbf{C}, T)$ the referential thermal conductivity tensor, $\mathbf{F}$ the deformation gradient tensor, and $J$ the determinant of $\mathbf{F}$. Note that $\mathbf{C} = \mathbf{F}^T \cdot \mathbf{F}$. For finite strains, referring the conductivity tensor to the reference configuration simplifies material symmetry considerations. For either small strains ($\mathbf{F} \approx \mathbf{I}$, $J \approx 1$) or heat conduction in a rigid solid ($\mathbf{F} \equiv \mathbf{I}$, $J \equiv 1$), equation 2 reduces to the classical relation, $\mathbf{q}(\mathbf{x}, t) = -\mathbf{k}(T) \cdot \nabla T(\mathbf{x}, t)$, where $\mathbf{k}(T)$ is the spatial conductivity (independent of deformation) and $\mathbf{x}$ is the position of a material particle in the current state; most often, $\mathbf{k}(T) = k(T)\mathbf{I}$, that is the conductivity is assumed to be isotropic. In general, however, $\nabla_0 T = \mathbf{F}^T \cdot \nabla T$ and hence, $\mathbf{k} = (1/J)\mathbf{F} \cdot \mathbf{K} \cdot \mathbf{F}^T$.

Part of a general thermomechanical analysis is the evaluation of the temperature field. Inserting equation 2 in the conservation of energy equation yields the thermal diffusion equation

$$\frac{\partial T(\mathbf{X}, t)}{\partial t} = \alpha_0(\mathbf{C}, T) : \nabla_0 \left( \nabla_0 T(\mathbf{X}, t) \right) \tag{3}$$

where $\alpha_0(\mathbf{C}, T) = \mathbf{K}(\mathbf{C}, T) / (\rho_0 c_F)$ is the referential thermal diffusivity tensor, in which $\rho_0$ is the referential mass density and $c_F(\mathbf{C}, T)$ the referential constant deformation specific heat. In terms of $\psi$ and $T$, $c_F(\mathbf{C}, T)$ may be written as

$$c_F(\mathbf{C}, T) = -T \frac{\partial^2 \psi(\mathbf{C}, T)}{\partial T^2} \tag{4}$$

Thus, $\mathbf{K}(\mathbf{C}, T)$ determines $\mathbf{q}$ for a given temperature gradient and deformation, but $\alpha_0(\mathbf{C}, T)$ establishes the time history of the temperature field, given appropriate boundary conditions. The method described here measures $\alpha_0(\mathbf{C}, T)$. Once the appropriate from of $\psi(\mathbf{C},T)$ is determined, $c_{\mathrm{F}}(\mathbf{C},T)$ may be calculated from equation 4, and $\mathbf{K}(\mathbf{C},T)$ thus established.

The heat transfer properties of polymers are sensitive to the molecular lattice structure of the material in the undeformed state (Parrott and Stuckes [2]) and changes in this structure induced either by deformation or temperature level. Tissue is a common example of an elastomer with complex internal structure, which may display directional variations in thermal conductivity in an undeformed state. Valvano et al. [3] measured the axial and radial components of $\mathbf{k}(T)$ in undeformed bovine aorta ($\mathbf{C} = \mathbf{I}$). Using the steady-state, hot-plate-cold-plate method, the axial thermal conductivity of bovine aorta was about 27 per cent higher than the radial conductivity at the two temperature levels examined, 25 and 50°C.

Initially isotropic polymers may exhibit directional changes in thermal conductivity when highly deformed. Choy et al. [4] demonstrated that for highly drawn polymers, the thermal properties change significantly under loading due to rearrangement of polymeric chains: the in-plane component of thermal conductivity was measured to be as large as 59 times greater than the out-of-plane component, for draw ratios of 25. Temperature changes are often considered in measurement of thermal properties. These changes have the effect of changing the diffusion equation to be solved for the temperature field to a nonlinear equation. Dashora [5] examined thermal conductivity data for several linear polymers and a cross-linked polymer as they vary from below the glass transition temperature to above the glass transition temperature. The temperature variation of the thermal conductivity was linked to the structure of the polymer.

In this paper, a novel apparatus is described that measures, for the first time, directionally dependent thermal diffusivity of elastomers as a function of finite strain and temperature level. The temperature range of interest lies above the glass transition temperature and below the rubber flow region. In addition to the usual difficulties encountered in measuring moderate thermal conductivities, such as those of elastomers, the apparatus must impose well-defined multiaxial finite deformations at temperatures from 20 to 100°C. For the heat transfer measurements, a method with minimal contact with the specimen is necessary to avoid disturbing the well-defined mechanical deformation. The need to measure the moderate $\alpha_0(\mathbf{C},T)$ of elastomers while the samples are simultaneously subjected to well-defined biaxial loads favors the flash method rather than a steady-state method; in the later, rigorous thermal insulation interferes with the application of multiple loads and subsequent measurement of finite strains. Biaxially-loaded elastomers may exhibit deformation induced changes in material symmetry (e.g., from isotropy to orthotropy); hence, at least three components of $\alpha_0(\mathbf{C},T)$ must be measured, which in turn requires at least three temperature measurements, along orthogonal axes. In the flash method, the time for the temperature on the non-illuminated face to reach half its ultimate value is of the order of 10 seconds, thus, small thermocouples can deliver the needed temporal resolution. The temperature level, deformation and load measurements are made while the specimen is at thermal and mechanical equilibrium. Data are reduced using a Marquardt-type regression to determine the corresponding thermal diffusivities, from which $\mathbf{K}(\mathbf{C},T)$ is calculated via independent measurements of $c_F$ and $\rho_0$.

## METHODS

A modified *flash method* has been developed in conjunction with a load frame that applies in-plane biaxial loads to a thin planar specimen. The flash

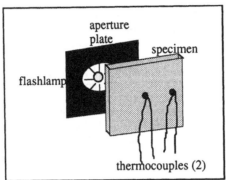

FIGURE 1. ANISOTROPIC FLASH METHOD, MAILLET ET AL.[7]

method, as developed by Parker et al. [6], measures thermal diffusivity in materials of moderate thermal diffusivity with minimal contact. However, only the component of the diffusivity normal to the dominant plane of the specimen was measured. Maillet et al. [7] and Lachi and Degiovanni [8] modified the flash method to measure two components of $\alpha(\mathbf{I},T)$, an in-plane component and the out-of plane component, of thin, undeformed specimens. Instead of illuminating the entire face of the specimen, a circular region on the center of a specimen face is illuminated and the temperature at the center of the circle, on the opposite face, and temperature on the opposite face, just in the shadow, are measured (Figure 1). Then, an iterative parameter fitting scheme is employed to determine the radius of the circle that is illuminated and the two components of $\alpha(\mathbf{I},T)$. Extending the flash method further and using the data reduction method outlined by Sawaf and Özisik [9], all three components of $\alpha_0(\mathbf{C},T)$ may be determined.

Here, the specimens have undeformed dimensions of approximately $30 \times 30 \times 1$ mm. The specimens are suspended between 4 load carriages using 0.2 mm diameter Kevlar® fibers, five to seven per side (Figure 2). A finite element analysis of the specimen suggested that a central square region, 1/4 of the specimen length on a side, is in a state of uniform stress when five or more fibers are used to load each side of the specimen. Load cells mounted on two of the orthogonally placed load carriages determine the in-plane applied loads, or changes therein that are induced by other effects. The specimen is maintained at a uniform temperature of between 20 and 100°C by an insulated environmental chamber equipped with electric heaters. As with the method of Lachi et al. [7], only part of the front face of the specimen is illuminated. An EG&G lamp delivers a flash in <135 µs, through an aperture plate, of approximately 1 joule to a square part of the central uniform stress region of the surface. A quartz window in the environmental chamber allows the flash to illuminate the specimen. The illuminated face of the specimen is coated with India ink to insure consistent and uniform absorption of the light. The temperature is measured at 3 locations on the non-illuminated face of the specimen: one at the center of the uniform stress region and two outside the projection of the lighted part of the front face (Figure 2a), but within the uniform stress region. Three E-type thermocouples, with nominal bead sizes of 0.2 mm, are mounted on an L-shaped bracket that can be placed in contact with the rear face using a micrometer adjustment. A port provides access for the thermocouple probes into the environmental chamber.

These thermocouples are brought into contact with the specimen after it reaches mechanical equilibrium. The response time of the thermocouples is <0.05 s,

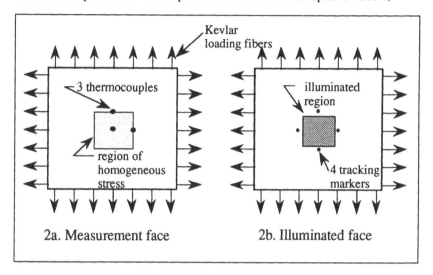

FIGURE 2.  SCHEMATIC OF A BIAXIALLY LOADED SPECIMEN

which is fast enough to resolve the response of the temperature on the non-illuminated (that is, measurement) face of the specimen, which takes of the order of 10 s to reach the maximum at the center thermocouple.

Deformations are measured by tracking the motion of four 50-100 μm diameter markers affixed to the illuminated face of the specimen surface in the form of a quadrilateral (Figure 2b). These markers are photographed at each mechanical equilibrium state for off-line strain analysis. Standard resistance strain gauges are inappropriate for finite strains in elastomers due to their low extensibility versus the relatively high extensibility of elastomers. Non-contact deformation measurements are thus preferred. Since the specimens are maintained in an environmentally controlled chamber and deformed without removing them, a method of thickness measurement is needed that allows the sensor to remain in the chamber throughout the test cycle. Thickness in the preliminary measurements presented here is determined based on the assumption of material incompressibility at a constant temperature level. From the initial thickness measurement, subsequent thicknesses may be determined from the measured in-plane deformation and the constant volume of the incompressible material. In future studies at multiple temperature levels, thicknesses will be measured directly.

Sawaf and Özisik [9] suggested the use of a Marquardt based method for estimating multi-directional conductivities. Their model problem was to provide a constant flux to the three positive faces of a cube and measure the temperature response on the three negative faces, each one an insulated boundary. A numerical model of the heat transfer was used to calculate the temperature of the specimen using estimated thermal diffusivities. Then, these values are compared with the measured values and the results are used to update the estimate of the diffusivities. This continues until the estimated and measured temperatures are within acceptable limits. Here, a finite difference analog of equation 3 with boundary conditions appropriate for the flash system is substituted for the diffusion equation and boundary conditions used by Sawaf and Özisik [9]. Note that for an incompressible solid, the equations are identical. This system of equations is solved by finite differences as a subprogram in the Marquardt parameter estimation scheme. Data reduction for the preliminary results described below is based on the method from Parker et al. [6]

$$\alpha_3 = \frac{1.38h^2}{\pi^2 t_{\frac{1}{2}}}$$                    (5)

where $h = \lambda_3 H$ is the current thickness of the specimen and $t_{\frac{1}{2}}$ is one half of the time required for the non-illuminated surface to reach its maximum temperature.

## RESULTS

Here, initial measurements demonstrate the change in $\alpha_3$, the out-of-plane component of the spatial thermal diffusivity, for a natural gum rubber subject to uniaxial stretch, $\lambda_1$, up to 1.51 and equal biaxial stretch, $\lambda_1 = \lambda_2$, up to 1.13. The stretch in the $i$-direction, $\lambda_i$, is the ratio of the current length in the $i$-direction to the reference length in that direction. Here, the dimensions of the undeformed specimen are the reference lengths. For an incompressible material, a good assumption for isothermal deformations of rubber, $\lambda_1\lambda_2\lambda_3 = 1$. The measurements were made at room temperature, which is well below the vulcanization temperature of the rubber. In general, reversible behavior is expected for rubbers that are tested at temperatures lower than their vulcanizing temperature.

Figure 3 shows the variation of $\alpha_3$ with $\lambda_1$. Note that $\lambda_1 = 1$ represents the reference state for both uniaxial and equal biaxial loading. As $\lambda_1$ increases $\alpha_3$ decreases for both uniaxial and biaxial loading. The out-of-plane direction of $\alpha_3$ is orthogonal to the direction of loading in both loading configurations. Thus, these results are analogous to the results of Choy et al. [4], wherein there was a decrease in the thermal conductivity in the out-of-plane direction of drawn and rolled polyethylene. Note that the effects are reversible in rubber at the moderate stretch ratios examined here and that the effects reported by Choy et al. [4] are irreversible due to the permanent deformation of drawn and rolled polyethylene. Thus, though the mechanisms causing the change in thermal transport may be similar for the rubber and polyethylene, additional irreversible effects must be considered when examining the results of Choy et al. [4].

## DISCUSSION

A primary obstacle to finding $\alpha_0(C,T)$ in elastomers has been the lack of an appropriate device, one that can measure the thermal diffusivity for specimens at well defined multiaxial load states. This paper describes such a device by which measurements of $\alpha_0(C,T)$ may be made while thin planar specimens are subjected to well defined uniaxial or biaxial loading. Based on the flash technique, the measurements are made with a minimum of contact. The device is well suited for elastomers, such as rubber and tissue, with high extensibility. The relatively short duration of the measurement, once a specimen has reached mechanical and thermal equilibrium, is highly desirable for materials, such as

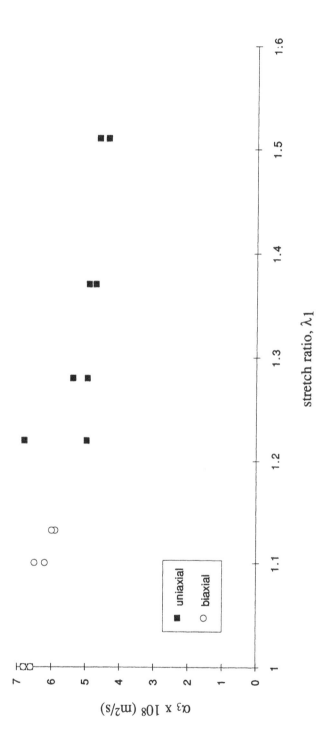

FIGURE 3. SPATIAL THERMAL DIFFUSIVITY VARIATION WITH STRETCH, $\lambda_1$

tissues, which can degrade quickly. An environmentally controlled chamber further enhances the viability of the specimens throughout the measurements.

Preliminary measurements of the out-of-plane component of thermal diffusivity in uniaxially and equally biaxially loaded commercially available natural gum rubber specimens have been made at room temperature. The component $\alpha_3$ decreases with increasing $\lambda_1$, in both the uniaxial and biaxial loading. These data follow a similar trend to that observed for other highly deformed polymers.

Measured thermal properties are vital to the complete thermomechanic modeling of non-linear heat conducting materials, such as elastomers. The determination of complete constitutive relations requires description of the thermal properties, as well as mechanical properties. Further, just as the 2nd Piola-Kirchhoff stress, defined in the reference configuration, is the most useful stress in constitutive formulation, but is without clear physical interpretation, so too, $\alpha_0(\mathbf{C},T)$ may be most convenient for constitutive formulation, but with little direct physical meaning. Conversely, whereas the Cauchy stress, defined in the spatial configuration, is physically meaningful, but of less use in constitutive formulation, the spatial thermal diffusivity will likely be the most useful in physical modeling. By the methods describe above, the thermal diffusivity tensor in both the referential and spatial configurations may be determined as a function of finite strain and temperature.

## ACKNOWLEDGMENTS

This work has been sponsored by the Army Research Office under grant ARO-DAAH04-95-2-2 to UMBC as part of the Center of Excellence for Materials Research.

## REFERENCES

1.  Coleman, B.D. and Gurtin, M.E. 1967. "Thermodynamics with Internal State Variables", *Journal of Chemical Physics*, 47(2): 597-613.
2.  Parrott, J.E. and Stuckes, A.D. 1975. *Thermal Conductivity of Solids*, London: Pion Limited, pp. 100-101.
3.  Valvano, J.W., Dalvi, V.P. and Pearce, J.A. 1993. "Directional Thermal Conductivity Measured in Bovine Aorta", ASME HTD-Vol. 268, Advances in Bioheat and Mass Transfer: Microscale Analysis of Thermal Injury Process, Instrumentation, Modeling and Clinical Applications, ASME Winter annual Meeting.
4.  Choy, C.L., Luk, W.H., and Chen, F.C., 1978, *Polymer*, 19: 155-162.
5.  Dashura, P., 1994, "A Study of Variation of Thermal Conductivity of Elastomers", *Physica Scripta*, 49: 611-614.
6.  Parker, W.J., Jenkins, R.J., Butler, C.P., and Abbott, G.L., 1961, *J. Applied Physics*, 32(9): 1679-1684.
7.  Maillet, D., Lachi, M., and Degiovanni, A., 1990, *Thermal Conductivity 21*, Cremers & Fine, ed., Plenum Press.
8.  Lachi, M., and Degiovanni, A., 1991, *Journal de Physique III*, 12: 2027-2046
9.  Sawaf, B. and Özisik, N. 1995. "Determining the Constant Thermal Conductivities of Orthotropic Materials by Inverse Analysis", *International Communications of Heat and Mass Transfer*,

# Author Index

Ahuja, S., 311
Altmann, H., 152
Alviso, C. T., 407
André, S., 72, 564
Ashworth, E., 598
Ashworth, T., 598
Aubuchon, S. R., 66

Beck, J. V., 29, 107, 183, 299
Bell, M. G., 231
Berger, H., 266
Bernnat, A., 481
Besant, R. W., 350
Biljaković, K., 266
Bilusić, A., 266
Blaine, R. L., 66
Booth, J. R., 325
Boulet, P., 442
Breidenich, F., 183

Cahill, D. G., 145
Capriglione, G. S., 243
Caps, R., 373, 407
Chen, G., 195
Childs, P. W., 395
Cusco, L., 13
Czarnetzki, W., 492

da Silva, M. G., 639
de Dianous, P., 442
Degiovanni, A., 72, 564
Diehl, K. C., 627
Dinwiddie, R. B., 107, 119, 466
Doss, D. J., 639
Dowding, K. J., 299

Ebert, H.-P., 407
Eilers, L., 299
Ellingson, W. A., 311

Flynn, D. R., 46
Franke, Th., 129, 162
Fricke, J., 152, 373, 383, 407

Gaal, P. S., 41, 119
Gembarovič, J., 95
Goerz, T. R., 29
Gorthala, R., 46
Graves, R. S., 325
Grunewald, J., 338
Gustafsson, S. E., 56
Gustavsson, M., 56

Hastings, I. J., 579
Häupl, P., 338
Häussler, P., 162
Henderson, J. B., 530
Hetfleisch, J., 373
Humphrey, J. D., 639
Hyldgaard, P., 172

Jason Lo, S. H., 288
Jeandel, G., 442

Karawacki, E., 56
Keene, B. J., 519
King, S., 311
Klemens, P. G., 209, 579, 590
Korder, S., 383

Latini, G., 613
Lee, S.-M., 145
Lévy, F., 266
Louvigné, P. F., 279
Lucuta, P. G., 579, 590

Madhusudana, C., 553
Mahan, G. D., 172
Maillet, D., 72, 564

Matzke, H. J., 579
McCurdy, A. K., 221, 231, 243, 254
McElroy, D. L., 3
McGrath, J. J., 183
Milano, G., 85, 362
Milcent, E., 279
Mills, K. C., 519
Mitchell, D. R., 350
Monaghan, B. J., 519

Nelson, G. E., 466
Nieto de Castro, C. A., 13
Nilsson, O., 152

Overfelt, R. A., 538

Pajić, D., 266
Papadaki, M., 481
Passerini, G., 613
Pekala, R. W., 407
Perkins, R. A., 13
Petrie, T. W., 395
Pincemin, F., 442
Polonara, F., 613

Ramires, M. L. V., 13
Relyea, H. M., 183
Rettelbach, Th., 373, 383, 407
Righini, F., 85
Riko, U. S., 419
Roetzel, W., 492

Salmon, D. R., 431

Sator, D., 383
Saxena, N. S., 56
Scarpa, F., 85, 362
Schmidt, R., 162
Serra, J. J., 279
Shutov, F., 455
Smontara, A., 266
Soldner, J., 504
Starešinić, D., 266
Steckenrider, J. S., 311
Stephan, K., 504
Stopp, H., 338
Strobel, A., 530

Tao, Y.-X., 350
Taylor, R. E., 29, 95, 538
Timmermans, G., 362
Tye, R. P., 419

van Gelder, M. F., 627
Villanueva, E., 553
Vitali, G., 613
Voelklein, F., 129

Wakeham, W. A., 481
Wang, H., 119, 288
Weaver, C. E., 466
Wilkes, K. E., 395
Wright, N. T., 639

Yarbrough, D. W., 325
Yu, X. Y., 195

Zhang, L., 195

# Subject Index

Ablation, 279
Acoustics, 504
Aerogels, 383, 407
Air filtration, 350
Alkanes, 613
Alloys, 519, 538
Aluminum, 288, 519, 598
Aluminum oxide, 311
Amorphous solids, 162
Angstrom bar technique, 152
Apparatus, 41, 46, 492
Apparent thermal conductivity, 3, 66, 362, 442, 455, 598
Argon, 481, 504
Aromatics, 613
ASTM, 3, 419

Building structures, 338

Calcium silicate, 431
Carbon, 29, 466
Ceramics, 311, 598
CFC, 3, 311
CFC-11, 325
Chemical vapor deposition (CVD), 183
Coatings, 553
Composites, 279, 299
Composites, carbon fiber, 29, 299, 466
Composites, carbon matrix, 29, 299
Composites, ceramic fiber, 311
Composites, ceramic matrix, 311
Composites, metal matrix, 288
Composites, polymer matrix, 455
Conduction, 29, 72, 107, 129, 299, 538
Conduction, electronic, 129
Conduction, phonon, 129, 172, 209
Contact conductance, 553, 564
Contacts, 553

Copper, 243, 519
Cut-bar apparatus, 41

Density, 598
Diamond, 152, 183
Differential scanning calorimetry, 3
Directionally-solidified superalloys, 530
Dispersion effects, 221, 231, 243

Effective thermal diffusivity, 325
Electrical conductivity, 266
Electrical conductivity, alloys, 519
Electrical resistivity, 129, 455
Electrical resistivity, alloys, 519
Energy-flow reversal, 254
Ethanol, 492

Fiber board, 373
Fibers, carbon/graphite, 466
Fibers, glass, 350, 373, 442
Fibrous insulation, 3, 338, 350, 373, 442
Finite difference analysis, 338
Finite element analysis, 152, 338
Fission products, 579
Flash method, 95, 107, 119, 311, 466, 519, 530, 639
Fluids, 13
Foams, 455
Foams, aging, 3, 325, 455
Foams, blowing agents, 325, 455
Foams, insulation, 338, 455
Foils, 553
Food products, 627
Frost, 350

Gas mixtures, 325
Gases, 481, 504
Grain boundaries, 129

649

Grain-size effects, 129, 209
Graphite, 466
Gravimetric measurement, 325
Group velocities, 221, 231, 243, 254
Guard design, 41
Guarded hot-plate, 46, 338, 373, 383, 407, 419, 431

HCFC, 3
Heat capacity, 3, 56, 66, 530, 538
Heat flow meter, 3, 338, 419
Heat loss corrections, 66
Heat pulse method, 85, 95, 119
Heat transfer coefficient, 72, 338
Heat treatment, 288
High flux, 279
High temperature, 85, 419, 431, 442, 455, 466, 627
Hot wire method, 407
Hyperstoichiometry, 579

Imaging, 311
Infrared detector, 311
Infrared opacifiers, 383
Infrared thermography, 183, 311
Insulation, 46, 350, 431
Insulation, evacuated panels, 373
Insulation, fibrous, 3, 350, 362, 419, 442, 466
Insulation, foam, 455
Insulation, glass fiber, 350, 373, 442
Insulation, high temperature, 419, 431, 466
Insulation, load bearing, 373, 383
Insulation, pipe, 3
Insulation, powder, 383
Interstitial materials, 553
Inverse heat conduction, 72
Iron, 3, 519

Joints, 553

Kalman filtering, 85, 362
Kinetic theory, 129
Knudsen number, 172

Lattice, thermal conductivity, 129, 209
Lattice, vibrations/waves, 221, 231, 243
Liquid metals, 530
Lorenz number, 266
Lorenz ratio, 519
Low temperature, 129, 162, 195, 383

Mattiessen's rule, 519
Mean free path, 129, 383
Metallic powders, 598
Metals, 129, 530, 538

Microhardness, 553
Mie theory, 442
Modeling, 338, 466, 564, 598
Modulated DSC, 66
Moisture content, 338, 350, 455
Moisture transport, 350
Molten materials, 530, 538
Multilayer systems, 338, 492

*n*-Butane, 13
*n*-Heptane, 492
Nickel, 519
Nickel based alloys, 538
Nitrogen, 481
Non-destructive testing, 311
Normal processes, 254
Nuclear fuel, 579, 590

Organic, liquids, 613
Organic, rubbers, 639
Organic, solids, 407
Oxides, 209, 311

Parameter estimation, 72, 85, 107, 183, 299, 362
Peltier effect, 492
Perlite, 383
Phase change, 266
Phenolics, 455
Phonon focusing, 231, 243
Phonon scattering, 129, 209, 590
Phonon transport, 129, 172, 209
Phonon-electron scattering, 129
Phonons, 221
Photoacoustic method, 504
Point defects, 209
Polymer, 66, 455
Porosity, 407, 455
Powder, 383
Pressure, effect of, 362
Propane, 13
Pyroceram 9606, 419

Quasi one-dimensional compounds, 266

Radial heat flow apparatus, 492
Radiation damage, 579
Radiative transport, 209, 373, 383, 407, 466
Reference materials, 431
Refractories, 431
Refrigerants, 619
Regularization, 72
Relaxation phenomena, 129

Seebeck coefficient, 129
Semiconductors, 129, 172, 266

Silica, 383
Silicon, 195
SIMFUEL, 579, 590
Single crystals, 530
Sodium, 231
Sol-gel process, 407
Solidification, 530, 538
Specific heat, 3, 56, 162, 266
Stabilization, 209
Standard reference material, 66, 119, 419
Steady-state method, 129
Steel, 3
Steel, austenitic, 3
Step heating, 95, 627
Strain, 639
Superalloy, 530, 538
Superlattice, 195
Surface tension, 538

Temperature oscillation method, 66, 492
Thermal aging, 325, 455
Thermal barrier coating, 209
Thermal contact resistance, 72, 553, 564
Thermal diffusivity, 29, 95, 107, 119, 152,
    183, 195, 279, 288, 311, 466, 492, 504,
    519, 530, 627, 639
Thermal expansion, 455
Thermal resistance, 46, 338, 419
Thermal wave method, 195

Thermistors, 627
Thermocouples, 29, 492
Thermoelastic, 639
Thermoelectric homogeneity, 3
Thermography, 311
Thermomechanical, 639
Thickness effect, 129, 195
Thin films, 129, 152, 162, 183, 195
Three-phonon processes, 209, 254
Titanium, 519
Transient effects, 564
Transient high flux, 279
Transient hot strip method, 56
Transient hot wire method, 481, 519
Tungsten, 231

Umklapp process, 129, 254, 266
Uranium dioxide, 579, 590

Vacancies, 209
Viscosity, 538

Water, 492
Wet porous media, 338, 455
Wiedemann-Franz law, 129

Young's modulus, 383

Zirconia, 209

Printed and bound by CPI Group (UK) Ltd, Croydon, CR0 4YY

23/10/2024

01777686-0020